T0320939

WATER-ROCK INTERACTION

PROCEEDINGS OF THE 13TH INTERNATIONAL CONFERENCE ON WATER-ROCK INTERACTION
WRI-13, GUANAJUATO, MEXICO, 16–20 AUGUST 2010

Water-Rock Interaction

Editors

Peter Birkle
Instituto de Investigaciones Eléctricas, Gerencia de Geotermia, Cuernavaca, Mexico

Ignacio S. Torres-Alvarado
Centro de Investigación en Energía, Universidad Nacional Autónoma de México, Temixco, Mexico

CRC Press is an imprint of the
Taylor & Francis Group, an **informa** business
A BALKEMA BOOK

Front cover:
Cauliflower-shaped, siliceous sinter deposits around the edge of an evaporation pond, which were formed by cooling of geothermal water priot to its reinjection into the Los Azufres geothermal reservoir (Michoacán State, Mexico) (Photo taken by P. Birkle).

CRC Press/Balkema is an imprint of the Taylor & Francis Group, an informa business

Typeset by Vikatan Publishing Solutions (P) Ltd., Chennai, India
Printed and bound in USA by Edwards Brothers, Inc, Lillington, NC

Published by: CRC Press/Balkema
P.O. Box 447, 2300 AK Leiden, The Netherlands
e-mail: Pub.NL@taylorandfrancis.com
www.crcpress.com – www.taylorandfrancis.co.uk – www.balkema.nl

ISBN: 978-0-415-60426-0 (Hbk + CD-rom)
ISBN: 978-0-203-83404-6 (eBook)

Water-Rock Interaction – Birkle & Torres-Alvarado (eds)
© 2010 Taylor & Francis Group, London, ISBN 978-0-415-60426-0

Table of contents

Water-rock interactions in geothermal systems

Role of mineral surfaces in kinetics of weathering

 ISBN 978-0-415-60426-0

Preface

In 1803, when Alexander von Humboldt arrived in Mexico, the colonial site of Guanajuato represented the third largest town in "Spanish" America, just overwhelmed in population by Mexico City and Havana. During the 18th century, the Guanajuato Mining District was reportedly producing one-third of the world's silver. Estimates of historic production range from 700 million to 1.5 billion ounces of silver, as well as 4 to 7 million ounces of gold. Remarkably, some areas are still today under exploitation. Geomorphological and environmental consequences from hundreds of years of mining activity can still be observed in the surroundings of this charming UNESCO World Heritage community.

The 13th International Symposium on Water-Rock Interaction (WRI-13) was held in Guanajuato, Mexico, from August 16 to 20, 2010. The symposium addressed main aspects of Water-Rock Interaction, including recent advances in understanding of principal geochemical processes in isotopic studies (Hotoshi Sakai memorial session), from high-temperature reservoirs to low-temperature systems (vadose zones, watersheds, karst systems), water-rock interaction through magmatic and ore forming processes, reconnaissance and concerns about groundwater quality and environmental aspects, recent advances in numerical modeling, specific interaction between fluids and minerals, sequestration of CO_2, geological hazards and organic processes. Special emerging topics of the present WRI-13 proceedings were environmental impacts by mining activities, the geochemical response through the injection of CO_2 into the subsoil, natural and anthropogenic sources for arsenic anomalies in groundwater and surface systems, the application of novel isotope and tracer methods (CFC, SF_6, $^{11}B/^{10}B$, $^{44}Ca/^{40}Ca$, $^{53}Cr/^{52}Cr$, $^{56}Fe/^{54}Fe$, $^{207,208}Pb/^{206}Pb$, ^{226}Ra, ^{228}Ra), and advanced geochemical methods for geothermal exploration.

The symposium was organized by participants from the Centro de Investigación Científica y de Educación Superior de Ensenada (CICESE, Ensenada, Baja California), Centro de Investigación en Matemáticas, A.C. (CIMAT, Guanajuato), Universidad de Guanajuato (UG, Guanajuato), Centro de Enseñanza Técnica y Superior (CETYS, Guanajuato), Instituto de Investigaciones Eléctricas (IIE, Cuernavaca, Morelos), Centro de Investigación en Energía (CIE, UNAM, Temixco, Morelos), Instituto Mexicano de Tecnología del Agua (IMTA, Jiutepec, Morelos), Centro de Geociencias (CGeo, UNAM, Juriquilla, Queretaro), Instituto de Geofísica (IGeof, UNAM, Mexico City), Centro de Investigación Científica de Yucatán (CICY, Mérida, Yucatán), and the Universidad Autónoma de Baja California Sur (UABCS, La Paz, Baja California Sur).

Sponsors of WRI-13 include the Consejo Nacional de Ciencia y Tecnología (CONACYT), Centro de Investigación Científica y de Educación Superior de Ensenada (CICESE), the International Association of GeoChemistry (IAGC), WRI-7/USGS, Universidad de Guanajuato (UG), Schlumberger Water Services, Instituto Mexicano de Tecnología del Agua (IMTA), Deutsche Gesellschaft für Technische Zusammenarbeit (GTZ), and Deutscher Akademischer Austauschdienst (DAAD).

Over 250 manuscripts were submitted for presentation in both oral and poster sessions of this symposium. Following a comprehensive review process, 231 manuscripts were accepted and included in the present volume. The articles were submitted by scientists, engineers, specialists, professionals, decision-makers, and students from 35 countries (Argentina, Austria, Australia, Belgium, Canada, Chile, China, Czech Republic, Denmark, Egypt, France, Germany, Hungary, Iceland, India, Israel, Italy, Japan, Mexico, Morocco, Palestine, Peru, Poland, Portugal, Russia, Senegal, Slovenia, Spain, Sweden, Switzerland, Thailand, Netherlands, Turkey, U.S.A. and U.K.).

The papers were divided into 16 major session topics, although many of them represent multi-disciplinary studies which incorporate a variety of advanced methodologies. The volume begins with the Plenary Lectures, followed by contributed papers for the specific topic sessions as follows:

- Hitoshi Sakai memorial session: Measurements and applications of stable and radiogenic isotopes and other tracers
- Water-rock interactions in geothermal systems

- Water in petrogenetical, magmatic and ore forming processes
- Water-rock interactions in watersheds
- Solute interactions during transport in vadose zone
- Water-rock interactions in karst and pore water chemistry in sedimentary rocks
- Water-rock interaction controlling groundwater quality
- Environmental Geochemistry
- Water-rock interactions in mine tailings
- Significance of water-rock interaction for reconnaissance and remediation of contaminated sites
- Characterization of mineral surfaces and water/mineral interfacial processes
- Role of mineral surfaces in kinetics of weathering
- Advances in numerical modeling of water-rock interaction processes
- Role of water-rock-gas interaction in sequestration of CO_2
- Geological hazards related to water-rock interaction
- Water-rock interactions in biogeochemical processes and genesis of petroleum

All of the manuscripts have undergone peer-review of their scientific content. Generally, two independent reviewers were asked to evaluate each of the submitted articles. The high quality of the present volume would not have been reached without the time-consuming and profound expertise of more than 260 volunteering colleagues listed below. We wish to express our sincere gratitude to:

Subhrangsu Acharyya
Werner Aeschbach-Hertig
Hind Al-Abadleh
Alessandro Aiuppa
Ma. Catalina Alfaro
Peter Alt-Epping
Jorge Alberto Andaverde-Arredondo
Laurent André
Muriel Andréani
Halldór Armannsson
Ma. Aurora Armienta
Stefan Arnórsson
Carlos Ayora
Tarik Bahaj
Rosa María Barragán
R.L. Bassett
Hamdallah Bearat
Pascal Benezeth
Peter Berger
Zsolt Berner
Prosun Bhattacharya
Jianmin Bian
Michel Bickler
Riccardo Biddau
Olivier Bildstein
Peter Birkle
David W. Blowes
Philipp Böning
Geneviève Bordeleau
Svetlana B. Bortnikova
Sylvie C. Bouffard
Susan Brantley
Jean-Jaques Braun
Kevin Brown
George Breit
Roberto Briones
Patrick Browne
Benjamin Brunner
Thomas D. Bullen
Jochen Bundschuh
Eurybiades Busenberg
Carles Canet
Valérie Cappuyns
Lucia Capra
Antonio Cardona
Bill Carey
Alejandro Carillo-Chávez
Ian Cartwright
Silvia Castillo Blum
Javier Castro Larragoitia
Dornadula Chandrasekharam
H.H. Chen
Benito Chen-Carpentier
Oleg Chudaev
Rosa Cidu
Jaime Cuevas
George Darling
Damien Daval
Jordi Delgado
Jo De Waele
Enrico Dinelli
Wolfgang Dreybrodt
Alan Dutton
Mike Edmunds
Derek Elsworth
Sara Eriksson
Ma. Vicenta Esteller
Williams C. Evans
Luca Fanfani
Matthew S. Fantle
Julianna E. Fessenden-Rahn
Sean Frape
Franco Frau
Christopher J. Gabelich
Yiqun Gan
Octavio Garcia-Valladares
Mel Gascoyne
Irina Gaus
Liliana Gianfreda
Pierre Glynn
José Marcus Godoy
Fraser Goff
Bibi R.N. Gondwe
Luis C. González Márquez
Eduardo González Partida
Sibylle Grandel
Steve Grasby
Peter Gratwohl
Bill Gunter
Huaming Guo
Carlos Gutiérrez
Alexandra Hakala
Russel J. Hand
Jeff Hanor
Anne Hansen
Henrik K. Hansen
Hao Hanzou
Russell Harmon
J.T. He
Roland Hellmann
George R. Helz
Bill Herkelrath
Alfredo Hernández
Graciela Herrera
Andrew Herzceg
Matthias Hinderer
Brian Hitchon

Jun Hong
Brian Horsfield
Vladimir Hristov
Xueyu Hu
Hudson-Edwards
Miguel Angel Huerta-Diaz
Charlotte Hurel
Shaul Hurwitz
Ivana Jačková
Andrew Jacobson
Valerie Jacquemet
Shao-Yong Jiang
Qian Jiazhong
Clark M. Johnson
Thomas M. Johnson
Blair F. Jones
Michel Jullien
Yousif Kharaka
Wolfram Kloppmann
Claus Kohfahl
Jiri Kopacek
Christian Koeberl
Pavel Kram
Thomas Kretzschmar
Hrefna Kristmannsdóttir
Martin Kubai
Jurate Kumpiene
Yoshihiro Kuwahara
Daniel Larsen
Pierfranco Lattanzi
Olesya Lazareva
Ma Lei
Wanting Ling
Marcelo Lippmann
Marta Litter
Dameng Liu
Fei Liu
Xiao-li Liu
Zaihua Liu
Jon Lloyd
Dinggui Luo
Andreas Luttge
Urs Mäder
Clara Magalhães
Frank D. Mango
Bruce Manning
Luigi Marini
Gudrun Massmann
Arash Massoudieh
M. Alisa Mast
Jörg Matschullat
Martin Mazurek
Enrique Merino
Angelo Minissale
Filip Moldan
Craig Moore
Mike Mottl
Gavin Mudd
Satoru Nakashima
Bibhash Nath
José Miguel Nieto Liñán
D. Kirk Nordstrom
Eric H. Oelkers
Atsushi Okamoto
Neus Otero
Nevzat Özgür
David Ozsvath
Tomas Paces
Kailasa Pandarinath
Zhonghe Pang
Marc Parmentier
Giovannella Pecoraino
Karsten Pedersen
Asaf Pekdeger
Albert Permanyer
Eugene Perry
Zell E. Peterman
Thomas Pichler
Helmut Pitsch
Maria D.R. Pizzigallo
Maxime Pontié
Simon Poulson
Rosa María Prol-Ledesma
Andrew Putnis
Clemente Recio
Marc H. Reed
François Renard
François Risacher
Brian Robinson
Dimitri Rouwet
Harold Rowe
Thomas R. Rüde
Brian Rusk
Boris Ryzhenko
Javier Sánchez
Bjorn Sandström
Edgar Santoyo
Hiroshi Sasamoto
Miklas Scholz
Robert R. Seal
S. Sevinc Sengor
Kotaro Sekine
Huimei Shan
Paul Shand
James B. Shanley
Barbara Sherriff
Xiao-Qing Shi
Orfan Shouakar-Stash
Vladimir Shulkin
Cristina Siebe
Adam Skelton
Pauline Smedley
A.R.M. Solaiman
Nic Spycher
Ondra Sracek
Lisa S. Stillings
Ken Stollenwerk
Simcha Stroes-Gascoyne
Mario César Suárez
Shulin Sun
Zhanxue Sun
Arny E. Sveinbjörnsdóttir
Yuri Taran
James Thordsen
Ignacio S. Torres-Alvarado
Eva-Lena Tullborg
Akira Ueda
Gene Ulmer
Manfred van Afferden
Theo W.J. van Asch
David Vaughan
América R. Vázquez Olmos
Mario Villalobos
Lukas Vlcek
Minghua Wang
Bronwen Wang
Y.X. Wang
Yanxin Wang
Richard B. Wanty
Mary H. Ward
Bodo Weber
Jenny Webster-Brown
Goeff Wheat
Tim White
Richard T. Wilkin
Jobst Wurl
Tianfu Xu
Echuan Yan
Norio Yanagisawa
Yossi Yechieli
Gaoke Zhang
Chaosheng Zhang
Cuiyun Zhang
Weimin Zhang
Zhanshi Zhang
Yan-Chun Zhang
Guodong Zhen
Liange Zheng
Chen Zongyu
Shengzhang Zou
Pierpalo Zuddas

We are especially grateful to Halldór Ármannsson, Thomas D. Bullen, Ian Cartwright, Mike Edmunds, Williams C. Evans, Brian Hitchon, Yousif Kharaka, Craig Moore, Tomas Paces, Kailasa Pandarinath, Zhonghe Pang, Robert R. Seal, Barbara Sherriff, Gene Ulmer, and Richard B. Wanty by their major efforts to reviewing 5 or more of the manuscripts. We would also like to thank Mirna Guevara García for the formatting of some manuscripts and to Oscar Alonso and Gema Alín for the graphical design of the WRI-13 logotype.

Peter Birkle & Ignacio S. Torres-Alvarado
Editors
WRI-13 Proceedings

Thomas Kretzschmar
Secretary General (WRI-13)

Water-Rock Interaction – Birkle & Torres-Alvarado (eds)
© 2010 Taylor & Francis Group, London, ISBN 978-0-415-60426-0

Organization

IAGC EXECUTIVE COMMITTEE

President	Russell S. Harmon (USA)
Vice-President	Clemens Reimann (Norway)
Past President	John Ludden (UK)
Treasurer	W. Berry Lyons (USA)
Secretary	Thomas D. Bullen (USA)
Editor (*Applied Geochemistry*)	Ron Fuge (UK)
Newsletter Editor/ Business Office Manager	Mel Gascoyne (Canada)
Council Members	Norbert Clauer (France) Shaun Frape (Canada) Zhonghe Pang (China) Andrew Herczeg (Australia) Nancy Hinman (USA) Martin Novak (Czech Republic) Harue Masuda (Japan) Andrew Parker (UK) Alakendra N Roychoudhury (South Africa) Rona J. Donahoe (USA)

WATER-ROCK INTERACTION WORKING GROUP

Tomas Paces (Czech Republic)
Yves Tardy (France)
Brian Hitchon (Canada)
Halldór Ármansson (Iceland; Chairman)
Mike Edmunds (UK)
Yousif Kharaka (USA)
Oleg Chudaev (Russia
Brian Robinson (New Zealand)
Luca Fanfani (Italy)
Susan Brantley (USA)
Yanxin Wang (China)
Thomas Kretzschmar (Mexico)

WRI-13 ORGANIZING COMMITTEE

Secretary General	Thomas Kretzschmar (CICESE)
Editors, Proceedings Volume	Peter Birkle (IIE), Ignacio S. Torres-Alvarado (CIE)
Scientific Committee	Ma. Aurora Armienta (UNAM), Peter Birkle (IIE), Carles Canet (UNAM), Gilberto Carrasco (Centro de Geociencias), Gilbert Carreño (UGto), Anne Hansen (IMTA), Ulises López (CETYS),

	Miguel Angel Moreles (CIMAT), Yann Renee Ramos (UGto), Ignacio S. Torres-Alvarado (CIE), Bodo Weber (CICESE)
Field trips	Carles Canet (UNAM, Coordinator), Gilberto Carrasco (Centro de Geociencias), Eduardo González Partida (Centro de Geociencias), Marie-Noelle Guilbaud (UNAM), Ulises López (CETYS), Luis Marín (UNAM), Yann Renee Ramos (UGto), Mario Rebolledo (CICY), Ignacio Reyes (UACH), Claus Siebe (UNAM), Bodo Weber (CICESE), Jobst Wurl (UABCS)
Assistants to Secretary General	Cecilia Hirata, Alicia Tsuchiya, Daniel Peralta, Horacio Sánchez, Jorge Milanez, David Pringle (CICESE)
Logistics (Ensenada)	Elena Enríquez (CICESE, Coordinator)
Logistics (Guanajuato)	Gilberto Carreño, Rocío Morales, Josefina Ortiz, Yann Renee Ramos (UGto), Miguel Angel Moreles (CIMAT)
Sponsors	CONACYT, CICESE, IAGC, WRI-7/USGS, Universidad de Guanajuato, Schlumberger Water Service, IMTA, GTZ, DAAD

Plenary lectures

Water-Rock Interaction – Birkle & Torres-Alvarado (eds)
© 2010 Taylor & Francis Group, London, ISBN 978-0-415-60426-0

The San Juan del Grijalva landslide

J. Aparicio
Instituto Mexicano de Tecnología del Agua, Jiutepec, Morelos, Mexico

ABSTRACT: On October and November, 2007, extraordinary precipitation events occurred in the Grijalva watershed, Mexico. Such events produced in turn flooding of about 80% of the Tabasco State territory, with water depths of about 4 m in some sites. Nearly 1.2 million people were affected and severe material and economic damages were produced. No human losses were recorded. These events also produced a landslide at the right margin of the Grijalva River in the river stretch between the Peñitas and the Malpaso dams—part of the Grijalva Hydroelectric complex—, which caused an obstruction in the river and a wave around 50 m in height. This wave devastated the San Juan del Grijalva town, causing 20 human deaths. This landslide is among the most important occurred in Mexico. Two concurring types of phenomena caused the landslide: one of a geological-geotechnical character and the other, of a meteorological character. Both contributed to the intensity and damages caused. The failure mechanism is described. As a consequence, the Mexican Federal Electricity Commission suspended power generation in the hydroelectric system and built a relief channel in a record time.

1 INTRODUCTION

1.1 *The Grijalva River system*

The extensive Tabasco flatlands, which have been forming for thousands of years due to large deposits of water borne sediment, have resulted in an intricate network of streams, lagoons, and flood areas. The nutrient-rich silt deposited by the floods enabled ancient cultures, such as the Olmecs and the Mayas, to flourish along the river courses and create highly sophisticated societies, which acquired profound hydraulic and hydrologic knowledge. These flatlands in southeastern Mexico are subjected nowadays to frequent flooding. There are references as old as the times of Hernán Cortés that give account of such floods (Álvarez 1994). The Tabasco flatlands lie in the confluence and deltas of several rivers, such as the Grijalva, Mezcalapa, Carrizal, Samaria, De la Sierra, and Usumacinta, some of which are among the greatest rivers in Mexico. Part of the Grijalva River is controlled by a series of dams (Angostura, Chicoasén, Malpaso, and Peñitas) forming the Grijalva Hydroelectric Complex, which also have a flood control function. The rest of the rivers flow uncontrolled through the flatlands into their mouth in the Gulf of Mexico. A considerable population lives in this region. In particular, Villahermosa, capital city of the State of Tabasco, lies at the confluence of the De la Sierra and Carrizal rivers (see Figure 1). According to the National Water Commission of Mexico, the annual precipitation in the region (2750 mm in the coastal zone and up to 4000 mm in the foothills) is one of the highest in the world and the highest in the Mexican Republic (CNA 1996).

1.2 *The October–November 2007 meteorological events*

From 28 to 30 October 2007, exceptional rainfall fell in the lower Grijalva River basin, generating runoff, which in turn caused floods over an area equivalent to 70% of the Tabasco State territory with water depths of up to 4 m in some areas. According to information from the Tabasco Government and the Mexican Ministry of Interior (Secretaría de Gobernación, 2007), about 1.2 million people were affected and there was extensive material and economic damage. No loss of human life was recorded in the flooded areas in the floodplains.

Several unfavorable effects converged during this event. During the early days of October, important rainfall events were recorded in the Peñitas basin and in the northern Chiapas range. These antecedent events produced high soil moisture content, which reduced considerably the infiltration capacity and therefore increased even more the runoff produced by the October–November events. The high moisture content in the zone of the landslide was also determinant for the massive soil and rock movement. October rainfall events, especially those occurring on the 23rd and subsequent days, produced a succession of floods with short time

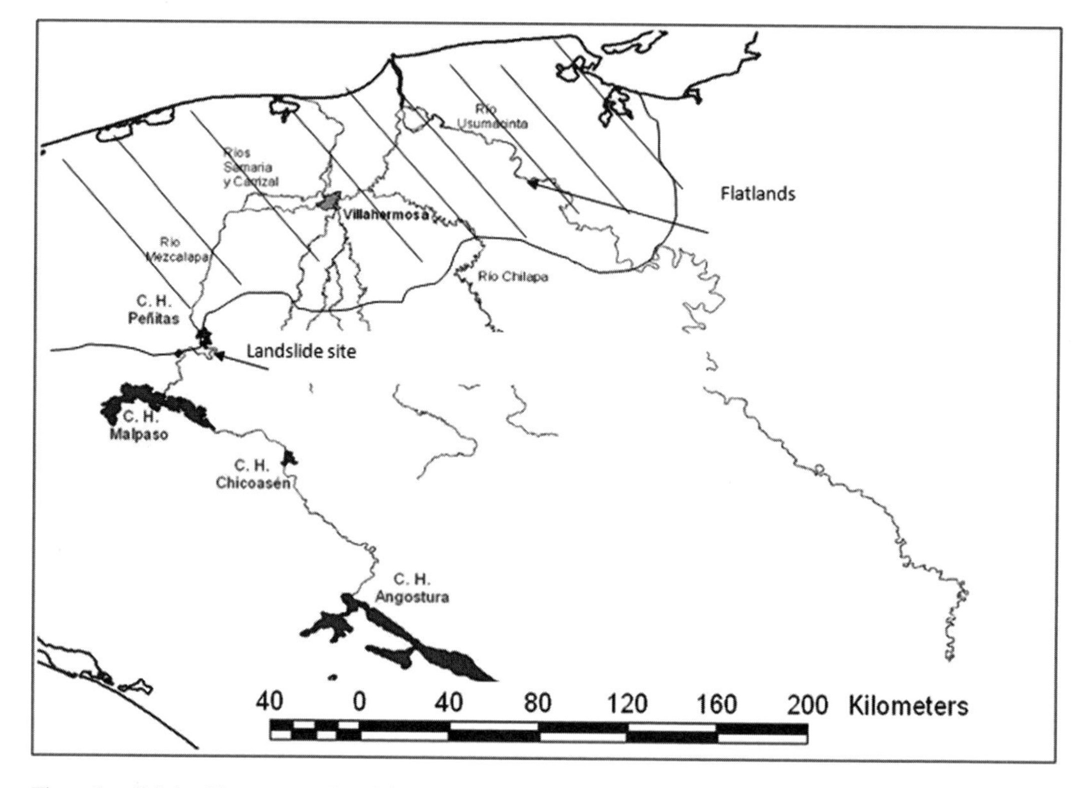

Figure 1. Grijalva River system (Aparicio *et al.*, 2008).

intervals between each. Extreme event simultaneity was notorious in this case. Precipitation in the middle Grijalva river basin was especially intense from 28 October to 1 November, particularly in the Peñitas river basin, where the accumulated precipitation depth in 72 h was 946 mm. These 24 h precipitation events have return periods of around 20 years (Aparicio *et al.*, 2009) although they are not the maximum recorded (which occurred in year 1990 with a precipitation depth of 645 mm in 24 h). However, if a 72 h duration is considered, return period exceeds 60 years. Simultaneity of these events was crucial for the hazard situation (Aparicio *et al.*, 2009).

2 THE LANDSLIDE

2.1 *Antecedents*

Landslides are a very common phenomenon in the world and are one of the most destructive phenomena of nature that cause damage to both property and life every year. While landslides are a natural landscape building mechanism, there is strong evidence that the frequency of such hazards has been increasing in the last few years (Haritashya *et al.*, 2006). Such increase in frequency is at least in part due to anthropogenic influence, in the form of deforestation and changes in land use. Climate change could as well have an important part in this incremental frequency.

There are a number of reported landslides in the oceans (Chaytor *et al.*, 2009, Lee 2009) and even in Mars (Anonymous 2008). Large landslides have occurred in history (vgr., Salvati *et al.*, 2009). Landslides can be produced by a number of causes: earthquakes, glacier melting, extraordinary precipitation, etc. However, high rainfall rates are perhaps the most frequent causes. Road networks and housing areas are in many cases the most frequently damaged elements (Petrucci *et al.*, 2009) and, furthermore, landslides usually block river courses, producing dams typically comprising unconsolidated and poorly sorted material, which make them vulnerable to rapid failure and breaching, in turn resulting in significant and sudden flood risk downstream (Davies *et al.*, 2007).

2.2 *Geology*

The zone surrounding the landslide area is characterized by a series of mountain ranges and valleys.

The sedimentary rock stratification age can be set in the middle- and upper Tertiary, specifically in the Eocene and Miocene epochs. The youngest rocks in the region are silty and clayey lutite layers intermingled with expansive clay loam (Arvizu 2009).

Several monoclinals, anticlines and synclines can be found in the region. The most important in this case are the Malpaso Synclinal to the W, the La Pera monoclinal y and the La Unión Anticlinal to the E. An important geological feature is the Chichonal Vulcano, which erupted in 1982, causing by that time a minor blockage of the Grijalva River as well.

Locally, the surface lithology consists mostly of alternate sandstones, lutites and limonites. In some outcrops near the landslide it is observed that the base of the lithologic sequence is made by conglomerate sandstone and polymictic conglometrates constitued by fragments of igneous rocks (Arvizu 2009). The zone of the landslide is limited to the north and south by parallel geologic faults having lateral displacements with 60° NE strike. To the NE there is a regional normal fault called La Laja.

Hernández-Madrigal *et al.*, (2010) report evidence of four large and older landslide areas (paleolandslides) near the 2007 event. To the SE, a rock flow with an area of 0.87 km^2 and an escarpment with direction perpendicular to the La Pera monoclinal; to the SW, two rotational landslides, each with an area of 0.13 km^2, on the left bank of the Grijalva River; and to the NW, a probable debris flow 750 m long covering 0.24 km^2, which has moved perpendicular to the Grijalva River. According to these authors, there are also several smaller (secondary) landslides in the area. The presence of these other, older landslides suggest that the area has a high susceptibility to deep-seated landsliding, probably linked to changes in groundwater level caused by the Grijalva River and to the steep terrain, geological structural conditions, acting together with the above mentioned high precipitation of the region. According to local inhabitants, the Cerro La Pera, where the landslide took place, is known as the "enchanted mountain" because of the frequent underground noises heard there. These noises may have been caused by incipient, precursory ground movements that preceded the landslide.

2.3 *Failure mechanism*

On November 4th 2007, the landslide occurred on the right margin of the Grijalva River between Malpaso and Peñitas dams (Figure 2). The landslide caused the development of a main scarp at the head and lateral scarps on both flanks of up to

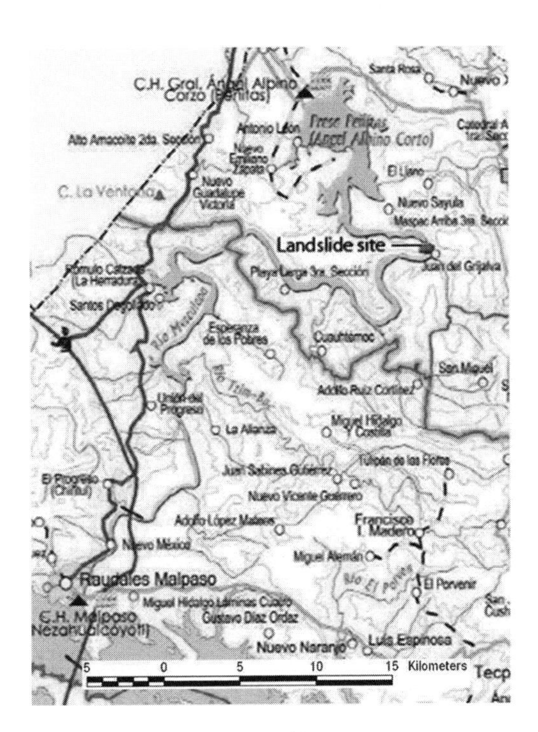

Figure 2. Location of landslide.

600 m in length and 50 m in height. The maximum width and length of the displaced mass is 1,170 and 1,570 m, respectively. Its total area was of 1.11 km^2, of which 0.76 km^2 corresponded to the landslide itself and 0.35 km^2 corresponded to the dam formed across the Grijalva River. (Hernández-Madrigal *et al.*, 2010). It is calculated that around 55 million cubic meters were displaced (Arvizu 2009, Arvizu *et al.*, 2008). This landslide produced an obstruction of the river course as well as a 50 m height wave which devastated the San Juan del Grijalva town, and caused 20 human deaths. The resulting loose material formed a plug 800 m long and 200 m wide, which temporarily blocked the river flow. Two phenomena were involved in the Juan del Grijalva landslide. One of them is geological and geotechnical in character and has to do with lithological, structural and geomechanical properties of the rocks; the other, meteorological, is related to the meteorological events described previously, which produced variations in the phreatic level and, therefore, ascending pressure variations opposed to the weight of the rock mass.

According to Alcántara-Ayala & Domínguez-Morales (2009) and based on field observations and the stratigraphic and geological setting of the area, the San Juan del Grijalva landslide was a translational mass movement and the slide surface took place on a lutite layer (Figures 3 and 4).

Figure 3. Landslide. View from upstream.

Figure 4. Landslide. View from downstream.

The rock mass slid down this soft lutite layer, composed of plastic clay notoriously weaker than the underlying sandstone with an angle of about 10° towards the river. The clay, upon saturation, led by the high precipitation, behaved like a lubricant, thus facilitating gliding. Then, the rock mass got loose in the NE portion, limited by the La Laja fault and flanked by a couple of lateral displacement faults to the North and South of the moving body (Arvizu 2009). The movement was stopped when the sliding mass found the left margin of the Grijalva River.

The magnitude of the landslide was so earsplitting that the vibrations produced during its occurrence were registered in a seismic station located at a distance of 16 km on the flanks of Chichonal volcano. Such records indicated that the mass movement took place during approximately 80 seconds (Alcántara-Ayala & Domínguez-Morales 2009).

The following main controlling factors of the San Juan del Grijalva landslide are considered by Domínguez-Morales (2008):

a. Structural geology: faults and fractures.
b. Water level changes and suction regime in the rock layers.
c. Mechanical properties of materials, expressly of lutites, which lowered resistance when saturated.
d. Spatial distribution and stratigraphic character of the rock masses.
e. Rock dipping.
f. Local topography, although the slope gradient before the landslide was slightly higher than 10 degrees.
g. Intense precipitation.
h. River bank erosion.
i. Deforestation.

The mass moved downslope exposing the sliding surface, on which triturated rock, large fragments and fractured rocky stacks could be observed (Ibid.).

On the other hand, according to Hernández-Madrigal *et al.*, (2010), the landslide mechanism consisted in both translational and rotational movements. They establish that the instability that developed at the base of the slope was possibly initiated by inferred large changes of the pore-water pressure due to the rising and lowering of water level in the upper dam reservoir at the base of the slope and heavy rainfall. The rotational movement of the toe of the landslide is inferred by the exhumation of well-cemented conglomerates and sands from the bottom of the river, that is, the rotational movement that dragged and raised material from the base of the slope and the bottom of the river, exposing it on the surface through the tilting of the sliding body. The translational movement was identified by Hernández-Madrigal *et al.*, (2010) by the translational sliding, in a SW direction and with counterclockwise horizontal rotation, of three large bodies or blocks of sandstones in alternation with clays, well-cemented coarse sandstones, and lateritic. The translational movement originated with the loss of confinement at the base of the slope that occurred in the preceding (initial) rotational movement at the base of the slope. The consequent increase of the shear stress suddenly accelerated the plastic deformation in the clayey layers, thus defining the weakness plane over which the landslide developed. The thrust exerted by these blocks on the rotational deposits at the base of the slope intensified the exhumation of the conglomerates that were raised to heights of 80 m above the water level of the Grijalva River.

2.4 *Relief works*

To avoid a possible failure and consequently a dam-brake flood which could result from erosion of the rock mass produced by the landslide and blocking the Grijalva River, the Federal Electricity

Commission, responsible for the operation of the Grijalva Hydroelectric Complex, stopped power generation in the system and started excavations with the aim of building a relief channel in two stages. The first stage considered a 6 m wide channel at an elevation 100 meters above mean sea level (mamsl) as the emergency work and the second stage the channel was 70 m wide at 85 mamsl. These works were performed in record time. Additionally, to avoid a second slide likely to be produced during the upcoming rainy season, the channel margins were stabilized, some relief wells were drilled and other drainage works were performed. All of the works were thoroughly instrumented to assess displacements, inclination variations, seismicity, phreatic levels and flow through the relief channel, among others. The opportunity and quality of these works deserved the awarding of the Edison Prize, given to Mexico, specifically to the Federal Electricity Commission and other institutions, such as the Mexican Institute of Water Technology, by the Edison Electric Institute.

Finally, to have a definitive solution, two tunnels 14 m in diameter are presently under construction.

3 CONCLUSIONS

Landslides are a common and frequent phenomenon all around the world. Hazards produced by landslides appear to be more frequent in recent years due to anthropogenic influence. The Juan del Grijalva landslide is one of the major recorded landslides in Mexico and constituted a threat in the security of the region, as well as a serious obstacle to Power generation in the important Grijalva River Hydroelectric Complex. The landslide was produced by both extraordinary meteorological conditions and a particular geological composition of the moving rock masses. The landslide mechanism consisted in both translational and rotational movements mainly caused by large changes of the pore-water pressure, exacerbated by the weakness of the soft lutite layer upon which the slide took place. The opportunity and quality of the relief works prevented further damages and they can be taken as an example to other actors involved in similar situations. Further research is needed in order to characterize and prevent hazards and damages in zones prone to landslides.

REFERENCES

Alcántara-Ayala, I. & Domínguez-Morales, L. 2009. The San Juan de Grijalva Catastrophic Landslide, Chiapas, Mexico: Lessons Learnt. http://150.217.73.85/wlfpdf/03_Alcantara-Ayala.pdf

Álvarez, J.R. (ed.) 1994. *Diccionario enciclopédico de Tabasco*. Villahermosa: Gobierno del Estado de Tabasco.

Anonymous 2008. Four Martian Landslides Caught in the Act. *Sky & Telescope* 115(6): 15.

Aparicio, J., Martínez-Austria, P.F., Güitrón, A. & Ramírez, A.I. 2009. Floods in Tabasco, Mexico: a diagnosis and proposal for courses of action. *J Flood Risk Management* 2: 132–138.

Arvizu, L.G. 2009. El deslizamiento en el río Grijalva, Chis. Conference, Academia de Ingeniería, México, 13 de noviembre de 2009.

Arvizu, G., Dávila, M. & Alemán, J. 2008. Landslide in Grijalva River, México, Abstract Paper No. 177–2. Joint Meeting of The Geological Society of America, Soil Science Society of America, American Society of Agronomy, Crop Science Society of America, Gulf Coast Association of Geological Societies with the Gulf Coast Section of SEPM, Houston, Texas.

Chaytor, J.D., Brink, U.S., Solow, A.R. & Andrews, B.D. 2009. Size distribution of submarine landslides along the U.S. Atlantic margin. *Marine Geology* 264(1, 2): 16–27.

CNA (Mexican National Water Commission) 1996. Diagnóstico de la Región XI Frontera Sur. Informe del contrato GRSP 96–01-I. Informe Ejecutivo.

Davies, T.R., Manville, V., Kunz, M. & Donadini, L. 2007. Modeling Landslide Dambreak Flood Magnitudes: Case Study. *J. of Hydraulic Engineering* 133(7): 713–720.

Domínguez-Morales, L. 2008. El deslizamiento del 4 de noviembre de 2007 en la comunidad Juan de Grijalva, municipio de Ostuacán, Chiapas, y su relación con el frente frío no. 4. Centro Nacional de Prevención de Desastres (CENAPRED), Internal Report.

Haritashya, U.K., Singh, P., Kumar, N. & Singh, Y. 2006. Hydrological importance of an unusual hazard in a mountainous basin: Flood and landslide, *Hydrol. Process* 20: 3147–3154.

Hernández-Madrigal, V., Mora-Chaparro, J. & Garduño-Monroy, V. 2010. Large block slide at San Juan Grijalva, Northwest Chiapas, Mexico. *Landslides*, published online 04 may 2010.

Kanungo, D., Arora, M.K., Sarkar, S. & Gupta, R. 2006. A comparative study of conventional, ANN black box, fuzzy and combined neural and fuzzy weighting procedures for landslide susceptibility zonation in Darjeeling Himalayas. *Engineering Geology* 85(3, 4): 347–366.

Lee, H.J. 2009. Timing of occurrence of large submarine landslides on the Atlantic Ocean margin. *Marine Geology* 264(1, 2): 53–64.

Petrucci, O., Polemio, M. & Pasqua, A.A. 2009. Analysis of Damaging Hydrogeological Events: The Case of the Calabria Region (Southern Italy). *Environmental Management* 43: 483–495.

Salvati, P., Balducci, V., Bianchi, C., Guzzetti, F. & Tonelli, G. 2009. A WebGIS for the dissemination of information on historical landslides and floods in Umbria, Italy. *GeoInformatica* 13(3): 305–322.

Secretaría de Gobernación 2007. Comparecencia del Secretario de Gobernación ante la Comisión de Asuntos Hidráulicos de la Cámara de Senadores, el 5 de diciembre de 2007 (Unpublished).

Water-Rock Interaction – Birkle & Torres-Alvarado (eds)
© 2010 Taylor & Francis Group, London, ISBN 978-0-415-60426-0

Drainage water—mine tailings interaction: Environmental risk and origin of secondary metal deposit

S.B. Bortnikova & Yu.A. Manstein
Trofimuk Institute of Petroleum Geology and Geophysics SB RAS

O.L. Gaskova & E.P. Bessonova
Sobolev Institute of Geology and Mineralogy SB RAS

N.I. Ermolaeva
Institute of Water and Ecological Problem SB RAS

ABSTRACT: The distribution of chemical elements (Zn, Cu, Fe, Pb, Cd, As, Sb, Be) in the water and bottom sediments of the Belovo swamp-settler was investigated in our integrated study of geochemistry, geophysics, and hydrobiology. This swamp collects drainage escaping from clinker heaps made up of waste of pyrometallurgical smelting of sphalerite concentrate. Water in the swamp has high TDS with extremely high contents of toxic elements. Bottom sediments in the swamp are the mixture of hydrogenic secondary Cu, Zn, Fe and other elements minerals. High metal concentrations lead to drastic changes in biota: phytoplankton, zooplankton, and bacteria communities. About 90% of zooplankton individuals have a genetic mutation expressed in morphological deformations. Infiltration of the high TDS swamp water into ground waters was detected by vertical electric sounding. Contouring of settler volume and preliminary estimation of useful component resources were done. The settler could be considered as a secondary deposit. Extraction of some metals from solutions and sediments can decrease costs on area remediation.

1 INTRODUCTION

Some concentrated drainage streams interacting with sulfide strongly altered tailings can be considered not only as dangerous for the environment but also as secondary raw material for extraction of useful components. These streams can be called "liquid ore" due to high concentrations of dissolved metals. One of the examples of such streams is drainage escaping from clinker heaps of the Belovo processing zinc plant (Kemerovo region, South-Western Siberia, Russia). This drainage is the product of seasonal streams and clinker interaction; it flows into swamp through special ditch. This swamp is bounded by fill dam and allows accumulation and settling of drainage. The mineral composition of the clinker heaps, zonality, content of metals and their speciation (water-soluble, exchangeable) were described in detail (Sidenko et al. 2001, Bortnikova et al. 2006). The aims of this study were: 1) to estimate the element concentrations in the water of the drainage streams and swamp-settler; 2) to determine chemical and mineral composition of the bottom sediments; 3) to identify the solution toxicity for biota; 4) to detect the solution dissemination depth in the settler.

2 STUDY AREA

The clinker heaps are located in the area of Belovo zinc processing plant in the town of Belovo (Fig. 1). The plant extracted zinc by smelting from a sphalerite concentrate which has been obtained from barite-sulfide ores mined at the Salair ore field. The sphalerite concentrate was of low quality and contained high amount of impurities because of Salair ore consist of a fine-grained intergrowth of different sulfide minerals. Since the mid-1990s the plant has ceased operation. About 1 million tons of clinker containing significant amount of sulfuric acid were left in the plant area as heaps. Large amounts of fine-grained coke dust (15–25%) were also present in the waste. The clinker was stored in the heaps of approximately 15 m height with flat top and a steep slope. This waste was affected by spontaneous ignition of the coke dust and subsequent burning of waste in the heaps, thus causing an acceleration of oxidizing processes. Oxidation

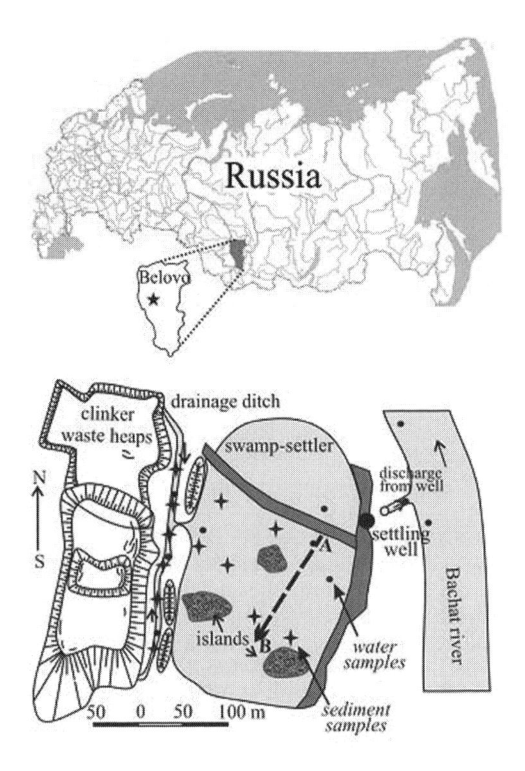

Figure 1. Schematic map of the study area with location of the sampling points.

rate is shown in occurrence of abundant secondary minerals on the waste surface. The western border of the waste is bounded by the swamp—settler in which the drainage from the waste discharges. Water from the swamp flows firstly into settling well and then—in the Bachat river. Water of drainage stream is a bright—dark blue solution, and bottom sediments have white, blue, green and yellow color.

3 METHODS

3.1 *Field sampling*

Various sampling campaigns were carried out in ten years (1999–2008). A total of 22 sampling points were monitored; they comprise surface water (drainage ditch, swamp-settler, discharge into the Bachat river), ground water (settling well, wells and water pumps in neighboring settlement, underground drainage into the Bachat river) and bottom sediments from ditch and swamp (Fig. 1). Sampling was carried out in summer and autumn, in 2008—in winter under from ice. The water samples were filtered through 0.45 μm filters using an all plastic equipment and collected into acid-cleaned polyethylene bottles. Values of pH, Eh, T were measured at the sampling site. Bottom sediments (hydrogenic flocs) were sampled into polyethylene packages and then were dried. Samples for zooplankton study were collected in the winter 2008 by filtration of 300–400 liters of swamp-settler water through Dzhedy net with cells of 62 μm.

3.2 *Laboratory analyses*

Anion species in solutions were determined by ion chromatography. Metals were analyzed by ICP-AES. In addition to the method, dry salt residues were received by evaporation and analyzed by the SR-XRF method. It allows us to decrease limits of element detection. Bottom sediments were analyzed by SR-XRF. Mineral composition was determined by X-ray powder diffraction. Zooplankton was studied using optical microscope according to standard methods.

3.3 *Geophysical investigations*

Estimation of solution extent at depth in the swamp volume was carried out in the winter 2008 by geoelectric sounding. The sounding was carried out using multielectrode resistivity meter. The electrodes penetrated through the holes drilled on the ice. The physical prerequisite for the use of the electric sounding in this case is the high mineralization of the swamp-settler solutions. The problem of solution area contouring in the swamp is reduced to the detection of the electrolyte with relatively low specific electric resistivity in the accommodating media (Manstein et al. 2000). Electric resistivity tomography was performed using Schlumberger array, spacing between electrodes 5 m, total electrodes—48.

4 RESULTS

4.1 *Hydrogeochemistry*

Water of drainage ditch from the clinker heaps shows high salinity and extremely high metal concentrations (Table 1). Solution compositions vary in time but the main trend is an increase of metal concentrations from 1999 to 2008.

Values of pH in the ditch range from 3.3 to 5.8. It is important that metals (Zn and Cu) account for about 50 equ-% of the cationic composition. Main metal species in drainage solutions are aqua-ions and sulfate complexes.

In the swamp-settler pH values of the solutions vary from 3.4 up to 6.9. Total dissolved solids (TDS) is not so high, but metals Zn and Cu account for the greater share of the cations.

Table 1. Solution composition in the drainage ditch and the swamp-settler, mg/L.

Periods	June 2005		February 2008		
Elements	Ditch	Settler	Ditch	Settler1*	Settler2
pH	5.8	4.81	3.30	3.4	6.95
SO_4^{2-}**	1.7	3.7	20	23	1.9
Cl^-	120	190	42	70	170
K+	21	76	120	110	170
Na+	210	530	920	910	360
Mg^{2+}	110	360	700	670	290
Ca^{2+}	250	510	280	390	300
Fe_{tot}	0.18	0.4	15	8.8	0.17
Zn**	0.15	0.37	1.7	1.6	2.1
Cu**	0.15	0.46	3.0	2.5	2.1
Al	2.5	6.7	200	130	6.0
Mn	4.6	22	83	75	18
Cd	1.5	4.2	9.2	9.4	4.4
Co	1.0	3.9	18	16	4.6
Ni	1.2	4.5	21	18	3.2
Be***	ld	2.1	13	17	2.5

* Settler1 is the south part of the swamp-settler, settler2—north one; ** SO_4^{2-}, Zn, Cu—in g/L; *** Be in μg/L.

In 2008 when the sampling occurred in the winter time, TDS of the solutions was higher than in summer solutions and concentrations of metals were sharply increased: Mg, Na, K 4–6 times, Fe and Al by 2 orders, Zn, Cu, Co and Ni by more than order. Besides extremely high concentrations of very toxic elements (Cd and Be) were observed. Obviously principal reason to this was freezing and concentration of metals under ice. Nevertheless occurrence of such high concentrations of dissolved metal species indicates more serious danger for ecosystem, than it was observed in the analysis of the summer solutions.

Discharge into the river Bachat carried out through settling well is not completely cleaned from metals. As a result metal concentrations are significantly higher in the downstream water than in the upstream water (Table 2).

4.2 *Bottom sediments*

Bottom sediments in the drainage ditch and swamp-settler are hydrogenic secondary minerals formed by solution supersaturation. Sediments are stratified substance. Generally, the upper layer is blue-green and the lower one is yellow-green or yellow-orange; thin black lenses occur in both layers. Blue-green color is caused by high concentration of Cu-compounds, concentration of this element reaches up to 27% (Table 3). Higher concentrations of Ca, Zn, Cd, Ni were determined in blue-green layers in comparison with yellow-orange ones

Table 2. Concentration of some elements in surface and ground waters as a result of the waste inflow, mg/l (Zn, Cu, Fe) and μg/l (Pb, Cd, As).

Samples	Zn	Cu	Fe	Pb	Cd	As
Run-off into the Bachat river						
Surface	6.8	0.19	1.0	5.2	230	17
Underground	11	0.32	2.7	11	430	30
Water from the Bachat river						
30 m upstream	0.76	0.087	0.4	3.5	4.6	4.1
300 m downstream	3.8	0.18	6.4	1200	91	17
Water in the settlement						
Pipeline	0.04	0.003	ld	2.4	0.13	Ld
Column in 100 m	0.56	0.032	0.04	ld	1.0	Ld
Wells in 100 m	0.76	0.17	0.12	2.0	3.0	3.0
in 150 m	0.1	0.0063	ld	0.52	0.33	Ld
MAC*	1	1	0.5	10	1	10

* Maximum allowable concentrations [GN 2.1.5.1315–03].

Table 3. Averaged element content in bottom sediments from drainage ditch and swamp-settler, ppm and % (Cu, Fe, Zn, Ca).

	Blue-green layers		Yellow-orange layers	
Elements	Ditch	Swamp	Ditch	Swamp
Cu	19	14	2.4	5.3
Fe	7.3	6.2	20	12
Zn	1.6	0.97	0.76	0.59
Ca	2.0	1.1	0.9	0.71
Pb	470	180	1400	300
Cd	90	46	32	47
Ni	500	400	92	160
Cr	74	81	350	230
Ag	79	25	190	40
As	100	67	610	130
Sb	200	43	570	88
Te	0.33	0.25	0.80	0.26
Se	11	10	8.4	2.2

where Fe concentration reaches up to 19% and Sb, As, Cr are elevated. Wide spectrum of chemical elements was determined both in blue-green and in yellow-orange sediments.

In mineral composition sediments consist of Cu, Zn, Fe, Ni—water hydroxylsulfates (Table 4). Reduced minerals such as native Cu deposited on Fe-scrub, sulfides and arsenides of Ni, Cu, and Fe (rammelsbergite, maucherite semseyite, cubanite) which can be formed at solution-coke dust interaction were also found out in the ditch.

In swamp-settler carbonates (malachite, azurite, rosasite) were also determined. They can be

Table 4. Secondary minerals formed hydrogenic flocs in drainage ditch and swamp-settler.

Minerals	Formula
Orthoserpierite	$Cu_3(SO_4)_2(OH)_2 \times 2H_2O$
Posnjakite	$Cu_4SO_4(OH)_6 \times H_2O$
Ramsbeckite	$Cu_{15}(SO_4)_4(OH)_{22} \times 6H_2O$
Schulenbergite	$(Cu,Zn)_7(SO_4)_2(OH)_{10} \times 3H_2O$
Guildite	$CuFe^{3+}(SO_4)_2(OH)$
Honessite	$Ni_6Fe_2(SO_4)(OH)_{16} \times 4H_2O$
Gerhardite	$Cu_2(OH)_3NO_3$
Cu-amorphous mineral	
Jarosite	$KFe_3(SO_4)_2(OH)_6$
Gypsum	$CaSO_4 \times 2H_2O$
Takanelite	$Mn^{2+}Mn^{4+}4O_9 \times H_2O$
Goethite	$FeO(OH)$
Rancierite	$(Ca,Mn^{2+})Mn^{4+}4O_9 \times 3H_2O$
Ellestadite	$Ca_5(OH)(SO_4)(SiO_4)(PO_4)_3$
Hydroxylellestadite	$Ca_{10}(SiO_4)_3(SO_4)_3(OH,Cl,F)_2$
Arsenosiderite	$Ca_6(H_2O)_6[Fe_9O_6(AsO_4)_9]$
Glauberite	$Na_2Ca(SO_4)_2$
Klaprothite	Cu_3BiS_3
Cubanite	$CuFe_2S_3$
Semseyite	$Cu_9Sb_8S_{21}$
Rammelsbergite	$NiAs_2$
Maucherite	$Ni_{11}As_8$
Skutterudite	$CoAs_5$
Malachite	$Cu_2CO_3(OH)_2$
Azurite	$Cu_3(CO_3)_2(OH)_2$
Sjogrenite	$Mg_6Fe^{3+}2(OH)_{16}(CO_3) \times 4H_2O$
Sergeevite	$Ca_2Mg_{11}(CO_3)_9(HCO_3)_4(OH)_4$
Nakauriite	$Cu_8(SO_4)_4(CO_3)(OH)_6 \times 48H_2O$
Rosasite	$(Cu,Zn)_2CO_3(OH)_2$

formed under drainage neutralization by the bed rocks of swamp. Stratification of sediments and physico-chemical modeling indicate that Al, Pb, and Ca sulfates are precipitated from solutions earlier than Cu and Zn sulfates. In the settler solutions metal species and their ratios are the same as in the ditch solutions.

4.3 *Zooplankton*

Seven groups of zooplankton have been found in winter samples: Cladocera *Bosmina longirostris* O.F. Mull.; Rotatoria—*Filinia longiseta longiseta* (Ehrb.), *Keratella quadrata* (O.F.Mull.), *Keratella cochlearis* (Gosse), *Br. angularis angularis* (Gosse), *Brachyonus quadridentatus*, *Testudinella patina*. 90% of detected zooplankton organisms have morphological deformity complicating specific definition considerably (Fig. 2). Judging by attributes *Bosmina longirostris* (O.F.Mull.) resembles *Bosmina obtusirostris arctica* Lill., but has typical for *B. longirostris* postabdomen. The majority

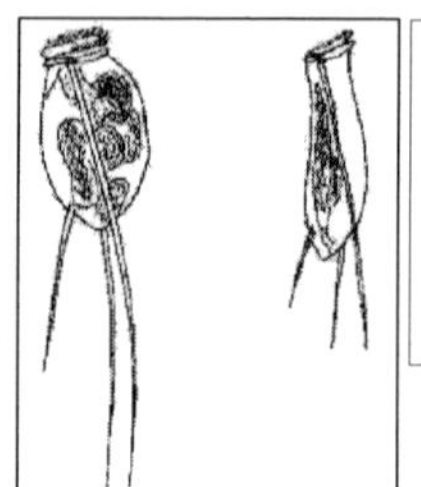

Filinia longiseta (Ehrb.)

Bosmina longirostris (O.F.Mull.)

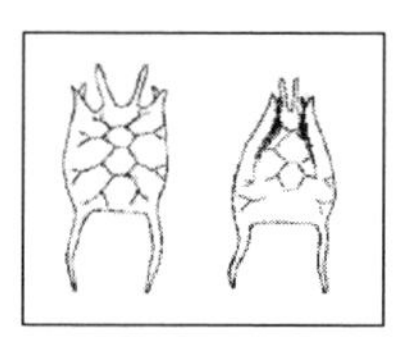

Keratella quadrata (O.F.Mull.)

Figure 2. Variations in morphology of some zooplankton organisms from the swamp-settler. Normal unit are shown on the left pictures, zooplankton from samples—on the right.

of Rotatoria have considerably deformed armour. Obviously influence of high metal concentrations leads to genetic mutation so long as all individuals are deformed identically. In summer and autumn samples zooplankton was absent. It seems that temperature increase and acceleration of metabolism lead to accumulation of lethal dose of toxic elements from the water.

The features of phytoplankton and bacteria are evidence to the extreme toxic conditions for biota in the swamp-settler and on the surrounding area.

4.4 *Electric tomography*

The result of geophysical sounding was the geoelectric cross-section up to 14 m depth on which several zones with different specific resistivity of media were revealed (Fig. 3). Each of them corresponds to particular hydrogeochemical condition.

1. Horizon with specific resistivity 3.5–4.2 Ohm · m. It is surface layer represented by concentrated waters in which metal concentrations (Cu, Zn, Cd, Ni, Co, and others) reach up to high toxicity. Solution composition is caused by discharge of the drainage under from clinker heaps;
2. Bottom sediments formed by secondary water Cu, Zn, Fe, Ni (and other metals) sulfates with combination of silt and organic matter; pore space is filled by solutions of the same composition as the surface layer; specific resistivity is 4.2–6.8 Ohm · m;

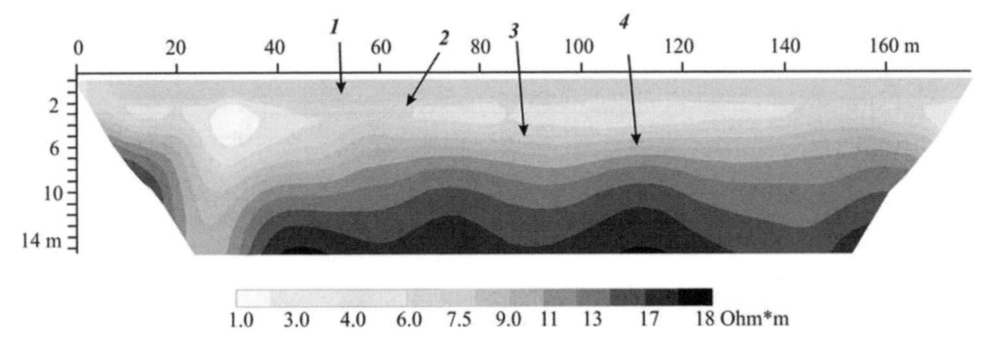

Figure 3. Vertical zoning of Belovo swamp-settler according to the data of geophysical sounding.

3. Horizon of ground water; according to values of specific resistivity, the infiltration of high toxic solutions into ground water can be inferred;
4. Impermeable clay horizon having resistivity of 8–18 Ohm · m is located at 6.5–10 m depth.

Infiltration of the concentrated solutions into ground waters is confirmed by analysis of water samples taken from the wells and water pumps situated on the area of the settlements in 60–100 m far from waste heaps and swamp-settler (Table 2). On the base of integrated results it can be arguable that waste heaps and swamp-settler are the sources of large-scale environment pollution, and metal dissemination occurs uncontrolled and in different directions.

At the moment regulation of these processes is practically impossible. In comparison with similar drainage solutions described in literature the Belovo swamp-settler corresponds to one of the most mineralized reservoir (Fig. 4). Even in comparison with well-known drainage within the Richmond mine (Iron Mountain) where solutions had pH values ranging to negative (Alpers et al. 1992, Nordstrom & Alpers 1999), solutions within Belovo waste heap have higher metal concentrations in some cases. Meanwhile in the Richmond mine solutions the main share in TDS belongs to iron, but in Belovo swamp-settler leading metals are Cu and Zn. The solutions in ditch and settler can be considered as potential row material for economic extraction of useful components which can decrease remediation costs.

Estimated resources of swamp-settler for area of 100 × 100 m (only south part of settler) are: 40 tons of Zn and 50 tons of copper are in dissolved species in settler solution plus 800 tons of Zn and 8000 tons of Cu are in secondary soluble minerals in bottom sediments.

Total estimation of settler resources will show more amounts of metals which are in easily extracted species. This task can be addressed by comparing reclamation and exploitation costs with respect to benefits for the ecological system.

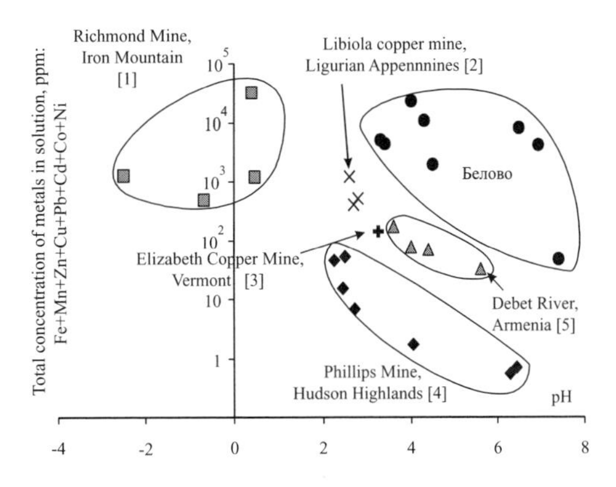

Figure 4. Comparison of pH values and metal concentrations in some drainage solutions and water from Belovo settler (1—Alpers et al. 1992; Nordstrom & Alpers 1999; Robbins et al. 2000, 2—Dinelli et al. 2001; 3—Balistrieri et al. 2007, 4—Gilchrist et al. 2009; 5—Kurkjan et al. 2004).

5 CONCLUSIONS

Water of the Belovo swamp-settler is concentrated strong toxic solutions with the extreme concentration of hazardous and very dangerous elements. The level of concentration and mobility of the elements is determined by previous ore processing as well as waste storage conditions. Remediation actions are urgently required because the heap and settler are located within the city's range and houses are in close proximity to these sources of pollution.

The depth of the settler solutions dispersion shows that there is intensive dispersion of pollutants into groundwater, which significantly complicates the choice of environmental measure. Costs of minimizing environmental damage can be reduced by processing the easily extracted Zn and Cu compounds from the settler solutions and bottom sediments.

ACKNOWLEDGEMENTS

The researches were financially supported by RFBR (Grant No 08–05–00688). We are grateful to operator of analytical devices—S. Nechepurenko, Yu. Kolmogorov, N. Palchic.

REFERENCES

Alpers, C.N., Nordstrom, D.K. & Burchard J.M. 1992. Compilation and interpretation of water-quality and discharge data for acidic mine waters at Iron Mountain, Shasta County, California, 1940–91. *U.S. Geol. Surv. Water-Resources Inv. Rept.* 91–4160: 173.

Balistrieri, L.S., Seal, R.R., Piatak, N.M. & Paul, B. 2007. Assessing the concentration, speciation, and toxicity of dissolved metals during mixing of acid-mine drainage and ambient river water downstream of the Elizabeth Copper Mine, Vermont, USA. *Applied Geochemistry* 22: 930–952.

Bortnikova, S.B., Gaskova, O.L. & Airijants, A.A. 2003. *Waste lakes: origin, development and influence on the environment*. Novosibirsk: "Geo".

Dinelli, E., Lucchini, F., Fabbri, M. & Cortecci, G. 2001. Metal distribution and environmental problems related to sulfide oxidation in the Libiola copper mine area (Ligurian Appennines, Italy). *Journal of Geochemical Exploration* 74: 141–152.

Gilchrist, S, Gates, A., Szabo, Z. & Lamothe, P.J. 2009. Impact of AMD on water quality in critical watershed in the Hudson River drainage basin: Phillips Mine, Hudson Highlands, New York. *Environmental Geology* 57: 397–409.

Kurkjian, R, Dunlap, C. & Flegala, A.R. 2004. Long-range downstream effects of urban runoff and acid mine drainage in the Debed River, Armenia: insights from lead isotope modeling. *Applied Geochemistry* 19: 1567–1580.

Manstein, A.K., Epov, M.I., Voevoda, V.V. & Sukhorukova, K.V. 2000. RF Patent Certificate no. 2152058, G 01 V 3/10 24.06.98. *Byull. Izobret. 18*.

Maximum allowable concentrations (MAC) of chemical substances in water GN 2.1.5.1315-03. 2003. Moscow.

Nordstrom, D.K. & Alpers, C.N. 1999. Negative pH, efflorescent mineralogy, and consequences for environmental restoration at the Iron Mountain Superfund site, California. *Proceed. from the National Acad. of Sciences*. 96: 3455–3462.

Sidenko, N.V., Giere, R., Bortnikova, S.B., Cottard, F. & Palchik, N.A. 2001. Mobility of heavy metals in self-burning waste heaps of the zinc smelting plant in Belovo (Kemerovo Region, Russia). *Journal Geochemical Exploration* 74 (1–3): 109–125.

Water-Rock Interaction – Birkle & Torres-Alvarado (eds)
© 2010 Taylor & Francis Group, London, ISBN 978-0-415-60426-0

Using CFCs and SF_6 for groundwater dating: A SWOT analysis

W.G. Darling, D.C. Gooddy & B.L. Morris
British Geological Survey, Wallingford, UK

A.M. MacDonald
British Geological Survey, Edinburgh, UK

ABSTRACT: Knowing the residence time of groundwater can be of importance in understanding key issues in the evolution of water quality. Chlorofluorocarbons (CFCs) and sulphur hexafluoride (SF_6) offer a convenient way of dating waters up to ~60 yrs old. Although any one of these gases can in principle provide a groundwater age, if two or more are measured on a water sample the potential exists to distinguish between different modes of flow or mixing. As with all groundwater dating methods, caveats apply. Recharge temperature and elevation must be reasonably well-constrained. Mainly for SF_6, the phenomenon of 'excess air' requires consideration. Mainly for the CFCs, local sources of contamination need to be considered, as do redox conditions. For both SF_6 and the CFCs, the nature and thickness of the unsaturated zone need to be factored into residence time calculations. This paper examines the pros and cons of the trace-gas dating method.

1 INTRODUCTION

Important hydrogeochemical issues such as the effects of climate change on water quality, or the origin of high-arsenic groundwater, involve the study of water–rock interaction processes occurring effectively at or close to the present day. As part of such investigations, it is important to have a readily-available way of dating young (i.e. up to decades old) waters. The atmospheric trace gases CFC-11, CFC-12, CFC-113 and SF_6 (sulphur hexafluoride) are increasingly being used as tracers of groundwater residence time (IAEA 2006). Large-scale production of CFC-12 began in the early 1940s, followed by CFC-11 in the 1950s and by CFC-113 in the 1960s. CFC-11 and CFC-12 were used mainly for refrigeration and air-conditioning, while CFC-113 was used as a solvent. Inevitably they leaked into the environment, with atmospheric concentrations rising until the 1990s, when production was cut back to protect the ozone layer as a result of the Montreal Protocol. SF_6, another industry-derived gas, has been detectable in the atmosphere since the early 1960s and is still rising strongly in concentration.

Theoretically the concentrations of these gases in a groundwater can be matched to a particular year of recharge. In practice there may be complications, either with the tracers themselves, or related to the fact that groundwater as sampled is probably rather rarely of a single recharge age, owing to a combination of wellbore and within-aquifer mixing processes. This review takes a 'SWOT' (Strengths–Weaknesses–Opportunities–Threats) approach to this and other aspects of CFC and SF_6 dating.

2 PRINCIPLES OF THE METHOD

The use of CFCs and SF_6 as indicators of residence time is based on the known rise of their atmospheric concentrations over the past 60 yrs, the observation that they are well-mixed in the atmosphere (unlike tritium), and the assumption that they dissolve in water according to their Henry's Law solubilities at the temperature of recharge (Plummer & Busenberg 1999). Figure 1 shows the concentrations of CFC-11, CFC-12, CFC-113 and SF_6 to be expected in groundwater recharged between 1950 and 2010 at a temperature of 10°C, typical of much of the UK. Higher or lower recharge temperatures would tend towards lower or higher concentrations respectively.

3 STRENGTHS

3.1 *Sampling and analysis*

Sampling by the bottle-in-can method of Oster (1994), or more recently the single bottle method of the USGS (IAEA 2006), is straightforward and requires no specialist expertise.

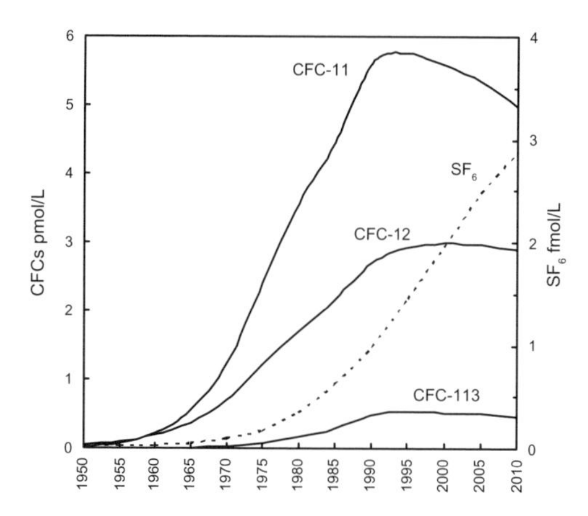

Figure 1. Variation over time of the concentrations in groundwater of the CFCs and SF_6 assuming equilibrium with the Northern Hemisphere atmospheric mixing ratios at a recharge temperature of 10°C (typical of the UK). Based on data from http://water.usgs.gov/lab/software/air_curve/.

CFCs and SF_6 are measured by gas chromatography using an electron capture detector (GC-ECD) following cryogenic pre-concentration (IAEA 2006). The detection limit for CFC concentrations in water is 0.01 pmol/L, while for SF_6 it is 0.1 fmol/L. Both CFC and SF_6 analyses should ideally be calibrated to a bulk air standard from an AGAGE atmospheric monitoring station (http://agage.eas.gatech.edu/). Analysis is rapid and cost-effective compared to tritium and other radioisotope techniques.

3.2 *Interpretation*

Measured CFC and SF_6 concentrations can be interpreted in terms of age simply by reading off the year of recharge from the curves in Figure 1 (adjusted if necessary for local recharge temperature (RT)). However, this presupposes that the measured water is the result of piston flow, i.e. as if along a simple tubular flowline from recharge to discharge. In reality, groundwaters are just as likely to be mixtures of young and older waters, or to be the result of mixing of waters arriving along flowlines of different lengths and therefore different ages. This last category in particular may be a consequence of mixing in the wellbore when a borehole is unlined.

Two basic ways of resolving this exist. One is to plot one CFC versus another. Figure 2a shows a plot of CFC-11 vs. CFC-12, with the piston flow curve and the modern–old binary mixing line. Given the error on the analysis, not to mention fairly ubiquitous small enhancements due to contamination

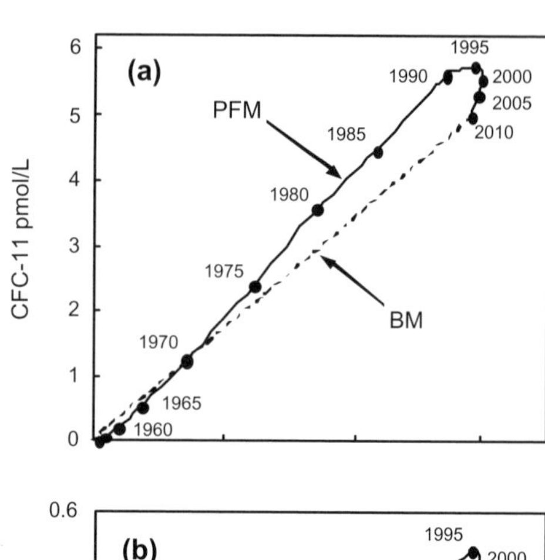

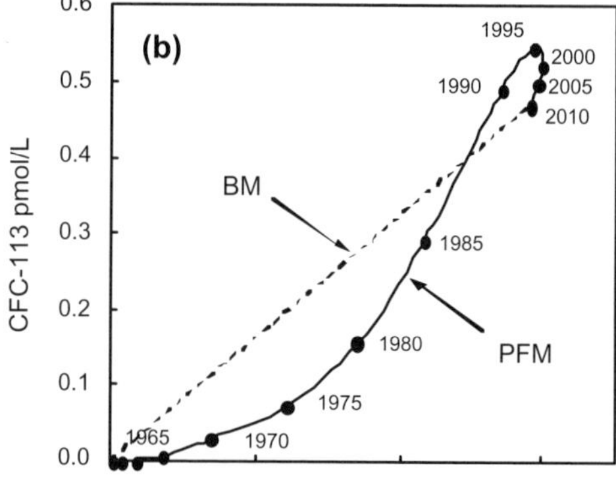

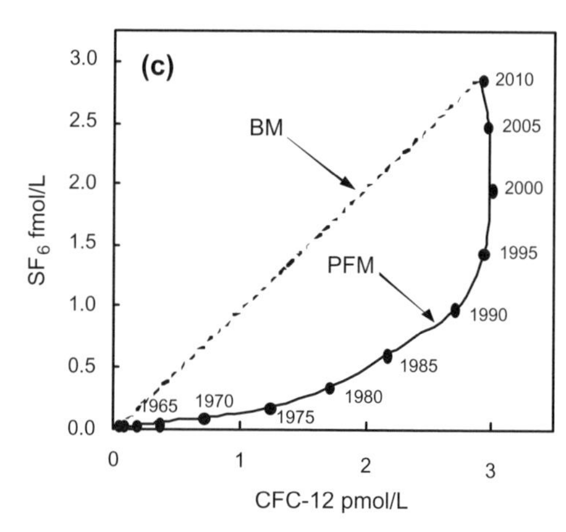

Figure 2. Plots of (a) CFC-11 vs. CFC-12, (b) CFC-113 vs. CFC-12, and SF_6 vs. CFC-12 showing the piston flow curve (PFM) and the binary mixing line (BM) between modern and old groundwater. RT = 10°C. Year of recharge shown for PFM.

(see 4.4 below), it is clear that that resolving piston flow from mixing will usually be difficult. The situation is somewhat improved by plotting CFC-113 versus CFC-12 (Fig. 2b) but there is still room for some ambiguity, particularly around the crossover in the late 1980s. An effective way of resolving this is to plot SF_6 versus CFC-12 (Fig. 2c). The resulting 'bow-shape' has good separation over most of the field, and no crossover. The other CFCs can substitute for CFC-12.

The simple piston flow and binary mixing model (PFM and BM) scenarios outlined by the SF_6–CFC bow are really opposite extremes of groundwater behaviour. In reality, most groundwater flow may be more complicated, which is why some intermediate lumped-parameter flow models such as the exponential mixing (EMM) and exponential piston flow (EPM) models have been proposed (Maloszewski & Zuber 1982).

Figure 3 compares the concentration curves for the three types of model, with residence times on the PFM, EPM and EMM curves, and amount of modern water expressed in fractional terms ('modern fraction') on the BM line. It is important to note that information derived from plotting analyses on this type of diagram must be considered in relation to hydrogeological information, not least regarding borehole construction. This is because groundwater mixing may occur just as readily (if not more so) in boreholes as in aquifers, giving misleading information if interpreted in isolation.

It will be noted that the EMM line actually extends back beyond the ~60 yr limit of the PFM. This is a consequence of the exponential nature of the putative mixing and is the reason why tracers of the post-war period can under appropriate conditions

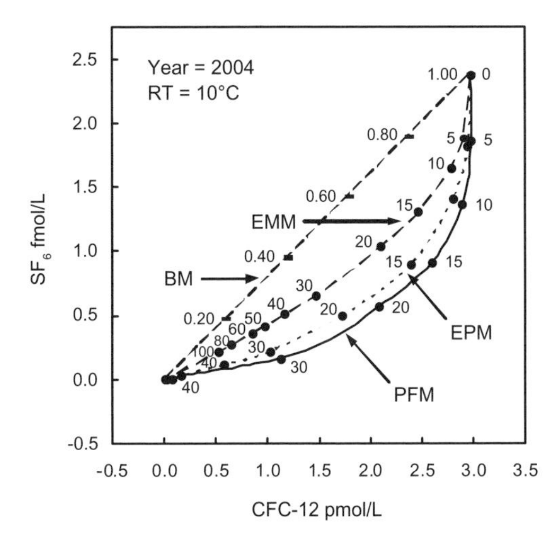

Figure 3. Example plot of SF_6 vs. CFC-12 showing the PFM, EPM and EMM age curves, with groundwater residence times in years. Also shown is the binary mixing line BM with amounts of modern water expressed as a fraction of unity. Based on Gooddy et al. (2006).

be used to infer groundwater residence times of up to ~100 yrs (Maloszewski & Zuber 1982).

4 WEAKNESSES

4.1 *Unsaturated zone*

The vadose or unsaturated zone (USZ) poses certain problems for the interpretation of CFCs and SF_6. The movement of atmospheric trace gases through the USZ occurs both in the dissolved and gas phases. When the USZ is thin, the trace-gas composition of moisture maps on to that of the atmosphere (Cook & Solomon 1995, Engesgaard et al. 2004). In the deeper USZ of simple porous aquifers there is a time lag for the diffusive transport of CFCs and SF_6, a function of the tracer diffusion coefficients, tracer solubility in water, and moisture content (Weeks et al. 1982, Cook and Solomon 1995). In the deepest USZs this suggests the tracer age could be greater than the mean advection time of recharge. In fractured aquifers however, the situation would be reversed meaning that CFC or SF_6 ages obtained from groundwaters effectively represent residence time only since recharge reached the water table, i.e. within the saturated zone (Darling et al. 2005). However, whatever the type of aquifer, provided there is some knowledge of rock properties the effect of the USZ can be factored in to age calculations.

4.2 *Elevation effects*

Atmospheric pressure falls almost linearly with rise in altitude, so that at ~5000 m above sea level it has declined to half the value. Therefore some knowledge of the average elevation of recharge is desirable if relief of the study area exceeds approximately 1000 m. In very high-altitude areas, such as the Altiplano of the Andes, the elevation of the discharge area would also have to be factored in. Taking a simple case of recharge at 1000 m and discharge at sea level (0 m), a correction factor of 1.13 would need to be applied to the measured CFC and SF_6 data to allow comparison with a calibration curve based on sea-level atmospheric pressure of approximately 1000 mb.

Temperature typically falls with rise in altitude, with an average 'lapse rate' of 6.5°C per 1000 m, though this is highly dependent on local factors. For the 1000 m to 0 m scenario considered above, this would introduce a correction factor of approximately 0.7 (it would vary slightly for the different gases). Thus the altitude-based pressure and temperature correction factors work in opposite directions, with the latter exceeding the former, and therefore tend to cancel out rather than reinforce each other.

Strictly speaking pressure and temperature are simply factors that require correcting for, rather than weaknesses. They only tend to become so in mountainous terrain where recharge elevation is poorly constrained.

4.3 *Excess air*

Excess air (EA) arises from the forcible dissolution of air bubbles that inevitably occurs during the recharge process. This is a process which supplements the dissolved gas content due to simple atmospheric equilibration, on which trace gas dating depends, and therefore requires correction. The impact of EA on a particular gas is in inverse proportion to the solubility of that gas.

EA is present in all groundwaters to some extent, related to factors such as the nature of matrix porosity, amount of fracturing, and size of seasonal fluctuations in water table elevation. It is usually present to the extent of a few cc/L (STP) but may be higher, sometimes significantly so, in fractured aquifers (Wilson & McNeill 1997). For 'normal' groundwaters with an EA component of up to ~5 cc/L, there is a need to correct measured SF_6 concentration but not CFC-12 as other uncertainties are of the same order. At greater concentrations of EA, both SF_6 and CFC-12 should be corrected. Recharge temperature affects the correction, which increases with temperature (Fig. 4). Figure 4 strictly only applies to simple unfractionated EA; partial or closed-system equilibration of

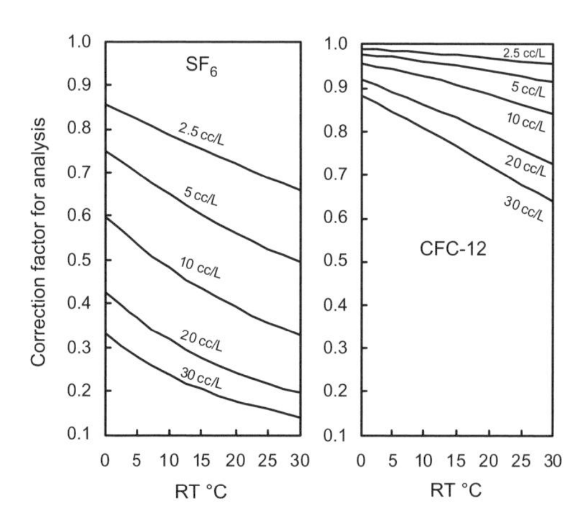

Figure 4. The effect of recharge temperature (RT) on the excess air (EA) correction that sample analyses require to give reliable ages. Owing to its lower solubility, SF_6 needs correction at all levels of EA while for CFC-12 only at >5 cc/L (STP).

EA can also occur and may need to be considered (Aeschbach-Hertig et al. 2000).

Likely EA values for particular aquifers can be found in the literature (e.g. Wilson & McNeill 1997) or derived from measuring N_2, Ar or other noble gases on a representative suite of samples. The correction can be accomplished by a simple iterative procedure that takes account of various factors (Busenberg & Plummer 2000).

4.4 *Contamination*

The CFCs in particular are likely to reach values in excess of atmospheric equilibrium owing to inputs from sources such as industrial activities and leaking landfills. Contamination may be atmospheric or in-ground (Höhener et al. 2003), and is not too surprising: for example it has been calculated that less than one-tenth of the amount of CFC-12 present in a single domestic refrigerator (of older design) could theoretically contaminate a moderately sized aquifer to more than ten times current atmospheric equilibrium levels (Morris et al. 2006), i.e. giving the water a modern fraction value of 10. Contamination tends to affect urban/peri-urban and fractured aquifers the most. Owing to their extremely low atmospheric equilibrium levels, CFC concentrations can reach several hundred times modern values but still be below drinking water guideline concentrations. But even very low amounts of contamination can be problematic to groundwater dating: with reference to Figure 3 it is clear that a small enhancement in CFC-12 could make a binary-mixed water appear to be the product of exponential mixing or piston flow.

SF_6 reaches contaminated values only rarely. Anthropogenic sources include high-voltage electricity supply equipment, Mg and Al smelting, and landfills (e.g. Fulda & Kinzelbach 2000, Santella et al. 2007).

Terrigenic sources include rocks containing fluorite, fine-grained volcanic rocks such as rhyolite, and areas of metallic mineralisation (e.g. Harnish & Eisenhauer, 1998; Koh et al. 2007).

4.5 *Microbial degradation*

There is ample evidence that CFCs are affected by microbial breakdown under low-O_2 conditions. This tends to affect CFC-11 and CFC-113 more rapidly than CFC-12, as reported in a number of studies (e.g. Khalil & Rasmussen 1989, Oster et al. 1996, Sebol et al. 2007, Horneman et al. 2008). Figure 5 shows an example of this in relation to a groundwater with redox layering. Typically reduction effects are revealed by major differences in

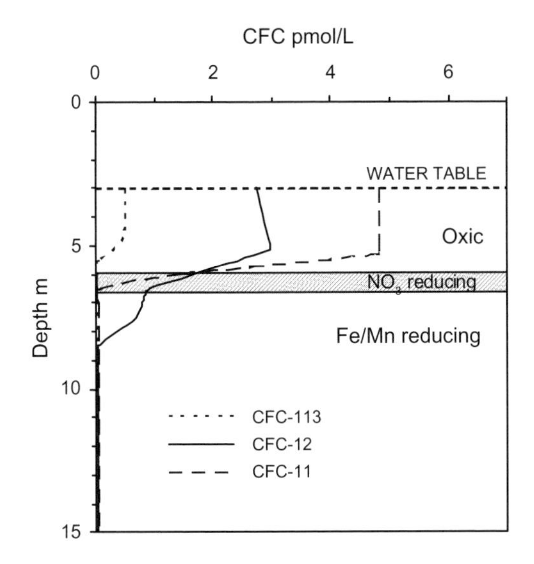

Figure 5. The effect of redox conditions on the bacterial degradation of the CFCs (after Sebol et al. 2007).

Table 1. Evidence for a possible thermal effect on CFC-12, which is below detection while other species remain detectable.

Site	CFC-12 pmol/L	CFC-11 pmol/L	CFC-113 pmol/L	SF_6 fmol/L
Bath, UK, temp 47°C, analysis by Spurenstofflabor				
Cross Spring	<0.01	0.26	0.04	-
Stall Street BH	<0.01	0.17	0.02	-
Chaudfontaine, Belgium, temp 37°C, analysis by BGS				
Chaud A	<0.02	0.55	-	0.18
Chaud B	<0.02	0.54	-	0.13

apparent age between the different CFCs. However, if conditions are highly reducing it is possible that all three CFCs will have been reduced to concentrations below the detection limit, thus showing a measure of agreement while not reflecting the true age. It is therefore important to be aware of redox conditions when interpreting CFC data.

While not bacterially mediated, sorption will also have the effect of lowering CFC concentrations. It is only likely to be a problem in aquifers where the matrix has a high organic matter content (IAEA 2006).

4.6 *Temperature*

Whereas CFC-12 is the most robust CFC under low-O_2 conditions (see previous section), there is some limited evidence that it may be more subject to temperature effects. Table 1 gives results obtained from two warm springs, in Belgium and the UK. Two different laboratories found CFC-12 below detection whereas CFC-11, CFC-113 and SF_6 were all present at detectable concentrations. The mechanism by which CFC-12 would be affected in preference to the other trace gases remains to be established, and indeed a larger database is required to determine whether the effect is real or a contamination-related artifact.

5 OPPORTUNITIES

5.1 *CFCs as flow tracers*

When groundwaters contain 'over-modern' concentrations of the CFCs (i.e. modern fraction values >1) clearly no quantitative residence time information can be obtained from them, beyond the fact that they must contain at least a proportion of post-war recharge. However, over-modern concentrations can sometimes be used to fingerprint water bodies and therefore shed light on processes such as infiltration, mixing and dilution (Busenberg & Plummer 1992, Böhlke et al. 1997).

Three CFCs are commonly measured on groundwater samples, which allows the use of trilinear plots. Such plots are commonly used in the interpretation of inorganic hydrochemical data as a way of detecting patterns of groundwater mixing or evolution (e.g. Hem 1992), though they require a certain amount of care in interpretation since they depict ratios rather than absolute concentrations. As with all such plots, concentrations of the three CFCs need to be converted to percentages. It is most convenient to do this using modern fraction values (see above), which are themselves directly proportional to the originally-measured concentrations. On this basis, modern air-equilibrated water sits in the centre of the plot. An example of such a plot is shown in Figure 6, which is based on a quarried and landfilled area in the southern UK (Darling et al. 2010).

In this case, neighbouring groundwaters have different CFC fingerprints so that probable flowpaths can be identified. The observed contrast between the proportions of individual CFCs presumably arises from the disposal of different kinds of waste material in the various landfills now occupying the several small-scale excavations remaining as a legacy of chalk quarrying operations.

5.2 *Waters of mixed age*

When CFCs and SF_6 are unaffected by contamination, they are as outlined earlier capable of discriminating between piston flow and mixed waters.

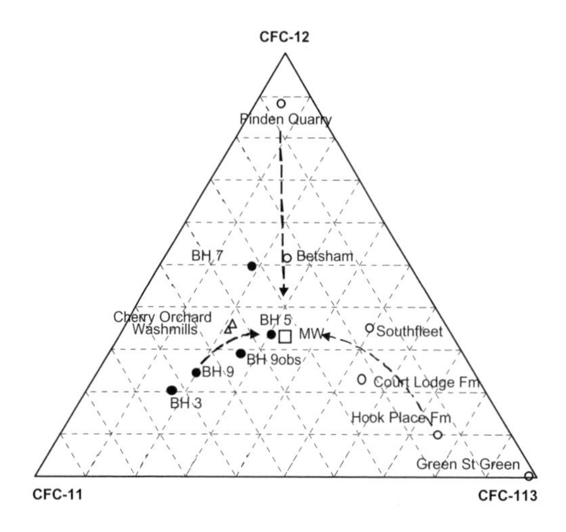

Figure 6. Trilinear plot of the CFCs using over-modern groundwater concentrations to determine flow/dilution trends in a landfilled area of the Chalk, southern UK. From Darling et al. (2010).

This makes them a rapid and cost-effective way of screening to determine if an old component is present, and in what proportion. This can help to decide whether or not to have the water dated by radiocarbon or other costly long-term age indicator.

The sensitivity of CFC and SF_6 analysis also offers a way of testing apparently old groundwaters for modern inputs which, though small in volume may have water-quality implications. In the past, tritium has performed this role but the CFCs and SF_6 can give more quantitative information and provide a degree of 'redundancy' that a single tracer cannot. Table 2 shows the results of CFC and ^{14}C measurements on groundwaters from a block-faulted, till-covered Triassic sandstone aquifer in the Lagan Valley of Northern Ireland. This shows firstly good agreement between the small percentages of modern water calculated from CFC-11 and CFC-12 concentrations in individual boreholes, and secondly the heterogeneity of this type of faulted aquifer both in terms of the proportion of modern water present and the calculated model age of the old component.

5.3 *New tracers*

Other atmospheric trace gases have been proposed as residence time indicators (Busenberg & Plummer 2008). CFC-13 (CF_3Cl), currently present in the atmosphere at about the same concentration as SF_6, has been rising steadily presumably because at this very low concentration it is not yet seen as a threat to the ozone layer. It also has the advantage that it has not been widely used industrially and therefore groundwaters are less likely to be contaminated with it than the conventional CFCs. Trifluoromethyl sulphurpentafluoride (SF_5CF_3) is also increasing in the atmosphere. Its advantage over SF_6 appears to be a lack of terrigenic production. To date it has reached only a few percent of the current SF_6 value so analytical precision is problematic. Both these new tracers seem likely to start featuring over the next few decades if atmospheric concentrations keep rising.

Table 2. Small amounts of modern water detected in Lagan Valley (Northern Ireland) groundwaters with Holocene bulk radiocarbon ages, showing the good agreement between CFC-11 and CFC-12. The right-hand column gives the model age for the old components.

Borehole	CFC-12	CFC-11	CFC av.	^{14}C model age in yrs	
	Modern Fraction			Bulk	Old comp.
CL	0.00	0.02	0.01	390	1550
CH	0.07	0.05	0.06	3900	4600
E3	0.00	0.01	0.01	3150	3400
KH	0.04	0.04	0.04	4700	4950
LB	0.09	0.04	0.07	2500	2990
LC	0.07	0.02	0.04	300	350
LP1	0.13	0.11	0.12	4200	5150
LP2	0.08	0.03	0.06	4450	5200
ML	0.06	0.05	0.05	7850	9200
MV	0.03	0.03	0.03	1950	2000
MO	0.00	0.04	0.02	1100	1300

6 THREATS

It is clear from Figure 1 that CFC concentrations in the atmosphere have started to decline. Study of that figure shows that concentrations similar to those of the present day (2010) also occurred 16–23 yrs ago, depending on the CFC. In theory, because the ratios between the CFCs are slightly different now from what they were in the 1987–94 period, it should still be possible to use CFCs to resolve the different ages (at least until they reach pre-industrial baseline values). In practice, measurement precision is unlikely to be high enough to do this unambiguously.

The lowering of atmospheric CFC concentrations is however much less of an issue if the SF_6 co-plot approach (Figs. 2c & 3) is used, because the rise of SF_6 so far shows no sign of significant slowing. Clearly for the next few decades the effect of lower CFCs will merely cause the CFC–SF_6 piston flow line to become more curved, with the benefit that it will become easier to distinguish between mixed and PFM or EPM groundwaters over much of the age range.

7 CONCLUSIONS

Like all groundwater dating methods, the CFCs and SF_6 have advantages and disadvantages. A SWOT analysis reveals the following.

Sampling and analysis is reasonably straightforward. Under ideal conditions the gases can be used to date groundwaters up to ~60 yrs old and resolve piston flow from simple mixing. Excess air is more of an issue than recharge elevation, though normally only for SF_6. Degradation of CFCs generally occurs only under low-oxygen conditions, but possibly also under thermal conditions. Contamination of CFCs (anthropogenic) or SF_6 (terrigenic) may occur, but then be usable as a flow rather than age tracer.

The conventional CFC tracers are declining in the atmosphere and will be unusable for dating in several decades' time. However, new trace gases are likely to replace them in this role.

ACKNOWLEDGMENTS

We thank Eurybiades Busenberg and Werner Aeschbach-Hertig for helpful reviews of the original manuscript.

REFERENCES

Aeschbach-Hertig, W., Peeters, F., Beyerle, U. & Kipfer, R. 2000. Paleotemperature reconstruction from noble gases in ground water taking into account equilibrium with entrapped air. *Nature* 405: 1040–1044.

Böhlke, J.K., Revesz, K., Busenberg, E., Deak, J., Deseo, E. & Stute, M., 1997, A ground-water record of halocarbon transport by the Danube River. *Environmental Science and Technology* 31: 3293–3299.

Busenberg, E. & Plummer, L.N. 1992. Use of Chlorofluorocarbons (CCl_3F and CCl_2F_2) as hydrologic tracers and age dating tools: the alluvium and terrace system of central Oklahoma. *Water Resources Research* 28: 2257–83.

Busenberg, E. & Plummer, L.N. 2000. Dating young groundwater with sulfur hexafluoride: Natural and anthropogenic sources of sulfur hexafluoride. *Water Resources Research* 36: 3011–3030.

Busenberg, E. & Plummer, L.N., 2008, Dating groundwater with trifluoromethyl sulfurpentafluoride (SF_5CF_3), sulfur hexafluoride (SF_6), CF_3Cl (CFC-13), and CF_2Cl_2 (CFC-12), *Water Resources Research*, 44: W02431, doi:10.1029/2007 WR006150.

Cook, P.G. & Solomon, D.K., 1995. Transport of trace gases to the water table: Implications for groundwater dating with chlorofluorocarbons and krypton-85. *Water Resources Research* 31: 263–270.

Darling, W.G., Gooddy, D.C., Riches, J. & Wallis, I. 2010. Using environmental tracers to assess the extent of river–groundwater interaction in a quarried area of the English Chalk. *Applied Geochemistry* doi: 10.1016/j.apgeochem. 2010.01.019.

Darling, W.G., Morris, B.L., Stuart, M.E. & Gooddy, D.C. 2005. Groundwater age indicators from public supplies tapping the Chalk aquifer of Southern England. *Water and Environment Journal* 19: 30–40.

Engesgaard. P., Højberg, A.L., Hinsby, K., Jensen, K.H., Laier, T., Larsen, F., Busenberg, E. & Plummer, L.N., 2004. Transport and time lag of chlorofluorocarbon gases in the unsaturated zone, Rabis Creek, Denmark. *Vadose Zone Journal* 3: 1249–1261.

Fulda, C. & Kinzelbach, W. 2000. Sulphur hexafluoride (SF6) as a new age-dating tool for shallow groundwater: methods and first results. *Proc. International Conference on Tracers and Modelling in Hydrogeology, Liège, Belgium.* IAHS Publication no. 262, 181–185. Wallingford, Oxfordshire, UK: IAHS Press.

Gooddy, D.C, Darling, W.G., Abesser, C. & Lapworth, D.J. 2006. Using chlorofluorocarbons (CFCs) and sulphur hexafluoride (SF_6) to characterize groundwater movement and residence time in a lowland Chalk catchment. *Journal of Hydrology* 330: 44–52.

Harnish, J. & Eisenhauer, A. 1998. Natural CF_4 and SF_6 on Earth. *Geophysical Research Letters* 25: 2401–2404.

Hem, J.D. 1992. *Study and interpretation of the chemical characteristics of natural water.* United States Geological Survey Water-Supply Paper 2254, 3rd edn, 263 pp.

Höhener. P., Werner, D., Balsiger, C. & Pasteris, G. 2003. Worldwide occurrence and fate of chlorofluorocarbons in groundwater. *Critical Reviews in Environmental Science and Technology* 33: 1–29.

Horneman, A., Stute, M., Schlosser, P., Smethie, W., Santella, N., Ho, D.T., Mailloux, B., Gorman, E., Zheng, Y. & van Geen, A. 2008. Degradation rates of CFC-11, CFC-12 and CFC-113 in anoxic shallow aquifers of Araihazar, Bangladesh. *Journal of Contaminant Hydrology* 97: 27–41.

IAEA (International Atomic Energy Agency), 2006. *Use of Chlorofluorocarbons in Hydrology: A Guidebook*, STI/PUB/1238, 277 pp. Obtainable from http://www-pub.iaea.org/MTCD/publications/PDF/Pub1238_web.pdf.

Khalil, M.A.K. & Rasmussen, R. 1989. The potential of soils as a sink of chlorofluorocarbons and other man-made chlorocarbons. Geophysical Research Letters 16: 679–682.

Koh, D.C., Plummer, L.N., Busenberg, E. & Kim, Y. 2007. Evidence for terrigenic SF_6 in groundwater from basaltic aquifers, Jeju Island, Korea: implications for groundwater dating. Journal of Hydrology 339: 93–104.

Maloszewski, P. & Zuber, A. 1982. Determining the turnover time of groundwater systems with the aid of environmental tracers. 1, Models and their applicability. *Journal of Hydrology* 57: 207–231.

Morris, B.L., Darling, W.G., Cronin, A.A., Rueedi, J., Whitehead, E.J. & Gooddy, D.C. 2006. Assessing the impact of modern recharge on a sandstone aquifer beneath a suburb of Doncaster, UK. *Hydrogeology Journal* 14: 979–997.

Oster, H. 1994. *Datierung von Grundwasser mittels FCKW: Voraussetzungen, Möglichkeiten und Grenzen.* Dissertation, Universität Heidelberg.

Oster, H, Sonntag, C. & Münnich, K.O. 1996. Groundwater age dating with chlorofluorocarbons. *Water Resources Research* 37: 2989–3001.

Plummer, L.N. & Busenberg, E. 1999. Chlorofluorocarbons. In: P.G. Cook & A.L. Herczeg (eds), *Environmental Tracers in Subsurface Hydrology*, 441–478. Dordrecht: Kluwer.

Santella, N., Ho, D.T., Schlosser, P. & Stute, M. 2007. Widespread elevated atmospheric SF_6 mixing ratios in the Northeastern United States: Implications for groundwater dating. *Journal of Hydrology* 349: 139–146.

Sebol, L.A., Robertson, W.D., Busenberg. E., Plummer, L.N., Ryan, M.C. & Schiff, S.L. 2007. Evidence of CFC degradation in groundwater under pyrite-oxidizing conditions. *Journal of Hydrology* 347: 1–12.

Weeks, E.P., Earp, D.E. & Thompson, G.M. 1982. Use of atmospheric fluorocarbons F-11 and F-12 to determine the diffusion parameters of the unsaturated zone in the southern high plains of Texas, *Water Resources Research* 18: 1365–1378.

Wilson, G.B. & McNeill, G.W., 1997. Noble gas recharge temperatures and the excess air component. Applied Geochemistry 12: 747–762.

Water-Rock Interaction – Birkle & Torres-Alvarado (eds)

The unsaturated zone as an observatory for the African Sahel

W.M. Edmunds
School of Geography and the Environment, Oxford University, UK

C.B. Gaye
University CA Diop, Dakar, Senegal

ABSTRACT: Using immiscible liquid displacement techniques it is possible to directly investigate the geochemistry of moisture from unsaturated zone materials. A profile from N Senegal is used to illustrate the scope of these techniques. Using pH, major and trace elements and comparing with isotopic data a better understanding is gained of timescales of water movement, aquifer recharge, hydrological environmental records and climate history as well as water-rock interaction. This approach has strong potential for the monitoring of recharge processes and rapid environmental change in areas of water stress such as the African Sahel.

1 INTRODUCTION

It is well known that the chemical properties of shallow groundwater are established overwhelmingly by reactions taking place in the unsaturated (vadose) zone. Atmospheric inputs from rainfall and dry deposition, with some increments from anthropogenic sources and industrial activities are also important as solute sources. Soil processes and land use strongly influence the reactivity through biogeochemical reactions, notably the regulation of the carbon and nitrogen cycles. Yet the intensity of soil and vegetation activity diminishes below the top couple of meters, and evapotranspiration activity slows or stops. Below this level water and solutes move slowly towards the water table to be stored as groundwater, carrying with them for some elements a memory of preceding events and for others a record of the water-rock interaction.

Unsaturated zone Cl records have been used widely for recharge and water balance studies (Scanlon et al. 2006) and isotopes (^{3}H, δ^2H, δ^{18}O) for investigation of age and hydrological processes (Edmunds 2005). Apart from lysimeters, little work has been conducted directly on the geochemistry of pore fluids. This is mainly due to the difficulties of extraction of moisture from unsaturated material with low water contents (typically 2–6 wt%).

Large areas of the African Sahel are comprised of unconsolidated sands or sandstones sometimes overlying low permeability basement rocks. Thicknesses of the unsaturated zone are variable but may frequently reach several tens of meters. Water movement is predominantly vertical under gravitation and capillary forces and with low recharge rates water may take decades or centuries to reach the water table.

The objective of the present study is to illustrate the scope for the use of a spectrum of unsaturated zone tracers for a range of applications using an example from unconsolidated sands in Senegal. New results are presented for chemical tracers, especially trace elements which expand the possibilities for using the unsaturated zone as an observatory.

2 FIELD AREA

The profile used for illustration is typical of much of the African Sahel. In Senegal, the Quaternary aquifer near Louga [15°37′N, 16°13′W] forms an extensive area of fixed dune sands which have been the subject of detailed study where some 20 profiles have been drilled (Cook et al. 1994; Edmunds & Gaye 1994). The current recharge in this region (290 mm mean average annual rainfall) has been estimated using the Cl mass balance approach at between 1 and 34 mm yr^{-1}. The unsaturated zone is 35 m thick (Figure 1) but may reach up to 45 m.

3 FIELD AND EXPERIMENTAL PROCEDURES

Areas of flat higher ground were selected as representative of the sandy terrain, as far as possible areas on natural grassland. Drilling was carried out using a hollow-stem hand augur with a string comprised of 1.5 m lengths. Sampling was possible

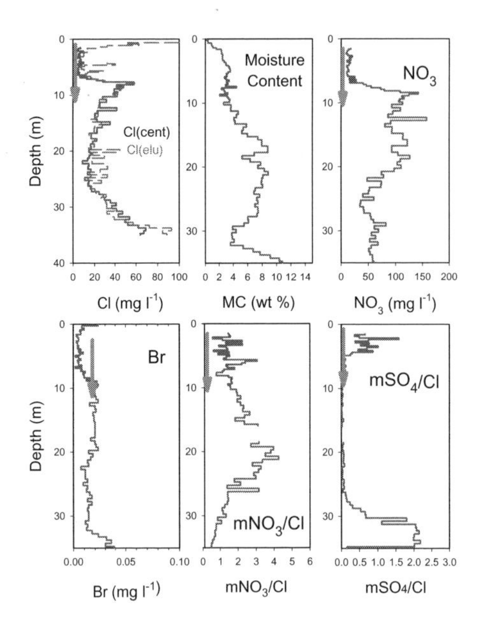

Figure 1. Chloride profile of unsaturated zone moisture (L18 site) with data obtained from centrifugation and elutriation. Also shown are MC, Br and NO_3 and SO_4, normalized to Cl.

to 35 m recovering unconsolidated sediments at intervals of 10–25 cm which were transferred to plastic bags and homogenised before rapidly transferring to sealed glass jars to avoid moisture loss. Any carry over of sediment from overlying material was usually visible as a colour change at the top of the sample interval and discarded. Moisture pH was measured in the field using standard soil analysis procedures with 0.01M $CaCl_2$ in distilled water slurry (ASA 1982).

Moisture contents were determined on a wet weight basis; procedures for isotopic analysis are described in Edmunds & Gaye (1996). Chemical analysis was carried out on samples obtained either by elutriation and/or by centrifugation using immiscible liquid displacement. For elutriation, 50 g of moist sand was stirred for one hour with 30 ml distilled and demineralised water and supernatant water recovered following centrifugation. This sample was only used for analysis of inert tracers—chloride and nitrate since the surface chemical equilibria were likely to be disturbed during this dilution procedure.

Interstitial water was extracted using a heavy immiscible liquid displacement technique using Arklone (trichloro-trifluoro-ethane). Approximately 80 g moist sand was placed in a 250 ml polypropylene bottle with 320 g of the heavy liquid and centrifuged at high speed (12000 rpm) for one hour. Moisture contents ranged from 1.6 to 8.8% and some water was recovered from all samples. Yields ranged from 24–63%, so that at low moisture contents up to 320 g were needed to obtain sufficient sample for analysis. Actual recoveries ranged from 0.9 to 8.7 ml. 10 ml were required for analysis and for this dilution (between 2–4 times) with demineralised water was generally required followed by filtration through 0.45 um membrane filters. A 5 ml aliquot was acidified with HNO_3. The unacidified sample was used for analysis of Cl, NO_3 and Br by colorimetry, SO_4, Fe, Mn, Si, Ba and Sr were measured using ICP-OES on acidified samples and remaining trace elements measured using ICP-MS.

4 RESULTS

The moisture contents in the top 2 m are influenced seasonally but below this depth in proportion to grain size. Interpretation of the interstitial water geochemistry is restricted by two factors—the need for indirect sampling and dilution (elutriation techniques and pH) and the small sample volumes obtained (centrifugation). Under the semi-arid conditions and in low organic carbon sandy soils and sediments considered here aerobic conditions prevail. Only pH and two chemical parameters, Cl and NO_3 were considered suitable for analysis using dilution/elutriation, other ions being prone to modifications of the moisture film. Centrifugation releases virtually undisturbed,

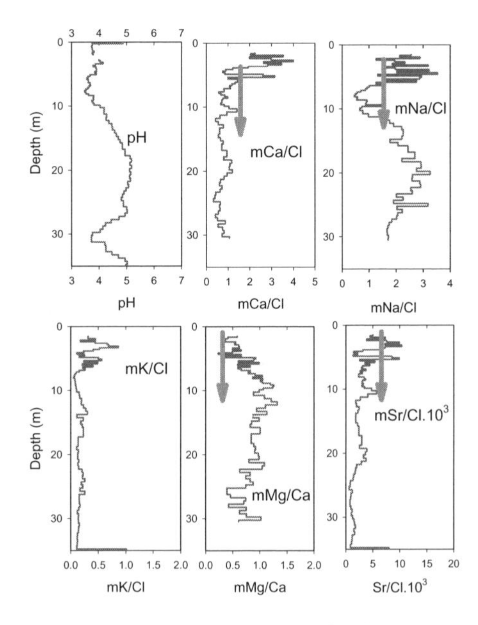

Figure 2. pH and major ion evolution in unsaturated zone moisture.

if small volume, water samples. It was possible to compare Cl in elutriate and centrifuged water samples (Figure 1) with a good correlation (r = 0.84). The noise in the data is clearly a reflection of the dilution process and unfortunately it was not possible to conduct replicate analyses. Chloride is chosen as the master variable since it is an inert element and its variations in the profile are related to evaporative concentration and recharge processes. Thus, ratios of other elements to chloride are sometimes used to understand the depth variations of reactive elements. Bromide, geochemically similar to Cl, is used to help follow possible changes in input conditions (Figure 1); nitrate and NO_3/Cl serve as indicators of inputs and also outputs from the soil horizons.

pH is low and the mean ionic balance (within 0.36%) also indicates that HCO_3 also must be negligible. Major and trace element profiles are plotted in Figures 1 to 4. In Figure 2 the concentrations of major cations Na, K and Ca are shown, as well as SO_4 normalised to Cl, to highlight the likely water-rock interaction as distinct from the trend relative to evaporative concentration. Rainfall compositions, from Louga (1990) are used to indicate likely input chemistry on subsequent diagrams.

Selected minor and trace elements in pore waters are shown in Figures 3 and 4. These have not been normalised to Cl but the principal trends which overprint evaporation trends are evident. These can be interpreted in relation to water-rock interaction (pH, redox environment) relative to depth in the profile.

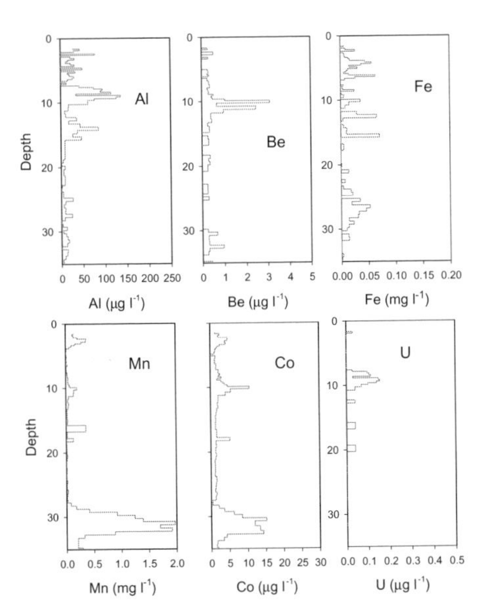

Figure 3. Trace element profiles in L18 unsaturated zone moisture.

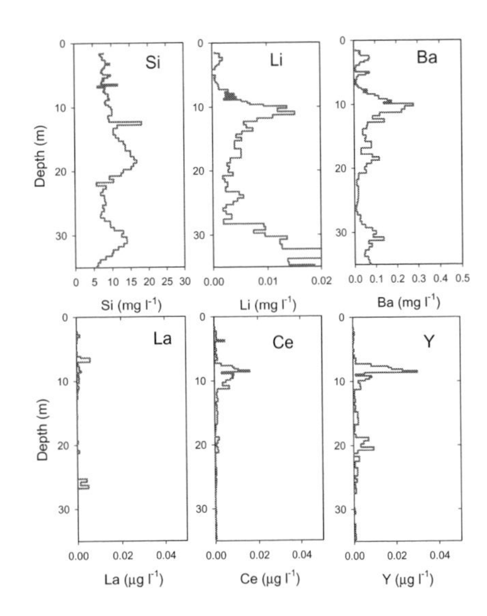

Figure 4. Trace element profiles in unsaturated zone moisture.

5 TRACERS IN THE UNSATURATED ZONE

5.1 *Unsaturated flow and groundwater age*

Tritium was used in the late 20th century to study the natural movement of water through unsaturated zones. In Senegal profiles from the unsaturated zone show well-defined 1960s tritium peaks several metres below surface, indicating homogeneous movement (piston flow) of water through profiles at relatively low moisture contents (2–4 wt%). These demonstrate that low, but continuous rates of recharge occur in many porous sediments. The usefulness of tritium as a tracer has now largely expired following cessation of atmospheric thermonuclear testing and radioactive decay. Providing the rainfall Cl, moisture contents and bulk densities of the sediments are known, then Cl accumulation can be substituted to estimate timescales. At the Louga research site this profile shows a Cl accumulation equivalent to 74 years storage in the unsaturated zone.

5.2 *Groundwater recharge rates*

The L18 profile indicates a mean annual recharge of 34.4 mm (average annual rainfall 290 mm and mean average rainfall Cl 2.8 mg l^{-1}) The spatial variability has also been determined using further profiles of Cl in local shallow wells (Gaye & Edmunds 1996).

5.3 *Climate records*

The oscillations in the 74-year Cl profile act as an evaporation record and therefore of alternating

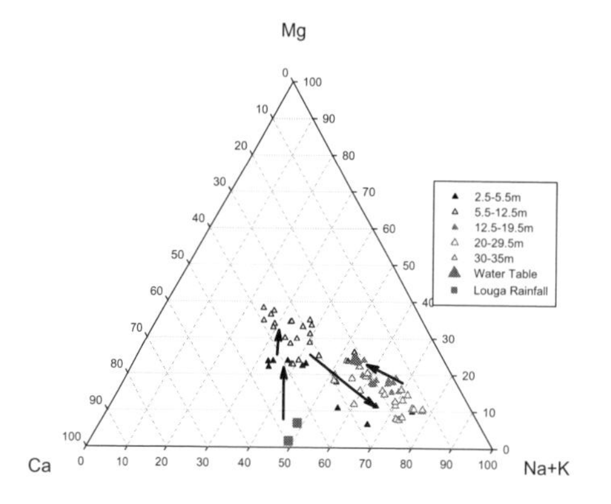

Figure 5. Cation evolution in unsaturated zone of profile L18.

periods of wet and dry years. The peak around 10 m coincides with the major period of Sahel drought which began in the 1970s. Subsequent profiles in porous sediments especially in China (Gates et al. 2008) confirm the value of the Cl profile as a high resolution proxy of climate variability over the decade to millennial scale.

5.4 *Water-rock-biosphere interactions*

Large increases in NO_3/Cl ratio above rainfall are found throughout the profile. This increase in the ratio has been interpreted widely here and across the Sahel region as due to N- fixation and release by natural leguminous vegetation (Edmunds & Gaye 1996). Bromine concentrations are similar to rainfall indicating an atmospheric source.

These sandy soils (Figure 2) are acidic and major ion variations are linked closely to pH which shows an increase from 4 to 5 over a 7 m interval below 10 m; the overall changes are summarized in Figure 5. There is a net gain in Na in the top 5 m and this is followed by a rise in the mMg/Ca ratio to around 1 and then mNa/Cl increases in the 10–20 m depth range. Silica is high throughout the profile. In the absence of detailed mineralogy and at the low pH it is inferred that exchange reactions followed by silicate mineral dissolution control the chemistry during the infiltration.

The overall low Fe concentrations (mean 0.016 mg l^{-1}) indicate an oxidizing environment; high Al with a distinct peak at 10 m is consistent with Al mobilization at low pH. The low pH also favours the mobilization of a number of trace metals which peak around 10 m before the pH rises and then decline; these include Be, Co, U and rare earth elements (Ce, La and Y). Lithium and Ba show also significant increases around 10 m and below, consistent with silicate mineral weathering.

The 8 m interval above the water table is also a zone of low pH and marked by an increase in SO_4, Mn, Co and Li. This coincides with lower moisture contents and coarser sediments. Two explanations are offered: i) enhanced water-rock interaction related to water table fluctuation—this interval marking a former higher level and/or ii) oxidation of sulphide minerals.

6 CONCLUSIONS

The chemistry of moisture in low moisture sands from a 35 m profile were extracted in mid-1990s using Arklone, an immiscible CFC liquid which was subsequently banned in the 1990s, subject to the Montreal protocol. The same work is now possible using a new liquid HFE-7100 (Li & Edmunds, in press). Extraction of moisture by different techniques, now including direct extraction allows a comprehensive investigation of the unsaturated zone geochemistry and provides a new tool for observing and monitoring environmental change over decadal scales or longer. As well as the applications shown, it is also possible to follow anthropogenic pollution.

REFERENCES

American Society of Agronomy (ASA) 1982. Chemical and microbiological properties. In A.L. Page (ed.), *Methods of soil analysis. Part 2*: 200–209. Madison: ASA.

Cook, P.G., Edmunds, W.M. & Gaye, C.B. 1992. Estimating palaeorecharge and palaeoclimate from unsaturated zone profiles. *Water Resources Research* 28: 2721–2731.

Edmunds, W.M. 2005. Contribution of isotopic and nuclear tracers to study of groundwaters. In P.K. Aggarwal, J. Gat, & K. Froehlich (eds.), *Isotopes in the Water Cycle: past, present and future of a developing science*: 171–192. Springer.

Edmunds, W.M & Gaye, C.B. 1994. Estimating the spatial variability of recharge in the Sahel using chloride. *Journal of Hydrology* 156: 47–59.

Edmunds, W.M. & Gaye, C.B. 1996. High nitrate baseline concentrations in groundwaters from the Sahel. *Journal of Environmental Quality* 26: 1231–1239.

Gates, J.B., Edmunds, W.M., Jinzhu, Ma. & Sheppard, P.R. 2008. A 700-year history of groundwater recharge in the drylands of NW China and links to East Asian monsoon variability. *The Holocene* 18: 1045–1054.

Li, J. & Edmunds, W.M. (in press). A new immiscible displacent for extracting interstitial water from unsaturated sediments and soils. *Vadose Zone Journal*.

Scanlon, B.R., Keese, K.E., Flint, A.L., Flint, L.E., Gaye, C.B., Edmunds, W.M. & Simmers, I. 2006. Global synthesis of groundwater recharge in semi-arid and arid regions. *Hydrological Processes* 20: 3335–3370.

Water-Rock Interaction – Birkle & Torres-Alvarado (eds)
© 2010 Taylor & Francis Group, London, ISBN 978-0-415-60426-0

Chemical weathering of silicate minerals: Linking laboratory and natural alteration

R. Hellmann
Environmental Geochemistry Group, LGIT-CNRS, Grenoble Cedex 9, France

R. Wirth
GeoForschungsZentrum Potsdam, Experimental Geochemistry, Potsdam, Germany

J.-P. Barnes
CEA, LETI, MINATEC, Grenoble, France

ABSTRACT: The interdiffusion-preferential cation release mechanism, commonly known as the leached layer theory, has been traditionally invoked to explain the prevalent occurrence of amorphous, silica-rich surface layers on weathered minerals. Based on a new approach using ultrathin Transmission Electron Microscopy (TEM) foils prepared in cross section from weathered mineral surfaces, it is possible to directly examine surface altered layers with analytical probes at the nanometer scale. Results show that the chemical and structural transitions from altered layers to parent minerals are spatially coincident and very sharp, which is not compatible with an interdiffusion mechanism. We argue that a coupled dissolution-reprecipitation mechanism can better explain the formation of surface altered layers in the context of both laboratory and natural chemical weathering.

1 INTRODUCTION

In order to understand the physico-chemical properties, the reactivity, and the metal and CO_2 sequestration capacity of soils, precise knowledge of nanometer-scale processes occurring at fluid-mineral interfaces is of primary importance. To understand chemical weathering processes, researchers have traditionally turned to laboratory-based experiments. However, a dichotomy has always existed between chemical weathering reactions simulated in laboratory experiments with chemical weathering reactions observed in the field. Because of the inherent complexity of natural weathering, which includes the effects of plants, lichens and microbial organisms, oxy-hydroxide metal coatings and/or clays, aqueous organic compounds, higher solution saturation states, and reaction time periods that are orders of magnitude greater than in the laboratory, it has commonly been assumed that chemical weathering in the laboratory and in the field occur by different mechanisms (Nesbitt & Muir 1988, Seyama & Soma 2003, Zhu et al. 2006).

Based on many decades of laboratory experiments, it is widely accepted that alumino-silicate minerals undergo chemical weathering via the 'leached layer' mechanism (Chou & Wollast 1985, Hellmann 1995, Casey et al. 1988, 1993, Petit et al. 1989, 1990). This mechanism accounts for incongruent and preferential cation release that characterizes the alteration of aluminosilicate minerals. A leached layer represents a surface altered layer that is structurally and chemically distinct from the parent mineral. It forms by way of an interdiffusion process between preferentially released cations from the parent mineral with protons from the surrounding aqueous bulk solution; this occurs at a diffusion front located at the inner interface between the leached layer and the mineral. The amorphous leached layer, which represents a structural relict of the parent mineral, is composed of silica that is 'leftover' after the preferential release of cations by diffusion (silica composition for chemical alteration conditions at acid to circum-neutral pH). The leached layer does not attain an infinite thickness because chemical hydrolysis reactions breakdown and release silica into solution at the outer interface of the leached layer.

The leached layer mechanism is based on two simultaneously occurring processes: interdiffusion and hydrolysis. Supporting evidence for the leached layer theory has been provided by diffusion-like, sigmoidal cation depth profiles obtained by surface sensitive analytical techniques that rely on surface bombardment of altered mineral surfaces with ion, electron or X-ray beams (e.g. SIMS, RBS, RNRA, XPS; see review in Chardon et al. 2006). These techniques, however, are all characterized by poor lateral spatial resolution (i.e. due to μm to mm-sized beam footprints).

Thus, the pertinent question that needs to be addressed is whether natural chemical weathering occurs via the same mechanism of interdiffusion and preferential cation release, or by way of another process. Alternatively, one can also ask the question as to whether laboratory and natural chemical weathering are controlled by an altogether different mechanism.

2 METHODS

Previous investigations of the surface chemistry of weathered minerals have nearly all been based on surface sensitive techniques based on bombarding the surface with ion, electron or X-ray beams that have μm to mm-sized beam diameters; the resultant chemical depth profiles represent indirect measurements at low spatial (lateral) resolution. This analytical methodology results in an artificial broadening of the individual cation depth profiles, and can thus lead to a misinterpretation of the operative mechanism (Hellmann et al. 2004).

A new approach to examine fluid-mineral interfaces is based on cutting or milling weathered mineral surfaces using either ultramicrotomy or focused ion beam milling (FIB). Both techniques provide a cross-sectional ultrathin section (electron transparent) that extends from the outer weathered surface to the unaltered parent. The dimensions of typical FIB thin sections are 15 μm (length) $*$ 5 μm (height), and ≤100 nm in thickness. The principal advantage of investigating surface altered layers in cross section is that the chemical depth profiles can be measured directly with high spatial resolution probes.

Both ultramicrotomy and FIB can produce artifacts, which is something to be aware of. Thin sections cut by ultramicrotomy have a tendency to be characterized by lamellar fracturing and spatial disordering of the cut mineral (Hellmann et al. 2003). However, this does not necessarily influence the quality of the analytical measurements, provided that large enough intact zones are located for subsequent analyses. In comparison, thin sections are more easily obtainable with the FIB technique. However, this technique has a major weakness in that artifacts are very easily produced. Two fundamental steps of FIB are based on the deposition of a Pt 'strap' (i.e. layer) on the surface area chosen, followed by ion milling of the thin section using a focused Ga beam. Both operations can result in the emplacement of measurable quantities of, respectively, Pt and Ga into the target mineral, which generally results in amorphization. Therefore, one of the most important preliminary steps is the application of a protective Au or C coating (≈100 nm in thickness) before the deposition of Pt and subsequent Ga milling (e.g. see Lee et al. 2007). For a thorough description of FIB applied to Earth sciences studies, see Wirth (2004, 2009).

A large suite of silicate minerals has been examined, subject to both laboratory and natural chemical weathering. TEM foils, prepared both by ultramicrotomy and FIB, were examined by various analytical techniques incorporated in transmission electron microscopes (TEM): high resolution TEM (HRTEM), energy filtered TEM (EFTEM), high angle annular dark field TEM (HAADF), electron energy loss spectroscopy (EELS), and energy dispersive X-ray spectroscopy (EDX). All of the above techniques provide either structural or chemical information at the nanoscale.

3 RESULTS

Here we describe two examples in detail. Figure 1 is a bright field (BF) TEM image of a FIB produced TEM foil showing, in cross section, a thick (400–500 nm) amorphous surface altered layer on unaltered crystalline wollastonite. The altered layer formed in response to laboratory chemical weathering. The wollastonite was altered in a flow through stirred reactor at ambient temperature at pH3. The distinctly lighter shade of grey associated with the altered layer indicates an average atomic

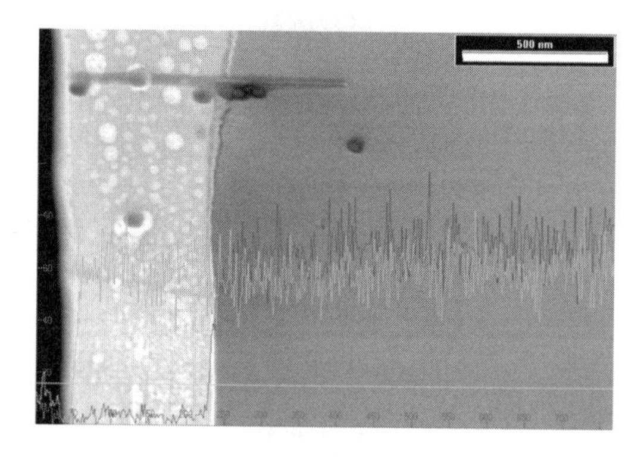

Figure 1. Bright field TEM image of FIB thin section showing cross sectional view of chemically weathered wollastonite; surface altered layer on left (light grey) and unaltered wollastonite on right (darker grey). The two EDX chemical profiles indicate no change in Si and significant Ca depletion in the altered layer. Note the sharp Ca gradient at the interfacial boundary. The bubbles and pits in the image are due to electron beam-solid interactions. The scale bar is 500 μm.

density that is less than that of the wollastonite. The superposed EDX chemical profiles reveal that the Si counts are approximately equivalent in both the unaltered wollastonite and the surface altered layer. On the other hand, the Ca counts reveal that the altered layer is depleted in Ca. Note that the change in Ca is abrupt and very sharp (i.e. similar to a step function), and spatially coincides with the interfacial boundary between the wollastonite and the altered layer. The Ca gradient at the interfacial boundary is only a few tens of nm wide. The second example is based on natural chemical weathering of a K-feldspar that was retrieved from a soil immediately adjacent and below a granite boulder.

Figure 2 is a high resolution TEM image that shows in cross section three distinct zones: the unaltered feldspar, a 20 nm-thick amorphous surface altered layer, and a several μm-thick layer containing abundant phyllosilicates oriented roughly parallel with the interfacial boundary separating the feldspar from the altered layer. The 20-nm thick altered layer appears to be made up of two equally thick zones with different densities. EFTEM analyses indicate that altered layer immediately adjacent to the feldspar is composed primarily of silica and is depleted in potassium. As was the case for the laboratory-weathered wollastonite, the chemical and structural boundaries are spatially coincident and very sharp. The width of the K gradient at the K-feldspar-altered layer boundary is estimated to be on the order of 3 nm.

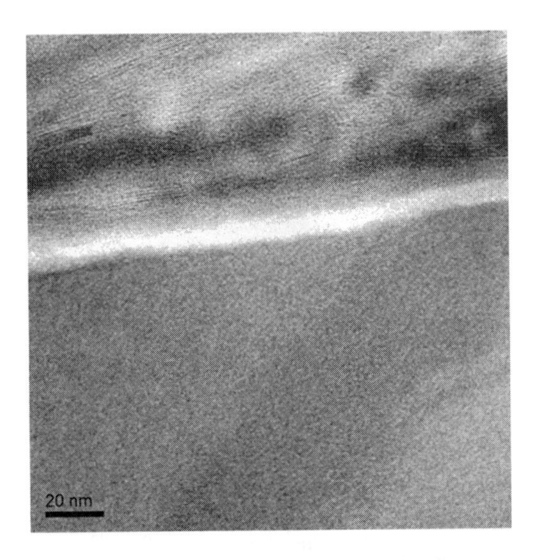

Figure 2. High resolution TEM image showing naturally weathered K-feldspar in cross section. The distinct and sub-horizontal 20 nm-thick band (white and medium grey) is an amorphous surface altered layer that separates the crystalline K-feldspar (below) from a zone comprised of abundant secondary phyllosilicates (above).

4 DISCUSSION

Even though we only have described two examples of silicate chemical weathering, one from the laboratory and one from the field, other silicate minerals that we have investigated are also characterized by similar surface altered layers. In all cases, the structural and chemical boundaries between the parent crystalline phase and the adjacent altered layer are spatially coincident and sharp. The layers are in general composed of silica and depleted in interstitial cations, as well as in Al (when originally present in parent mineral). The main difference between the various weathered minerals is the thickness of the altered layers. The thickness is dependent on the composition and structure of the parent mineral, as well as on the conditions of chemical weathering, i.e., acid pH conditions favor thicker layers compared to circum-neutral pH. The overall physical, chemical, and mineralogical character of the layers is also dependent on whether chemical weathering took place in the laboratory or under natural conditions (i.e. comparison of Figs. 1 and 2). Chemical weathering in the field occurs under conditions that are more complex (numerous abiotic and biotic processes), and over time scales that are generally orders of magnitude greater.

The results that we have shown, in particular the extremely sharp and coincident chemical and structural boundaries, are at odds with a chemical weathering process controlled by interdiffusion and preferential cation release (i.e. leached layer theory). If the surface altered layers had been formed by interdiffusion, the measured cation depth profiles are predicted to be sigmoidal, with much broader chemical gradients. Moreover, diffusion modeling of our results confirm that abrupt changes in chemistry in altered layers greater than 50 nm in thickness (such as in Fig. 1) are incompatible with interdiffusion.

We propose that laboratory and natural surface altered layers form by a process that involves coupled interfacial dissolution-reprecipitation. This process is compatible with the sharp chemical and structural boundaries that we measured. Dissolution-reprecipitation implies that the intrinsic dissolution process is characterized by the stoichiometric release of all elements at the reactive interface of the parent mineral. The release of elements from this interface is controlled by chemical reactions, and not by a diffusion process. The release of all constituent elements is followed by the nearly synchronous precipitation of silica within a thin fluid film that contacts the reactive interface of the parent mineral. It is important to note that this process is not directly dependent on the fluid saturation state of the bulk solution. Silica-rich surface altered layers formed on all laboratory weathered minerals, even though the bulk solutions were completely undersaturated with respect to all polymorphs of silica (crystalline and amorphous).

These novel results call into a question a large body of previously published studies on chemical weathering. The well-entrenched leached layer theory has already been challenged by previous chemical weathering studies that advocate this mechanism (see, e.g. Hellmann et al. 2003), as well as by more recent studies involving carbonation reactions (e.g. Daval et al. 2009a, 2009b). In addition, reactions involving mineral replacement (a subcategory of chemical weathering) may also be controlled by dissolution-reprecipitation (see pioneering study by O'Neil & Taylor 1967, see also reviews in Putnis 2002, 2009). If the idea of dissolution-reprecipitation is validated by further research, then this will completely change existing conceptual ideas concerning how chemical weathering really works, both in the laboratory and in the field.

The presence of precipitated silica-rich surface layers has important implications with respect to coupled chemical weathering-carbonation reactions. According to recent studies (e.g., Daval et al. 2009b) carbonation of Ca-bearing silicates such as wollastonite leads to the precipitation of calcium carbonate microcrystallites and phyllosilicates on and within the precipitated silica layer. There is evidence that this leads to partial passivation, thereby slowing down the weathering reaction of the primary mineral. This, of course, has important implications with respect to the efficacy of CO_2 sequestration, both in the field as well as with respect to industrial processes.

REFERENCES

Casey, W.H., Westrich, H.R. & Arnold, G.W. 1988. Surface chemistry of labradorite feldspar reacted with aqueous solutions at pH = 2, 3, and 12. *Geochimica et Cosmochimica Acta* 52: 2795–2807.

Casey, W.H., Westrich, H.R., Banfield, J.F., Ferruzzi, G. & Arnold, G.W. 1993. Leaching and reconstruction at the surfaces of dissolving chain-silicate minerals. *Nature* 366: 253–255 (1993).

Chardon, E.S., Livens, F.R. & Vaughan, D.J. 2006. Reactions of feldspar surfaces with aqueous solutions. *Earth-Science Reviews* 78: 1–26.

Chou, L. & Wollast, R. 1985. Steady-state kinetics and dissolution mechanisms of albite. *American Journal of Science* 285: 963–993.

Daval, D., Martinez I., Corvisier, J., Findling, N., Goffé, B. & Guyot, F. 2009. Carbonation of Ca-bearing silicates, the case of wollastonite: Experimental investigations and kinetic modeling. *Chemical Geology* 265: 63–78.

Daval, D., Martinez, I., Guignier, J.-M., Hellmann, R., Corvisier, J., Findling, N., Dominici, C., Goffé, B. & Guyot, F. 2009. Mechanism of wollastonite carbonation deduced from micro- to nanometer length scale observations. *American Mineralogist* 94: 1707–1726.

Hellmann, R. 1995. The albite-water system Part II. The time-evolution of the stoichiometry of dissolution as a function of pH at 100, 200 and 300°C. *Geochimica et Cosmochimica Acta* 59: 1669–1697.

Hellmann, R., Penisson, J.-M., Hervig, R.L., Thomassin, J.-H. & Abrioux, M.-F. 2003. An EFTEM/HRTEM high-resolution study of the near surface of labradorite feldspar altered at acid pH: evidence for interfacial dissolution-reprecipitation. *Physics and Chemistry of Minerals* 30: 192–197.

Hellmann, R., Penisson, J.-M., Hervig, R.L., Thomassin, J.-H. & Abrioux, M.-F. 2004. Chemical alteration of feldspar: a comparative study using SIMS and HRTEM/EFTEM. In R.B. Wanty & R.R. Seal II (eds.), *Water Rock Interaction*: 753–756. Rotterdam: Balkema.

Lee, M.R., Brown, D.J., Smith, C.L., Hodson, M.E., MacKenzie, M. & Hellmann, R. 2007. Characterization of mineral surfaces using FIB and TEM: A case study of naturally weathered alkali feldspars. *American Mineralogist* 92: 1383–1394.

Nesbitt, H.W. & Muir, I.J. 1988. SIMS depth profiles of weathered plagioclase, and processes affecting dissolved Al and Si in some acidic soil conditions. *Nature* 334: 336–338.

O'Neil, J.R. & Taylor, H.P.J. 1967. The oxygen isotope and cation exchange chemistry of feldspars. *American Mineralogist* 52: 1414–1437.

Petit, J.-C., Dran, J.-C. & Della Mea, G. 1990. Energetic ion beam analysis in the Earth sciences. *Nature* 344: 621–626.

Petit, J.-C., Dran, J.-C., Paccagnella, A. & Della Mea, G. 1989. Structural dependence of crystalline silicate hydration during aqueous dissolution. *Earth Planetary Science Letters* 93: 292–298.

Putnis, A. 2002. Mineral replacement reactions: from macroscopic observations to microscopic mechanisms. *Mineralogical Magazine* 66: 689–708.

Putnis, A. 2009. Mineral replacement reactions. In E.H. Oelkers & J. Schott (eds.), *Thermodynamics and Kinetics of Water-Rock Interaction*: 87–124. Washington, D.C.: Mineralogical Society of America.

Seyama, H. & Soma, M. 2003. Surface-analytical studies on environmental and geochemical surface processes. *Analytical Sciences* 19: 487–497.

Wirth, R. 2004. Focused Ion Beam (FIB): A novel technology for advanced application of micro- and nanoanalysis in geosciences and applied mineralogy. *European Journal Mineralogy* 16: 863–876.

Wirth, R. 2009. Focused Ion Beam (FIB) combined with SEM and TEM: Advanced analytical tools for studies of chemical composition, microstructure and crystal structure in geomaterials on a nanometre scale. *Chemical Geology* 261: 217–229.

Zhu, C., Veblen, D.R., Blum, A.E. & Chipera, S.J. 2006. Naturally weathered feldspar surfaces in the Navajo Sandstone aquifer, Black Mesa, Arizona: electron microscopic characterization. *Geochimica et Cosmochimica Acta* 70: 4600–4616.

Water-Rock Interaction – Birkle & Torres-Alvarado (eds)
© 2010 Taylor & Francis Group, London, ISBN 978-0-415-60426-0

Integrated CO_2 sequestration and geothermal development: Saline aquifers in Beitang depression, Tianjin, North China Basin

Z. Pang
Key Laboratory of Engineering Geomechanics, Institute of Geology and Geophysics, Chinese Academy of Sciences, Beijing, China

F. Yang, Y. Li & Z. Duan
Key Laboratory of Engineering Geomechanics, Institute of Geology and Geophysics, Chinese Academy of Sciences, Beijing, China
Graduate School, Chinese Academy of Sciences, Beijing, China

ABSTRACT: This paper presents a conceptual model named CO_2-EATER (Enhanced Aquifer Thermal Energy Recovery) to integrate geological sequestration of CO_2 with geothermal energy development and explains how this can be achieved using the Beitang depression of North China Basin as an example. First results of chemical and isotopic characteristics of the saline aquifers are discussed, where field scale tests are underway to verify the model. The water is dominated by Na-HCO_3 type with TDS of 0.7–1.5 g/L. Isotopic composition shows that the formation water is of meteoric origin that has slightly reacted with the formation matrix. Chemical geothermometry indicates that the Guantao formation contains geothermal waters with temperatures around 80°C. Due to the low salinity, CO_2 solubility can be rather high, which implies high storage capacity, favorable for CO_2 sequestration. The results show that this site is suitable for the planned field scale test from a geochemical point of view.

1 INTRODUCTION

China is one of the largest users of geothermal energy in the world for direct heating. Saline aquifers in the North China Basin, which have been used to supply hot water for space heating and bathing for decades, are typical examples of geothermal development of this kind. However, re-injection of waste water is limited (Pang 2007). Field tests have shown that re-injection into Guantao sandstone formation is rather difficult and injectivity is generally very low, and decreases with time, likely due to clogging problems (Minissale et al. 2008). As a consequence, the water table has dropped rapidly. A 60 m drop has generally been recorded in areas of heavy extraction of geothermal water.

In order to achieve sustainable development of geothermal resources, it is important to improve the injectivity of the formation. On the other hand, Tianjin is a major urban centre, and an industrial as well as a port city, where economic development has been booming in recent years. CO_2 emissions from the use of fossil fuels have become a major concern of the government and general public. Plans are underway to test the possibility of using some of the saline aquifers for geological sequestration of CO_2 (GSC).

In this context, it is necessary to assess the storage capacity and integrity of the various saline aquifers for GSC purposes. In order to use the saline aquifers for GSC and at the same time maintain the geothermal energy development, it is necessary to integrate the two operations to achieve both objectives. Pang et al. (2008) proposed a conceptual model called CO_2-EATER (Enhanced Aquifer Thermal Energy Recovery) that suggested injection of CO_2 to improve injectivity of the targeted saline aquifers through increasing their porosity and permeability. The advantage of this model is one stone two birds: improving geothermal energy recovery while offering the opportunities for CO_2 sequestration. This model is considered a different perspective as compared to that proposed by Pruess et al. (2006) in which CO_2 is used as a working fluid in heat extraction in an enhanced geothermal system. Our model has an emphasis on water-rock-gas interactions in which CO_2 is used as a reactant. A field test has been planned in the study area for the purpose of model verification. This paper presents first results of chemical and isotopic characterization of the saline aquifers to understand the water-rock interaction processes as relevant to both objectives, which will shed light on the viability of the site for such tests.

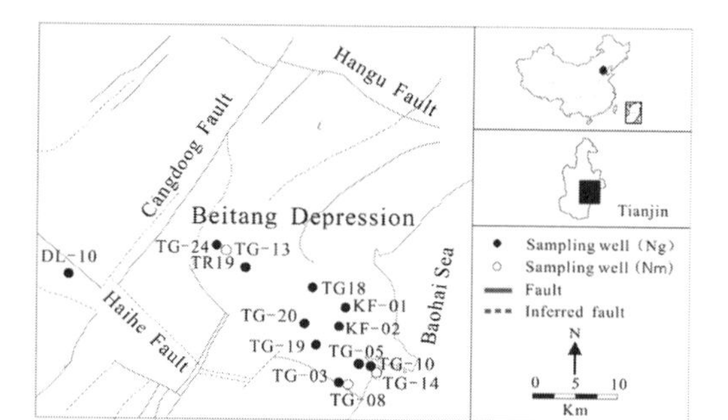

Figure 1. Location of water sampling wells in the Beita Depression, the North China Basin (for cross section see Li et al., this volume).

2 REGIONAL HYDROGEOLOGY

The Beitang Depression is a tectonic depression in the North China Basin (Fig. 1) with an area of 1200 km^2. It is bordered by the Hangu, Cangdong and Haihe faults and the Bohai Sea in the four directions, respectively.

The Upper Tertiary Guantao and Minghuazhen formations lie unconformably above the highly folded and tectonized Mesozoic to Precambrian crystalline basement. Being typical of terrestrial fluvial sediments, the Guantao formation is formed by a series of sandstone interbedded with clay layers. The upper and lower limits are 1300 m and 1750 m deep in the Beitang Depression, with major regional variations.

Due to its higher porosity (up to 35%) and favorable temperature (40–90°C) the Guantao formation has become the main target of geothermal development. The Minghuazheng formation lies between 520 m to 1300 m on top of the Guantao formation, occurring in a thick stratum of mudstone and silt sandstone, the lower part of which can be viewed as cap-rock of the Guantao formation.

We have sampled the Guantao formation and performed XRD analysis. The reservoir mineralogy is mainly composed of quartz, feldspar (albite and anorthite), calcite, dolomite and minor mica and chlorite. Regional groundwater generally flows from northwest to southeast. Groundwater temperatures of the upper Tertiary formations measured at wellhead range from 40 to more than 90°C, and, due to its greater depth, the Guantao formation is usually hotter than the Minghuazheng formation.

3 SAMPLING AND ANALYSIS

We conducted a sampling campaign in December, 2009 during which 16 water samples were collected from the geothermal production wells sunk in the upper Tertiary formations in the Beitang depression. Chemical and isotopic analyses of water samples as well as a semi-quantitative analysis of mineralogy have been completed so far. The well depth ranges from 1400 m to 2100 m.

For metallic elements and cation analysis, samples were acidified with HNO_3 to adjust pH to less than 2. For dissolved and free CO_2 and bicarbonate ions, titration with 0.05 M NaOH and 0.02 M HCl were conducted on site. Unstable hydrogeochemical parameters like pH, temperature, conductivity, Eh and dissolved oxygen as well as ferrous ion were determined in the field, using a HACH field set. Samples were filtered using 0.45 μm membranes in the laboratory before determinations.

The chemical analysis of the water was performed in the Analytical Laboratory, Beijing Research Institute of Uranium Geology, where main anions (F^-, Cl^-, SO_4^{2-}, NO_3^-) were determined using DIONEX-500 ion chromatograph and HCO_3^- was determined by a 785DMP titrator and cations by an OPTIMA2X00/1500 ICP-OES, trace elements by ICP-MS, all within one week of sampling.

Besides, three core samples from the Guantao formation and two from the Minghuazhen formation were collected from a new drillhole in the depression and analyzed for major and minor mineral components by XRD and SEM in laboratories of the Institute of Geology and Geophysics, Chinese Academy of Sciences. The locations of the wells are shown in Figure 1 and field measurements and analytical data for the water samples in Table 1.

4 RESULTS AND DISCUSSION

4.1 *Hydrogeochemistry*

The chemical features of water samples collected from Guantao formation are typical of HCO_3-Cl-Na type water with TDS ranging from less than 0.7 g/L to about 15 g/L. The wellhead temperatures measured are between 60 and 70°C with three samples below 40°C. pH determined in-situ indicates that geothermal water is neutral to alkaline with an average value of 7.7.

As shown in the Piper diagram (Fig. 2), Na^+ and HCO_3^- are the predominant ions in all 16 samples while Cl^- and SO_4^{2-} are secondary anions, Ca^{2+} and Mg^{2+} are minor cations with less than 6% of the total in meq/L. The concentration of trace elements, such as Li, Sr, and F in the water are relatively high and the average values for all the water samples are 1.2, 0.76 and 5.6 mg/L, respectively, mainly due to water- rock interactions.

Table 1. Chemical and isotopic composition of water samples from the Upper Tertiary saline aquifers in the Beitang Depression, the North China Basin.

Code	Aquifer	T °C	pH_f	F^- mg/L	Cl^- mg/L	NO_3^- mg/L	SO_4^{2-} mg/L	HCO_3^- mg/L	Ca^{2+} mg/L	Mg^{2+} mg/L	Na^+ mg/L	K^+ mg/L	TDS g/L	$\delta^{18}O$ ‰VSMOW	δ^2H ‰VSMOW
TG08	Nm	40	8.2	3.2	52	0.1	53	596	5.1	0.8	309	4.5	0.7	−9.26	−73.0
TG10	Nm	46	8.4	4.3	21	0.0	28	669	4.6	0.7	293	4.1	0.7	−10.00	−72.4
TG13	Nm	43	8.6	4.0	85	0.8	74	590	10.0	2.0	300	11.8	0.8	−9.36	−72.6
TG05	Ng	70	7.4	5.7	316	1.1	220	640	12.7	1.0	546	10.1	1.5	−9.21	−72.7
TG03	Ng	72	8.0	5.8	306	1.5	192	644	12.0	1.0	531	11.4	1.4	−9.18	−72.0
TG20	Ng	63	7.4	4.5	277	1.5	185	641	14.7	1.4	509	10.5	1.4	−9.16	−72.6
TG19	Ng	63	7.5	6.7	325	1.1	209	651	14.0	1.4	557	10.9	1.5	−8.90	−72.2
TG14	Ng	70	7.6	4.9	273	2.2	172	670	10.8	1.1	519	11.5	1.4	−9.45	−72.2
KF02	Ng	66	7.6	6.9	258	0.1	175	635	10.9	0.9	496	9.8	1.3	−9.95	−71.1
KF01	Ng	60	7.7	5.3	226	1.3	163	645	11.6	0.9	477	10.0	1.3	−9.64	−71.8
TG18	Ng	60	7.7	6.4	285	1.2	197	620	12.9	1.3	491	10.4	1.4	−9.08	−71.0
DL10	Ng	–	7.2	7.2	317	1.3	203	610	34.3	6.8	467	39.2	1.4	−9.38	−73.3
TG24	Ng	46	7.6	5.6	205	1.0	144	613	23.1	4.8	397	32.1	1.2	−9.40	−72.9
TR19	Ng	54	7.7	6.2	259	0.0	174	611	18.8	3.1	465	17.7	1.3	−9.36	−73.3
TR19B	Ng	36	8.0	6.9	357	0.8	234	602	19.5	3.2	523	18.0	1.5	−9.34	−72.1
TR20	Ng	36	7.8	5.2	283	1.5	183	606	19.0	3.1	444	18.0	1.3	−9.43	−73.1

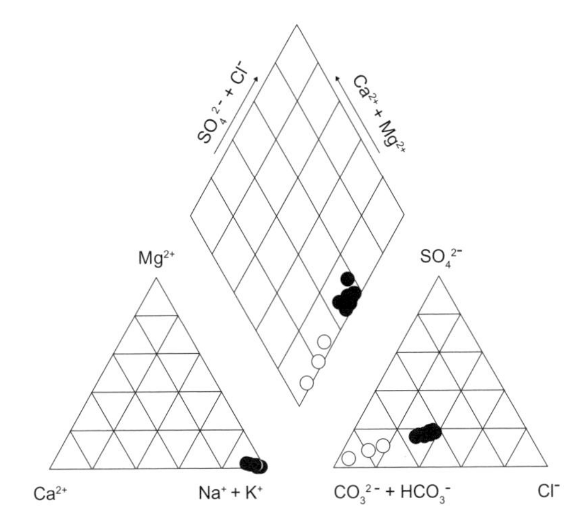

Figure 2. Piper diagram for the Tertiary formation water of Tianjin, China. The solid circles represent water from the Guantao formation while open circles represent water from the Minghuazhen formation.

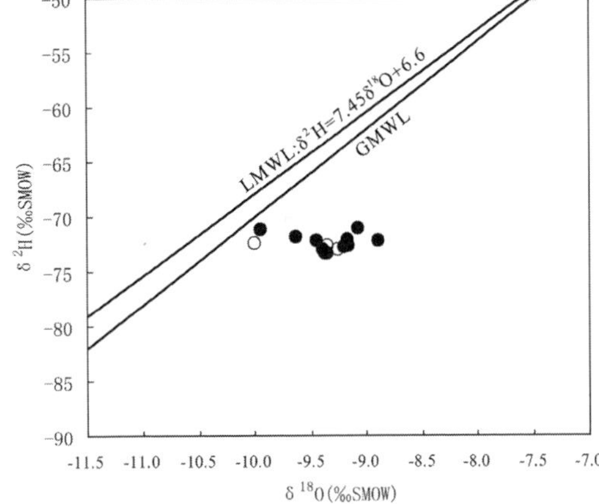

Figure 3. Isotopic composition of the Tertiary formation water of Tianjin, China. The solid circles represent water from the Guantao formation (Ng) while open circles represent water from the Minghuazhen formation (Nm).

4.2 *Isotopic composition*

Stable isotopes in the water samples show a quite uniform pattern with deuterium identical to that of meteoric water, but Guantao formation waters are enriched in Oxygen-18. Thus, they are shifted towards the right in the δ^2H versus $\delta^{18}O$ plot (Fig. 3), which can be explained either by the temperature effect in water-rock exchange of isotopes or the signature of paleo-water, considering the fact that the water has been dated to be recharged about 24 ka before present (Chen 1988, Gao et al. 2009).

4.3 *Geothermometry and reservoir temperature*

Many geothermometers have been established and proposed to calculate geothermal reservoir temperature in the past decades including both empirical chemical geothermometers and theoretical geothermometers. Reservoir temperature can

Table 2. Calculated reservoir temperatures according to cation and silica geothermometers.

Code	T (°C)	Na-K	Na-K-Ca	Quartz	Chal.	Li-Mg
TG08	40	55	106	52	38	26
TG05	70	65	116	78	64	120
TG03	72	72	122	81	67	119
TG20	63	70	120	74	60	134
TG19	63	68	118	76	62	128
TG14	70	73	123	86	71	94
TG10	46	54	105	59	45	25
KF02	66	68	118	77	63	126
KF01	60	70	121	77	63	130
TG18	60	71	121	76	61	134
TG24	46	141	182	70	55	121
TG13	43	100	146	65	50	100
TR19	54	99	146	69	55	13

be obtained with the assumption of water-rock equilibrium in most geothermal systems. Na-K (Giggenbach et al. 1988), quartz (Fournier 1977) and Na-K-Ca (Fournier & Truesdell 1973) as well as Li-Mg (Kharaka & Mariner 1989) were used in this case and reservoir temperatures obtained from the three methods are listed in Table 2.

It can be seen that, estimated values using quartz and chalcedony range from close to the measured wellhead temperatures to about 10 degrees higher than that. Reservoir temperatures obtained from the Na-K and Na-K-Ca geothermometers are much higher than measured temperatures and are inconsistent and thus unlikely to be realistic values.

Temperature calculated through Mg-Li geothermometer method are almost 30 to 60°C higher than those measured at wellhead. This discrepancy may have been caused by the difference in water chemistry between continental sedimentary basins and those of marine origin on which the geothermometer equation was calibrated. It seems that the reservoir temperature is not likely to be significantly higher than 80°C.

4.4 *Aqueous speciation*

The saturation state of the geothermal water with respect to common minerals has been calculated using the code PHREEQC 2.15 (Parkhurst & Appelo 1999). It refers to temperatures measured at the wellhead. The water is slightly supersaturated with calcite (SI = 0.2–0.9), which is usually due to CO_2 degassing during fluid ascent in the borehole, though calcite equilibrium in reservoir conditions of geothermal systems is hypothesized and the computed reservoir CO_2 pressure P_{CO_2} ($10^{-1.3}$–$10^{-2.8}$ Pa) is higher than the atmospheric CO_2 partial pressure of $10^{-3.5}$ Pa. The results for free CO_2 and bicarbonate ion as well as carbonate ion determined on site in our sampling campaign also confirm this hypothesis.

4.5 *Potential capacity for CO_2 storage*

Four trapping mechanisms have been proposed by IPCC (2005). Among which the solubility trapping mechanism has been suggested to be the major one (80% of total CO_2) through investigations on natural analogs of CO_2 gas fields (Gilfillan et al. 2009). As it is more complicated to evaluate other forms of trapping, here we make a first analysis of our test site by considering the solubility trapping alone and use the model of Bachu et al. (2007).

$$M_{CO_2} = A \cdot h \cdot \phi \cdot (\rho_s \cdot X_s^{CO_2} - \rho_0 \cdot X_0^{CO_2})$$

where A and h are aquifer area and thickness, Φ is the porosity, ρ the density of formation water, X^{CO_2} is the CO_2 content (mass fraction) in formation water and the subscripts *0* and *S* stand for initial CO_2 content and CO_2 content at saturation, respectively. Water chemistry analysis shows that the TDS of Guantao formation water ranges from less that 1.0 g/L to more than 1.5 g/L which indicates a relatively low salinity from 0.10 to 0.15‰. Since CO_2 saturation increases with decreasing salinity of the formation water, the Guantao formation is definitely an excellent CO_2 sequestration reservoir. In addition, the porosity of the Guantao formation is 33.6 to 38.7% with an average porosity of about 35%, which also provides favorable conditions for CO_2 sequestration. Therefore, the CO_2 storage potential of the Guantao formation is very promising.

4.6 *Integrated GSC and geothermal development*

Although the formation is favorable for GSC in terms of accessibility, impermeable cap-rock and high storage capacity, it would be a substantial cost in terms of renewable energy if geothermal development has to give way to GSC completely in the region and elsewhere. As a matter of fact, studies on gas-water-rock interactions in sandstone reservoirs have flourished in the last decade due to the need for enhanced oil recovery (CO_2-EOR). The fact that deep saline aquifers in sedimentary basins are promising candidates for geological sequestration of CO_2 further encouraged the research efforts in the experiments at laboratory and well field scales, plus numerical modeling at various scales. It has been shown by these studies that porosity and permeability of the sandstone reservoirs would increase as a consequence of CO_2 injection (Xu et al. 2004, Kharaka et al. 2006). The conceptual model CO_2-EATER proposed by

Pang et al. (2008), aims to achieving two objectives at the same time using an integrated technological design. The model suggests that injection of large amounts of CO_2 would first dissolve the carbonates in the cement/fillings between the pores and thus clean up the clogging materials responsible for the reduced injectivity of the formation. When the water is over-saturated with respect to carbonate minerals the latter would precipitate, but at a certain distance, outside of the radius of disturbance of geothermal wells; if well regulated, an ideal situation for CO_2 sequestration and geothermal development. Depending on PT and salinity conditions some injected CO_2 will dissolve in formation water, but most injected CO_2 initially would be present as supercritical CO_2 that will replace and push the geothermal water into the production wells. Producing geothermal water allows large volumes of CO_2 to be sequestrated without causing high fluid pressures and induced seismicity. As the porosity is high in such formations, pressure build-up is limited, so safety concerns arising from the cap-rock integrity are minor. Further tests at field scale are underway to verify this concept as an attempt to overcome the bottle-neck technological barrier in sustainable development of geothermal energy from sandstone reservoirs while offering the chance for GSC.

5 CONCLUSIONS

1. A concept for integrating geological sequesteration of CO_2 and geothermal development has been presented using the Beitang area in Tianjin, China as an example. Aspects of the model are described and the isotopic and chemical characterization discussed against geothermal energy potential and CO_2 storage capacity of the Tertiary Guantao and Minghuazhen formations.
2. The water type is dominated by HCO_3-Na and TDS is less than 2.0 g/L. High CO_2 storage capacity can be expected due to the high solubility of the water and the high porosity of the target formation.
3. As geothermal reservoirs, the Guantao and Minghuazhen formations in the Beitang Depression do not possess a high potential for geothermal resource development compared to other units, though thermal waters in the temperature range between 40 and 80°C are rather abundant.
4. Isotopic composition confirms the meteoric origin of the formation water in a terrestrial environment. Age data previously obtained indicates that the Guantao formation is an open to semi-open system, in which the water has been recharged and renewed during an extended period of time, 20–25 thousand years. Safety issues need to be carefully evaluated with respect to CO_2 sequestration.
5. Results show that the Guantao and Minghuazhen formations are viable for field scale tests to verify the proposed model of integrated geological sequestration of CO_2 and geothermal development.

ACKNOWLEDGEMENTS

This study has been supported financially by China National High-Tech R&D (863) Program (2208AA062303). Comments of Brian Hitchon and Gene C. Ulmer helped to improve the earlier version of the manuscript, and those by Halldór Ármannsson and Yousif Kharaka did the later extended version.

REFERENCES

Bachu, S., Bonijoly, D., Bradshaw, J., Burruss, R., Holloway, S, Christensen, N.P. & Mathiassen, O.M. 2007. CO_2 storage capacity estimation: methodology and gaps. *International Journal of Greenhouse Gas Control* 1: 430–443.

Chen, M. 1988. *Geothermics of North China Basin.* Bejing: Science Press, in Chinese.

Fournier, R.O. & Truesdell, A.H. 1973. An empirical Na-K-Ca geothermometer for natural waters. *Geochimica et Cosmochimica Acta* 37: 1255–1275.

Fournier, R.O. 1977. Chemical geothermometers and mixing models for geothermal systems. *Geothermics* 5: 41–40.

Gao, B., Nei, R., Li, X. & Mu, C. 2009. Application of environmental isotopic techniques to studying Tianjin geothermal fluid origin and flow. *Ground water* 31(4), in Chinese.

Giggenbach, W.F. 1988. Geothermal solute equilibria. derivation of Na-K-Mg-Ca geoindicators. *Geochimica et Cosmochimica Acta* 52: 2749–2765.

Gilfillan, S.M.V., Sherwood Lollar, B., Holland, G., Blagburn, D., Stevens, S., Schoell, M., Cassidy, M., Ding, Z., Lacrampe-Couloume, G., Zhou, Z. & Ballentine, C.J. 2009. Solubility trapping in formation water as dominant CO_2 sink in natural gas fields. *Nature* 458: 614–618.

Intergovernmental panel on Climate Change (IPCC). 2005. Special Report on Carbon Dioxide Capture and Storage. Working Group III of the Intergovernmental Panel on Climate Change (B. Metz, O. Davidson, H.C. de Coninck, M. Loos & L.A. Meyer (eds.), Cambridge University Press, Cambridge, UK & New York, USA, 442 p.

Kharaka, Y.K., Cole D.R., Hovorka, S.D., Gunter, W.D., Knauss, K.G. & Freifeld, B.M. 2006. Gas-water-rock interactions in Frio Formation following CO_2 injection: Implications for the storage of greenhouse gases in sedimentary basins. *Geology* 34(7): 577–580.

Kharaka, Y.K. & Mariner, R.H. 1989. *Chemical geothermometers and their application to formation waters from sedimentary basins. Thermal history of sedimentary basins*. Springer-Verlag, New York, 99–117.
Li, Y., Pang, Z., Yang, F. & Duan, Z. 2010. Modelling the geochemical response of CO_2 injection into the Guantao Formation, North China Basin. *Proc. WRI-13*, in this volume.
Minissale, A., Borrini, D. & Montegrossi, G. 2008. The Tianjin geothermal field (north-eastern China): Water chemistry and possible reservoir permeability reduction phenomena. *Geothermics* 37(4): 400–428.
Pang, Z. 2007. Geothermal resources and key technical issues in China. *Keynote presentation at IEA-GIA NEET Network Meeting 2007, Beijing, China*, http://www.iea.org.
Pang, Z. 2008. Advances in the study of geological sequesteration of CO_2. *Keynote speech, Sec. Nat. Symp. on Geological Disposal of Waste, Dunhuang, China, September* 2008.
Pang, Z., Yang, F., Duan, Z. & Huang, T. 2008. Gas-water-rock interactions in sandstone reservoirs: implications for enhance re-injection into geothermal reservoirs and CO_2 geological sequestration. *Proc. Workshop for Decision Makers on Direct Heating Use of Geothermal Resources in Asia, Tianjin, China, 11–18 May 2008.*
Parkhurst, D.L. & Appelo, C.A.J. 1999. User's guide to PHR-EEQC (version 2)—A computer program for speciation, batch reaction, one dimensional transport, and inverse geochemical calculations. U.S. Geological Survey Water-Resources Investigations Report 99–425.
Pruess, K. 2006. Enhanced geothermal systems (EGS) using CO_2 as working fluid-A novel approach for generating renewable energy with simultaneous sequestration of carbon. *Geothermics* 35 (2006): 351–67.
Xu, T., Apps, J.A., Pruess, K. & Yamamoto, H. 2007. Numerical modeling of injection and mineral trapping of CO_2 with H_2S and SO_2 in a sandstone formation. *Chemical Geology* 242: 319–346.

Water-Rock Interaction – Birkle & Torres-Alvarado (eds)
© 2010 Taylor & Francis Group, London, ISBN 978-0-415-60426-0

Ascent and cooling of magmatic fluids: Precipitation of vein and alteration minerals

M. Reed & J. Palandri
University of Oregon, Eugene, Oregon, USA

ABSTRACT: Fluids expelled from magma ascend into fractured host rock and react with the wall rock, producing hydrothermal alteration. Computer model calculations of magmatic fluid reaction with granite over a temperature range from 600°C to 200°C show that a single fluid composition is able to produce the full range of observed wall rock alteration assemblages including potassic (biotite, K-feldspar, muscovite, anhydrite) at high temperature, and quartz-sericite-pyrite, advanced argillic and propylitic at intermediate and low temperatures. The key reaction that enables a fluid saturated with feldspar and biotite at 600°C to produce sericitic and advanced argillic assemblages at lower temperature is the strongly temperature dependent disproportionation of aqueous SO_2, which yields strongly acidic compositions at lower temperatures. Additional calculations of P-T changes along adiabatic fluid ascent pathways yield various assemblages of quartz, sulfides, magnetite, and sulfates that enable us to understand the causes of vein mineral precipitation.

1 INTRODUCTION

The ascent and cooling of a magmatic hydrous phase originating at great depth and high temperature (600°C) is a common hydrothermal process that has received little attention in numerical chemical modeling. Systems of this kind are exemplified by porphyry copper deposits and analogous deposits of tin and molybdenum, wherein granitic magma yields fluids of moderate salinity as in the B35 fluid inclusions of Butte, Montana (5 wt% NaCl equiv, Rusk et al. 2008) and similar ones described for Bingham Canyon (Redmond et al. 2004). We modeled processes of wall rock alteration by computing reaction of Butte Granite with a magmatic fluid closely resembling that analyzed in Butte B35 inclusions (Rusk et al. 2004) over a series of temperatures ranging from 600°C to 200°C at a pressure of 1 kb. The effects of adiabatic fluid ascent on vein mineral assemblages were modeled for isentropic and isenthalpic pathways, starting at 600°C and 2.5 kb, with additional runs for isobaric temperature drop and isothermal pressure drop. The goal is to understand the thermodynamic constraints on what mineral assemblages form in veinlets and alteration envelopes.

2 VEINS AND ALTERATION

High-temperature veinlets in magmatic hydrothermal systems are of two common varieties–those with wall rock alteration, and those without (Fig. 1). Modeling of diffusion and reaction in the Butte vein system (Geiger et al. 2002), show that centimeter-scale alteration envelopes likely need tens of years to form, during which time the fracture apparently remains open to fluid movement at either lithostatic or hydrostatic pressure, as indicated by fluid inclusions (Rusk et al. 2008). The fracture subsequently fills with quartz, sulfides, ±oxide, forming the vein with an alteration envelope (Fig. 1). Alteration envelope mineral assemblages differ depending on temperature and pH, and constitute the well known series of assemblages including potassic, sericitic and propylitic.

Veins dominated by quartz commonly accompanied by molybdenite but lacking alteration envelopes apparently form where pressure drop in newly opened fractures causes quartz to precipitate, thereby blocking fluid movement and precluding significant wall rock reaction. An alternative possibility is that veins lacking alteration form from non-reactive fluids, but analyses of fluid

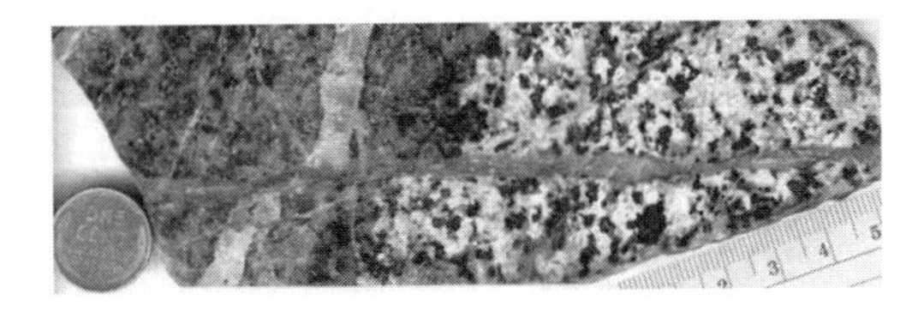

Figure 1. Near-vertical quartz vein with a dark alteration envelope of biotite, K-feldspar, sericite, quartz cut by a quartz-molybdenite vein (horizontal) that lacks an alteration envelope.

inclusions (Rusk et al. 2004) show that there are no systematic differences in fluid composition among vein types, and reaction modeling (below) shows that one fluid type is capable of producing the full range of alteration assemblages –potassic, sericitic, advance argillic, and propylitic.

3 METASOMATIC PETROLOGY OF MAGMATIC HYDROTHERMAL ALTERATION

To explore the hydrothermal alteration mineral geochemistry we map onto P-T-X space the stability fields of all minerals in the system, as well as the compositions of solid solutions, and the composition of the aqueous phase (X refers to 20 compositional variables). Results are given for the presence or absence of many of the minerals in the system displayed in Figure 2. The X coordinate is w/r (water/rock ratio), which expresses the extent of mixture along a pseudo-binary between the fluid and rock compositions, given in Table 1.

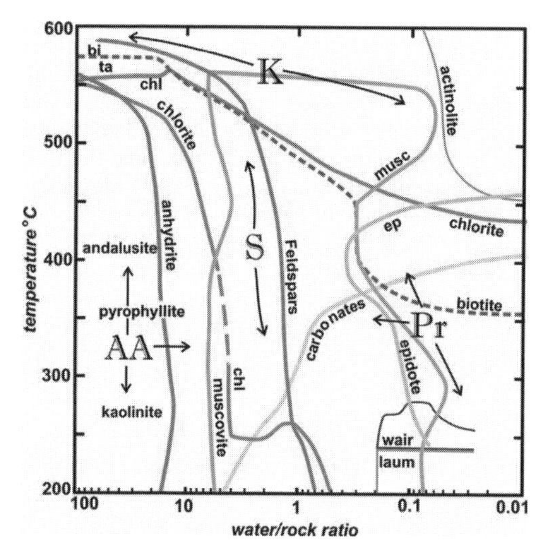

Figure 2. Phase diagram showing the distribution of major alteration minerals computed for reaction of granite with a single magmatic fluid. Quartz is present throughout. Sulfides, sulfur, Al-silicate boundaries, garnet and solid solution features are omitted for simplicity. The curves are labeled with mineral names on the side of the curve where the mineral is present. The large letters with arrows indicate the extent of alteration types common in magmatic-hydrothermal deposits: K, potassic; S, sericitic; AA, advanced argillic; Pr, propylitic. The diagram shows that the fluid produces potassic alteration assemblages spanning all w/r ratios at near-magmatic temperatures, but cooling yields a highly acidic fluid capable of producing sericitic and advanced argillic assemblages. Abbreviations: bi, biotite; chl, chlorite; ep, epidote; laum, laumontite; ta, talc; wair, wairakite. "Feldspars" includes albite, K-feldspar, and plagioclase (an25-an35).

The phase distribution shown in Figure 2 was computed using computer program CHIM-XPT (Reed 1998) by calculating reaction of the Butte Granite with a fluid composition (Table 1) constrained by LA-ICP-MS fluid inclusion analyses (Rusk et al. 2004).

The reaction was executed at 50-degree temperature intervals between 600°C and 200°C at 1kb pressure. Additional calculations were run over short intervals of composition and temperatures to refine many of the phase boundaries where the 50° spacing was not sufficient to define them. Aqueous species thermodynamic data are largely from Shock et al. (1997) and mineral data are largely from Holland & Powell (1998).

The T-w/r diagram differs from conventional phase diagrams because each line marks the edge of a mineral stability field, not a transition from one mineral to another, although some overlapping boundaries do show a transition between phases.

At water/rock less than unity and temperatures less than 550°C, feldspar-bearing assemblages dominate, which we call "potassic" (K, Figure 2) at high temperature and "propylitic" (Pr) at lower temperature where epidote and chlorite are significant (Fig. 2). At w/r between one and five, chlorite, muscovite and quartz are dominant, as in "sericitic" alteration (S), and at larger w/r aluminum silicates and quartz dominate, as in advanced argillic alteration (AA, Fig. 2).

The cause of the particular mineral distribution pattern described above is that the magmatic

Table 1. Magmatic fluid and rock compositions.

Species	Moles	Rock oxide	wt%
H^+	0.58315E+01	SiO_2	65.38
H_2O	0.51900E+02	Al_2O_3	15.30
Cl^-	0.67063	Fe_2O_3	2.11
SO_4^{2-}	0.73150	FeO	2.71
HCO_3^-	0.26561E+01	MnO	0.09
HS^-	0.17108E+01	MgO	2.12
SiO_2	0.43861E–01	CaO	3.81
Al^{3+}	0.13090E–03	Na_2O	3.09
Ca^{2+}	0.11493E–03	K_2O	4.00
Mg^{2+}	0.74562E–05	P_2O_5	0.17
Fe^{2+}	0.84923E–05	BaO	0.97
K^+	0.71378E–01	CuO	0.0037
Na^+	0.34556	PbO	0.0024
Mn^{2+}	0.69967E–03	ZnO	0.0074
Zn^{2+}	0.71840E–03		
Cu^+	0.24832		
Pb^{2+}	0.15532E–03	fluid pH (590°C): 5.9	
Ag^+	0.87008E–05	TDS (wt%): 13.4	
Ba^{2+}	0.26301E–05	(excluding	
HPO_4^{2-}	0.72956E–06	C species)	

fluid is pH-neutral at high temperature but acidic at temperatures less than 550°C, i.e. the same fluid that is in equilibrium with feldspars and biotite at 600°C where pH is 5.9 is so acidic at 500°C (pH 3) that it forms quartz and aluminum silicates. The fundamental reason for the neutral-to-acidic pH change with decreasing temperature is that $SO_2(aq)$, or its equivalent, $H_2SO_3(aq)$, is the dominant sulfur species at high temperature, but it disproportionates with decreasing temperature, forming sulfuric acid and H_2S, as in the following reaction:

$$SO_2(aq) + H_2O(aq) = 1\tfrac{1}{2}\, H^+(aq) + \tfrac{3}{4} SO_4^{2-}(aq) + \tfrac{1}{4}\, H_2S(aq) \qquad (1)$$

Between 600°C and 100°C, the equilibrium constant for reaction (1) increases by 17 orders of magnitude (10^{-15} to 10^2), shifting the reaction strongly to the right at lower temperature, yielding hydrogen ion.

The principal significance of the modeling results is that a single fluid (Table 1) reacting at different temperatures and in variable water/rock ratios produces all of the alteration types known from magmatic-hydrothermal systems: potassic, sericitic, advanced argillic and propylitic and sub-types within those. Thus, it is possible that a single fluid actually does produce all of the alteration types. If so, our focus in understanding the zoning of vein and alteration type in magmatic-hydrothermal systems must shift from an examination of magma evolution to examination of fluid separation, storage, cooling and expulsion from plutons and their cupolas.

In addition to rationalizing the large-scale alteration patterns in porphyry copper and similar ore systems, the calculations show that, depending on the details of cooling and rock contact, a single fluid parcel may produce potassic alteration at depth then yield sericitic then propylitic alteration as it ascends, cools, and reacts along the way. Another ascending parcel of the same fluid that finds a more direct vertical pathway may produce potassic assemblages at depth then advanced argillic alteration as it moves upward. In this way, the many variations known in alteration zoning and cross-cutting patterns (Seedorf et al. 2005) may develop.

4 PRESSURE-TEMPERATURE ASCENT PATHS AND VEIN FORMATION

The preceding discussion addresses wall rock alteration bordering veins, but what about the vein mineral content, itself? As fluids ascend through fractures above a pluton they depressurize and cool, but the particular pressure-temperature path of a given fluid parcel depends on the thermal history of the pathway. For fluids that start at lithostatic pressure and ascend toward hydrostatic pressure (e.g. Rusk & Reed 2002), we can frame the P-T conditions starting from an adiabatic path. If a substantial quantity of fluid had previously ascended, it is likely that an adiabatic thermal gradient would exist, the particulars of which depend on the conditions of ascent, but it would lie somewhere between an isenthalpic (free expansion) and isentropic (reversible expansion) path, as shown in at (b) in Figure 3.

Early in the system history, ascending fluids would encounter not-yet-heated rock, and would thus be cooler than an adiabatic path, as in region (a). Later in the system history and depending on how open the system at depth has been to hydrostatic pressure, an incomplete transition from lithostatic to hydrostatic pressure (but one changing in that direction) could yield conditions in region (c). The low-pressure ends of the curves are limited by hydrostatic pressure for fluids that begin at 2.5 kb at depth. Lower pressures are not addressed here, but they could be controlled by a hydrostatic pressure gradient, with temperature depending on other details such as adiabatic ascent, with or without fluid phase separation.

To examine the consequence to vein mineral assemblages of various P-T ascent paths, we modeled simultaneous cooling and depressurization of the fluid given in Table 1 along the two limiting adiabatic pathways and along isobaric and isothermal paths shown in Figure 3. All ascent paths produce quartz as the dominant mineral (Fig. 4), however secondary minerals differ for the various paths.

Simple isobaric temperature decrease yields a decrease in pH as reaction (1) proceeds, resulting in precipitation of pyrite, and minor muscovite and andalusite (Fig. 4). pH changes little along the isentropic path, in which pyrite forms with small amounts of muscovite, copper sulfides and sulfate minerals. The isenthalpic path results in a modest increase of pH by about one unit, which favors magnetite over pyrite for most of the P-T range, with additional sulfides, sulfates and feldspar (Fig. 4). In the isothermal pressure drop (not shown), pH increases by two units, yielding magnetite, sulfides, sulfates and one carbonate,

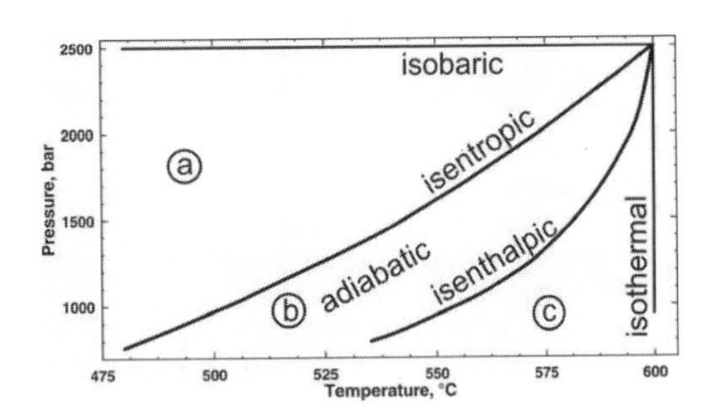

Figure 3. Possible P-T paths for fluid ascending from a pluton. See text for explanation of labeling.

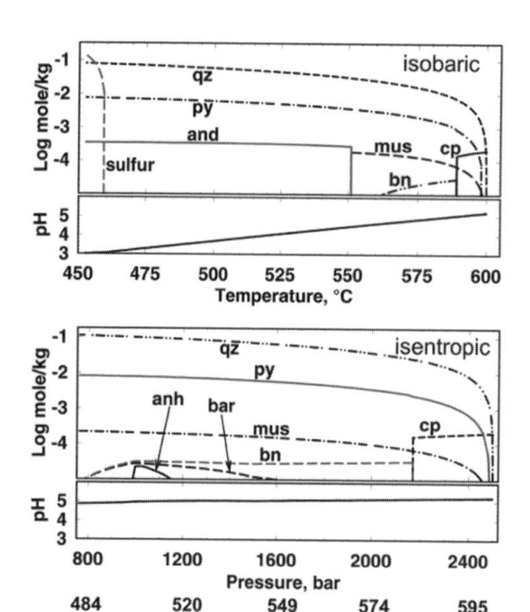

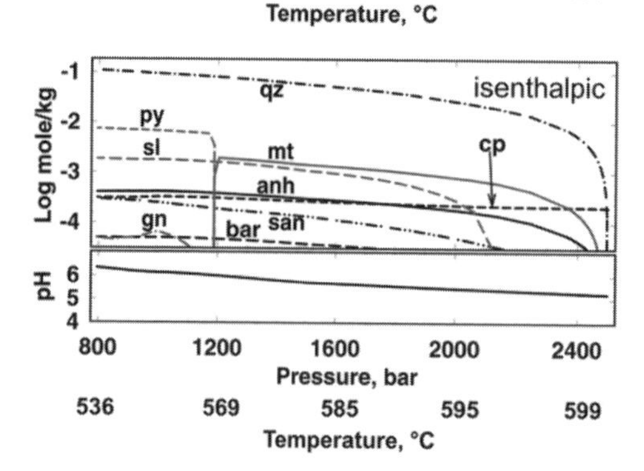

Figure 4. Mineral precipitates and pH change for three of the four fluid P-T ascent paths shown in figure 3 (isothermal excluded). Depending on the P-T path, the pH becomes either acidic or basic, which affects the mineral assemblages. The temperature scale numbers are paired with the corresponding pressures on the scale. Abbreviations: and, andalusite; anh, anhydrite; bar, barite; bn, bornite; cp, chalcopyrite; gn, galena; mt, magnetite; mus, muscovite; py, pyrite; qz, quartz; san, sanidine; sl, sphalerite.

rhodonite. pH increases upon pressure decrease in the isenthalpic and isothermal cases are driven by the association of HCl, consuming $H^+$and Cl^-.

With the exception of sulfur and andalusite, all of the minerals computed in the ascent paths (Fig. 4) occur in the deep vein assemblages of Butte and other porphyry copper systems. Natural veins display differences in relative quantities of pyrite, magnetite, and copper sulfides and the presence or absence of anhydrite, as shown in these results.

However, we are at the preliminary stages of modeling such simultaneous P-T change processes, and more complete modeling using alternate fluid compositions and examining simultaneous wall rock reaction is needed to refine our ability to relate observed assemblages to particular fluid ascent histories. It is clear from these results that adiabatic ascent yields common vein assemblages and that the thermal history of the individual fluid ascent pathways distinctly affects the vein mineral assemblages. It is also apparent that a single fluid yields multiple vein mineral assemblages depending on its P-T ascent pathway (Figure 4), just as a single fluid yields the full range of alteration assemblages, depending on its P, T and reaction history (Fig. 2).

This study was supported by US National Science Foundation grant EAR-0440198.

REFERENCES

Geiger, S., Haggerty, R., Dilles, J.H., Reed, M.H. & Matthai, S.K. 2002. New insights from reactive transport modelling: the formation of the sericitic vein envelopes during early hydrothermal alteration at Butte, Montana. *Geofluids* 2: 185–201.

Holland, T.J.B. & Powell, R. 1998 An internally consistent thermodynamic data set for phases of petrological interest. *Journal of Metamorphic Geology* 16: 309–343.

Redmond, P.B., Einaudi, M.T., Inan, E.E., Landtwing, M.R. & Heinrich, C.A. 2004. Copper deposition by fluid cooling in intrusion-centered systems: New insights from the Bingham porphyry ore deposit, Utah. *Geology* 32: 217–220.

Reed, M.H. 1998. Calculation of simultaneous chemical equilibria in aqueous-mineral-gas systems and its application to modeling hydrothermal processes, Chapter 5. In J. Richards & P. Larson (eds), *Techniques in Hydrothermal Ore Deposits Geology, Reviews in Economic Geology* 10: 109–124.

Rusk, B.G. & Reed, M.H. 2002. Scanning electron microscope-cathodoluminescence of quartz reveals complex growth histories in veins from the Butte porphyry copper deposit, Montana. *Geology* 30: 727–730.

Rusk, B., Reed, M., Dilles, J.H., Klemm, L. & Heinrich, C.A. 2004. Compositions of magmatic hydrothermal fluids determined by LA-ICP-MS of fluid inclusions from the porphyry copper-molybdenum deposit at Butte, Montana. *Chemical Geology* 210: 173–199.

Rusk, B., Reed, M. & Dilles, J., 2008. Fluid Inclusion Evidence For Magmatic-Hydrothermal Fluid Evolution in the Porphyry Copper-Molybdenum Deposit at Butte, Montana. *Economic Geology* 103: 307–334.

Seedorf, E., Dilles, J.H., Proffett, J.M., Einaudi, M.T., Zurcher, L., Stavast, W.J.A., Johnson, D.A. & Barton, M.D. 2005. Porphyry deposits: characteristics and origin of hypogene features. *Economic Geology 100th Anniversary Volume*: 251–298.

Shock, E.L., Sassini, D.C., Willis, M. & Sverjensky, D.A. 1997. Inorganic species in geologic fluids: Correlations among standard molal thermodynamic properties of aqueous ions and hydroxide complexes. *Geochimica et Cosmochima Acta* 61: 907–950.

Water-Rock Interaction – Birkle & Torres-Alvarado (eds)
© 2010 Taylor & Francis Group, London, ISBN 978-0-415-60426-0

FTIR/ATR characterization of TiO_2/aqueous carboxylate solution interfaces undergoing chemical changes

F. Roncaroli, P.Z. Araujo, P.J. Morando, A.E. Regazzoni & M.A. Blesa
Gerencia Química, Comisión Nacional de Energía Atómica, San Martín (Buenos Aires), Argentina
Instituto de Ingeniería e Investigación Ambiental and Instituto Sabato, Universidad Nacional de San Martín
Consejo Nacional de Investigaciones Científicas y Técnicas

ABSTRACT: FTIR/ATR is used to (a) structurally characterize surface complexes formed upon the adsorption of simple carboxylates onto metal oxides; (b) derive the (Langmuir) stability constants of the these complexes, and discuss the stability trends in a series of related surface complexes; (c) study the kinetics of adsorption of the ligands onto metal oxide films; (d) study the kinetics of photochemically induced ligand transformation (essentially, oxidation by dissolved oxygen). Titanium dioxide is used as a model for minerals with high stability in water and high photoactivity.

1 INTRODUCTION

The sediment / water interface is not only the locus of adsorption of solutes; it is also the site in which chemical and photochemical transformations of adsorbates may take place. Many structural probes of static interfaces became available in the last part of the XXth Century (Brown 1990) and the results were used to model sediment/water interfaces (Davis & Kent 1990). Description and modeling of interfaces undergoing chemical changes is more limited. A very simple and versatile structural tool is infrared spectroscopy, in the Attenuated Total Reflection mode (FTIR/ATR). The structural characterization of many systems using FTIR/ATR has been reported since the seminal paper by Tejedor-Tejedor & Anderson (1986).

In this paper we describe the use of FTIR/ATR to (a) structurally characterize surface complexes formed upon the adsorption of simple carboxylates onto metal oxides; (b) derive the (Langmuir) stability constants of the these complexes, and discuss the stability trends in a series of related surface complexes; (c) study the kinetics of adsorption of the ligands onto metal oxide films; (d) study the kinetics of photochemically induced ligand transformation (essentially, oxidation by dissolved oxygen). Titanium dioxide is used as a model for minerals with high stability in water and high photoactivity.

These experimental studies are supplemented with theoretical calculations of the stability of various possible surface species.

2 EXPERIMENTAL

ATR-FT-IR measurements were performed in a Nicolet Magna IR 560 spectrometer equipped with a MCT-A detector cooled by liquid N_2. The spectrometer was operated by the OMNIC 5.0 program from Nicolet. A 12 reflection trapezoidal ZnSe ATR crystal ($0.3 \times 7.2 \times 1.0$ cm) was placed inside a horizontal ATR flow cell Spectratech ARK 0056–303. The ATR flow cell was mounted on a Spectratech ARK accessory. For the preparation of TiO_2 thin films, 200 μL of a water suspension of TiO_2 (60 mg/10 mL) were spread on one of the faces of the ATR crystal (7.2×1.0 cm) and allowed to evaporate overnight. After flowing 100 mL of water to remove impurities and equilibrate the surface, carboxylic acid solution were flowed through the cell; as soon as it reached the ATR cell spectra acquisition was started. Spectra were recorded from 1000 – 2000 cm^{-1}, usually every 3 minutes (250 scans, resolution 2 cm^{-1}).

3 RESULTS AND DISCUSSION

Figure 1 shows the results of the spectral study of the adsorption of glycolic and thioglycolic acids (hydroxy- and mercapto-carboxylic acids) onto a titanium dioxide sample composed mainly of anatase at pH 4. Two independent components are required to fit the glycolic spectral data, and only one in the case of thioglycolic acid. Band assignments are rather straightforward (not shown).

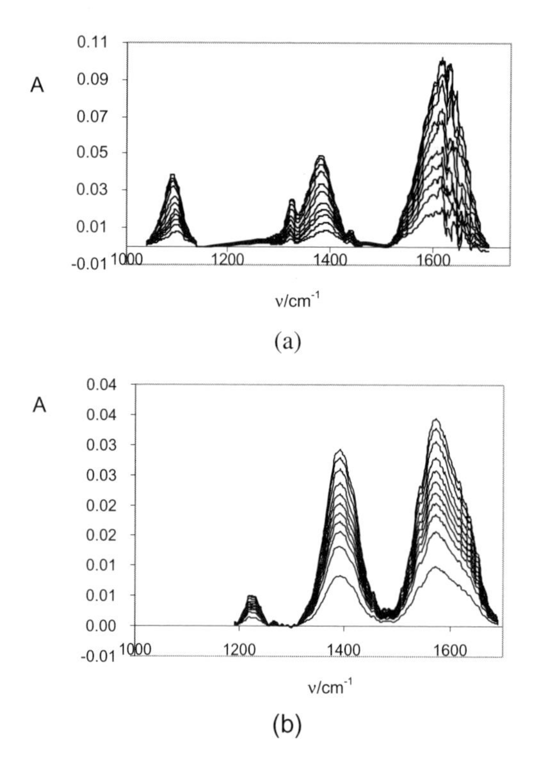

Figure 1. FTIR/ATR spectra of adsorbed glycolic (a) and thioglycolic (b) acids onto TiO_2 at various solution concentrations. Absorbance (A) is plotted against wavenumber ν.

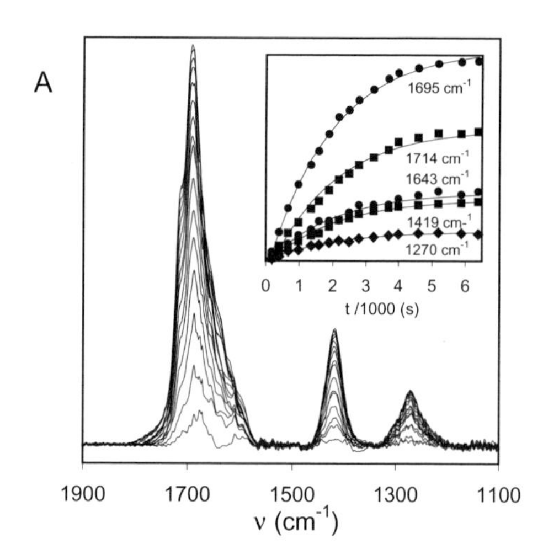

Figure 2. IR spectral changes obtained during the adsorption of oxalic acid a TiO_2 film. Inset:selected traces at the most relevant wavenumbers. $k_{obs} = 5.5 \pm 0.6 \times 10^{-4}$ s^{-1}. [Oxalic acid] = 1.67×10^{-6} M, pH 4.0, T 25.0°C, I 0.01 M (NaCl). Absorbance (A, in arbitrary units) is plotted against wavenumber ν.

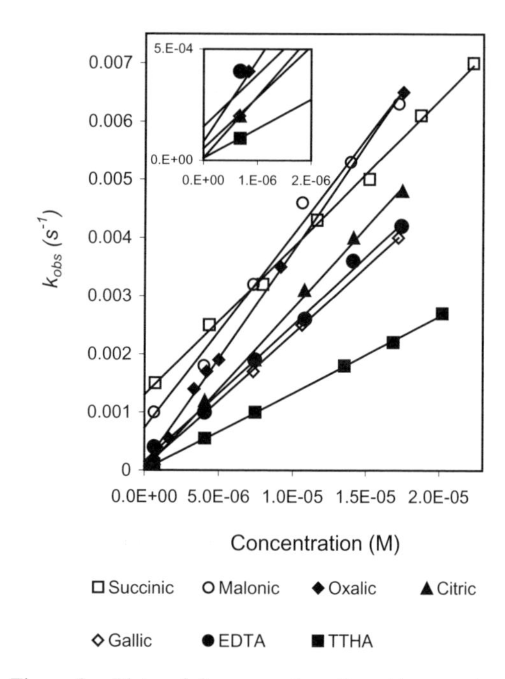

Figure 3. Plots of k_{obs} vs. carboxylic acid concentration for different carboxylic acids under study. pH 4.0, T 25.0 °C, I 0.01 M (NaCl). Inset: zoom of the same figure at the origin showing intercepts. Values of the slopes (a) and intercepts (b) are listed in Table 1.

The conditional Langmuir stability constants at pH 4 derived from the same data are 8.7×10^5 and 8.6×10^3 mol^{-1} L for glycolic acid, and 4.9×10^5 and 2.9×10^3 mol^{-1} L for thioglycolic acid (two adsorption isotherms are required even though only one spectral component is found). The former complexes are thus somewhat more stable (by a factor *ca* 2) at this pH. It is well known that Ti(IV) has a large affinity for –OH groups, and it is thus reasonable that the surface chelates of glycolic acid are more stable. In a similar way, it can be shown that the stability of aminocarboxylates formed in the surface of TiO_2 is lower than the stability hydroxocarboxylates. A good example is provided by the lower stability of the EDTA complex as compared to that of the citrate complex. The zwitterionic nature of aminocarboxylates in a wide pH range hinders somewhat the surface complexation reaction, most probably because of the decrease in the negative charge of the ligand due to the positive $-NH_3^+$ group.

Kinetics of adsorption can also be followed by FTIR/ATR. Figure 2 shows the time evolution of the spectrum of adsorbed oxalic acid, and Figure 3 shows the dependence on ligand concentration of the pseudo-first order rate constant derived from a study of a series of carboxylic acids. The data in Figure 3 can be fitted by the simple expression:

$$k_{obs} = a[\mathrm{L}] + b \tag{1}$$

It can be shown that *a* measures the rate of adsorption, and *b* measures the rate of desorption of the outer sphere complex, formed by dissociation of the inner sphere species. Furthermore, the data demonstrate that *a* is diffusion controlled; *b* is determined by the composite action of the back diffusion rate and the thermodynamic stability of the inner sphere complexes.

As an example of photo-induced changes in the surface, we shall discuss the case of gallic acid (3, 4, 5-trihydroxybenzoic acid), which is a very simple model for natural organic matter (NOM), adsorbed on TiO_2.

Gallic acid is totally oxidized as described in the following equation, in a process that require a series of successive transformations, equivalent to 24 one-electron steps:

$$C_7O_5H_6 + 6O_2 \rightarrow 7CO_2 + 3H_2O$$

FTIR/ATR spectra do not provide evidence of intermediates, indicating a low steady state concentration in the interface.

Figure 4(a) shows the time evolution of the spectra and Figure 4(b) shows the first order kinetic plot for two types of experiments, under continuous and under intermittent illumination in the near UV region. The reaction is first order at all times under intermittent illumination, and at short times under continuous irradiation. The deviation at long times is indicative that partially oxidized molecules consume photons.

In earlier work, quantum-chemical calculations using semiempiral MSINDO method was used to calculate the configurations corresponding to the most stable surface species. In the case of oxalate, it could thus be demonstrated that at high degrees of coverage, in the presence of bulk water, the complex with a free pendant carboxylate group is among the most stable ones, together with chelates formed involving both carboxylate groups.

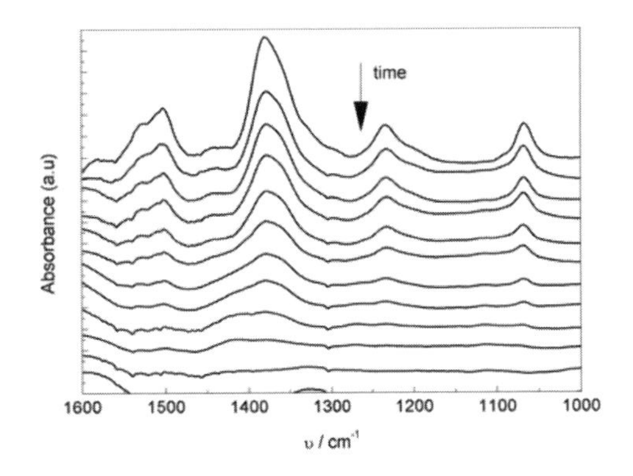

(a)

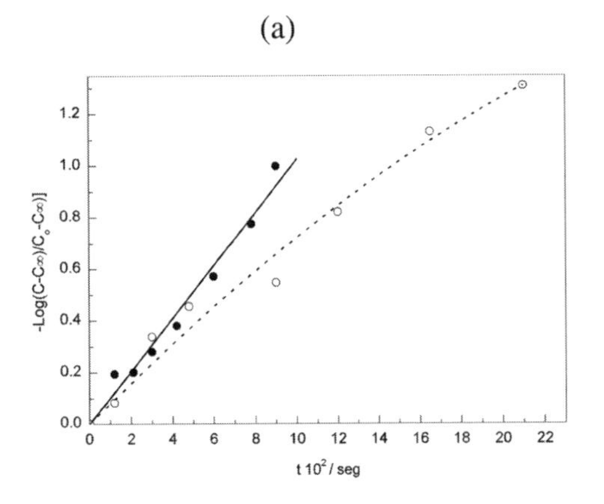

Figure 4. (a) ATR-FTIR spectral time evolution upon photolysis of 1800 μL GA 1×10^{-4} mol L^{-1} on the film prepared with 150 μL of 20 g L^{-1} TiO_2 suspension. From the top to bottom 0, 2, 3.5, 5, 7, 10, 13, 15, 20, 25, 35 and 45 min; (b) First order kinetic plot of the heterogeneous photocatalytic degradation of GA from IR main band intensity: (●) Single film experiment. (O) Continuous irradiation experiment. Films were deposited from 150 μL of 20 g L^{-1} TiO_2 suspension. Initial GA concentration, 1.0×10^{-4} mol L^{-1}.

4 CONCLUSIONS

FTIR/ATR is useful not only to characterize surface species under equilibrium conditions. The evolution of dynamic interfaces can also be followed *in situ*, as illustrated in this paper by the behavior of adsorption and photocatalytic oxidation of carboxylic acids onto TiO_2.

REFERENCES

Brown, G.E. Jr. 1990. Spectroscopic Studies of Chemisorption Reaction Mechanisms at Oxide-Water Interfaces. In M.F. Hochella Jr. & A.F. White (eds.), *Mineral-Water Interface Geochemistry*. Reviews in Mineralogy vol. 23: chap. 8. Washington D.C: Mineralogical Society of America.

Davis, J.A. & Kent, D.B. 1990 Surface Complexation Modeling in Aqueous Geochemistry. In M.F. Hochella Jr. & A.F. White (eds.), *Mineral-Water Interface Geochemistry*. Reviews in Mineralogy vol. 23: chap. 5. Washington D.C: Mineralogical Society of America.

Mendive, C.B., Bredow, T., Feldhoff, A., Blesa, M.A. & Bahnemann, D. 2009. Adsorption of Oxalate on Anatase (100) and Rutile (110) Surfaces in Aqueous Systems: Experimental Results vs. Theoretical Predictions. *Physical Chemistry Chemical Physics* 11: 1794–1808.

Tejedor-Tejedor, M.I. & Anderson, M.A. 1986 In situ Attenuated Total Reflection Fourier Transform infrared studies of the goethite (α-FeOOH)-aqueous solution interface. *Langmuir* 2: 203–210.

Hitoshi Sakai memorial session: Measurements and applications of stable and radiogenic isotopes and other tracers

Water-Rock Interaction – Birkle & Torres-Alvarado (eds)

 ISBN 978-0-415-60426-0

Evolution of groundwater in volcanic aquifer of Axum, Ethiopia: Isotopic and hydrochemical evidence

T. Alemayehu & M. Dietzel
Graz University of Technology, Graz, Austria

A. Leis
Joanneum Research, Graz, Austria

ABSTRACT: Hydrochemical and isotopic data are used to examine the decisive processes controlling groundwater chemistry and to reveal sources of groundwater. Groundwater from Axum area shows distinctive chemical and isotopic variation suggesting spatial diversity of the aquifer and different geochemical evolution. Individual compositions appear to be controlled by the type of host rock and intensity of water-rock interaction. Weathering of silicates and to lesser extent of calcite containing veins as well as ion-exchange explains most of the observed spatial variability in groundwater composition. The δ^2H_{VSMOW} and $\delta^{18}O_{VSMOW}$ values of the groundwaters indicate a meteoric origin. The $\delta^{13}C_{VPDB}$ values of dissolved inorganic carbon (DIC) range from −12 to +1‰ revealing magmatic CO_2 and soil CO_2 as predominant sources for DIC.

1 INTRODUCTION

Groundwater from a volcanic aquifer is the main source of drinking water supply for the town of Axum. The groundwater is mainly hosted by fractured basalt rocks (Fig. 1). The lateral variation in water yield and water quality observed in the study area reflects the heterogeneous nature of the aquifer. Poor groundwater quality from high yielding bore holes has been a major problem for the water supply of the town. The local volcanic environment shows remarkable variation e.g. in terms of major ion concentrations (Ayenew et al. 2008).

Variation of groundwater composition generally reflects the mineral content within the aquifer and the duration and intensity of the water-rock interaction (Elango & Kannan 2007). Kinetics of water-rock interactions can greatly affect the chemical composition of groundwater starting from the recharge to discharge zone. Characterization of groundwater chemistry is required to identify the geochemical evolution of groundwater chemistry along the flow path.

In the present study, the water chemistry of the main groundwater fields of the Axum area is investigated to identify mechanisms which are responsible for groundwater variability. The overall aim is to apply hydrochemical and isotopic tools for groundwater characterization and evolution for improved groundwater resource management.

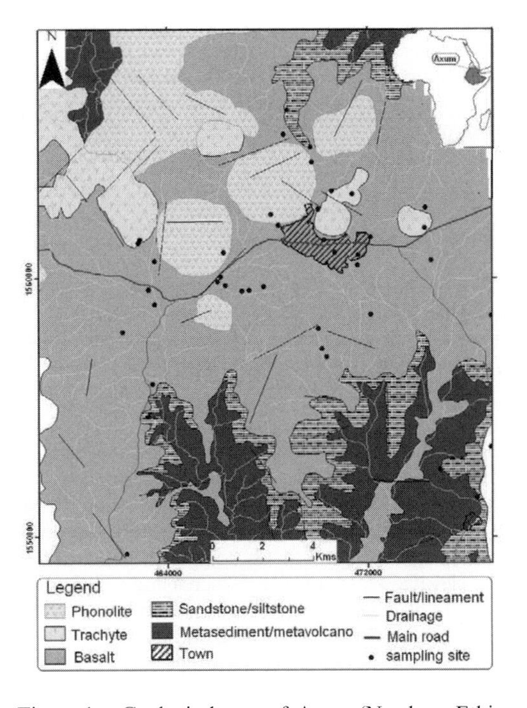

Figure 1. Geological map of Axum (Northern Ethiopia) and sampling point.

2 SAMPLING AND ANALYSIS

Water samples were collected from springs, wells and boreholes at various depths. All samples were filtered through 0.45 μm membranes in the field and the samples were separated in different aliquots. Field parameters, including pH, electrical conductivity (EC) and temperature were recorded during sampling. Total alkalinity was measured in-situ by titration with diluted HCl. Major and trace elements were analysed using a coupled plasma-optical emission spectrometry (Perkin Elmer 4300) and ion chromatography (Dionex 600).

Samples for the stable isotope analyses of the water ($^{18}O/^{16}O$ and $^{2}H/H$) were collected in HDPE bottles and for stable carbon isotopes ($^{13}C/^{12}C$) of dissolved inorganic carbon (DIC) in 10 ml gas tight vials initially flushed with Helium and preloaded with six droplets of phosphoric acid.

The isotopic composition of DIC was analyzed using a fully automated peripheral continuous-flow gas preparation device (Gasbench II), which was connected to a Finnigan Delta^{plus} XP Mass Spectrometer. The oxygen isotopic composition of the water was measured by the classic CO_2–H_2O equilibrium technique with a fully automated device coupled to a Finnigan DELTA^{plus} Mass Spectrometer. The isotopes of hydrogen were analysed using a continuous flow Finnigan DELTA^{plus} XP Mass Spectrometer coupled to HEKAtech high-temperature oven by chromium reduction.

3 RESULTS

3.1 *Groundwater chemistry*

The chemistry of the groundwater progressively evolved from Ca^{2+}-HCO_3^- to Mg^{2+}-HCO_3^- and Na^+-HCO_3^- water types. The total dissolved solid (TDS) concentration indicates large differences between Ca^{2+}-HCO_3^- and Mg^{2+}-HCO_3^- types with a maximum value of 2160 mg/L in deep groundwater (down to 150 m). The pH ranges from 6.8 to 8.5, with mean pH 7.4. The estimated internal partial pressure of CO_2 (pCO_2; see Fig. 2a) is significantly higher than that of the Earth's atmosphere indicating the presence of soil CO_2 and/or an additional CO_2 source. The elevated pCO_2 results in increasing weathering capacity which determines the degree of solute concentration leaching out of mineral phases.

The groundwater passes through a basaltic rock aquifer that influences its hydrochemical signature. The local basalt is mainly composed of plagioclase feldspar, olivine and pyroxene minerals. Accordingly, weathering of such primary silicate minerals leads to enrichment of respective major cations and occasionally to the formation of clay minerals. Groundwaters tend to be of Na–HCO_3^- type in feldspar-rich rocks reflecting albite dissolution.

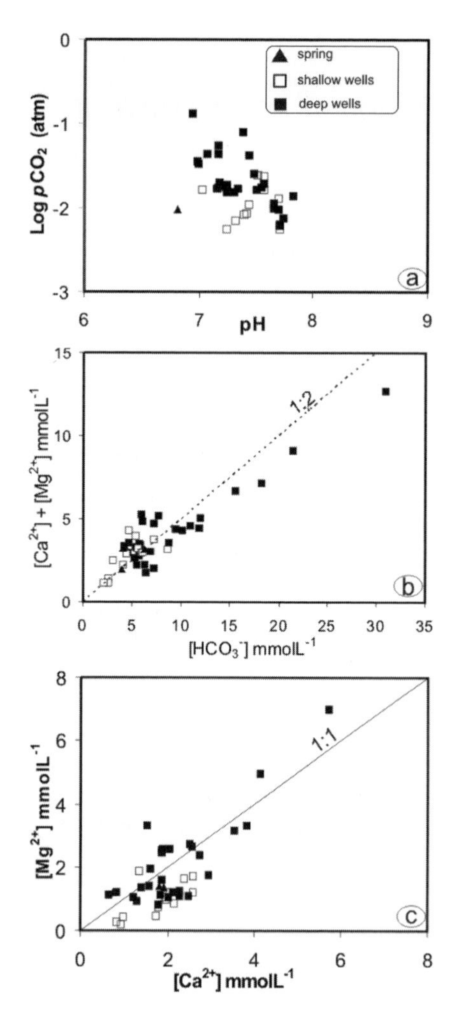

Figure 2. Plots of (a) pCO_2 vs. pH (b) $[Ca^{2+}] + [Mg^{2+}]$ vs. $[HCO_3^-]$ (c) $[Mg^{2+}]$ vs. $[Ca^{2+}]$ for Axum groundwaters.

In volcanic aquifers, high HCO_3^- concentrations can be caused by the incongruent dissolution of silicate minerals (Appello & Postma 2005). However, a significant increase of HCO_3^- concentrations may account processes other than silicate weathering (Fig. 2b). Additional HCO_3^- may be released into the solution by carbonate dissolution as calcite, along with other hydrothermal phases, is locally present as filling veins.

Magnesium enrichment of deep groundwaters with high bicarbonate content is typical for intense dissolution of mafic rocks (Figs. 2b and 2c). However, most shallow wells (depth of less than 40 m) are indicated by elevated Ca^{2+} content.

As most of the groundwaters are supersaturated with respect to calcite at elevated pCO_2, calcite may be precipitated due to ongoing CO_2 degassing from the solutions.

3.2 *Stable isotope composition*

The isotopic data of water (δ^2H and $\delta^{18}O$) indicate that the groundwaters are of meteoric origin. Most δ^2H and $\delta^{18}O$ values plot along or close to the global and local meteoric water line (see Fig. 3). In two samples, a positive $\delta^{18}O$ shift is attributed to oxygen exchange with the host rock. Similar features have been described by Federico et al. (2002) in volcanic aquifers. The presence of rather isolated aquifers support the possibility that such fluids may interact with surrounding rocks more strongly. δ^2H and $\delta^{18}O$ values for surface water and spring water are distinct from the other solutions. The shift is due to evaporation of the original local precipitation (see evaporation trend in Fig. 3).

The $\delta^{13}C_{DIC}$ values in the analysed groundwater are in the range from −12 to +1‰ (Fig. 4). Such range is mostly referred to the origin of CO_2 from two sources. The isotopically heavier group with $\delta^{13}C_{DIC}$ from −5 to +1‰ indicates the presence of magmatic CO_2 (deep groundwater). Typical $\delta^{13}C$ values of CO_2 from magmatic source range from −8 to −5‰ (e.g. Hoefs 1997). The equilibrium $^{13}C/^{12}C$ fractionation between CO_2 and DIC varies with temperature and pH (e.g. Dietzel & Kirchhof 2002, Federico et al. 2002). In nearly closed system with respect to magmatic CO_2, the initially CO_2 loaded solution may yield a pH value close to 6.5. At this pH, fractionation between CO_2 gas and DIC is about −4‰. Taking this fractionation value, $\delta^{13}C_{CO_2}$ values between −9 and −3‰ are obtained from isotopic composition of the DIC, which is in general agreement with the isotopic composition of CO_2 from magmatic origin. However, the isotopic composition of DIC is likely to be influenced by dissolution of calcite filling veins deposited from hydrothermal fluids and aqueous solution during volcanic activities.

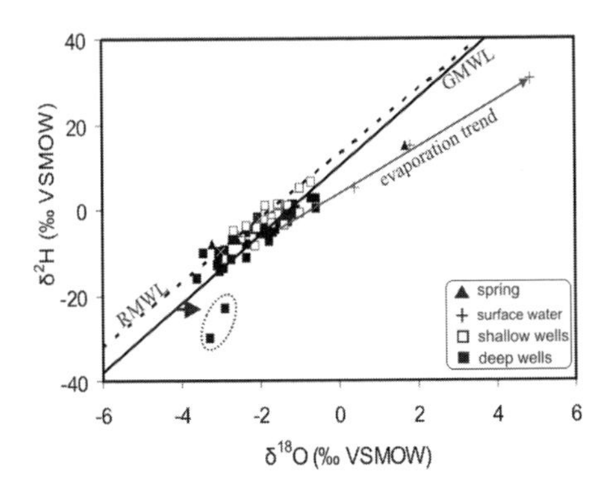

Figure 3. Relationships between δ^2H and $\delta^{18}O$ values for all sampled solutions. GMWL (solid line) refers to Global Meteoric Water Line of Craig (1961). RMWL denotes the regional meteoric water line for Addis Ababa ($\delta^2H = 7.14\ \delta^{18}O+11.98$).

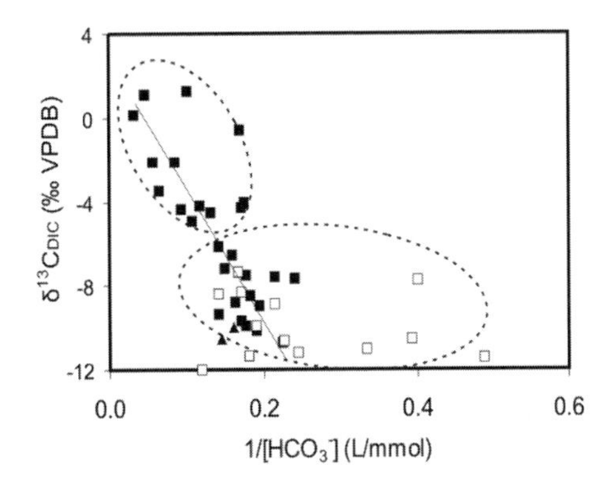

Figure 4. $\delta^{13}C$ values versus the reciprocal of HCO_3^- concentrations. The $\delta^{13}C_{DIC}$ values indicate the likely source of magmatic CO_2 (enriched in ^{13}C) and in the presence of soil CO_2 (enriched in ^{12}C). Symbols are given in Fig. 1.

In contrast, the isotopically lighter group with $\delta^{13}C_{DIC}$ from −12 to −5‰ is mostly referred to shallow wells where the CO_2 is mainly derived from the soil. The stable carbon isotopic composition of soil CO_2 produced by respiration in the soil is approximately equivalent to $\delta^{13}C$ of the predominant organic matter. The average $\delta^{13}C$ value generated from C_3 and C_4 plants is about −27 and −13 ‰, respectively (Clark & Fritz 1997). At the given pH range (mean pH 7.5) of the shallow groundwaters, the carbon isotope fractionation between DIC and $CO_{2(gas)}$ is close to −8‰. Accordingly, the $\delta^{13}C_{CO_2}$ of soil falls in the range from −20 to −13‰, suggesting more inputs from C_4 plant type dominant in the region (Terwilliger et al. 2008). The dissolution of silicates in the shallow aquifers remains at open system conditions with respect to soil CO_2.

The intermediate carbon isotope signature of DIC can be interpreted as mixtures between the discussed two end members.

3.3 *Geochemical evolution*

The chemical evolution of the sampled groundwater of Axum is attributed to the host rock composition and the extent of water-rock interaction. The interaction behaviour can be evaluated by the degree of saturation with respect to mineral phases present in the aquifer. The computer code *PHREEC-2* (Parkhurst & Appelo 1999) was used to calculate the saturation indices (*SI*) for minerals contributing to water chemistry according to the equation

$$SI = \log\left(\frac{IAP}{K_{sp}}\right) \quad (1)$$

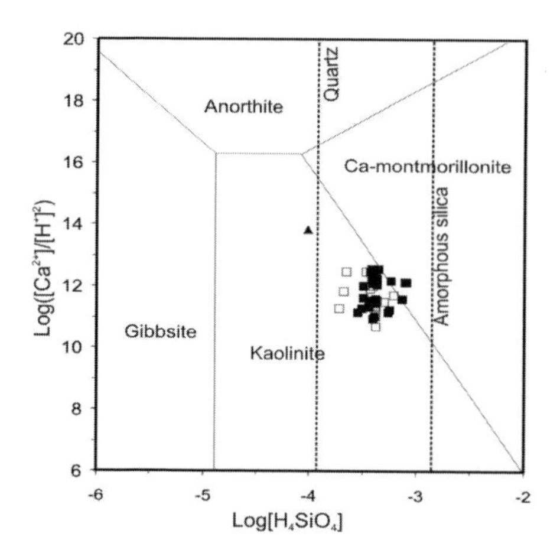

Figure 5. Mineral stability diagram for $CaO-Al_2O_3-SiO_2-H_2O$ systems and calculated activities of sampled solutions from Axum area. Solubility constants from Tardy (1971) were used for construction of equilibrium boundary (25°C and 1 atm). Symbols are given in Fig. 1.

where *IAP* is the ion activity product and K_{sp} is the equilibrium constant at given temperature.

The weathering of silicate minerals appears to be controlled by kinetics and precipitation of secondary silicate phases such as kaolinite or smectite. The chemical compositions of sampled solutions are plotted in Figure 5, where predominant area for silica containing solids in the $CaO-Al_2O_3-SiO_2-H_2O$ systems are given, calculated from solubility. Highest silica concentrations are caused by intensive weathering of silicate rocks at high pCO_2 (elevated cation content). The chemical composition of sampled solutions plots in the stability field of kaolinite and along the equilibrium line of kaolinite and Ca-montmorillonite. This is consistent with the occurrence of such clay minerals in the weathered rocks. Considering clay minerals, particularly montmorillonite, cation exchange reactions may have an additional impact on the observed chemical composition of the groundwaters in the study area.

4 CONCLUSIONS

The chemical variability of the groundwater in the Axum area reflects the individual behaviour for uptake of gaseous CO_2 and the extent of water-rock interactions. The chemical composition is controlled by complex reaction involving dissolution and precipitation of silicates and to lesser degree carbonates as well as cation exchange between groundwater and clay minerals. While the composition of shallow groundwaters is governed by the uptake of soil CO_2, weathering of silicate at higher pCO_2 is a relevant in the deep aquifer.

The geochemical trends in the groundwater evolution are supported by $\delta^{13}C_{DIC}$ data which indicate CO_2 from soil horizon and magmatic-derived CO_2 as predominant source for DIC. The δ^2H and $\delta^{18}O$ values clearly show that the solutions are of meteoric origin with apparent effect of evaporation and water-rock interactions.

ACKNOWLEDGEMENTS

We acknowledge the Austrian Agency for International Cooperation in Education and Research (OeAD) for a PhD grant for the first author. Moreover, the authors greatly appreciate support from Mekelle University and Graz University of Technology during sample collection and solution analyses, respectively.

REFERENCES

Appelo, C.A.J. & Postma, D. 2005. *Geochemistry, groundwater and pollution*. Rotterdam: Balkema.

Ayenew, T. Molla, D. & Wohnlich, S. 2008. Hydrogeological framework and occurrence of groundwater in the Ethiopian aquifers. *Journal of African Earth Sciences* 52(3): 97–113.

Clark, I.D. & Fritz, P. 1997. *Environmental Isotopes in Hydrogeology*. New York: Lewis Publishers.

Craig, H. 1961. Isotopic variations in meteoric waters. *Science* 133: 1702–1703.

Dietzel, M. & Kirchhoff, T. 2002. Stable isotope ratios and the evolution of acidulous groundwater. *Aquatic Geochemistry* 8: 229–254.

Elango, L. & Kannan, R. 2007. Rock-water interaction and its control on chemical composition of groundwater, In D. Sarkar, R. Datta & R. Hannigann (eds), *Concepts and Applications in Environmental Geochemistry*: 229–246. Amsterdam: Elsevier Science.

Federico, C. Aiuppa, A. Allard, P. Bellomo, S. Jean-Baptiste, P. Parello, F. & Valenza, M. 2002. Magma-derived gas influx and water-rock interactions in the volcanic aquifer of Mt. Vesuvius, Italy. *Geochimica et Cosmochimica Acta* 66: 963–981.

Hoefs, J. 1997. *Stable Isotope Geochemistry*. Berlin Heidelberg: Springer-Verlag.

Parkhurst, D.L. & Appelo, C.A.J. 1999. User's guide to PHREEQC (version 2)—a computer program for speciation, batch-reaction, one-dimensional transport, and inverse geochemical calculations. *Water-Resources Investigation Report* 99–4259, USGS, Denver, CO.

Tardy, Y. 1971. Characterization of the principal weathering types by the geochemistry of waters from European and African crystalline massifs. *Chemical Geology* 7(4): 253–271.

Terwilliger, V.J. Eshetu, Z. Colman, A. Bekele, T. Gezahgne A. & Fogel, M.L. 2008. Reconstructing palaeoenvironment from $\delta^{13}C$ and $\delta^{15}N$ values of soil organic matter: a calibration from arid and wetter elevation transects in Ethiopia. *Geoderma* 147: 197–210.

Water-Rock Interaction – Birkle & Torres-Alvarado (eds)
© 2010 Taylor & Francis Group, London, ISBN 978-0-415-60426-0

CFC and SF_6 concentrations in shallow groundwater: Implications for groundwater age determination

L. Aquilina, V. De Montety & T. Labasque
Caren - Geosciences Rennes, Université Rennes 1 - CNRS, Rennes, France

J. Molénat & L. Ruiz
Caren - Sols Agrohydrosystèmes Spatialisation, Agrocampus ouest - INRA, Rennes, France

V. Ayraud-Vergnaud
LADES, Rennes, France

E. Fourré
LSCE, CEA - CNRS - Université St Quentin en Yvelines, Saclay, France

ABSTRACT: CFC12, CFC-111, CFC-113 and SF_6 measurements have been carried out during two hydrological cycles in a granitic catchment, western France. The well network allowed the variability of the CFC and SF_6 concentrations to be investigated in the large, variably saturated zone from 2–3 m down to 6–8 m. All gases show an important variation range which is clearly related to the water table variation. A strong disagreement is observed between SF_6 and CFC, the later showing systematically older ages, especially before 1990. Laboratory experiments support potential SF_6 production and a slight CFC sorption and/or degradation.

1 INTRODUCTION

During the last two decades, dissolved CFC gases have been measured in groundwater in order to determine the groundwater residence-time, or 'age' (Busenberg & Plummer 1992, Cook et al. 1995). Following the Montreal protocol, CFC use has been prohibited and the CFC atmospheric concentrations have stabilized and begun to decline. The atmospheric evolution leads to higher uncertainties in the age determination for recent waters and the CFC use has thus been more limited during the last years.

Another anthropological gas, sulphur hexafluoride SF_6, has also been used for residence time determination (Bauer et al. 2001, Busenberg & Plummer 2000, Goody et al. 2006, Koh et al. 2007, Zoellmann et al. 2001). There are however relatively few studies which have recently compared SF_6 and CFC. Two recent publications have indicated a strong disagreement between CFC and SF_6 ages (Gooddy et al. 2006, Koh et al. 2007), which was also observed previously in a multi-tracer study in fractured aquifers (Plummer et al. 2001). This is clearly observed in Fig. 5 of Koh et al. (2007) which compares the apparent CFC-12 and SF_6 ages. No CFC ages more recent than 15 yrs are obtained whilst the SF_6 apparent ages are all more recent than 15 yrs, most of them being younger than 10 yrs. Potential SF_6 lithogenic production has been suggested by Koh et al. (2007), following previous observations (Busenberg & Plummer 2000). CFC decline and SF_6 potential local production make the groundwater residence time determination complex.

The aim of this study is to characterize the CFC and SF_6 concentration acquisition in the upper part of a fractured aquifer, western France, which presents a thick, variably saturated zone.

2 MATERIALS AND METHODS

2.1 *Site monitoring*

CFC and SF_6 concentrations have been monitored on site at the Kerrien site (47°57'N-4°8'W, Fig. 1), located close to the sea, in the South Western part of French Brittany (north-west of France). The bedrock is made of fissured and fractured granite, overlayed by weathered material, with mean thickness of about 20 m. The catchment is equipped with 53 piezometers ranging from 4 to 15 m (Fig. 1) (Ruiz et al. 2002; Martin et al., 2006). 13 campaigns have been carried out during the whole monitoring period (12/2004–03/2007) for CFC measurements, 9 campaigns for SF_6 (12/2005–03/2007). The CFC monitoring covers two hydrological cycles. Close to 100 samples have been collected and analysed for CFC and SF_6 content as well as physico-chemical

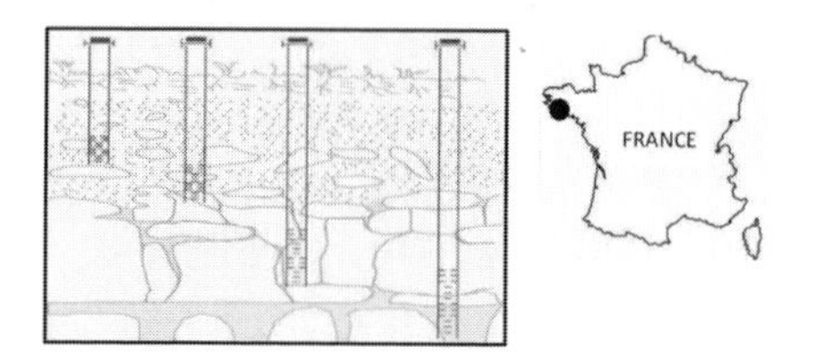

Figure 1. Site location and piezometer network.

parameters and chemical concentrations (Cl, SO_4, NO_3).

2.2 *Laboratory experiments*

In order to assess potential rock influence on CFC and SF_6 concentrations, batch experiments were carried out. Tap water was equilibrated with atmosphere and injected into 20 L plastic flasks. After 24 hrs, physico-chemical parameters were measured and the first sample taken (t_0). Then, about 8 kg of crushed granite alterite (diameter about 1 mm) were added in the flask. During the experiment, the water and dissolved gases were sampled, and physico-chemical parameters measured. Samples were also collected from the air above the water and rock, in order to make sure that no degassing occurred.

2.3 *Sampling and analytical methods*

CFC and SF_6 samples were taken in glass ampoules closed with two PTFE three ways valves. Water was pumped in boreholes with a MP1 Grunfos pump, connected to glass ampoules by nylon tubing. Sample was collected after stabilization of conductivity and pH, dissolved oxygen and temperature measured in the field (WTW 350i). 20 ml and 500 ml of water were stored in glass ampoules for CFC and SF_6 determination, respectively. Samples for determination of noble gas concentration were taken in 500 ml glass flasks, closed by a rubber cap and a metal ring. The flasks were placed in the bottom of 10 L water container. After a few minutes of water circulation, the flasks were filled and closed without atmospheric contact.

CFC concentrations are determined by Purge and trap (PT) extraction and analyzed with a gas chroma-tograph equipped with an electron capture detector (GC/ECD). Uncertainties for CFC determinations on equilibrated water with actual atmosphere are around 1% For SF_6 determinations, larger volumes implied modifications of the analytical system. The trap is 1/8 inch in diameter and 20 cm long, filled with a strong absorbant. Chromatographic column is a 30 m PLOT molecular Sieve 5 A.

Trapping is realized at −100°C. Uncertainties are around 20% for concentrations near 0.05 fmol/L and 5% for 2 fmol/L. Noble gas (Ne, Ar) were determined by headspace extraction and chromatographic analysis using a µGC chromatograph (GC 3000, SRA instruments). Detection is classically realised with a catharometer. Uncertainties (CV) are around 3% for Neon measurements (water equilibrated with atmosphere at 12°C) and less than 1% for argon. Ne and Ar concentrations are used to estimate recharge temperature and excess air since excess air correction can be very important for groundwater age determination with SF_6.

3 RESULTS

The histogram of the CFC-12 concentration (Fig. 2) shows that age distribution is not homogeneous at the scale of the aquifer. Many CFC concentrations correspond to the years before1995/1990.

The distribution presents a double peak, with a concentration range between 490–520, which corresponds to the highest sample number (23 for a total of 76 measurements). 84% of the concentrations

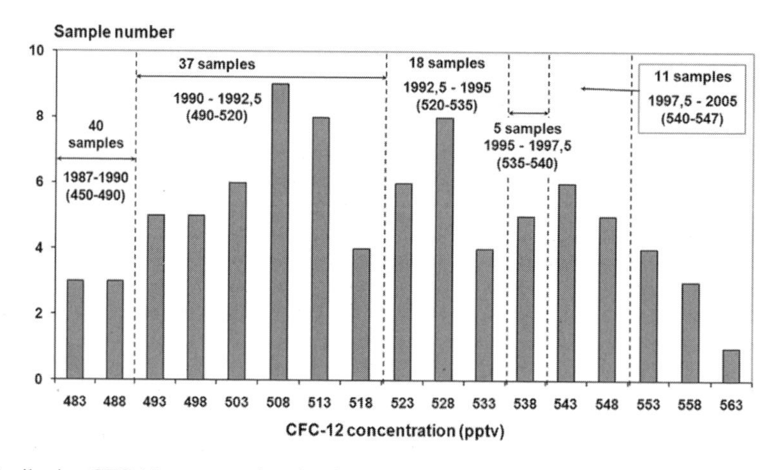

Figure 2. Distribution CFC-12 concentrations in piezometers.

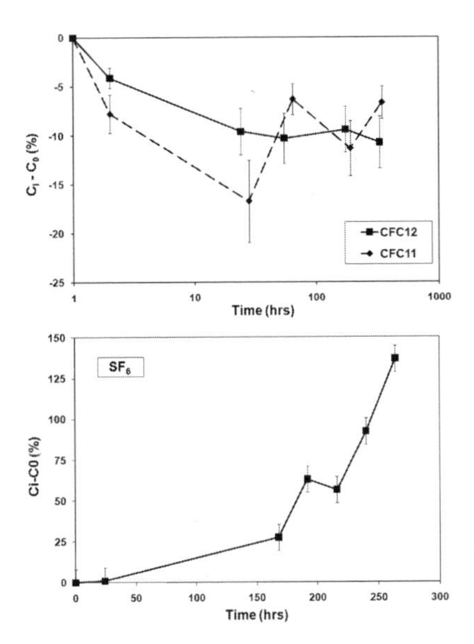

Figure 3. SF_6 and CFC-12 concentrations during experimental leaching.

encountered in the Kerrien aquifer correspond to the years 1976–1995. Only 4% correspond to the years 1995–1997,5 whilst 12% of the total concentrations measured represent the years 1997,5–2005 years. Conversely, most of the SF_6 concentrations correspond to the last 8 years. The whole Kerrien data set shows a great discrepancy between CFC and SF_6.

The results of the experimental leaching of the weathered granite is presented in Fig. 3. A clear increase of SF_6 concentrations, which become higher than the atmospheric equilibrium, is noticed during the whole experiment. After 250 hrs, the SF_6 enrichment reaches 100% of the initial concentration. The CFC concentrations, on the contrary show a decrease during the same experiments. The decrease is continuous and reaches about 10% after 250 hrs.

4 DISCUSSION

4.1 *CFC measurements*

Precise monitoring and tracer experiments in the Kerrien catchment led Legout et al. (2007) to describe the unsaturated zone in Kerrien as a double compartment system. At this site, water transfer in the soil and vadose zone is thought to occur as i) rapid (in the order of a day) transfer through macropores, ii) piston-like transfer with velocities in the order of m-yr, iii) as slower exchanges between the water and the non-mobile compartment. Microbial degradation, adsorption, and exchanges with the non-mobile water are the main processes which can lead to CFC concentration shifts. Such processes can act differently within the different water compartments and the different water-transfer ways.

The Kerrien measurements do not indicate important microbial degradation effects. They indicate that equilibrium with atmosphere is reached during the recharge process. On the contrary, systematic CFC concentrations are below equilibrium, whatever the depth, during the discharge period. Even at the top of the water table, a few months after recharge and equilibrium, CFC concentrations correspond to more than 10 yrs before present, which is unlikely based on tracer tests (Legout et al. 2007). The CFC concentration decrease does not seem to reflect residence-time but rather potential adsorption along the minerals and/or retention in the non-mobile water compartment. The experimental leaching of weathered granite agrees with the potential adsorption mechanisms and/or the exchange with the non-mobile compartment since a clear but slow CFC concentration decrease is observed.

4.2 *SF_6 measurements*

SF_6 concentrations reflect a seasonal variation which is dependent on the water table. However, contrary to CFC, the SF_6 concentrations remain relatively high and close to the present equilibrium. As regards the residence-times deduced from SF_6, they are much lower than the corresponding CFC ages. Most of the residence times range from 1997 to 2005, i.e. they are at least 10 yrs younger than the corresponding CFC age. During several campaigns, concentrations are clearly above the atmospheric equilibrium. SF_6 lithogenic production can explain high SF_6 concentrations.

Potential production of SF_6 from the rock has been evidenced (Busenberg & Plummer 2000, Koh et al. 2007). Such production does not seem to be restricted to granitic environments. No clear explanation has been given to the mechanisms responsible for the production. F is present in the rock incorporated in secondary minerals (such as fluorite) found in fault coatings. SF_6 production might be related to the energy produced by U, Th and K disintegration leading to fluorite breaking and bonding of F and S in the solution. The leaching experiments we have carried out give a strong confirmation of lithogenic SF_6 production. These experiments show that the SF_6 production is not related to a specific mineralogy. We favour the fact that SF_6 production occur in any rock containing F as a trace element in minerals and an energy source. Thus, SF_6 production might be a relatively common process.

4.3 *CFC and SF_6 piston ages*

The CFC and SF_6 monitoring and the leaching experiments indicate that two mechanisms may influence the gas concentrations, and modify the

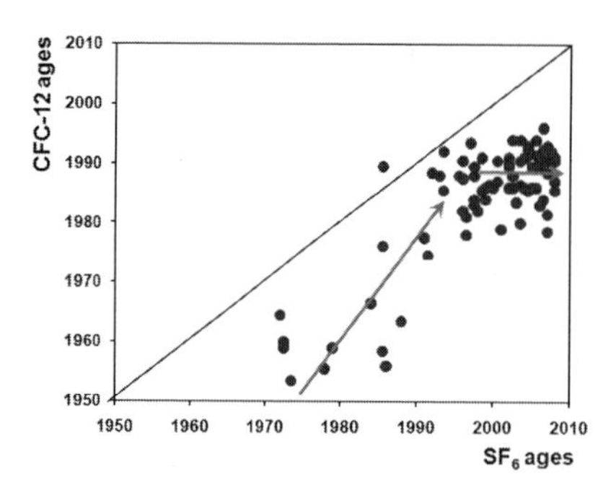

Figure 4. SF_6 and CFC-12 concentrations during experimental leaching.

residence-time determination: *i)* SF_6 lithogenic production may increase the SF_6 concentrations. Such mechanism does not seem to be related to a hydrogeological zone (soil, unsaturated zone etc.) nor to a water transfer context, but could occur all along the water transfer in the aquifer. *ii)* On the contrary, CFC concentrations are reduced through adsorption/concentration processes along the minerals and in the non-mobile compartment These mechanisms are limited to the vadose zone transfer, especially during the discharge period.

Such mechanism explains the double peak of the concentration distribution observed in the Kerrien catchments: during the recharge process, CFC concentrations are equilibrated with atmosphere in the unsaturated zone and rapidly transferred to the water table, on the contrary during the discharge period, CFC concentrations have been slightly modified during the transfer in the unsaturated zone and present concentrations below the atmospheric equilibrium.

The relative CFC and SF_6 ages do not have to be correlated and depend on the potential transfer mechanisms such as piston flow, exponential etc. However, since the samples correspond to very shallow depths, we compare the SF_6 to CFC piston ages in Fig. 4 in order to test the potential influences of the mechanisms deduced from the experimental leaching. CFC measurements show some scatter with relatively constant ages (1985–1995) for SF_6 ages ranging from 1990 to present. For older piston ages, SF_6 to CFC age correlation shows a difference which increases with the residence time. Such tendency is also evidenced in fractured rocks from the Blue Ridge Mountains (Plummer et al. 2001). It can be interpreted in the light of our experimental results which indicate rock production. Such production should induce an increase of SF_6 with time.

5 CONCLUSIONS

Investigation of CFC and SF_6 concentrations in the vadose zone of a granitic catchment indicates a strong disagreement between the two gases types. CFC always present residence times older than 1995, even at shallow depths, after recharge processes whilst SF_6 'piston ages' are much more recent and often over the equilibrium concentrations.

Analyses of these measurements and leaching experiments indicates that CFC may be affected by slight sorption and degradation processes during the recharge processes in the vadose zone whilst SF_6 lithogenic production increases the SF_6 concentrations.

REFERENCES

Bauer, S., Fulda, C., & Schäfer, W. 2001. A multi-tracer study in a shallow aquifer using age dating tracers ^3H, ^{85}Kr, CFC-113 and SF_6 - indication for retarded transport of CFC-113. *Journal of Hydrology* 248: 14–34.

Busenberg, E. & Plummer, L.N. 2000. Dating young groundwater with sulfur hexafluoride: Natural and anthropogenic sources of sulfur hexafluoride. *Water Resources Research* 36(10): 3011–3030.

Busenberg, E. & Plummer, N. 1992. Use of chlorofluorocarbons (CCl_3F and CCl_2F_2) as hydrologic tracers and age-dating tools: alluvium and terrace systems of central Oklahoma.: Water Resources Research 28: 2257–2283.

Cook, P.G., Solomon, D.K., Plummer, L.N., Busenberg, E. & Schiff, S.L. 1995. Chlorofluorocarbons as tracers of groundwater transport processes in a shallow, silty sand aquifer. *Water Resources Research* 31(3): 425–434.

Gooddy, D.C., Darling, W.G., Abessser, C. & Lapworth, D.J. 2006. Using chlorofluorocarbons (CFCs) and sulphur hexafluoride (SF_6) to characterize groundwater movement and residence time in a lowland chalk catchment. *Journal of Hydrology* 330: 44–52.

Koh, D.C., Plummer, L.N., Busenberg, E. & Kim, Y. 2007. Evidence for terrigenic SF_6 in groundwater from basaltic aquifers, Juju Island Korea: implications for groundwater dating. *J. of Hydrology* 339: 93–104.

Legout, C., Molenat, J., Aquilina, L., Gascuel-Odoux, C., Faucheux, M., Fauvel, Y. & Bariac, T. 2007. Solute transfer in the unsaturated zone-groundwater continuum of a headwater catchment. *Journal of Hydrology* 332 3–4: 427–441.

Martin, C., Molénat, J., Gascuel-Odoux, C., Vouillamoz, J.M., Robain H., Ruiz, L. & Aquilina, L. 2006. Modelling the physical and chemical characteristics of shallow aquifers on water and nitrate transfer in small agricultural catchments. *Journal of Hydrology* 326: 25–42.

Plummer, L.N., Busenberg, E., Böhlke, J.K., Nelms, D.L., Michel, R.L. & Schlosser, P. 2001. Groundwater residence times in Shenandoah National Park, Blue Ridge Mountains, Virginia, USA: a multi tracer approach. *Chemical Geology* 179: 93–111.

Ruiz, L., Abiven, S., Durand, P., Martin, C., Vertès, F. & Beaujouan, V. 2002. Effect on nitrate concentration in stream water of agricultural practices in small catchments in Brittany I-Annual nitrogen budgets. *Hydrology and Earth System Sciences* 6(3): 497–505.

Zoellmann, K., Kinzelbach, W. & Fulda, C. 2001. Environmental tracer transport (^3H and SF_6) in the saturated and unsaturated zones and its use in nitrate pollution management. *Journal of Hydrology* 240: 187–205.

Water-Rock Interaction – Birkle & Torres-Alvarado (eds)
© 2010 Taylor & Francis Group, London, ISBN 978-0-415-60426-0

An attempt to date saline groundwater in the multi aquifers system of the Dead Sea area, using carbon isotopes

N. Avrahamov & O. Sivan
The Ben Gurion University, Beer-Sheva, Israel

Y. Yechieli
Geological Survey of Israel, Jerusalem, Israel

B. Lazar
The Hebrew University of Jerusalem, Jerusalem, Israel

ABSTRACT: Dating of brines is a great challenge due to their complex chemical characteristic. This work presents an attempt to date hypersaline groundwater in the Dead Sea (DS) area, using carbon species and isotopes of groundwater. While most ions exhibit conservative behavior, the carbon system parameters indicate additional processes. The depletion in the $\delta^{13}C_{DIC}$ in saline groundwater, compared to the DS, is probably due to anaerobic oxidation of organic matter by sulfate reduction and methane oxidation. Methane was indeed found, in low concentrations, in many groundwater samples. The very low $^{14}C_{DIC}$ value of some of the saline groundwater samples (~14 pMC, >10000 years), much lower than both fresh and saline end–members of the system, suggests the presence of ancient brine source in the lower sub–aquifer. Other samples of saline groundwater, from the upper sub-aquifer, have $^{14}C_{DIC}$ values closer to that of the DS (~80 pMC, <150 years).

1 INTRODUCTION

The main dating tool in many groundwater systems is radiocarbon activity in the dissolved inorganic carbon (DIC). However, there is a suite of geochemical processes that affect the DIC and its isotopic composition, and therefore should be considered in $^{14}C_{DIC}$ age determination of groundwater (Mook 1980). Quantification of the DIC sources and processes that affect their contents is challenging. In carbonate aquifers the most prevalent process is dissolution of ^{14}C-free ancient marine $CaCO_3$. Methane (CH_4) production (methanogenesis) in groundwater can also contribute carbon to the DIC pool (Barker et al. 1979).

Few studies have dealt with the carbon system with respect to dating of brines. Due to uncertainties in geochemical histories, the radiocarbon groundwater dating may give questionable results. Dating of groundwater brine from Gorleben channel (Germany) was found to be problematic leading to the conclusion that a reliable ^{14}C groundwater dating is presently not possible in this region (Buckau et al. 2000).

The purpose of this study is to get a better understanding of the processes that affect ^{14}C dating and provide an estimate of saline groundwater in the DS area.

1.1 *Carbon isotopes in water in the Dead Sea area*

Measurements of $\delta^{13}C_{DIC}$ in the Dead Sea (DS) water column show that the $\delta^{13}C_{DIC}$ of the DS is mainly controlled by abiotic processes, and is strongly influenced by CO_2 escape due to the increasing P_{CO_2} of the water (Barkan et al. 2001). The $^{14}C_{DIC}$ content of the DS and the factors controlling it were investigated by Stiller et al. (1988) and Talma et al. (1997). Radiocarbon in the DS is affected by CO_2 exchange with the atmosphere, and contributions of rain, groundwater and floods from the Jordan River and Judea Mountains (Fig. 1). These sources are also affected by water–rock interaction. Preliminary $^{14}C_{DIC}$ analysis of some fresh and saline groundwater implied relatively old ages although actual dating was problematic (Yechieli et al. 1996).

1.2 *Hydrogeological setting*

The two main aquifers in the western part of the DS rift are the limestone and the dolomite layers of the Judea Group of Upper Cretaceous age and the alluvial aquifer of Quaternary age (Fig. 2). The alluvial aquifer is generally separated from the other aquifer by faults of the western margin

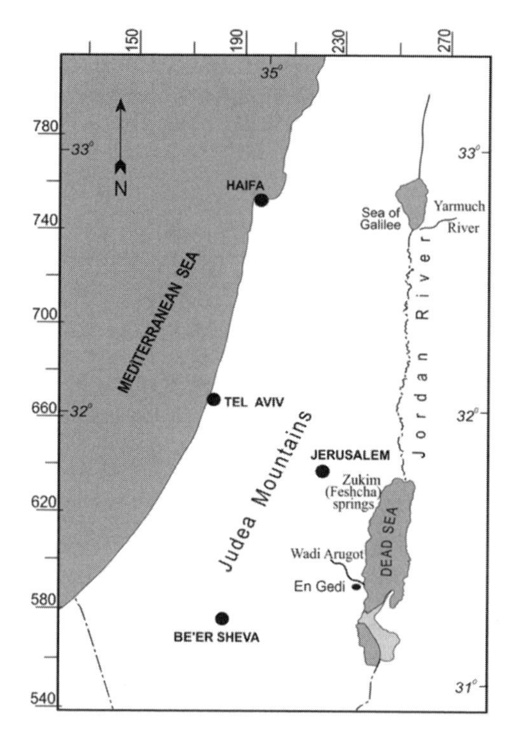

Figure 1. Location map of the Dead Sea area.

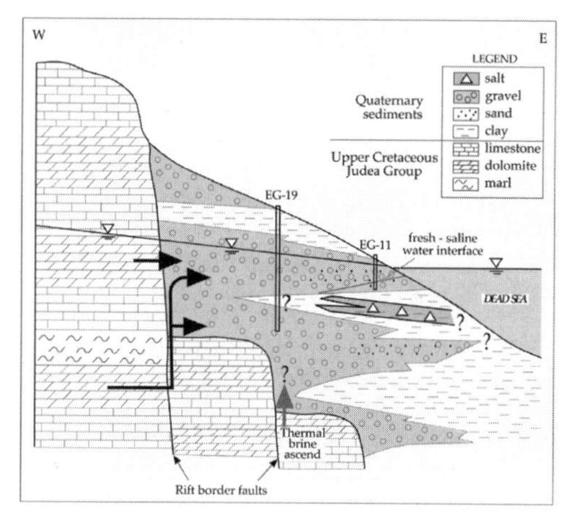

Figure 2. Generalized cross section of the Dead Sea coastal aquifer.

of the DS rift. The fresh water source to the alluvial aquifer is mostly from the Cretaceous aquifer through the previously mentioned fault zone. Due to the low precipitation and high evaporation rates in the DS area, only small amount of flood water penetrates directly to the alluvial aquifer. The alluvial aquifer is built of alternating clay and gravel layers. These alternations create hydrological condition of several sub–aquifers differing in their water levels, temperature, salinity and water chemistry. It was postulated that the DS brines are the result of sea water infiltration and evaporation from the Neogene Sea that have been captured along the Jordan Valley (Starinsky 1974). Many of the brine emerge as thermal springs, mostly in the Kedam-Shalem system (Gavrieli et al. 2001).

Water levels in the DS have decreased by 20 meters in the last 30 years at an average rate of 80 cm per year, and in the last years at 1 m per year. As a result, groundwater levels have also been decreasing. There is a recirculation of DS water flowing into the aquifer although the general trend is of groundwater level decrease (Kiro et al. 2008).

2 METHODS

Water samples were taken from boreholes along a transect in the Arugot alluvial fan (Fig. 1). Samples were taken with a bailer and a submersible pump.

Samples for DIC and carbon isotopes were immediately filtered through 0.45 μm filters and transferred into 20 mL pre poisoned syringes ($HgCl_2$ powder) to avoid bacterial activity. For CH_4 and DIC analyses, 5 mL was transferred from the syringe to a vacutainer for head space measurements. DIC and CH_4 concentrations were measured by gas chromatograph (GC).

For $\delta^{13}C_{DIC}$ and $^{14}C_{DIC}$ analyses, the inorganic carbon was stripped by phosphoric acid under a high vacuum and purged through a vacuum line at Weizmann Institute. The CO_2 used for ^{14}C analysis was reduced to graphite, and ^{14}C activity was determined using Accelerator Mass Spectrometer (AMS) at the University of Arizona Laboratory. The ^{14}C results are reported as pMC relative to an NBS oxalic acid standard, and the maximum error of the measurements is ±0.5 pMC.

3 RESULTS AND DISCUSSION

The dynamic hydrogeological DS alluvial system is expected to host several water bodies of different salinity and ages. The decrease in DS levels causes a decrease in groundwater levels, flushing of the original brines and movement of the fresh–saline water interface toward the sea.

Major ion composition of the groundwater samples show a mixing relation between the fresh spring water and the saline DS water (Fig. 3). However, carbon system parameters show non conservative behavior (Fig. 3) indicating that other processes besides mixing occurred. While the very

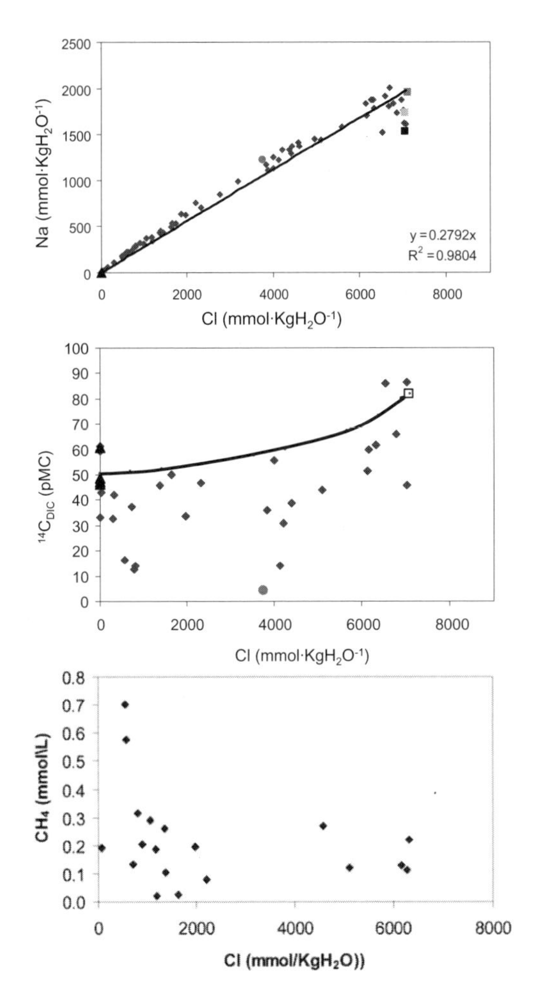

Figure 3. Concentrations of Na, methane and radiocarbon versus Cl concentration. The lines in the upper 2 graphs indicate mixing between the fresh and DS end-members. The Qedem-Shalem brines affect many data points in the Na-Cl graph plot.

high concentrations of the ions in the hypersaline water samples make the system harder to detect other processes than mixing, these processes are distinguishable with the carbon parameters.

Some of the saline groundwater has dissolved inorganic carbon (DIC) concentrations close to the DS but much lower values of $\delta^{13}C_{DIC}$. The depletion in the $\delta^{13}C$ in these groundwater samples is probably due to anaerobic oxidation of organic matter by sulfate reduction and methane oxidation. The gap between the $^{14}C_{DIC}$ values in DS and the groundwater samples with DS characteristic (salinity and most ion composition) can also be explained by the low $^{14}C_{DIC}$ contribution from oxidation of old organic matter in the sediment or methane oxidation.

The $^{14}C_{DIC}$ values in groundwater sample near the DS shore are higher than those of the recent DS (86 pMC in EG16/11 m compared to 82 pMC in the DS in 2003). In 1977 the DS had $^{14}C_{DIC}$ value of 107pMC (Talma et al. 1997), when the ^{14}C values in the atmosphere were much higher. During the 1980s the values decreased until in 1990 the DS $^{14}C_{DIC}$ value was 83 pMC (Talma et al. 1997). Therefore, we can assume that the groundwater in the vicinity of the DS percolated to the ground in the 1980s.

The samples from the lower sub–aquifer have salinity range from 800 mmol $Cl \cdot KgH_2O^{-1}$ at 50 m depth to 4700 mmol $Cl \cdot KgH_2O^{-1}$ at 59 m (borehole EG19), both with the same $^{14}C_{DIC}$ values (~14 pMC). This means that two different water bodies exist in the lower sub–aquifer, one above the other, which were trapped there thousands years ago (>10,000 years). The low ^3H concentrations in these samples (below 1 TU) support this concept. It can be speculated that these two different water bodies represent an ancient fresh-saline water interface zone of one of the precursor lakes which had higher water level than the modern DS.

Table 1. Chemical and isotopic values of groundwater from selected boreholes.

Borehole (depth, m)	Na (mmol · Kg H_2O^{-1})	CH_4 (mmol · Kg H_2O^{-1})	^{14}C pMC	$\delta^{13}C$ ‰	Cl (mmol · Kg H_2O^{-1})
EG19 (40)	1350				4200
EG19 (50)	280	0.29	14.2	–4.4	800
EG19 (59)	1520	0.13	13.9	–11.2	5200
EG11 (13)	441		45.7	–8.5	1380
EG11 (19)	1230	0.12	35.8	–12.6	4030
EG16 (11)	1620		86.1	–10.2	7000
Dead Sea	1660		82	2.8	7100

REFERENCES

Barkan, E., Lazar, B. & Luz, B. 2001. Dynamics of the carbon dioxide system in the Dead Sea. *Geochimica et Cosmochimica Acta* 65(3): 355–519.

Barker, J.F., Fritz, P. & Brown, R.M. 1979. C–14 measurements in aquifers with CH_4. In *Isotope Hydrology 1978. Proceeding Symposium. 19–23 June 1978, Neuherberg, Vol. II.* IAEA, Vienna: 661–678.

Buckau, G., Artinger, R., Geyer, S., Wolf, M., Fritz, P. & Kim, J.I. 2000. ^{14}C Dating of Gorleben groundwater. *Applied Geochemistry* 15: 583–597.

Gavrieli, I., Yechiel, Y., Halicz, L., Spiro, B., Bein, A. & Efron, D. 2001. The sulfur system in anoxic subsurface brines and its implication in brine evolutionary pathways: the Ca–chloride brines in the Dead Sea area. *Earth and Planetary Science Letters* 186: 199–213.

Kiro, Y., Yechieli, Y., Lyakhovsky, V., Shalev, E. & Starinsky, A. 2008. Time response of the water table and saltwater transition zone to a base level drop. *Water Resources Research* 44, W12442, doi:10.1029/2007 WR006752.

Mook, W.G. 1980. Carbon–14 in hydrogeological studies. In P. Fritz & J.C. Fontes (eds), *Handbook of Environmental Isotope Geochemistry*. Vol. 1, Chap. 9. Amsterdam: Elsevier.

Stiller, M., Carmi, I. & Kaufman, A. 1988. Organic and inorganic ^{14}C concentrations in the sediments of lake Kinneret and Dead Sea (Israel) and the factors which control them. *Chemical Geology* 73: 63–78.

Starinsky, A. 1974. *Relationship between Ca–chloride brines and sedimentary rocks in Israel* [Ph.D. Thesis]. Jerusalem: Hebrew University of Jerusalem (in Hebrew).

Talma, A.S., Vogel, J.C. & Stiller, M. 1997. The radiocarbon content of the Dead Sea. In T.M. Niemi, Z. Ben–Avraham & J.R. Gat (eds), *The Dead Sea: the lake and Its Setting*. Oxford: Oxford University Press.

Whiticar, M.J. 1999. Carbon and hydrogen isotope systematics of bacterial formation and oxidation of methane. *Chemical Geology* 161:291–314.

Yechieli, Y., Ronen, D. & Kaufman, A. 1996. The source and age of groundwater brines in the Dead Sea area, as deduced from ^{36}Cl and ^{14}C. *Geochimica et Cosmochimica Acta* 60: 1909–1916.

Water-Rock Interaction – Birkle & Torres-Alvarado (eds)
© 2010 Taylor & Francis Group, London, ISBN 978-0-415-60426-0

Isotopic exchange between thermal fluids and carbonates at offshore Campeche oil fields, Gulf of Mexico

P. Birkle
Instituto de Investigaciones Eléctricas, Gerencia de Geotermia, Cuernavaca, Morelos, Mexico

M.A. Lozada Aguilar, E. Soriano Mercado & J.J. Torres Villaseñor
PEMEX-PEP, Activo Integral Cantarell, Cd. del Carmen, Campeche, Mexico

ABSTRACT: Formation water and core samples (dolomite, breccia, halite) from the Cantarell and Ku-Maloob-Zaap oil reservoirs, Gulf of Mexico, have been analyzed for $^{11}B/^{10}B$, $\delta^{13}C$, and $^{87}Sr/^{86}Sr$. Differences in $^{87}Sr/^{86}Sr$-ratios between formation water (0.70796–0.70903) and adjacent salt domes (0.70707) exclude halite dissolution as major process to explain the chemical heterogeneity (TDS = 12–290 g/L) and partial hypersaline composition of formation water. Low Cl/Br ratios of fluids suggest sub-aerial evaporation of seawater as a primary mineralization process. Dolomitization and clay desorption represent the dominant processes altering the primary geochemistry and mineralogy of the carbonate host rock. Similar $^{11}B/^{10}B$ and $^{87}Sr/^{86}Sr$ ratios of dolomite and breccia host rock and reservoir fluids, as well as a short residence time (late Pleistocene) for formation water within the reservoir, suggest a relatively fast and ongoing chemical-isotopic exchange between reservoir fluids and host rock, also reflected by similar $^{87}Sr/^{86}Sr$-ratios for fracture and matrix dolomite.

1 INTRODUCTION

The Cantarell complex, located 85 km offshore in the Bay of Campeche, Gulf of Mexico, is one of the largest oilfields in the world. Upper Cretaceous carbonate breccia and highly dolomitized Jurassic limestone form the main host rock type of the highly fractured reservoir; upper Paleocene and middle Eocene calcarenite and sandstone formations are less important. Discovered in 1976, Cantarell´s production peaked with 2.1 million barrels per day (330,000 m^3/d) in 2003. Production declined rapidly after that, and by 2009 had fallen to 772,000 barrels per day (123,000 m^3/d). Due to the Cantarell decline, production at the adjacent Ku, Maloob and Zaap fields, which produce from Kimmeridgian, Lower Paleocene-Upper Cretaceous and Middle Eocene carbonates, was boosted in 2007. With an oil production of 802,002 barrels/day in November 2009, Ku-Maloob-Zaap recently became Mexico's most productive oil field.

In the last few years, the oil-water contact in the Cantarell field has been advancing, and this has considerably reduced the oil window zone. One likely explanation is that the natural fracture network, which provides most of the permeability in the field, favors production of water and gas over oil (Cruz & Sheridan 2009).

In order to reconstruct the enhanced trend of water invasion into production wells, hydrochemical-isotopic studies were performed to provide actualized concepts for the hydrogeological reservoir model. In the present paper, a variety of isotopic methods ($^{11}B/^{10}B$, $\delta^{13}C$, $^{87}Sr/^{86}Sr$) are applied to reconstruct the origin of chemical-isotopic diversity of brine composition at the Cantarell and Ku-Maloob-Zaap oil reservoirs. Samples from formation water are compared with carbonate reservoir rocks and adjacent halite domes to define the type and degree of reservoir alteration by formation fluids.

2 METHODS

A total of 92 water samples from Upper Jurassic to Eocene reservoir units from 43 oil production wells were taken from the Cantarell and Ku-Maloob-Zaap oilfields during three sample periods from March to November 2008. Partially low water contents of the samples emulsion (< 5%) and low API (American Petroleum Institute gravity) values from 14° to 22° of the hydrocarbons required the partial use of an organic de-emulsifier and centrifuge technique to separate the water phase from the oil phase. Thermal Ionization Mass Spectrometry (TIMS) was applied at the Saskatchewan Isotope Laboratory (University of Saskatchewan, Canada) and at Geochemical Technologies Corp. (Waco, Texas, U.S.) to measure $^{87}Sr/^{86}Sr$ and $^{11}B/^{10}B$ ratios, respectively. Sr isotope ratios were adjusted relative to a fixed value of 0.710250 for the NIST SRM 987 standard, and with a 1σ analytical uncertainty for the samples of 0.005%. In the case of ^{11}B, a precision 1σ of less than +/– 0.5 ‰ for replicates is given by the standard deviation from analysis of the NBS (NIST) Standard Reference Material SRM 951. δ^{13}C-ratios were determined at the NSF-Arizona AMS Facility (University of Arizona, Tucson). Similarly, ten core samples from Jurassic and Cretaceous dolomitized grainstone, dolomite and breccia, as well as one sample from an adjacent Jurassic salt dome were analyzed for Sr, B and C isotopic composition. Prior to the analytical procedure, core material was crushed, and matrix and fracture minerals were manually separated.

3 RESULTS

3.1 *Primary origin of brine salinity*

The origin of salinity in fluids from sedimentary basins has historically been attributed to sub-aerial evaporation of seawater (Carpenter 1978), shale membrane filtration (Graf 1982), and the dissolution of evaporates (Land & Prezbindowksi 1981). Sub-aerial evaporation of seawater in particular has been attributed to explain the partial hypersaline composition of brines in several US (Kharaka & Hanor 2004) and Mexican (Birkle et al. 2009, Birkle & Angulo 2005) oil reservoirs of the Gulf of Mexico.

For the present Cantarell reservoir, the comparison of the isotopic composition of a halite core sample (^{87}Sr/^{86}Sr = 0.70707; δ^{11}B = 27.3‰, δ^{13}C = 5.4‰), representative for the regional salt bodies and extracted from the well Ek-63 at a depth of 4,320 m.b.s.l., with reservoir formation water (^{87}Sr/^{86}Sr = 0.70796–0.70903; δ^{11}B = 18.6‰ to 43.4‰, δ^{13}C = –12.8‰ to 9.0‰), showed significant isotopic differences between these phases. This suggests that halite dissolution does not explain water mineralization (Fig. 1).

Additionally, low Cl/Br (135–560) and Na/Br (116–477) ratios confirm that evaporation of seawater beyond halite precipitation is the principal process to explain the hypersaline brine composition, as the brines reach TDS concentrations of up to 290,000 mg/L (well C-317). Increasing fluid salinity with reservoir depth, as shown for the Sr-trend of groundwater from shallow Paleocene horizons towards deeper Jurassic layers in Figure 1, confirms the gravity-related infiltration of the most saline water towards the deepest horizons of the reservoir.

3.2 *Dolomitization of the reservoir*

In general, marine carbonates preserve the ^{87}Sr/^{86}Sr-signature of seawater present during their sedimentation. In the present case, alteration of the primary isotopic composition of formation water and reservoir host rock (from the initial seawater composition) was caused by post-depositional

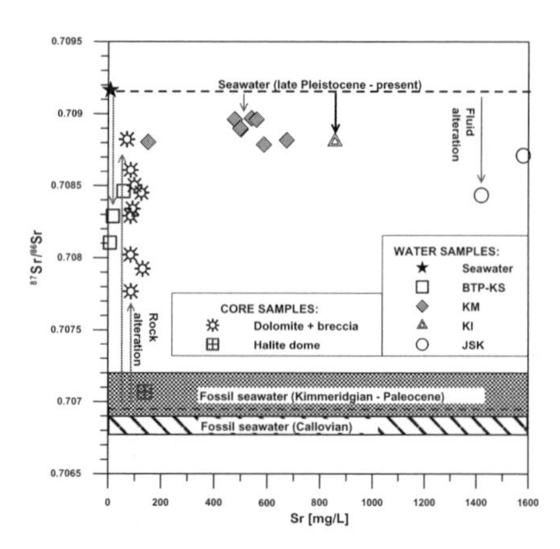

Figure 1. Converging trends of ^{87}Sr/^{86}Sr-ratios between formation water and reservoir host rock (Ku-Maloob-Zaap oilfield) by secondary dolomitization, initiating from a primary Kimmeridgian to Paleocene seawater composition of the reservoir carbonates, and a late Pleistocene isotopic seawater composition for formation fluids. Legend: BTP-KS = late Cenozoic (Paleocene) – late Cretaceous (Turonian); KM = late to early Cretaceous (Albian – Cenomanian), KI = early Cretaceous (Berriasian – Aptian), JSK = late Jurassic (Kimmeridgian). Also shown is Callovian seawater composition in comparison with local Jurassic halite dome.

exchange processes. Formation water with a maximum residence time of 35,000 years within the reservoir (determined by the ^{14}C method) has slightly lower ^{87}Sr/^{86}Sr ratios in comparison to Pleistocene-recent seawater (descending arrows in Fig. 1), whereas secondary dolomite and altered breccia host rock have greater ^{87}Sr/^{86}Sr in comparison to the original Kimmeridgian-Paleocene seawater composition (McArthur et al. 2001; ^{87}Sr/^{86}Sr = 0.7069–0.7072) (ascending arrows in Fig. 1). Considering the relatively short residence time of formation water within the reservoir, the observed isotopic converging trends between both phases imply a relatively fast (since Late Pleistocene) and ongoing chemical exchange process between reservoir fluids and host rock. Moreover, the ion balance between calcium enrichment and magnesium depletion in the present water samples supports the hypothesis of dolomitization as the principal secondary alteration process.

3.3 *Stages for fracture filling*

In order to reconstruct the sequential stages for dolomitization, secondary dolomite from host rock matrix and fracture fillings, extracted from one single drilling core (Well Zaap-8, Core No. 3, Early Cretaceous; Depth 3521–3526 m b.s.l.), was analyzed for ^{11}B/^{10}B, ^{13}C/^{12}C, and ^{87}Sr/^{86}Sr ratios (Fig. 2). The hypothesis of a complete isotopic homogenization of the mineral phases by co-genetic, secondary

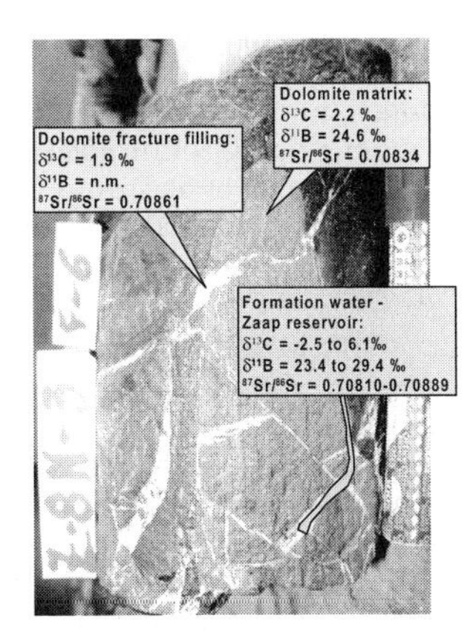

Figure 2. $\delta^{11}B$, $\delta^{13}C$ and $^{87}Sr/^{86}Sr$ isotopic composition of dolomite matrix and fracture filling from the well Zaap-8, and isotopic value range for formation water from the Zaap reservoir.

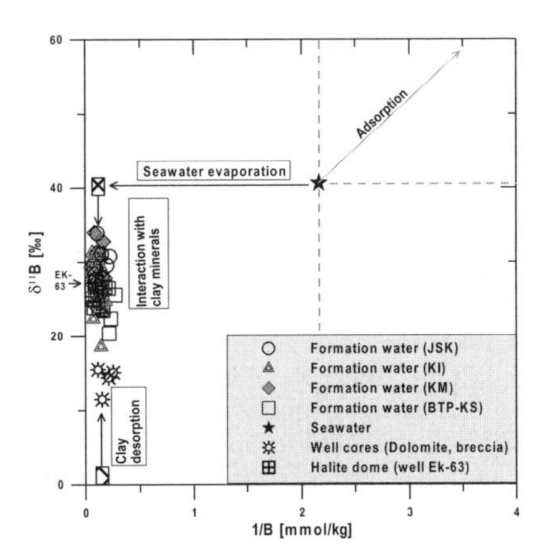

Figure 3. Boron and $\delta^{11}B$ composition of formation water from the Cantarell oilfield with present seawater and reservoir core samples (dolomite, breccia and halite). Legend details in Fig. 1.

dolomitization of the host rock—matrix as well as fracture fillings—by contact with infiltrating surface water during Late Pleistocene seems feasible. The vertical chemical-isotopic stratification from low-saline groundwater towards hypersaline aquifer units explains the wide isotopic range for these groundwater samples from the Zaap reservoir (value range is given in Fig. 2).

3.4 *Clay desorption*

Boron concentrations of Cantarell formation water are elevated (46.1–192 mg/L) in comparison to standard meteoric water (< 1.0 mg/L) and seawater (4.5 mg/L) (Fig. 3). In contrast, $\delta^{11}B$ values between +18.6‰ (well 432) and +34.0‰ (well 2072) are less than that of dissolved boron from present-day seawater from the Gulf of Mexico with a $\delta^{11}B$-value of +40.6‰. The most probable mechanisms that can affect both the B concentration and B isotope signature in formation fluids could be a) desorption of exchangeable B in clays, resulting in elevated B with low ^{11}B (in fluids), and/or b) a potential involvement of deep generated fluids. The second option can be excluded, as elevated ^{14}C-concentrations of the present reservoir fluids (1.6–30.0 pmC) and more than 10 km of accumulated sedimentary column exclude major contributions of deep, ^{14}C-free magmatic fluids. Because B partitions preferentially into vapor and $\delta^{11}B_{\text{vapor-liquid}}$ values are small for geothermal systems, ranging from ~3‰ to 1‰ at ~140 to 300°C (Leeman et al. 1992), effects by liquid-vapor separation would be negligible for Cantarell fluids that have average temperatures of 110 to 120°C.

$\delta^{11}B$-values between +11.5‰ and +24.6‰ for dolomite and breccias from the Jurassic-Cretaceous reservoir probably do not represent the total rock value for carbonate rock (with a global value range for modern marine carbonates of +22.1 ± 3‰, Hemming & Hanson 1992), but rather adsorbed-B on clay surfaces. Spivack et al. (1987) showed that adsorbed-B in marine sediments has an average $\delta^{11}B$ of +14‰, but it accounts for only 10 to 20% of the total-B in the sediments. The primary boron isotopic composition is not known for the Mexican reservoir sediments, but Williams et al. (2001) reported $\delta^{11}B$-values between -2‰ and +2‰ for pore filling clays from organic mudstone (with interbedded sandstones) from the Wilcox Formation in south-central Louisiana. Total clay concentrations of the analyzed carbonate cores (dolomite, breccias) reach up to 16.9% (M-456, N-2), whereas dolomitization degree ranges between 54.5% (M-456: N-2) and 99.4% (Z-36: N-2). Isotopic variations observed in formation water and carbonate rock are interpreted to result from the interaction between fluid B and surface adsorbed-B. $\delta^{11}B$ values of formation water decreased from its initial seawater composition for the Gulf of Mexico (+40.6‰) to the current values between +18.6 and +30.4‰. Elevated B/Cl ratios (0.001–0.025) for the Cantarell and Ku-Maloop-Zaap formation waters are beyond the evaporation trend of seawater, indicating secondary enrichment of reservoir fluids by an additional source other than evaporated seawater. In the case of U.S. sedimentary basins from the Gulf of Mexico, ^{11}B depletion accompanied by B-enrichment of formation water is explained by the progressive dissolution of ^{11}B-depleted crustal silicate minerals (Land & Macpherson 1992).

As for the $^{87}Sr/^{86}Sr$-ratios, the wide range of $\delta^{11}B$ values for formation water (18.6–34.0‰) and core samples (11.5–24.6‰) suggest spatial variations for the degree of isotopic exchange at the Cantarell reservoir. Interaction of brines with clay minerals controls their B content and isotopic composition.

3.5 *Origin of salt domes*

In general, evaporates preserve the $^{87}Sr/^{86}Sr$-signature of the respective marine water composition during the period of sedimentary deposition (McArthur et al. 2001). In southeastern U.S. and Mexican sedimentary basins, the evaporite deposits are known as Louann Salts and Istmo Salts, respectively (Salvador 1991). Most authors assigned a Middle Jurassic age (Callovian) for the saline column as a unique and extensive event for seawater evaporation and salt precipitation. Comparing the known strontium isotopic composition of seawater during Callovian ($^{87}Sr/^{86}Sr = 0.70687–0.70696$; Salvador 1991) with halite core material from the well EK-63 (depth = 4,315–4,324 mb.l.s.; $^{87}Sr/^{86}Sr = 0.70707$), the close relation between both values supports the Middle Jurassic as the principal depositional period for halite evaporates at the specific Ek reservoir site (Fig. 1).

4 CONCLUSIONS

Understanding the types and degree of water-rock interaction processes in oil reservoirs is essential to reconstruct ongoing groundwater mobilization by active petroleum extraction. The present study on secondary alteration processes in the Cantarell and Ku-Maloob-Zaap oil reservoirs postulates several geochemical reactions to explain abrupt changes in current aquifer mobility and pathway reactivity. An improved fracture permeability by extensional tectonic episodes from Pliocene to Holocene (Mitra et al. 2006), as well as the generation of porosity by active dolomitization (petrographic studies by Martínez 2009) could provide enhanced conditions for hydraulic conductivity between groundwater horizons. The abrupt invasion of oil reservoir intervals by formation water is explained by the formation of preferential fracture-related conduits, allowing accelerated dynamics for groundwater migration.

An elevated degree of alteration for reservoir dolomitization, as well as the abundance of secondary clay minerals (montmorillonite, smectite, illite) allow us to postulate desorption of exchangeable boron as a principal process that alters the primary isotopic composition of reservoir fluids. Probably, due to the lack of suitable thermal conditions (T = 110–120°C), the water-rock interaction process maintained partial equilibrium, reflected by the partial degree of isotopic homogenization between host rock ($\delta^{11}B_{Carbonates} = +11.5$ to $+24.6‰$; $\delta^{11}B_{Salt} = +27.3$ to $+34.3$ ‰) and Cantarell formation water ($\delta^{11}B_{Water} = +18.6‰$ to $+34.0‰$).

REFERENCES

Birkle, P. & Angulo, M. 2005. Hydrogeological model of Late Pleistocene aquifers at the Samaria-Sitio-Grande petroleum reservoir, Gulf of Mexico, Mexico. *Applied Geochemistry* 20: 1077–1098.

Birkle, P., Martínez, B.G. & Milland, C.P. 2009. Origin and evolution of formation water at the Jujo-Tecominoacán oil reservoir, Gulf of Mexico. Part 1: Chemical evolution and water-rock interaction. *Applied Geochemistry* 24: 543–554.

Carpenter, A.B. 1978. Origin and chemical evolution of brines in sedimentary basins. *Oklah. Geol. Surv. Circ.* 79: 60–77.

Cruz, L. & Sheridan, J. 2009. Relative contribution to fluid flow from natural fractures in the Cantarell field, Mexico. *SPE Latin American and Caribbean Petroleum Engineering Conference, 31 May-3 June 2009, Cartagena de Indias, Colombia*, Paper 122182-MS.

Graf, D.L. 1982. Chemical osmosis, reverse chemical osmosis, and the origin of subsurface brines. *Geochimica et Cosmochimica Acta* 46: 1431–1448.

Hemming, N.G. & Hanson, G.N. 1992. Boron isotopic composition and concentration in modern marine carbonates. *Geochimica et Cosmochimica Acta* 56: 537–543.

Kharaka, Y.K. & Hanor, J.S. 2004. Deep fluids in the continents: I. Sedimentary Basins. In H.D. Holland & K.K. Turekian (eds.), *Treatise in Geochemistry* 5: 499–540, Elsevier.

Land, L.S. & Prezbindowksi, D.R. 1981. The origin and evolution of saline formation water, Lower Cretaceous carbonates, south-central Texas, U.S.A. *Journal of Hydrology* 54: 51–74.

Land, L.S. & Macpherson, G.L. 1992. Origin of saline formation waters, Cenozoic section, Gulf of Mexico sedimentary basin. *AAPG Bulletin* 76(9): 1344–1362.

Leeman, W.P., Vocke, R.D. & Mc Kibben, M.A. 2002. Boron isotopic fractionation between coexisting vapor and liquid in natural geothermal systems. In Y.K. Kharaka & A.S. Maest (eds.) *Proc. 7th Int. Symp. WRI-7:* 1007–1010. Rotterdam: Balkema.

Martínez R.M. 2009. Dolomitzation and generation of vugular porosity in the K/T breccia of the Cantarell field, Marina-Campeche zone. *Ph.D. thesis, U.N.A.M.*, Mexico City, 224 p. (original in Spanish).

McArthur, J.M., Howarth, R.J. & Bailey, T.R. 2001. Strontium isotope stratigraphy: LOWESS version 3: Best fit to the marine Sr-isotope curve for 0–509 Ma and accompanying look-up table for deriving numerical age. *The Journal of Geology* 109: 155–70.

Mitra, S., Duran J.A.G., Garcia J.H., Hernandez, S.G. & Subhotosh, B. 2006. Structural geometry and evolution of the Ku, Zaap, and Maloob structures, Campeche Bay, Mexico. *AAPG Bulletin* 90(10): 1565–1584.

Salvador, A. 1991. Chapter 8: Triassic-Jurassic. In A. Salvador (ed.), *The Gulf of Mexico Basin. The Geology of America,* Vol. J: 131–180. Boulder, CO: The Geological Society of America.

Spivack, A.J., Palmer, M.R. & Edmond, J.M. 1987. The sedimentary cycle of boron isotopes. *Geochimica et Cosmochimica Acta* 51: 1939–1949.

Williams, L.B., Hervig, R.L., Wieser, M.E. & Hutcheon, I. 2001. The influence of organic matter on the boron isotope geochemistry of the gulf coast sedimentary basin, USA. *Chemical Geology* 174: 445–461.

Water-Rock Interaction – Birkle & Torres-Alvarado (eds)
 ISBN 978-0-415-60426-0

Interpreting Ca and Fe stable isotope signals in carbonates: A new perspective

T.D. Bullen
Water Resources Discipline, U.S. Geological Survey, Menlo Park, California, USA

R. Amundson
Department of Plant and Soil Biology, UC Berkeley, Berkeley, California, USA

ABSTRACT: Calcium and iron stable isotope compositions ($^{44}Ca/^{40}Ca$, $^{56}Fe/^{54}Fe$) of co-precipitated aragonite and siderite and co-located water, sampled along a 10 m stream transect down-gradient from a CO_2^- charged spring, are used to define mineral-liquid isotope fractionation factors for this natural system ($\Delta_{aragonite\ Ca\text{-}aqueous\ Ca} \sim -0.6‰$, $\Delta_{siderite\ Fe\text{-}aqueous\ Fe} \sim -1.7‰$). The factor for Ca is greater than that obtained in laboratory studies of aragonite crystallization, and close to the equilibrium fractionation factor deduced from field studies of long-term re-crystallization of calcium carbonate. The factor for Fe is less than that obtained in laboratory studies of siderite crystallization, and close to the calculated equilibrium fractionation factor based on theoretical models. While differences in siderite and aragonite crystallization rate may in part account for these differences, the results might likewise be explained in terms of the relative roles of processes leading to expression of either the theoretical or natural long-term equilibrium fractionation factors in a given mineral-water system.

1 INTRODUCTION

The field of metal stable isotope biogeochemistry has expanded rapidly over the past decade, fueled in part by the desire to use the stable isotope signals as indicators of paleo- and present-day environmental conditions. Of the metal stable isotope systems that have been found to have significant variation of isotope composition in nature and which have now been used in laboratory and field-based research studies, the calcium (Ca) and iron (Fe) systems have received the most attention. Calcium isotopes have been used mainly as indicators of ocean composition as reflected in the isotope composition of biocalcified marine organisms through the geologic record. Iron isotopes have been used mainly as indicators of redox conditions and controls based on the isotopic contrast between coexisting ferrous (Fe(II)) and ferric (Fe(III)) iron aqueous species.

Although there are differences in the geochemical behavior of Ca and Fe during water-rock interaction, there are also important similarities. For example, although the predominant Ca and Fe aqueous species are typically "hexaquo" complexes (i.e., the metal ions are positioned within inner hydration spheres formed by six water molecules), minor ligands such as carbonate, chloride and sulfate are important as intermediates in chemical reactions. In solids such as carbonates, Ca and Fe tend to be loosely bound to oxygen atoms of the CO_3^{2-} structural units of the crystal lattice, and as a consequence solids such as carbonates generally have lighter Ca and Fe isotope compositions than coexisting aqueous phases.

Here we use the results of an isotopic study of concomitant Ca^- and Fe carbonate precipitation at a natural CO_2^- charged groundwater spring to help identify the controls on solid-phase Ca and Fe isotope composition. The field site has previously been studied for chemistry, mineralogy, and carbon and oxygen isotope systematics (Amundson & Kelly 1987), providing a unique opportunity to refine our understanding of stable isotope behavior during partitioning of metals among Fe and Ca aqueous species and carbonates.

2 FIELD SETTING AND METHODS

2.1 *Salt Lick Spring, Clear Lake Highlands, CA*

Salt Lick Spring and its associated travertine deposits are located along the north-eastern border of the Clear Lake Volcanics in the Coast Range of California, USA. The geology of the Clear Lake Volcanics consists of Quaternary volcanic rocks of varying lithologies which overlie meta-sedimentary and sedimentary rocks of marine origin (Donnelly et al. 1977). In the immediate area of the study site and to the north is an extensive region consisting primarily of meta-sedimentary rocks with occasional outcrops of ultrabasic igneous rock. The spring itself is CO_2^- charged, and emerges on the steep side slope of a rugged mountainside where there are several other travertine depositing springs and travertine deposits. Salt Lick Spring is currently the most active spring in the area, and has the greatest flow rate (~60 L min^{-1}) and the largest travertine deposit.

Amundson & Kelly (1987) provide a description of the dynamics of spring-stream hydrology, water chemistry and travertine mineralogy and distribution along a ~50 m transect downstream from the spring. The CO_2^- charged (>1 atm PCO_2) groundwater is Mg-Na-Ca-HCO_3-Cl type, has pH~6.3 and contains ~300 pm Ca, ~4 ppm Fe(II) and ~120 ppm SiO_2. On exiting the spring, the water flows smoothly over older travertine deposits minimizing oxygenation. As a result, based on XRD analysis, Ca-substituted siderite ($FeCO_3$) and lesser aragonite ($CaCO_3$) are the main minerals deposited along the first 10 m of the stream in response to progressive CO_2 degassing. Downstream from this point, aragonite forms a greater proportion of the mineral precipitate as Fe(II) contents of the stream water decrease to a low level.

2.2 *Sample collection, processing and analysis*

In October, 2000, spring and stream water and the upper 1 cm of streambed mineral deposits were sampled along the first 10 m of the stream. Water samples were immediately filtered (0.2 µm syringe filter apparatus) into pre-rinsed plastic containers and acidified to pH = 2.0 with double Teflon-distilled HNO_3. Solids were dissolved using double Teflon-distilled HNO_3 in the laboratory. Chemical compositions of the water samples and mineral digests were determined using a Perkin Elmer Elan 6000 inductively-coupled plasma mass spectrometer.

To prepare samples for Ca isotope analysis, sufficient water sample or mineral digest solution to provide 1.5 µg of Ca was mixed with a ^{42}Ca–^{43}Ca "double spike" amendment and allowed to homogenize. Ca in this mixture was purified using quartz glass columns packed with AG-50-X12 cation exchange resin and HNO_3 as the eluent. To prepare samples for Fe isotope analysis, sufficient water sample or mineral digest solution to provide 1.5 µg of Fe was mixed with a ^{57}Fe–^{58}Fe "double spike" amendment and allowed to homogenize. Fe in this mixture was purified using small Teflon columns packed with Re-Spec resin (Eichrom Industries) and HNO_3 at various strengths as the eluent.

$^{44}Ca/^{40}Ca$ and $^{56}Fe/^{54}Fe$ ratios of water and mineral samples were determined by thermal ionization mass spectrometry (TIMS), using the double spike to allow for correction of stable isotope fractionation resulting from either sample processing or mass spectrometry. For both Ca and Fe, the isotope composition of the sample was obtained by subtracting the double spike component from the measured ratios of the sample-double spike mixtures using the algorithm employed at the Menlo Park Metal and Metalloid Isotope Facility for all double spike experiments. Details of the double spike TIMS approach are given in Fantle & Bullen (2009).

Total procedural replicates were performed for each sample. For Ca, column-processed seawater, La Jolla Ca and NIST 915 A Ca were analyzed at least four times each during the analytical sessions. Results are reported for $^{44}Ca/^{40}Ca$ in standard "delta" notation relative to seawater. All replicates agreed within 0.15‰. Relative to seawater, analyses of La Jolla Ca averaged –1.34‰, while analyses of NIST 915 A Ca averaged –2.03‰. For Fe, column-processed BIR basalt and IRMM-014 Fe were analyzed at least four times each during the analytical sessions. Results are reported for $^{56}Fe/^{54}Fe$ in standard "delta" notation relative to BIR, which is representative of "bulk earth" Fe. All replicates agreed within 0.10‰. Relative to BIR, analyses of IRMM-014 averaged –0.09‰.

3 RESULTS

Based on chemical compositions reported by Amundson & Kelly (1987), the travertine sampled along the studied stream reach is approximately 19% aragonite and 26% siderite (molar basis). The remaining material is probably a mixture of organic matter and silicate minerals not identified by XRD. Here, we assume that the majority of Ca in the mineral precipitate is associated with aragonite, while the majority of Fe is associated with siderite.

The concentration of dissolved Ca in stream water remains constant within analytical uncertainty (~100 ppm) along the first 10 m of stream reach (Amundson & Kelly 1987), due to the small amount of aragonite precipitated relative to the aqueous Ca pool. $\delta^{44/40}Ca$ of aqueous Ca is essentially constant over this stream reach, as shown in Figure 1. $\delta^{44/40}Ca$ of Ca in the aragonite precipitate is likewise essentially constant over this stream reach, and is on average 0.6‰ less than that of aqueous Ca. Thus, the fractionation factor $\Delta_{\text{aragonite Ca-aqueous Ca}}$ is –0.6‰.

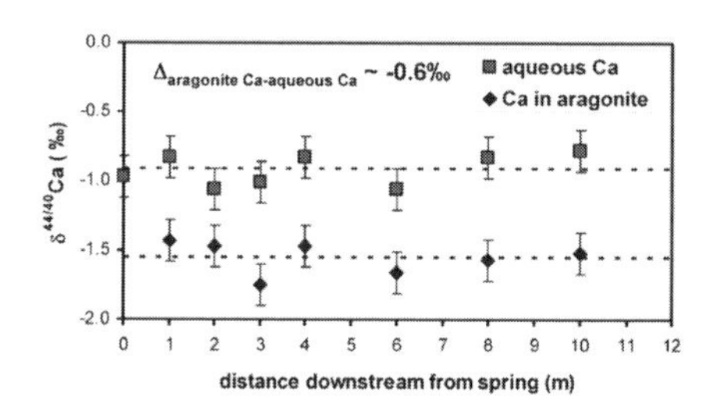

Figure 1. Ca stable isotope compositions of aqueous Ca and aragonite Ca for samples collected along the stream gradient at Salt Lick Spring. Error bars show maximum uncertainty of replicate measurements. $\Delta_{\text{aragonite Ca-aqueous Ca}}$, the average difference of $\delta^{44/40}Ca$ between solid and liquid phase Ca, is approximately –0.6 ‰.

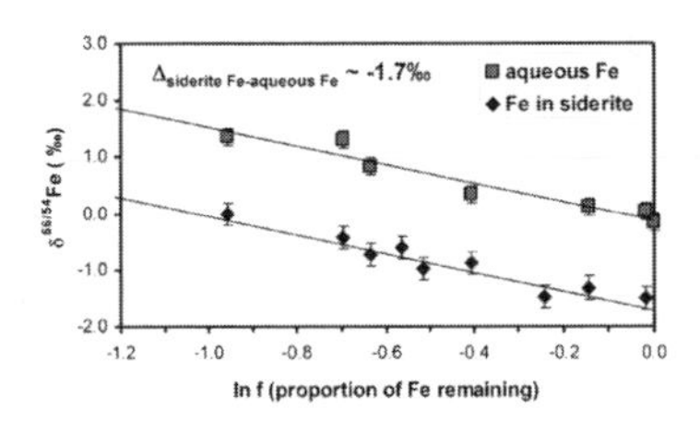

Figure 2. Fe stable isotope compositions of aqueous Fe and siderite Fe for samples collected along the stream gradient at Salt Lick Spring. Error bars show maximum uncertainty of replicate measurements. $\Delta_{siderite\ Fe-aqueous\ Fe}$, the fractionation factor between solid and liquid phase Fe defined by the slope of the lines through the data, is approximately −1.7 ‰.

The concentration of dissolved Fe decreases by an order of magnitude along the studied stream section, from 4.5 ppm at the spring to 0.5 ppm at 10 m distance along the flowpath. Precipitation of ~90% of the aqueous Fe pool as siderite during the ~30 second travel time down this portion of the stream reach provides the perfect condition to observe a Rayleigh fractionation pattern for the Fe isotopes, as shown in Figure 2. This diagram shows the $\delta^{56/54}$Fe of siderite Fe and aqueous Fe plotted versus the fraction of Fe remaining in solution relative to the Fe concentration at the spring on a logarithmic scale. In this diagram, the slope of a line through either the solid- or liquid-phase data defines the fractionation factor ε (or $\Delta_{siderite\ Fe-aqueous\ Fe}$) which is ~ −1.7‰.

4 INTERPRETATIONS AND PERSPECTIVES

To our knowledge, this is the first study to consider both Ca and Fe stable isotope compositions of coexisting carbonates together in order to understand processes during crystallization. Here we discuss our results in the context of previous attempts to determine metal stable isotope fractionation factors for carbonate-water systems.

There is a rapidly expanding literature concerning expected equilibrium isotope fractionation of Fe and Ca between coexisting carbonate and aqueous phases. On one hand, theoretical estimates can be obtained by calculating reduced function partition ratios (β-factors) for various metal complexes based on methods such as Mossbauer spectral analysis (e.g., Polyakov & Mineev 2000) and force field modeling of molecular vibrational frequencies (e.g., Schauble et al. 2001). Alternatively, estimates can be obtained based on experimental and natural mineral-water systems that are presumed to be at equilibrium based on independent geochemical and/or geologic factors. In Figure 3, we compare

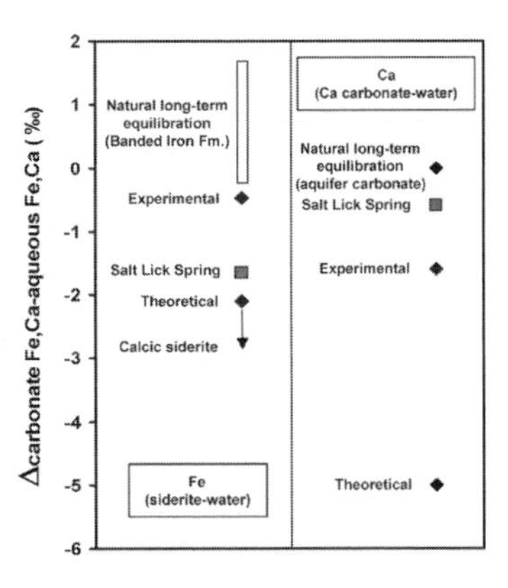

Figure 3. Fe and Ca isotope fractionation factors for Salt Lick Spring carbonates shown in relation to various estimates for equilibrium isotope fractionation factors. Fe: theoretical estimate from Johnson et al. (2003) based on β-factors from Polyakov & Mineev (2000) and Schauble et al. (2001), extension to more negative values for calcic siderite based on results of Johnson et al. (2005); experimental estimate from Wiesli et al. (2004); range for Banded Iron Formations (natural long-term equilibration) from Johnson et al. (2003). Ca: theoretical estimate from E. Schauble (pers. comm.) based on comparison of calculated β-factors for $Ca(H_2O)_{6\,aq}$ and aragonite; experimental estimate from Gussone et al. (2003); value for re-crystallized Madison aquifer carbonate (natural long-term equilibration) from Jacobson & Holmden (2008).

the fractionation factors for Ca and Fe measured in the Salt Lick Spring carbonate-water system with various equilibrium factors determined either theoretically or in laboratory/field studies.

The extremes of reported equilibrium isotope fractionation factors are defined by theoretical estimates on one hand (most negative values of the Δ scale), and isotopic analysis of rocks that have had sufficient time to re-crystallize in equilibrium with trapped pore fluids on the other (most positive values of the Δ scale). The latter are represented by Madison aquifer carbonates which have equilibrated with high-pCO_2 groundwater in the case of Ca (Jacobson & Holmden 2008), and Late Archean to Early Proterozoic Banded Iron Formation (BIF) mineral suites including siderite in the case of Fe (Johnson et al. 2003). The Ca and Fe isotope fractionation factors derived here for the Salt Lick Springs carbonate-water system have intermediate values, as do fractionation factors obtained in laboratory-based mineral synthesis experiments.

It is immediately apparent that the fractionation factor for Fe defined by siderite-water pairs at Salt Lick Spring is less than that obtained in experiments, and close to the theoretical equilibrium factor for the siderite-water system. In contrast, the fractionation factor for Ca defined by aragonite-water pairs at Salt Lick Spring is greater than that obtained in experiments, and close to the equilibrium factor reflecting long-term fluid-rock re-equilibration in the Ca carbonate (calcite)-water system. We note that apart from this difference in isotope composition relative to theoretical and experimental values, Ca and Fe have similar geochemical behavior in this system, including relative distribution of aqueous species and rate of mass loss from solution as implied by the similar proportions of siderite and aragonite in the near-spring travertine deposits. However, the concentration of aqueous Ca is nearly two orders of magnitude greater than that of aqueous Fe in Salt Lick Spring water, and this difference may play a role in promoting the isotopic contrasts.

A possible explanation for the relative lightness of Fe in siderite and relative heaviness of Ca in aragonite at Salt Lick Spring could be a difference in crystallization rate. For example, in their study of calcite crystallization in unstirred laboratory reactors, Lemarchand et al. (2004) demonstrated that $\Delta_{calcite\ Ca\text{-}aqueous\ Ca}$ became less negative with increasing crystallization rate, approaching 0‰ at the fastest rates due to Ca diffusion limitation. By analogy, slow precipitation of siderite at Salt Lick Spring would result in a relatively light Fe isotope composition, while rapid precipitation of aragonite would result in a relatively heavy Ca isotope composition. Differences in precipitation rate might result from differences in crystal seed size, i.e. smaller and more numerous seeds for siderite compared to those for aragonite. Although this explanation is speculative, we note that in the original study of this site Amundson & Kelly (1987) reported that mainly aragonite crystals were observed sticking to plastic tape placed in the stream bed in an attempt to capture recent carbonate precipitates. Siderite crystals may likewise have formed but may have been too small to provide a good visual or XRD record.

An alternative explanation arises from the fact that equilibrium isotope fractionation factors based on theory and those describing natural long-term equilibration are so different (Fig. 3), which suggests that different processes or environmental conditions may lead to expression of those possible equilibrium states in natural and experimental systems. Further, the fact that mineral-water fractionation factors observed in natural and experimental systems lie between these extremes could indicate that observed compositions represent any particular balance of those processes. For example, expression of theoretically light equilibrium solid compositions in the short term, apparently the dominant effect on siderite at Salt Lick Spring, requires a dynamic crystal-liquid interface where the metal can pass easily back and forth between hydrated and dehydrated states (i.e. in an efficient forward-backward isotope exchange scenario) before being incorporated into the crystal. This condition is most likely to occur at slow rates of crystallization. Expression of heavy equilibrium solid compositions are more difficult to explain, but at least in the long term may result from differences in metal cation hydration coordination at the environmental conditions (e.g. high pressure, complex porewater chemistry) prevailing during cycles of crystal dissolution and re-precipitation. Clearly there remains much experimental and field-based work to be done in order to fully understand the controls on metal stable isotope fractionation in complex mineral-water systems.

REFERENCES

Amundson, R. & Kelly, E. 1987. The chemistry and mineralogy of a CO_2-rich travertine depositing spring in the California Coast Range. *Geochemica et Cosmochimica Acta* 51: 2883–2890.

Donnelly, J.M., Hearn, B.C. Jr. & Goff, F.E. 1977. The Clear Lake Volcanics, California: Geology and field trip guide. In *Field Trip Guide to the Geysers-Clear Lake area. Cordilleran Section, Geol. Soc. Amer.* 2: 25–56.

Fantle, M.S. & Bullen, T.D. 2009. Essentials of iron, chromium and calcium isotope analysis of natural materials by thermal ionization mass spectrometry. *Chem. Geology* 258: 50–64.

Gussone, N., Eisenhauer, A., Heuser, A., Dietzel, M, Bock, B., Bohm, F., Spero, H.J., Lea, D.W., Bijma, J. & Nagler, T.F. 2003. Model for kinetic effects on calcium isotope fractionation ($\delta^{44}Ca$) in inorganic aragonite and cultured planktonic foraminifera. *Geochemica et Cosmochimica Acta* 67: 1375–1382.

Jacobson, A.D. & Holmden, C. 2008. $\delta^{44}Ca$ evolution in a carbonate aquifer and its bearing on the equilibrium isotope fractionation factor for calcite. *Earth and Planetary Science Letters* 270: 349–353.

Johnson, C.M., Beard, B.L., Beukes, N.J., Klein, C. & O'Leary, J.M. 2003. Ancient geochemical cycling in the Earth as inferred from Fe isotope studies of banded iron formations from the Transvaal Craton. *Contributions to Mineralogy and Petrology* 144: 523–547.

Johnson, C.M., Roden, E.E., Welch, S.A. & Beard, B.L. 2005. Experimental constraints on Fe isotope fractionation during magnetite and Fe carbonate formation coupled to dissimilatory hydrous ferric oxide reduction. *Geochemica et Cosmochimica Acta* 69/4: 963–993.

Lemarchand, D., Wasserburg, G.J. & Papanastassiou, D.A. 2004. Rate-controlled calcium isotope fractionation in synthetic calcite. *Geochemica et Cosmochimica Acta* 68:4665–4678.

Polyakov, V.B. & Mineev, S.D. 2000. The use of Mossbauer spectroscopy in stable isotope geochemistry. *Geochemica et Cosmochimica Acta* 64/5: 849–865.

Schauble, E.A., Rossman, G.R. & Taylor, H.P. Jr. 2001. Theoretical estimates of equilibrium Fe-isotope fractionations from vibrational spectroscopy. *Geochemica et Cosmochimica Acta* 65: 2487–2497.

Wiesli, R.A., Beard, B.L. & Johnson, C.M. 2004. Experimental determination of Fe isotope fractionation between Fe(II), siderite and "green rust" in abiotic systems. *Chemical Geology* 211: 343–362.

Water-Rock Interaction – Birkle & Torres-Alvarado (eds)
© 2010 Taylor & Francis Group, London, ISBN 978-0-415-60426-0

Environmental isotopes as indicators of aquitard effectiveness, Murray Basin, Australia

I. Cartwright
National Centre for Groundwater Research & Training and School of Geosciences, Monash University, Clayton, Australia

T. Weaver
URS Australia Pty Ltd., Southbank, Australia

D. Cendon
Australian Nuclear Science and Technology Organisation, Menai, Australia

I. Swane
Terrenus Pty Ltd., Wilston, Australia

ABSTRACT: The distribution of $\delta^{13}C$ values, $^{87}Sr/^{86}Sr$ ratios, and ^{14}C activities ($a^{14}C$) show that considerable inter-aquifer flow occurs in the Wimmera Region of the Murray Basin, southern Australia. Many of the potential aquitards, (e.g., Bookpurnong Beds and Ettrick Marl) are ineffective barriers to groundwater flow; only the Geera Clay forms an effective aquitard and restricts inter-aquifer leakage between the Loxton-Parilla Sands and/or the Murray Group and the underlying Renmark Formation. This study illustrates the utility of environmental isotopes in determining inter-aquifer flow and aquitard effectiveness in regions where direct studies of aquitards is lacking.

1 INTRODUCTION

Aquitards are important components of groundwater flow systems. Under ideal circumstances, their low hydraulic conductivities prevent significant groundwater flow between over- and underlying aquifers. This in turn protects deeper groundwater from near surface contamination and isolates shallow groundwater and surface water systems from the impacts of pumping of deeper aquifers. The effectiveness of aquitards depends on their thickness and integrity; in particular, the presence of fractures or interconnected coarser-grained units may allow interformational flow. It is possible to directly examine groundwater flow and solute transport within aquitards. However, as localised zones of high hydraulic conductivity may be critical in determining aquitard integrity, groundwater chemistry and environmental isotopes from the over- or underlying aquifers are a more effective method to study regional-scale leakage through aquitards. Understanding the effectiveness of aquitards in preventing inter-aquifer flow is essential to characterising groundwater flow systems and in assessing the threats to groundwater resources. Here we use environmental isotopes to constrain inter-aquifer mixing and aquitard effectiveness in the Wimmera region of the southern Murray Basin, Australia.

1.1 *Hydrogeological setting*

The Murray Basin is ~300,000 km^2 in area and contains up to 600 m of late Palaeocene to Recent sediments overlying Proterozoic to Mesozoic basement (Lawrence 1988, Evans & Kellett 1989: Fig. 1a). Except for the southwest where groundwater discharges to the Southern Ocean the Murray Basin is closed and regional groundwater flow is from the basin margins towards the basin centre (Fig. 1a) where discharge into salt lakes, playas, and the Murray River occurs.

The Murray Basin in the Wimmera region deepens irregularly northward to in excess of 300 m. The Loxton-Parilla Sands is a sequence of marine sands and silts that form the shallowest unconfined unit (Fig. 1c). In the west of the region, the Loxton-Parilla Sands are underlain by the Murray Group, which comprises marine and marginal marine limestone with interbedded calcareous sands, marls, and silts. The lowermost Renmark Formation is a confined aquifer of fluvial clays, silts, sands, and gravels that is up to 200 m thick in the north of the area. The Ettrick Formation, Bookpurnong Beds and Winnambool Formation comprise up to 35 m of calcareous clays with silt, sand, and locally limestone layers (Brown & Radke 1989); the Geera Clay consists of up to 75 m of massive clays with minor sand and silt

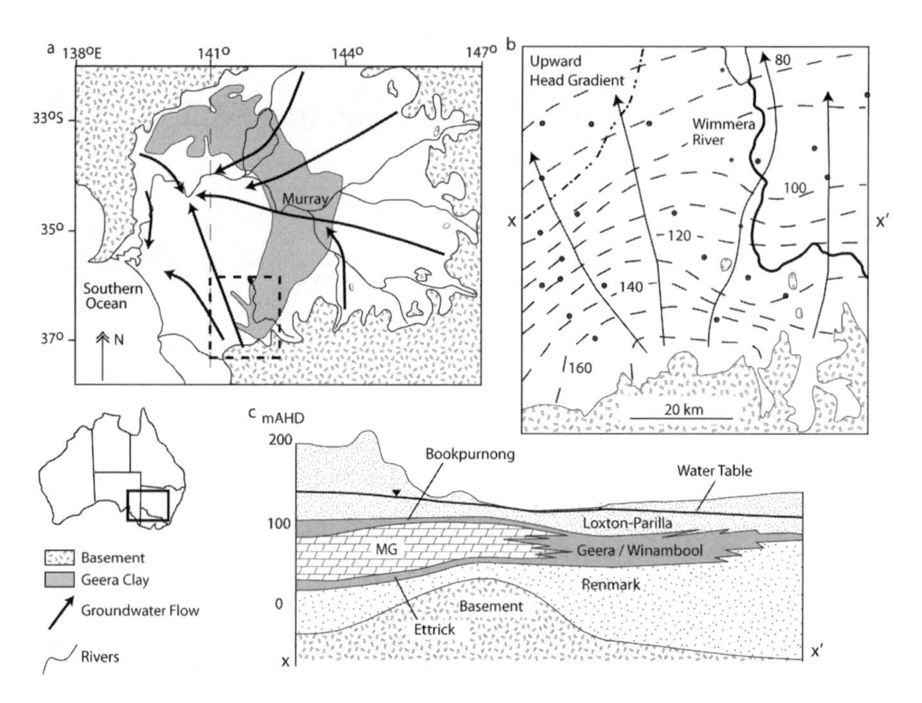

Figure 1. a) Map of the Murray Basin showing extent of the Geera Clay and location of Wimmera region (dashed box). b) Hydraulic heads (in m) and groundwater flow directions in the Loxton-Parilla Sands; flow directions in the other aquifers are similar. Vertical head gradients are downward except northwest of dash-dot line. Dots are sampling localities. c) Cross section across the Wimmera Region (x-x′in Fig. b); Mid Tertiary Aquitard units shown in grey. MG = Murray Group, AHD = Australian Height Datum (mean sea level).

layers. These units are collectively termed the Mid Tertiary Aquitard (Fig. 1), although their efficiency as confining units is not well constrained. Horizontal hydraulic conductivities determined from pump tests are $6 \times 10^{-6} - 6 \times 10^{-5}$ m/s in the Loxton-Parilla Sands, $1 \times 10^{-5} - 2 \times 10^{-4}$ m/s in the Murray Group, and $1 \times 10^{-5} - 3 \times 10^{-4}$ m/s in the Renmark Formation; vertical hydraulic conductivities of the Geera Clay are 10^{-10}–10^{-9} m/s (Lawrence 1988).

Groundwater flow in all aquifers is to the north or northwest (Fig. 1b). Lateral hydraulic gradients are $\sim 5 \times 10^{-4}$ and, except in the northwest of the area vertical hydraulic gradients are generally downward (typically 0.02 to 0.05). The shallow unconfined aquifers likely receive recharge across the region (Leaney et al., 2003); however, the confined units, especially the Renmark Formation and Murray Group, can only be recharged through the overlying units.

2 GEOCHEMICAL VARIATIONS

^{14}C activities ($a^{14}C$ in percent modern carbon, pMC) and corrected ^{14}C ages of groundwater (Fig. 2a) have the following distribution: 1) $a^{14}C$ of groundwater in the Murray Group and the Renmark Formation in the west are <15 pMC, corresponding to residence times of 6.5 to >30 ka. 2) Renmark groundwater in the east has $a^{14}C$ between 1 and 68 pMC and residence times of 2.3 to >30 ka. 3) $a^{14}C$ of Renmark groundwater in the centre are higher (40–67 pMC) than those in groundwater from the overlying Murray Group or Mid Tertiary Aquitard units (4.5–30 pMC). Groundwater residence times in the Renmark Formation in this area increase northward from 1.6 to 6.8 ka. 4) The Loxton-Parilla Sands groundwater has variable $a^{14}C$ (14–106 pMC) that show no consistent variation with either location or depth. Groundwater residence times in this aquifer are 14.5 ka to Modern and are always younger than those of the underlying units.

The pattern of groundwater residence times shows little correspondence to distance from the basin margin, implying that flow is not only subhorizontal along the flow paths inferred from the hydraulic heads (Fig. 1b). In turn this suggests that there may be considerable inter-aquifer mixing despite the presence of several possible aquitards.

$^{87}Sr/^{86}Sr$ ratios of groundwater show a similar pattern of spatial variation to $a^{14}C$ (Fig. 2b). $^{87}Sr/^{86}Sr$ ratios of Renmark Formation groundwater increase eastwards from 0.709–0.710 to 0.712–0.715. Discounting some high $\delta^{13}C$ values that are impacted by bacterial DIC reduction, there is also a general

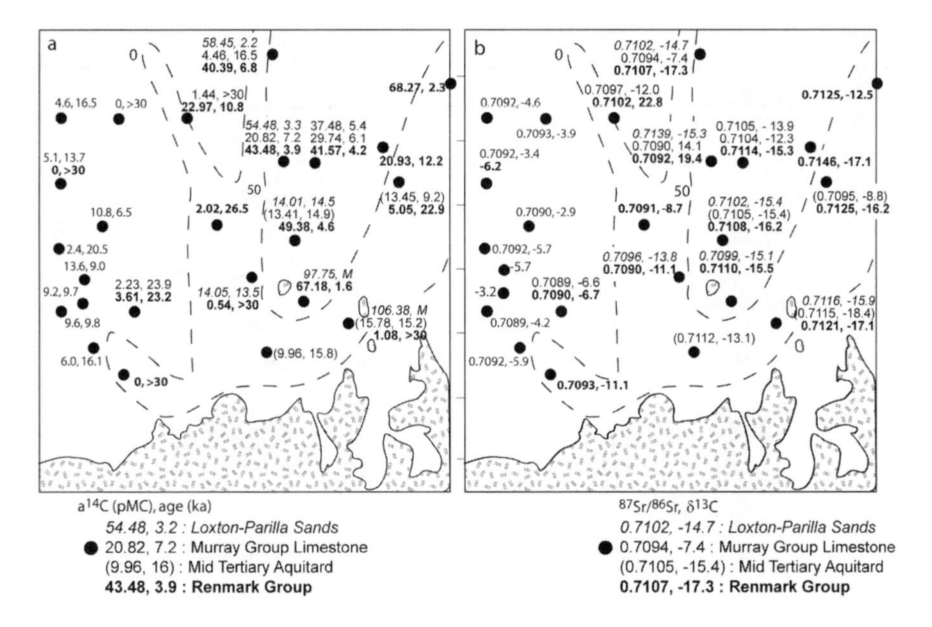

Figure 2. Distribution of a^{14}C & ^{14}C ages a) and $^{87}Sr/^{86}Sr$ ratios and $\delta^{13}C$ values b) in the Wimmera region. Dashed lines show extent and thickness of the Geera Clay.

westward decrease in $\delta^{13}C$ values from –7 to –6‰ where the Renmark Group is overlain by the Murray Group to –17 to –13‰ where the Murray Group is absent. Groundwater from the Murray Group generally has $^{87}Sr/^{86}Sr$ ratios of ~0.709, except in the centre of the region where $^{87}Sr/^{86}Sr$ ratios are 0.710–0.711. $\delta^{13}C$ values in the Murray Group groundwater also vary from –7 to -3‰ in the west to –12 to –14‰ in the centre of the area. Groundwater from the Mid Tertiary Aquitard units has $^{87}Sr/^{86}Sr$ ratios of 0.710–0.712 and $\delta^{13}C$ values of –16 to –14‰, while Loxton-Parilla groundwater has $^{87}Sr/^{86}Sr$ ratios of 0.710–0.714 and $\delta^{13}C$ values of –18 to –9‰.

Overall, the $^{87}Sr/^{86}Sr$ ratios and $\delta^{13}C$ values imply mixing between groundwater that has derived the majority Sr from silicate weathering and C from the soil zone (Silicate endmember in Fig. 3) and groundwater with both C and Sr derived dominantly from marine carbonate dissolution (Carbonate endmember in Fig. 3). Similar mixing trends are evident elsewhere in the southwest Murray Basin (Dogramaci & Herczeg 2002). The spatial distribution in Fig. 2b implies that leakage of water from the Murray Group into the Renmark Formation in the west of the region has occurred, producing locally low $^{87}Sr/^{86}Sr$ ratios and high $\delta^{13}C$ values in groundwater in that unit. Likewise, leakage from the Loxton-Parilla Sands into the Murray Group in the centre of the region results in the groundwater from the limestone having locally anomalously high $^{87}Sr/^{86}Sr$ ratios and low $\delta^{13}C$ values (Fig. 2b).

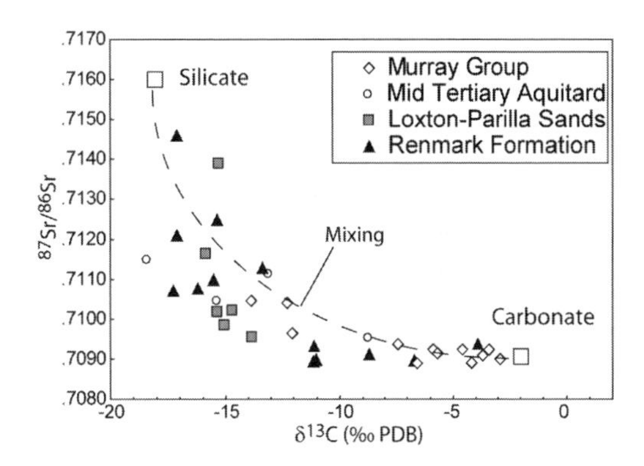

Figure 3. $^{87}Sr/^{86}Sr$ vs. $\delta^{13}C$ for Wimmera groundwater. The composition of the endmembers was calculated using $^{87}Sr/^{86}Sr$ ratios and $\delta^{13}C$ values of the aquifer minerals and soil zone CO_2 (c.f., Dogramaci & Herczeg 2002).

3 DISCUSSION

Environmental isotopes have allowed inter-aquifer mixing and the effectiveness of potential aquitards in the Wimmera Region to be assessed; this is vital information for understanding the hydrogeology of this region that was hitherto not well understood. There is little correlation between groundwater flow directions and groundwater residence times, implying that there is not simple lateral flow within the aquifers. The pattern of $^{87}Sr/^{86}Sr$ ratios, $\delta^{13}C$ values, and a^{14}C imply the following inter-aquifer flow has occurred. 1) Downward flow between the Murray Group and the Renmark Formation through the Ettrick Formation. This observation is consistent with the significant upward leakage between these aquifers across the Ettrick Formation in discharge areas of the Murray Basin (Dogramaci et al. 2001, Dogramaci & Herczeg 2002). 2) Downward flow between the Loxton-Parilla Sands and Murray Group through the Bookpurnong Beds. Both the Ettrick Formation and Bookpurnong Beds contain sand and silt layers that are locally extensive, and are not effective aquitards. The very variable a^{14}C in the Renmark groundwater in the east of the region implies that here too there is a significant component of leakage through the Loxton-Parilla Sands into the Renmark Formation. The inferred inter-aquifer leakage is dominantly downward, which is in agreement with the observed head gradients.

The relatively young ^{14}C ages in the Renmark Formation in the centre of the region coincide with where the Geera Clay is thickest (Fig. 2a) The observation that ^{14}C ages increase along the flow direction of flow in this region indicates that recharge occurs in the south of the region where the Geera Clay grades into the Renmark Formation and contains sand and silt layers, with subsequent northward lateral flow beneath the Geera Clay. The Geera Clay is an effective aquitard and a lack of leakage in this region is limited resulting in relatively simple groundwater flow paths. A check on the reasonableness of this conclusion may be made from the hydraulic conductivities required for such flow. The distribution of groundwater residence times implies northward lateral velocities of ~20 m/year, which for a porosity of 0.1–0.3 implies a flux of 2–6 m^3/m^2/year. Hydraulic gradients in the Renmark Formation are $\sim 5 \times 10^{-4}$ and thus it would require that hydraulic conductivities were $4 - 12 \times 10^3$ m/year ($1 - 3 \times 10^{-4}$ m/s). These are within the range estimated for the Renmark Formation (Lawrence 1988), implying that the proposed flow regime is plausible.

3.1 *Implications for groundwater use*

Renmark Formation groundwater beneath the Geera Clay locally has TDS <3500 mg/L and is thus a viable resource for irrigation or stock water. While the Geera Clay protects the deeper groundwater from salinisation or contamination from overlying units, protection of the recharge area is required to ensure the long-term viability of this resource. By contrast, the interconnectivity between the Loxton-Parilla Sands and the Murray Group implies that any changes to the salinity of surface water, as may happen with climate change or further landuse change, may impact the Murray Group groundwater in this area. Renmark Formation groundwater in the west of the Wimmera Region is receiving downward flow from the Murray Group and is not protected from salinisation or contamination; although the long residence times imply that any changes to groundwater quality will be slow. The irregular distribution of the ^{14}C residence times in the Renmark groundwater in east of the region implies that there are pathways of preferential flow and that this groundwater is vulnerable to contamination.

4 CONCLUSIONS

The combined use of major ion geochemistry and environmental isotopes has allowed groundwater flow in the Wimmera region of the Murray Basin to be constrained. In particular, the multi tracer approach employed here has permitted the assessment of inter-aquifer mixing and the effectiveness of aquitards. the study shows that many of the potential aquitards in this region are ineffective barriers to cross-formational flow, which has significant implications for the use and protection of water resources.

REFERENCES

Brown, C.M. & Radke, B.M. 1989. Stratigraphy and sedimentology of mid-Tertiary permeability barriers in the subsurface of the Murray Basin, southeastern Australia. *Bureau of Mineral Resources Journal of Australian Geology and Geophysics* 11: 367–386.

Bureau of Meteorology 2009. Commonweath of Australia Bureau of Meteorology. *www.bom.gov.au.*

Dogramaci, S.S. & Herczeg, A.L. 2002. Strontium and carbon isotope constraints on carbonate-solution interactions and inter-aquifer mixing in groundwaters of the semi-arid Murray Basin, Australia. *Journal of Hydrology* 262: 50–67.

Dogramaci, S.S., Herczeg, A.L., Schiff, S.L. & Bone, Y. 2001. Controls on $\delta^{34}S$ and $\delta^{18}O$ of dissolved sulfate in aquifers of the Murray Basin, Australia and their use as indicators of flow processes. *Applied Geochemistry* 16: 475–488.

Evans, W.R. & Kellett, J.R. 1989. The hydrogeology of the Murray Basin, southeastern Australia. *Bureau of Mineral Resources Journal of Australian Geology and Geophysics* 11: 147–166.

Lawrence, C.R. 1988. Murray Basin. In J.G. Douglas & J.A. Ferguson (eds.), *Geology of Victoria*: 352–363. Geological Society of Australia (Victoria Division), Melbourne.

Leaney, F.W., Herczeg, A.L. & Walker, G.R. 2003. Salinization of a fresh palaeo-ground water resource by enhanced recharge. *Ground Water* 41: 84–92.

Water-Rock Interaction – Birkle & Torres-Alvarado (eds)
© 2010 Taylor & Francis Group, London, ISBN 978-0-415-60426-0

Tracing the inorganic carbon system in the groundwater from the lower Jordan Valley basin (Jericho/Palestine)

S.K. Khayat & S. Geyer
UFZ Environmental Research Centre, Leipzig-Halle, Germany

A.M. Marei
Al-Quds University, Jerusalem, Palestine

ABSTRACT: The study aims at tracing the carbon system using carbon isotopes, rubidium and strontium to trace the different reactions in Jericho aquifer. The results refute the previous studies about the role of Jericho fault, which considered as a barrier between Mountain and the Pleistocene aquifers. The isotopes signatures from the Pleistocene groundwater adjacent to the fault system show the same signatures from mountain springs. There seems to be a fresh water leakage from mountain aquifer through the fault, with δ^{13}C~12 ‰ in both locations. The effect of different salts leachate inputs due to the mineralization and dissolution activities are found eastwards. Typical isotopic signature of δ^{13}C for brackish water from the eastern wells of the study area was ~–7%. The rest of the wells between the fault system in the West and the brackish wells in the East of Jericho plain area show a mixing δ^{13}C signature.

1 INTRODUCTION

The eastern aquifer of the Jordan valley is fed mainly from the drainage of the calcareous mountain runoff further to the West where the water quality is highly deteriorated as soon as it reaches the Pleistocene-Holocene formation in the Jordan valley rift, where the water salinity increases with distance to the East. Many previous studies pointed the problem of salinity in the study area, mainly the mineral compositions of the rocks and sediments, as well as an old deep brine that normally predominate in the eastern part (Khayat et al. 2006, Salameh 2002, Marie & Vengosh 2001, Rosenthal 1994). In this area, the effective management of groundwater resources is important not only for potable water supply and agriculture, but also it aids in creating a stable political situation in this area of conflict. Each use requires a good understanding for the groundwater system, but conventional hydrochemical techniques are not always successful for such environments. The calcareous mountain aquifer plays an important role in feeding the whole system through water runoff that infiltrated along at different places along the stream path. Thus tracing inorganic carbon system with carbon stable isotopes will be helpful in tracing the possible changes in the fresh water quality that deteriorated as soon as it enters the study area. The extent mineralization and dissolution is spatially variable within the aquifer system due to variability in the infiltration conditions with distance (Khayat et al. 2009).

2 HYDROLOGY OF THE STUDY AREA

The main local aquifer systems for the whole sub-basin from recharge point to the Jordan River were classified as shown in Figure 1 (Rof & Rafti 1963):

2.1 *Regional mountains aquifer: Upper Cretaceous aquifer*

Cenomanian to Turonian carbonate rocks (K, Figure 1) form most of the mountains outcrops west of the study area. This aquifer is Karsti-

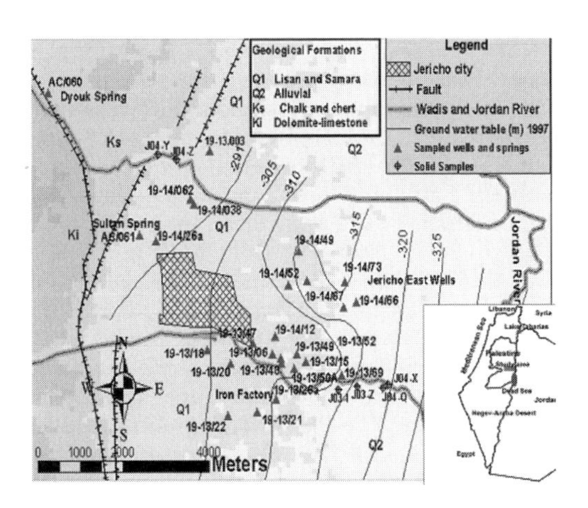

Figure 1. Study area, geological formations and sampling location.

fied and represents one of the most important water resources in the region. The outlets of this sub-aquifer are in the Jericho springs, Wadi Qilt Springs and Wadi Nu'emeh Springs. The aquifer in the upper mountain part fed directly through infiltration from precipitation in the upper mountain and from direct infiltration during stream runoff. During the flood events the excess runoff water drained further to the East to the Jericho plain area feeding the Jordan valley deposits.

2.2 *Jordan Valley deposits aquifer (Dead Sea)*

Two Dead Sea groups aquifers are composed of Pleistocene to recent age deposits. These deposits begin directly after the Jericho fault system that separates the Karstified mountain aquifer and the Jordan valley deposits. It is composed of: a) Holocene or sub-recent alluvial aquifer (Q2, Figure 1), which is distributed mainly in the Jordan Valley and neighboring areas. The alluvial aquifer, often directly overlying the Pleistocene gravel aquifer, is hydraulically interconnected with this aquifer. b) Pleistocene Samara Lisan aquifer/ aquiclude (Q1, Figure 1). This includes three pleistocenic members: Samara coarse clastic, Samara silt and Lisan. Samara aquifer is a lateral facies succession from terrestrial/fluvial, to deltaic/limnic and limnic/ brackish lake environments. They reflect the Plio-Pleistocene depositional conditions of the Lisan Lake. Lisan, the marl, gypsum and silt lacustrine unit, is generally considered an aquiclude, void of exploitable water. The natural recharge by rain is almost negligible. Therefore, the aquifer is mainly fed by inflows of excess water from neighboring aquifers. The aquifer supplies agriculture in the Jordan Valley between Jericho and Fari'a Graben.

2.3 *Hydraulic separation between the two aquifers*

As explained above, there are 2 main aquifer systems in the study area. The mountain calcareous aquifer with its fresh groundwater of $CaCO_3$ type, and Dead Sea group deposits that is highly affected with the Lake Lisan sediments which act as the source of salinity and water dissolution in the area. The groundwater quality from both aquifers is thus totally different. Both areas are separated by the Jericho fault that lies directly at the feet of the mountains. This fault act as blocks composed of impermeable Senonian chalk which prevent lateral flow of groundwater from the Upper Cretaceous aquifer to the Plio-Pleistocene shallow aquifer system (Golani 1972). However, groundwater from the Upper Cretaceous aquifer leaks into the Plio-Pleistocene aquifer along the Dead Sea fault (Yechieli et al. 1995).

3 SAMPLING AND ANALYTICAL METHODS

Groundwater samples for analysis of C stable isotopes were collected in 1.5 L bottles from 25 spring and wells (Figure 1), in which the dissolved inorganic carbon (DIC) was precipitated by treatment with alkaline $BaCl_2$. The resulting precipitates were washed with dionized water and dried. $^{13}C/^{12}C$ dissolution of the dried $BaCO_3$ precipitates was made with anhydrous H_2PO_4 used to yield CO_2 gas for analysis. Isotope ratio measurements of $\delta^{13}C$ were measured against V-PDB standard using Gas Source Isotope Ratio Mass Spectrometry (IRMS), while the trace elements were analyzed by ICP-mass spectrometry using both collision-cell quadrupole and high resolution sector-field based instrumentation. All measurement was carried out on the UFZ-Environmental Research Centre, Halle, Germany.

4 RESULTS AND DISCUSSION

In the study area, the $\delta^{13}C$ (DIC) values of the spring water were about –12‰ consistent with the carbon signature in calcareous rocks (Fritz & Kendal 1997). The HCO_3 content in these samples was the highest in proportion to other TDS (75% of TDS). The recharge mechanism varied according to the infiltration from one place to another along the flow path. "Samed" well 19–14/26a and the North Nuwe'ma wells, very adjacent to the spring's system in the West, also show similar results with higher HCO_3- and same $\delta^{13}C$ signatures to mountain aquifer springs (Figures 2–4). This emphasize that there is a leakage from the upper calcareous aquifer to the shallow Pleistocene aquifer. This means that the conclusions by Golani (1972) about the impermeable fault are not definitely correct. The samples from the shallow aquifer further to

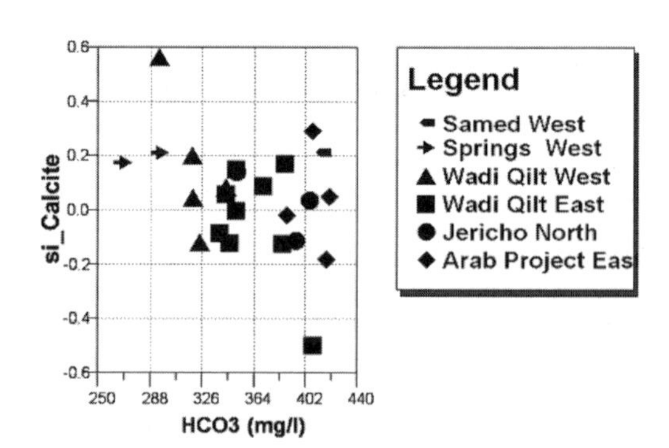

Figure 2. Relative increase in HCO_3 vs. a decrease in calcite saturation (si) from West to East.

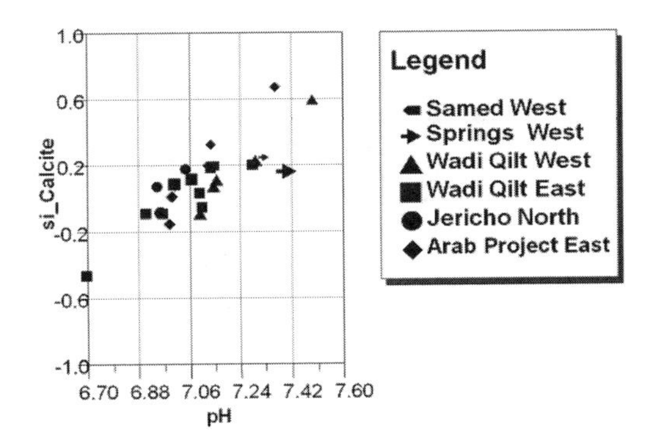

Figure 3. Dissolution of calcite decreases the pH values, due to CO_2 release from West to East.

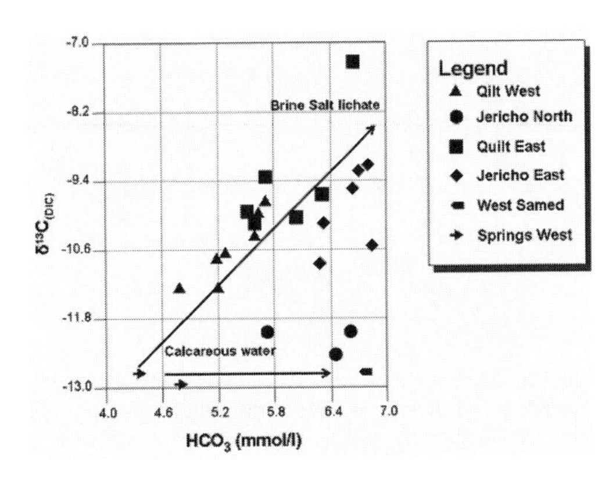

Figure 4. $\delta^{13}C$ vs. bicarbonate in mmol/l.

the East show different behavior with regard to the carbon system. Much more calcite dissolves due to continuous replenishment of CO_2, thus, the DIC concentration in these wells is much higher. This dissolution continues gradually along the flow path from the West to the East (Figure 2). The pH gradually gets lower due to CO_2 replenishment (Figure 3). An additional input of carbonates coming from Lisan deposits with $\delta^{13}C$ values between −9 and −7‰, rich in Mg, Na and Cl, initiates other mineralization forms, mainly dolomitization, in which some of the precipitated calcite undergoes a dolomitization process resulting in a surplus Ca release (Khayat et al. 2009).

The loss of some calcite by dolomitization has a minor effect on the $\delta\ ^{13}C$, the gain of dolomite add carbonate with $\delta^{13}C$~0‰ to the DIC pool, which is lead to dilution of DIC and resulting in the above mentioned typical $\delta^{13}C$ signature for the infiltrated water in the shallow aquifer (Figures 4–5).

Many factors play a role in the changes of the DIC signatures, these are mainly related to the lithological formation of Lisan and Samra with brackish flush out, as well as anthropogenic influences in this agricultural and populated area.

In this study, Rb and Sr were used to differentiate between the effects of carbonate minerals transformations and salt leachate. NO_3 was also used as indicator for the anthropogenic influences within the aquifer. Rb occurs in the minerals pollucite, carnallite, leucite and lepidolite as well as K-bearing minerals and brackish waters. These minerals occurs in small amount in the study area, however, they can give a more precise picture about the mineral role in changing the carbonate phase, rather than most common aragonite and anhydrite phases. Both Rb and Sr can also derived in groundwater by dissolution of Ca-silicates, this was the case in the spring water and adjacent wells to the West.

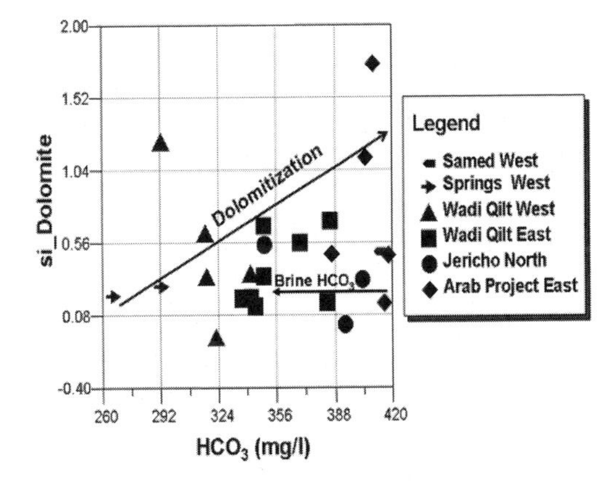

Figure 5. Increase of bicarbonate concentration vs. an increase of the dolomite saturation (si) from West to East.

The presence of higher free Rb in the groundwater to the East can give an indication of the dissolution of the Dead Sea basin minerals in the groundwater. Sr substitutes calcium in carbonate minerals due to their similar ionic size. Thus, the presence of high strontium concentrations in the groundwater can give an indication of dissolution of Sr-bearing carbonate minerals.

The Rb/CO_3, Sr/CO_3 molar ratios and $\delta^{13}C$ values were used to trace the effect of dissolute mineral leachates in the Jordan rift, which mainly has a highly Rb/CO_3 ratio with respect to these signatures resulted from carbonate minerals transformations (Figure 6).

The NO_3/Ca molar ratio vs. $\delta^{13}C$ was used to trace the anthropogenic influences on the carbon system of the groundwater (Figure 7). The increase of Ca concentration can also give an indication of the carbonate minerals dissolution, mainly calcite.

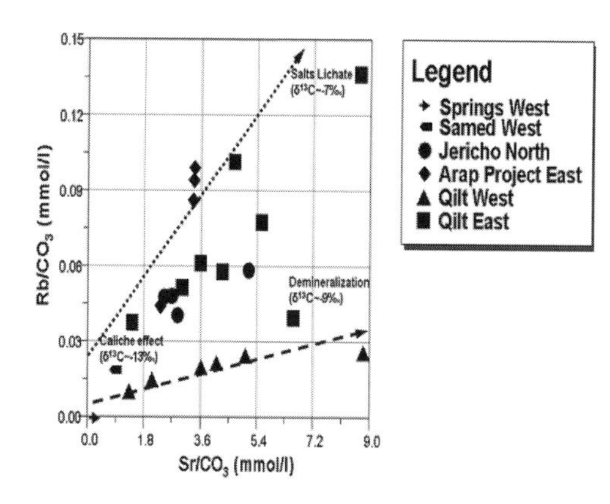

Figure 6. Sr/CO_3 vs. Rb/CO_3 molar ratios.

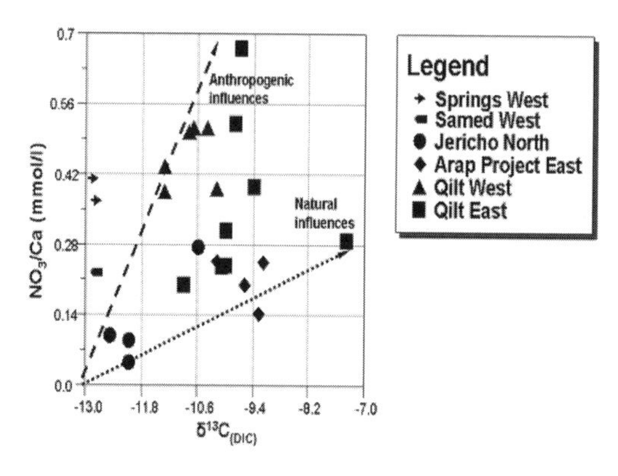

Figure 7. $\delta^{13}C$ vs. NO_3/Ca molar ratio. Note two different behaviors due to different NO_3 content.

4 main processes affect the carbon system in the Pleistocene aquifer:

1. The influence from calcareous rocks (Upper Cenemonian formation), which mainly affects the springs and adjacent well (Samed) in the West and the North. This is characterized by low Sr/CO_3 molar ratios and depleted $\delta^{13}C \approx -12‰$ (Figure 6).
2. The anthropogenic influence indicated by the low pH value, high NO_3/Rb molar ratio and $\delta^{13}C$ between –10 and –11‰ (Figures 3–7).
3. The salts leachate influence, mainly in the eastern wells, with a high Rb/CO_3 molar ratio and $\delta^{13}C \approx -7‰$ (Figure 6).
4. The mineral transformation influences (dissolution, dolomitization), which is characterized by high Sr/CO_3 molar ratio and $\delta^{13}C \approx -9‰$ (Figure 6).

REFERENCES

Golani, U. 1972. Groundwater resources in Jericho region. Technical report, *Tahal Report HR/72/016* (in hebrew).

Khayat, S. Ghanem, M. Tamimi, A. Haddad, M., Geyer. S., Hötzl, H., Ali, W. & Möller, P. 2009. Hydrochemistry and isotope hydrogeology in the Jericho area/Palestine. In H. Hotzl (ed.), *The water in the Jordan Valley:* 284–325. Springer Verlag.

Khayat, S., Hötzl, H., Geyer, S. & Ali, W. 2006. Hydrochemical investigation of water from the Pleistocene wells and springs, Jericho area. *Palestine Hydrogeology Journal* 14 (1–2): 192–202.

Marei, A. & Vengosh, A. 2001. Sources of salinity in groundwater from Jericho area, Jordan valley. *Journal of Groundwater* 39: 240–248.

Rofe & Roffety 1963. West Bank. *Hashemite Kingdom of Jordan.* Central Water Authority.

Rosenthal, E. & Vengosh, A. 1994. Saline Ground-water in Israel: Ist on the Water Crisis in the Country. *Journal of Hydrology* 156: 389–430.

Salameh, E. 2002. Sources of water salinities in the Jordan Valley area/Jordan. *Acta Hydrochimica et Hydrobiologica* 29: 329–362.

Yechieli, Y., Ronen, D., Berkovitz, B., Dershowitz, W. & Hadad, A. 1995. Aquifer characteristics derived from the interaction between water levels of a terminal lake (Dead Sea) and adjacent aquifer. *Water Recourses* 31: 893–902.

Water-Rock Interaction – Birkle & Torres-Alvarado (eds)
© 2010 Taylor & Francis Group, London, ISBN 978-0-415-60426-0

The application of radium isotopes in the evaluation of saline water circulation in the Dead Sea aquifer

Y. Kiro & A. Starinsky
Institute of Earth Sciences, Hebrew University of Jerusalem, Givat Ram, Jerusalem, Israel

Y. Yechieli
Geological Survey of Israel, Jerusalem, Israel

C.I. Voss
US Geological Survey, Reston, VA, USA

Y. Weinstein
Department of Geography, Bar-Ilan University, Ramat-Gan, Israel

ABSTRACT: Radium isotopes in the Dead Sea Lake and the surrounding aquifer were studied in order to define the processes controlling their activities in the lake and in the groundwater. ^{226}Ra activities in the groundwater show a significant removal of ^{226}Ra from the lake water as they enter the aquifer. Short-lived radium isotopes show a nonlinear relation with salinity which is caused by the effect of salinity on the adsorption of radium. Simulations of radium distribution in the aquifer, using the code of SUTRA-MS, were done in order to estimate the adsorption distribution coefficient of radium in the aquifer. ^{228}Ra activities in the Dead Sea water are much higher than the expected activities according to the radium sources to the Dead Sea. These observations together with the removal of ^{226}Ra in the Dead Sea groundwater may indicate a large-scale saline water circulation in the aquifer.

1 INTRODUCTION

The Dead Sea hypersaline water system is unique in terms of its unusual geochemical composition, rapid lake level changes and water composition of the brines discharging along its shoreline. The lake level has been dropping since the 1930's, reaching a rate of 1 m/yr in recent years. This rapid lake level drop affects the groundwater flow and its discharge to the lake. Previous studies showed that saline water circulates in the aquifer also during the lake level drop (Kiro et al. 2008).

The Dead Sea is highly enriched with ^{226}Ra (Stiller & Chung 1984) and ^{228}Ra (Somayajulu and Rengarajan, 1987). The present study focuses on the fate of the various radium isotopes during lake water circulation in the aquifer.

The Dead Sea is a hypersaline lake with a water salinity of 340 g/L. The activity of ^{226}Ra is ~140 dpm/L, and that of ^{228}Ra is ~1.5 dpm/L, which is more than 400 times and 100 times their activity in the ocean, respectively. The main sources of ^{226}Ra and ^{228}Ra to this terminal lake are brines (remnants from ancient lakes) discharging along the lake shoreline, with typical activities of 400–500 and 2–20 dpm/L, respectively (Moise et al. 2000).

The continuous lake level drop of the Dead Sea provides a unique opportunity to study the interaction between hypersaline brines and the aquifer in this extremely dynamic system. We studied the behavior of radium isotopes in groundwater whose salinity ranges between fresh water and the Dead Sea salinity.

Radium isotopes are commonly used as tracers for submarine groundwater discharge (e.g. Moore, 1996) and for estimating adsorption distribution coefficients (Krishnaswami et al. 1982). The possibility to study the Dead Sea water after it enters the aquifer and mixes with fresh groundwater offers a natural case, which demonstrates the effect of salinity on the activity of radium.

2 RADIUM IN THE GROUNDWATER

Recirculated Dead Sea water found in the aquifer adjacent to the lake shore carry reduced ^{226}Ra activities (60 dpm/L, Fig. 1) compared with the lake. Groundwater with lower salinities show mixing between the above 60 dpm/L Dead Sea water and Ra-free fresh groundwater (Fig. 1).

Unlike the ^{226}Ra, the short-lived radium isotopes ^{224}Ra and ^{223}Ra are strongly enriched in the aquifer

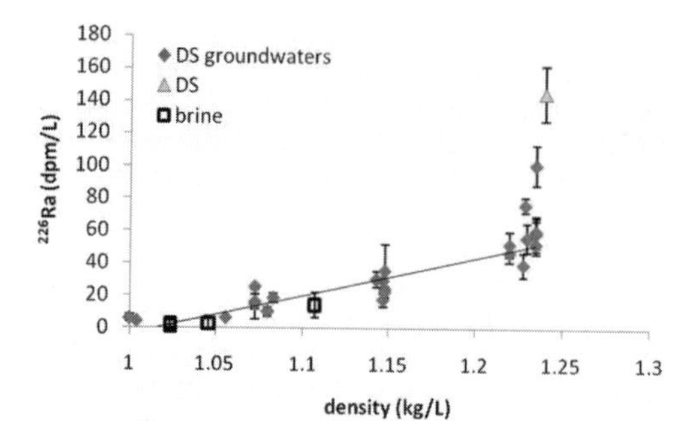

Figure 1. ^{226}Ra activities in groundwater near the Dead Sea. ^{226}Ra decreases from 140 dpm/L in the lake water to 60 dpm/L after the water enters the aquifer at a distance of less than 50 m from the shore.

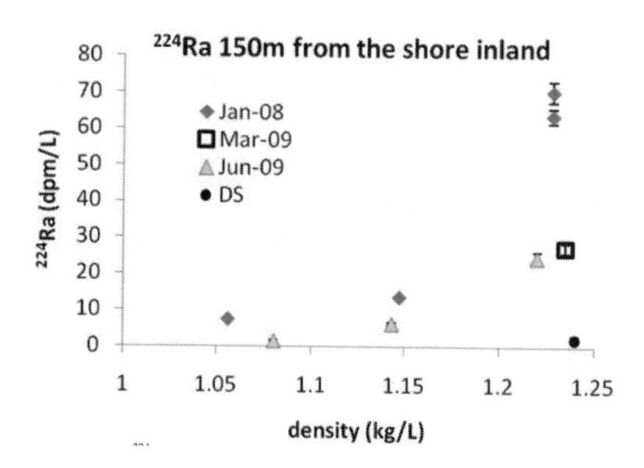

Figure 2. ^{224}Ra activities in the Dead Sea groundwater. The source of ^{224}Ra is the aquifer sediments and the relation between the salinity and ^{224}Ra activity demonstrates the effect of the salinity on the adsorption of radium.

compared with the lake (up to 70 and 35 dpm/L and up to 4 and 0.7, respectively, in groundwater and in the lake). Also here, the activities of the short-lived radium isotopes decrease with decreasing groundwater salinity, though in a nonlinear relation (Fig. 2).

Considering the dynamic conditions of the Dead Sea level drop, an approach of studying radium isotopes in a dynamic flow field was taken. Analysis and simulations with SUTRA-MS (Hughes & Sanford 2005) were performed in order to describe the processes controlling radium distribution in the Dead Sea aquifer.

The large decrease of ^{226}Ra on entering the aquifer cannot be explained by adsorption because of the high salinity of the water. This is also supported by the simulations, which show that a large adsorption distribution coefficient does not cause a response of the ^{226}Ra distribution, and thus cannot fit the field data. As with solute transport, where a large distribution coefficient causes a slow

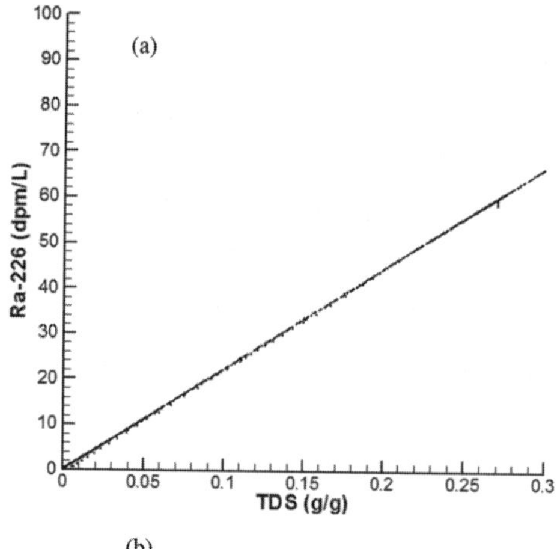

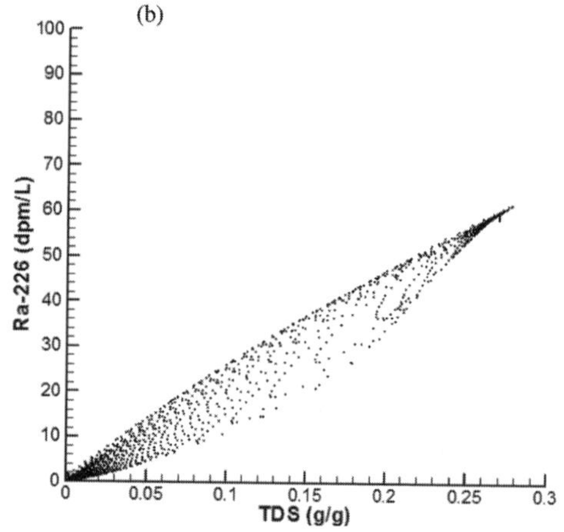

Figure 3. Simulation results of ^{226}Ra vs. salinity. (a) in steady-state and (b) after 3 years of a lake level drop. In steady-state the relation is linear and during the drop of the lake level, the relation changes with time.

response (retardation) to changes, the same holds for the response of radium to the Dead Sea lake level drop. A large distribution coefficient results in no response of ^{226}Ra which cannot account for the observed decrease of ^{226}Ra in diluted water.

Therefore, it is suggested that the removal of radium on entering the aquifer is caused by coprecipitation with barium in barite. The Dead Sea water is supersaturated with barite (by a factor of 15) and on entering the aquifer it encounters abundant nuclei, which favors crystallization. The crystallization of barite is indicated by the significantly reduced barium concentrations in the circulated Dead Sea water compared with the lake water (5 mg/L and 1.5 mg/L, respectively).

According to the simulations, the relation between ^{226}Ra activity and salinity over time changes during the lake level drop. The relation in steady-state is linear and it becomes non-linear during the lake level drop, which continues to change over time (Fig. 3). On the other hand, the relation between the salinity and the activity of short-lived radium isotopes is constant in time and is not affected by the Dead Sea level drop. The simulations show a relatively good fit with field data and may allow a good estimation of the radium adsorption distribution coefficient. The ^{226}Ra constrains the distribution adsorption coefficient to less than 6 m^3/kg in the fresh water and the short-lived radium isotopes define the relation between the salinity and the adsorption of radium.

3 RADIUM MASS BALANCES IN THE DEAD SEA—IMPLICATIONS ON LARGE-SCALE SALINE WATER CIRCULATION

The high activities of the relatively short-lived ^{228}Ra in the lake (Somayajulu & Rengarajan 1987, Stiller & Chung 1984) and the discrepancy between ^{226}Ra activity in the lake and its known sources imply that the circulation of Dead Sea water in the aquifer plays a major role in the Dead Sea radium mass balances.

The Dead Sea water loses ^{226}Ra after it enters the aquifer due to precipitation of barite and gains ^{228}Ra due to recoil and production (Fig. 4). Therefore, a mechanism of saline water circulation removes ^{226}Ra from the Dead Sea and contributes ^{228}Ra without changing the water mass balance.

Preliminary mass balance calculations using water fluxes from Lensky et al. (2005) show that the saline water circulation is around 1000 million m^3/yr which is three times greater than the whole water discharge to the lake. Additional sampling of the different sources is planned, in both sides of the Dead Sea, to re-examine these calculations.

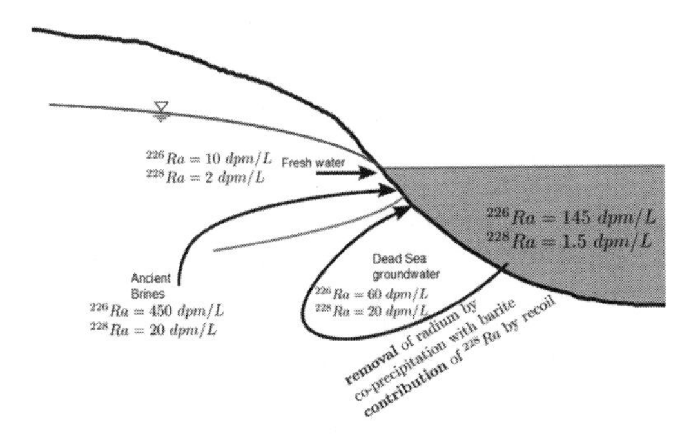

Figure 4. A conceptual model of the Dead Sea Lake-groundwater interaction that controls the radium distribution in the aquifer and affects the Dead Sea radium mass balance.

4 CONCLUSIONS

1. ^{228}Ra activities in the Dead Sea are higher than expected based on the conventional mass balance, suggesting an additional source of ^{228}Ra to the Dead Sea.
2. ^{226}Ra in circulated Dead Sea water is low compared with the lake, suggesting the removal of ^{226}Ra from the Dead Sea via precipitation of barite.
3. A mechanism of saline water circulation can explain the needed additional source of ^{228}Ra and the removal of ^{226}Ra from the Dead Sea Lake.

REFERENCES

Hughes, J.D. & Sanford, W.E. 2005. SUTRA-MS a version of SUTRA modified to simulate heat and multiple-solute transport. *U.S. Geological Survey, Open-File Report* 2004–1207, p. 141.

Kiro, Y., Yechilei, Y., Lyakhovsky, V., Shalev, E. & Starinsky, A. 2008. Time response of the water table and saltwater transition zone to a base level drop. *Water Resources Research* 44, doi: 10.1029/2007 WR006752.

Krishnaswami, S., Graustein, W.C., Turekian, K.K. & Dowd, J.F. 1982 Radium, Thorium and Radioactive Lead Isotopes in Groundwaters—Application to the Insitu Determination of Adsorption-Desorption Rate Constants and Retardation Factors. *Water Resources Research* 18: 1663–1675.

Lensky, N., Dvorkin, Y., Lyakhovsky, V., Gertman, I. & Gavrieli, I. 2005. Water, salt, and energy balances of the Dead Sea. *Water Resources Research* 41, doi: 10.1029/2005 WR004084.

Moise, T., Starinsky, A., Katz, A. & Kolodny, Y. 2000. Ra isotopes and Rn in brines and ground waters of the Jordan-Dead Sea Rift Valley: Enrichment, retardation, and mixing. *Geochimica et Cosmochimica Acta* 64: 2371–2388.

Moore, W.S. 1996. Large groundwater inputs to coastal waters revealed by ^{226}Ra enrichments. *Nature* 380: 612–614.

Somayajulu, B.L.K. & Rengarajan, R., 1987, Ra-228 in the Dead-Sea. *Earth and Planetary Science Letters* 85: 54–58.

Stiller, M. & Chung, Y.C. 1984. Radium in the Dead Sea—a possible tracer for the duration of meromixis. *Limnology and Oceanography* 29: 574–586.

Water-Rock Interaction – Birkle & Torres-Alvarado (eds)
© 2010 Taylor & Francis Group, London, ISBN 978-0-415-60426-0

Evaluation of seawater intrusion using Sr isotopes: An example from Ensenada, B.C., México

K.M. Lara & B. Weber
Departamento de Geología, Centro de Investigación Científica y de Educación Superior de Ensenada (CICESE), Ensenada, Baja California, México

ABSTRACT: To evaluate seawater intrusion into a coastal aquifer Sr isotopes and concentrations of water samples from wells of the Maneadero valley (Ensenada, Baja California) and from recharge feeder were analyzed. The NE feeder has $^{87}Sr/^{86}Sr$ of ~0.7076 that is significantly lower than seawater Sr (~0.7092). The SE feeder has $^{87}Sr/^{86}Sr$ as low as ~0.7063 but higher Sr concentrations. Water samples from the wells have $^{87}Sr/^{86}Sr$ ranging between the values of these two recharging creeks. Samples from wells close to the coast have the highest salinities and Sr concentrations, respectively, but none of them has $^{87}Sr/^{86}Sr$ higher than the NE feeder Sr. Mixing models suggest only minor amount of seawater is added to the aquifer. Hence, high salinities of water from some wells cannot be explained by seawater intrusion only. We suggest increasing salinity by evaporation and salt recycling from irrigation as a reasonable process to explain high salinities.

1 INTRODUCTION

Scientists become increasingly concerned about the effects of seawater intrusion on coastal aquifers. Due to β^- decay of ^{87}Rb to ^{87}Sr the $^{87}Sr/^{86}Sr$ depends on the Rb/Sr ratio and the age of a given rock or mineral. The $^{87}Sr/^{86}Sr$ of present-day oceans is constant around the world (0.70918 ± 0.00001; Faure & Mensing 2005) due to the relatively long residence time (5.0×10^6 years) of strontium in seawater and its complete homogenization within about 10^3 years. Considering that the Sr concentration of seawater (7.74 ppm) is about 10–100 times greater than that of average river water, mixing of seawater with continental water (by assuming the Sr isotopic composition of river water being sufficiently different from seawater) will cause an immediate response not only on the Sr concentration but also on its isotopic composition.

On the other hand, if salinity of water within a given aquifer rises due to reasons other than seawater intrusion (like evaporation and salt recycling from irrigation, Milnes & Renard 2004) then the Sr isotopic composition in the aquifer will remain unaffected. In samples from coastal aquifers that show relatively high salinity the analysis of element concentrations alone cannot always distinguish between increasing salinity that comes from intruded seawater or from increasing salinity due to evaporation at the surface and salt recycling. Strontium isotopic composition modeling, instead, is a useful tool not only for quantifying the amount of seawater that intruded the coastal aquifer but also to determine and to quantify sources of recharge (e.g. Jørgensen et al. 2008, Langman & Ellis 2010).

2 REGIONAL BACKGROUND

Coastal aquifers provide fresh water all over the world, especially in semiarid and arid zones. Due to their proximity to the ocean, coastal aquifers are highly sensitive to disturbances. Sometimes, when a coastal aquifer is overexploited, the phreatic level lowers below the sea level and, in order to restore the hydrologic balance, marine water moves towards the zones occupied previously by fresh water.

According to the CNA (for the Mexican National Water Commission), there are 17 coastal aquifers with seawater intrusion problems known in Mexico. The Maneadero aquifer, that provides fresh water to the city of Ensenada, Baja California, is one of those aquifers, having problems of decreasing water quality due to rising salinity of water in the wells. The Maneadero aquifer supplies about 70% of the water used to sustain extensive agriculture in the Maneadero Valley. However, recharge rates have not kept pace with extraction and they have generally been less than the total discharge of the aquifer (i.e. extraction, evaporation and losses to the ocean). Since 1968, the year after which extraction increased significantly, the discharge rates are higher than recharge rates (Daesslé et al. 2004).

Geophysical studies (e.g. Lujan 2006 and references therein) have shown increasing salinity of the Maneadero aquifer over past decades, which

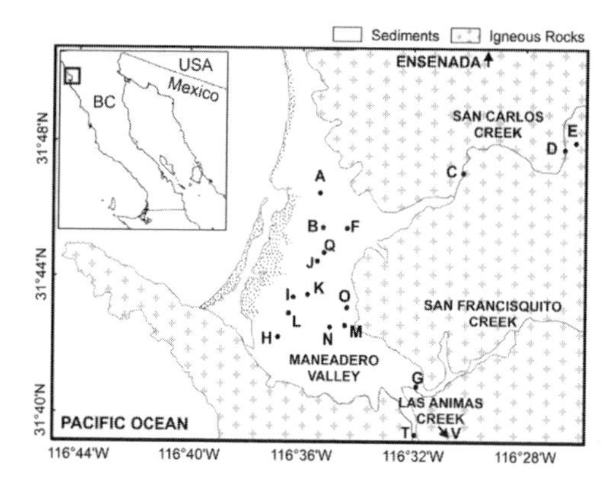

Figure 1. Simplified geologic map of the Maneadero area showing sample sites of wells in the sedimentary basin (white) and from creeks. Igneous rocks (cross filling) are Cretaceous volcanic rocks and granite. Note: Dotted area around estuary is partially flooded when tide is in.

was interpreted in terms of progressive seawater intrusion. The hydrogeochemical evolution of the Maneadero aquifer was studied by Daesslé et al. (2004) during a season that was dryer than average (from 2001 to 2002). These authors concluded that the rise in TDS levels (total dissolved solids) through time indicates the vulnerability of the aquifer to seawater intrusion. The comparison of TDS data from different wells revealed that the areas most affected by seawater intrusion were those close to the coast and in the center of the aquifer. Furthermore, the authors suggested a progression of seawater intrusion towards the wells into San Carlos creek (Fig. 1).

In this approach to evaluate and to quantify possible seawater intrusion into the Maneadero aquifer, we applied $^{87}Sr/^{86}Sr$ ratios and Sr concentrations obtained by isotope dilution analysis.

3 ANALYTICAL PROCEDURE

Water samples were collected from wells throughout the Maneadero Valley (Fig. 1) and from recharge feeder creeks, as well as from a thermal spring (D) at the end of the wet and dry seasons in March and October 2009, respectively. Conductivity was measured on-site, as a first estimation of aquifer salinity. Water samples were collected in pre-cleaned polyethylene bottles, filtered, and gently acidified with a few mL of double distilled HNO_3. In order to characterize the chemical composition of the samples and to obtain an estimate of the Sr content to enhance spiking, cation contents were measured with a ICP-OES Liberty 110 at CICESE.

Either 10, 20, or 50 ml of each water sample were weighed, spiked with a Sr tracer (99.89% ^{84}Sr), and evaporated to dryness. To remove any organic matter, 5–10 drops H_2O_2 were added to the samples and evaporated again. Then, the samples were dissolved in double distilled 8N HNO_3 and loaded onto ion exchange columns packed with ~300 µl of Sr-Spec® resin. Major and interfering elements were eluted first with 8M, then with 3M, and finally with 0.3M HNO_3. The final Sr cut was collected with ultrapure water and dried down in small Teflon® beakers. Chemical and chromatographic procedures were carried out in class 100 cleanlab facilities at the Geology Department, CICESE, Ensenada. Total procedure blanks were typically 0.1 ng/g Sr. Isotope ratios were analyzed with a Finnigan MAT 262 multicollector mass spectrometer at the Geophysics Institute, UNAM, México City by measuring simultaneously 88, 87, 86, 85 (for Rb monitor), and 84 masses on faraday collectors in static mode. Six blocks of ten 16s integrations each block were measured per run. $^{87}Sr/^{86}Sr$ were corrected for mass fractionation by normalizing to $^{86}Sr/^{88}Sr = 0.1194$. Values were adjusted by a –0.0085% correction factor obtained from repeated measurements of the NIST987 standard.

4 RESULTS

Measured conductivities, Sr concentrations obtained by isotope dilution, and isotope ratios of samples from sites shown in Figure 1 are listed in Table 1 and displayed in Figure 2 together with calculated hypothetical mixing lines. Conductivities and major element concentrations (Ca, Na, Mg, K; not shown) of all samples show linear correlations with respect to the Sr concentrations, indicating that Sr concentrations behave proportional to salinity.

San Carlos Creek, which carries water around the year, is the northeastern recharge feeder for the Maneadero aquifer (C, E; Fig.1). It has the highest $^{87}Sr/^{86}Sr$ (0.70757–0.70761) and moderate Sr concentrations (0.3–0.6 ppm). One sample collected from the San Carlos thermal spring (D) has the lowest $^{87}Sr/^{86}Sr$ (0.70527) and Sr concentration (0.08 ppm) of all analyzed samples, suggesting that thermal springs, which are linked to major active faults, have a different origin. Samples collected from the SE recharge feeder (San Francisquito Creek, Fig. 1) have $^{87}Sr/^{86}Sr$ of 0.70626 (G, wet) and 0.70659 (G, dry) with significantly higher Sr concentrations of ~1.8 ppm. Two samples from the southern feeder (Las Animas Creek) that were recently taken after the 2010 raining season plot close to a mixing line calculated from NE and SE feeder creek data (Fig. 2).

Table 1. $^{87}Sr/^{86}Sr$, Sr concentrations and conductivies of water samples from Maneadero Valley and adjacent creeks.

Sample		Conduct.	Sr		std. err
Site	Date	mS	ppm	$^{87}Sr/^{86}Sr$	$2\sigma_{(m)}$
A	Mar-09	4.51	1.66	0.707544	13
A	Oct-09	5.60	1.74	0.707519	11
B	Mar-09	1.70	0.56	0.707324	15
B	Oct-09	1.89	0.57	0.707347	15
C	Mar-09	1.62	0.49	0.707578	10
C	Oct-09	1.98	0.54	0.707574	07
C1*	Oct-09	1.98	0.60	0.707613	09
D	Mar-09	0.81	0.08	0.705273	12
E	Mar-09	1.47	0.45	0.707598	10
E	Oct-09	1.44	0.33	0.707565	08
F	Mar-09	1.52	0.46	0.707523	10
G	Mar-09	4.38	1.82	0.706259	11
G	Oct-09	5.44	1.76	0.706590	10
H	Oct-09	10.3	2.95	0.706810	09
I	Oct-09	45.9	9.54	0.707284	10
J	Oct-09	7.33	1.83	0.707367	10
K	Oct-09	7.25	2.21	0.706709	14
L	Oct-09	9.23	2.71	0.706751	09
M	Oct-09	2.72	0.71	0.706877	28
N	Oct-09	3.51	1.00	0.706756	11
O	Oct-09	3.71	1.00	0.706735	10
Q	Apr-09	7.42	2.49	0.707526	08
Q	Oct-09	9.19	2.53	0.707501	13
T	Mar-10	2.31	0.73	0.706996	09
V	Mar-10	2.43	0.82	0.706961	15

* Sample site C1 close to sample site C, on Fig. 1.

In the northern section of the valley, downstream from San Carlos Creek, samples A, F, and Q have similar Sr isotopic composition to San Carlos Creek (C, E). Sample A and Q, however, have significantly higher Sr concentration with respect to San Carlos Creek. Samples B and J both yielded $^{87}Sr/^{86}Sr$ slightly below NE feeder values (0.70732–0.70737), of which sample J has a significantly higher salinity and Sr concentration. Another 7 samples were taken from wells within the southern part of Maneadero Valley (Fig. 1). Samples M, N, and O are from wells at a greater distance from the coast. They have relatively low Sr concentrations and $^{87}Sr/^{86}Sr$ between 0.70674 and 0.70688. The results of the latter samples plot again close to the NE–SE recharge feeder mixing line (Fig. 2). Samples H, I, K, and L that were taken from wells relatively close to the coast within the main farmed zone of the Maneadero Valley have the highest salinities and Sr concentrations, respectively, ranging from 2.2 to 9.5 ppm Sr (note: average seawater has 7.74 ppm Sr). However, all samples have $^{87}Sr/^{86}Sr$ ratios between average NE and SE recharge feeder creek

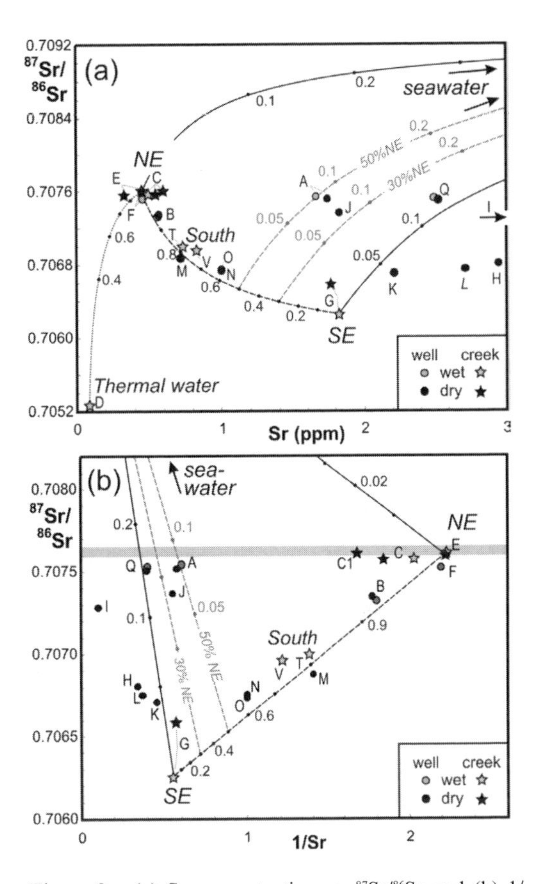

Figure 2. (a) Sr concentration vs. $^{87}Sr/^{86}Sr$ and (b) 1/Sr vs. $^{87}Sr/^{86}Sr$ diagrams. Stars are water samples from feeder creeks; dots are from wells within Maneadero Valley (see Fig. 1). Grey symbols are samples taken in the dry season, black symbols in the wet season (Note: From some sites samples of both seasons were analyzed.)—Mixing hyperbolas (a) and mixing lines (b) were calculated for mixtures of present-day seawater ($^{87}Sr/^{86}Sr$ = 0.70918, 7.74 ppm Sr; Faure & Mensing 2005; solid black lines) with (1) NE feeder (San Carlos Creek) and (2) with SE feeder (San Francisquito Creek) using sample site G in dry season, which lower $^{87}Sr/^{86}Sr$ (0.70626) and 1.8 ppm Sr (Note: Seawater composition lies off the diagrams).—Samples from the southern feeder creek, Las Animas (T, V), were taken after heavy rains in spring 2010. Using NE and SE feeder compositions as endmembers, the latter samples plot close to a mixing line (dashed black line) with ~80% NE and ~20% SE.—In addition, black dotted hyperbola illustrates theoretical mixing of San Carlos thermal water (only in Fig. 2a). Grey dashed mixing lines were calculated for mixtures of SE feeder with 30% NE and 50% NE feeder, respectively, which is additionally mixed with seawater (three components).—Heavy grey line (only in Fig. 2b) marks the $^{87}Sr/^{86}Sr$ value of the higher NE feeder, illustrating that downstream samples and wells have similar isotope ratios with higher Sr concentrations and none of the samples has $^{87}Sr/^{86}Sr$ above the grey line.

data. Samples H, K, and L (0.70671–0.70681) are indistinguishable from mixed samples (M, N, O); the extremely salty (46 mS/ 9.5 ppm Sr) sample "I" has a higher $^{87}Sr/^{86}Sr$ ratio of 0.70728.

5 DISCUSSION AND CONCLUSIONS

In order to evaluate possible seawater intrusion into the aquifer of the Maneadero Valley, we calculated mixing models for seawater and the different potential source compositions analyzed in this study (Fig. 2). None of the analyzed samples from the wells within Maneadero Valley have higher $^{87}Sr/^{86}Sr$ ratios than those from the NE recharge feeder (San Carlos Creek). In the northern section of the Maneadero aquifer, where it can be assumed that San Carlos Creek is the only water source, no seawater intrusion is indicated by our data.

Most of the samples from the southern part of the valley have significantly lower isotope ratios, indicating another source of water mixed with water from the NE feeder (San Carlos Creek). Another minor water source for the Maneadero aquifer lies at its SE edge. Our only sample site from this area has a lower isotope ratio but a higher Sr concentration with respect to San Carlos Creek. A mixture of both recharge feeder creek waters is indicated for several water samples from wells in the southern part of the aquifer away from the coast, where seawater intrusion is not indicated. Such mixtures of different source waters complicate the evaluation for Maneadero Valley in terms of adequate mixing models, as at least three component mixing needs to be considered.

In addition to mixing lines between the different endmembers, theoretical mixing lines of different mixtures of the NE and the SE recharge feeder creek waters with seawater (three component models) are shown in Figure 2. For some samples, such three component mixtures seem to be a reasonable solution, especially for sample J whose composition can be explained by adding ~8% of seawater to a mixture of about 35% SE and 65% NE recharge feeder creek water. However, such a mixing model is unlikely for samples from site A, which is located north of San Carlos Creek.

Contamination of the coastal aquifer with seawater can be estimated for the data from those wells that lie closer to the cost (H, L, K, I, Q), only if a single water source (i.e., from the SE recharge feeder creek) is assumed. In such a scenario samples H, L, and K may contain ~5% of seawater, whereas samples I and Q may contain more than 10% of seawater (Fig. 2). Considering the intermediate composition obtained from the southern feeder (Las Animas Creek) a three-component mixing should be assumed. Then, seawater contamination must be significantly lower. In view of their relatively high salinities and Sr contents another process than seawater intrusion has to be considered to explain increasing salinity in the wells of the Maneadero Valley.

On the basis of our preliminary data we conclude that seawater intrusion into the coastal aquifer of the Maneadero Valley is not the main reason for increasing salinity of water in the sampled wells. Instead, it is more likely that salinity increases by evaporation at the surface and salt recycling from irrigation (Milnes & Renard 2004), which in turn contaminates the water of the aquifer with Sr having the same isotopic composition as the water from the aquifer itself. Contamination with a minor seawater component is possible only in those wells that are located close to the coast. This interpretation is further confirmed by negative $\delta^{18}O$ (–5.2% to –6.9%) and δD (–31% to –42%) values that lie between the SE and NE feeder values for all except one (I) samples analyzed from wells.

ACKNOWLEDGEMENTS

This study was supported by CICESE internal project 644131. We are grateful to Gabriela Solís Pichardo and Peter Schaaf (both UNAM) for supporting isotope analysis. We want to thank Thomas Kretzschmar for his continuous assessment, Mario Vega for ICP analysis and Gabriel Rendón for lab assistance, all at CICESE. We also thank to Thomas Bullen (USGS) for his comments and Akira Ueda (Kyoto) for review. Thanks to Billy, Ismael, Roberto, and Sarah for their help during fieldwork.

REFERENCES

Daesslé, W. Sanchez, E.C., Camacho-Ibar, V.F., Mendoza-Espinosa, L.G., Carriquiry, J.D., Macias, V.A. & Castro, P.G. 2005. Geochemical evolution of groundwater in the Maneadero coastal aquifer during a dry year in Baja California, Mexico. *Hydrogeology Journal* 13: 584–595.

Faure, G. & Mensing, T.M. 2005. *Isotopes: Principles and Applications* (Third Edition), John Wiley & Sons, Inc.

Jørgensen, N.O., Andersen, M.S. & Engesgaard, P. 2008. Investigation of a dynamic seawater intrusion event using strontium isotopes ($^{87}Sr/^{86}Sr$). *Journal of Hydrology* 348: 257–269.

Langman, J.B. & Ellis, A.S. 2010. A multi-isotope (δD, $\delta^{18}O$, $^{87}Sr/^{86}Sr$, and $\delta^{11}B$) approach for identifying saltwater intrusion and resolving groundwater evolution along the Western Caprock Escarpment of the Southern High Plains, New Mexico. *Applied Geochemistry* 25: 159–174.

Lujan, B. 2006. *Utilización de ondas electromagnéticas para detectar la invasión de agua marina en el acuífero del Valle de Maneadero en Ensenada, BC.* Tesis MC, CICESE, Ensenada BC, México.

Milnes, E. & Renard, P. 2004. The problem of salt recycling and seawater intrusion. in coastal irrigated plains: an example from the Kiti aquifer (Southern Cyprus). *Journal of Hydrology* 288: 327–343.

Water-Rock Interaction – Birkle & Torres-Alvarado (eds)
© 2010 Taylor & Francis Group, London, ISBN 978-0-415-60426-0

Geochemistry of groundwater and its flow system in the Osaka Basin, Japan

H. Masuda, K. Makino, K. Matsui, T. Okabayashi, H. Yoshioka & Y. Kajikawa
Osaka City University, Osaka, Japan

U. Tsunogai
Hokkaido University, Sapporo, Japan

ABSTRACT: Groundwaters were geochemically characterized to determine the flow system in the Osaka Basin, Japan. The aquifers beneath the center of the basin can be divided into three groups: (1) unconfined and uppermost confined aquifers, <50 m depth, (2) confined aquifers between 50 and ~600 m depth, where the pore water would be squeezed from the intercalated less saline permeable layers, and (3) aquifers >600 m depth down to the basement rocks, in which highly saline waters probably having hydrothermal origin are occasionally found. Seawater younger than 20 years infiltrates into the aquifers <100 m depth in the western part of the basin, where the ground surface is below sea level. Anthropogenic pollutants remain more than 20 years even in the aquifer >100 m where the aquifer structure provide stagnant conditions.

1 INTRODUCTION

In Japan, until the middle 1970s, groundwater was extensively used as a water resource for industrial and agricultural purposes, causing serious subsidence in many Quaternary plains where the large cities are located. The Osaka Basin (Fig. 1) is the one such large Quaternary sedimentary basin. Depth to basement reaches >1500 m, and high quality groundwater aquifers are hosted in the basin sediments. Similar to the other sedimentary basins in the country, groundwater use from the unconsolidated aquifers has been strictly controlled to prohibit subsidence since the mid-1970s. However, recently increasing uptake of groundwater from various depths, sometimes using techniques beyond regulatory control, has arisen again with worrying associated hazards including ground subsidence and aquifer depletion. On the contrast, since the groundwaters in the unconfined and uppermost confined aquifers <50 m depth have not been used due to pollution, the excessive hydropressure of aquifers would cause other disasters such as floating of construction and liquefaction when earthquake occurs. In order to make mitigation plans, we have to know the whole system of groundwater flow in the basin.

In this paper, we document groundwater flow estimated from the geochemical features, including stable isotopes of groundwaters taken from the aquifers at various depths.

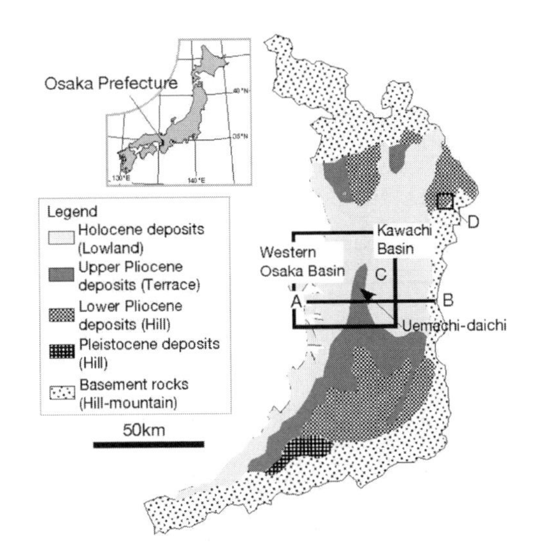

Figure 1. Surface geology of Osaka Prefecture, Japan (modified after Itihara 1991). Line A–B shows the vertical profile of geology and major element chemistry in Figure 5. Box C shows the area with topography and major element chemistry of groundwaters <300 m depth in Figure 4. Box D is the area studied for nitrate N and O isotopes of groundwaters.

2 GEOLOGICAL BACKGROUND

Figure 1 is a map of the surface geology of Osaka Prefecture, and typical depth profiles of geology beneath the center of the basin with the categories of groundwater aquifer as shown in Figure 2.

As shown in Figure 1, the Osaka Basin comprises two sub-basins; the western Osaka Basin and the Kawachi Basin, divided by Uemachi-daichi, a terrace uplifted by faulting. The basement rocks, comprising Cretaceous granitic rocks and Mesozoic-Paleozoic sedimentary formations, are exposed in the surrounding mountains. Plio-Pleistocene sedimentary rocks, the Osaka Group, cover the basement rocks and are exposed on the hilly areas surrounding the lowland plain, where the cover comprises thin Holocene sediments.

The Osaka Group may be roughly divided into two units; an upper formation comprising intercalated marine clays and non-marine sandy layers, and a lower formation comprising freshwater sediments.

Groundwater aquifers can be divided into three units (Fig. 2): 1) shallow aquifers that include unconfined and uppermost confined ones; 2) deep aquifers of the Upper Osaka Group between 50 and ~ 600 m depth at the center of the basin, which become shallower towards the surrounding area; and 3) deepest aquifers in the Lower Osaka Group at >600 m including shallow waters recovered from the basement rocks in the surrounding mountainous area. Groundwaters from the first unit are mostly used for gardening in suburban areas and for religious purpose in temples and monitoring of groundwater level in the urban area. Drinking water and water for washing and emergency purposes for industrial and large office buildings comes mainly from the second unit. The third unit supplies water mainly for bathing purposes.

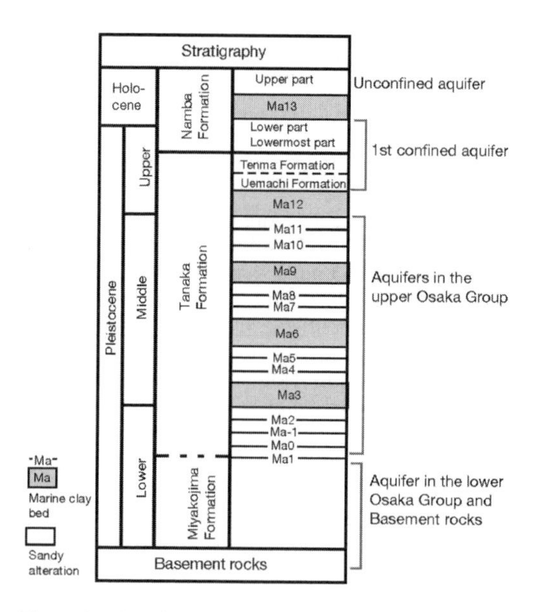

Figure 2. Stratigraphy beneath the centre of the Osaka Basin. Depth of the base of the sedimentary formation is about 1500 m.

3 METHODS

Groundwaters were taken from the wells installed at different depths. Water temperature, EC, pH, ORP, DO and alkalinity were measured in the field, and the water was filtered and sampled in separate plastic and glass bottles for the different analyses. Hydrochloric acid was added for cation analyses. Major chemical composition, hydrogen and oxygen isotopes of water, nitrogen and oxygen isotopes of nitrate ion, Si, Fe, Mn, As, Hg, and VOCs were analyzed in the laboratory. In this paper, we mainly report the analytical results of major element chemistry, stable isotope characteristics, and VOCs.

4 RESULTS

4.1 *Major element chemistry*

Groundwaters from the shallow aquifers are mostly Ca-HCO_3 types in the recharge zone areas; e.g., Uemachi-daichi is the recharge zone at the centre of the plain, and dilute Ca-HCO_3 type groundwater is dominant in the <10 m well waters (Fig. 3). This type of groundwater occurs in the shallow wells (<10 m) in the terraces and hills surrounding the lowland plain.

Two groundwaters from the shallow aquifers located on the eastern side of Uemachi-daichi are

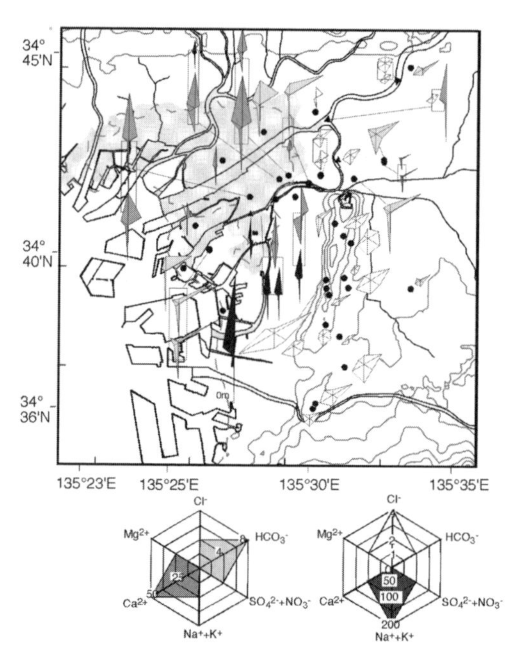

Figure 3. Major chemistry of groundwaters <300 m depths in and around Osaka City, indicated by box C in Figure 1, at the center of the Osaka Basin. The shaded area in the map is below sea level. Grey scales on the map correspond to the scale of concentrations of hexadiagram legends. Part of the data is from Makino et al. (in press).

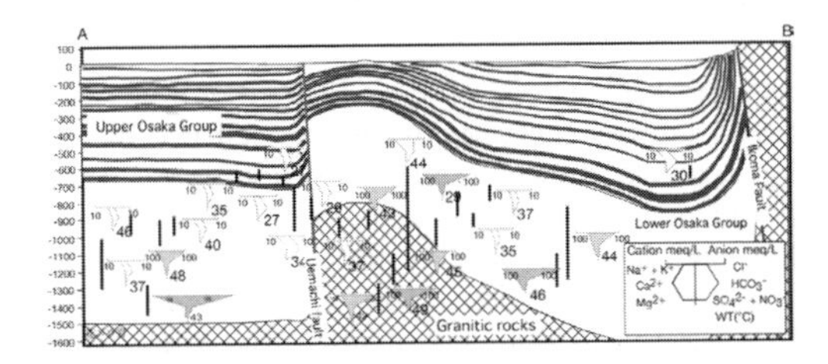

Figure 4. Depth profile of geology beneath the Osaka Basin along the line A–B in Figure 1 and major element chemistry of deepest groundwaters used for hot spas. The Upper Osaka Group comprises intercalated freshwater sediments and marine clay layers indicated by black lines. The Lower Osaka Group comprises freshwater sediments. The bars attached to the hexadiagrams give the depth of the screens of each well. (Geological profile after Uchiyama et al. 2001).

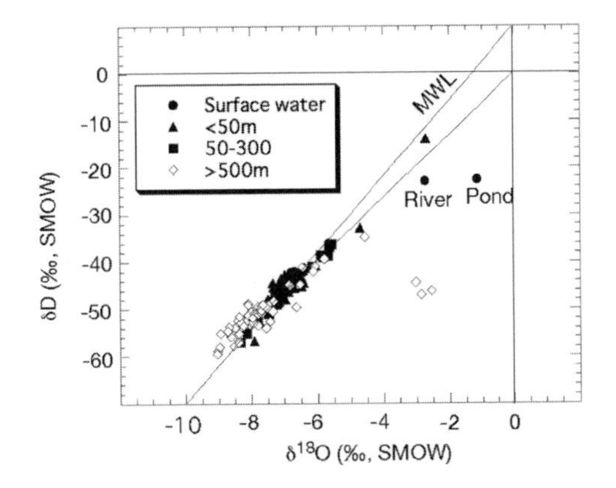

Figure 5. Relationship between oxygen and hydrogen isotope ratios of groundwaters and surface waters in the Osaka Basin. The groundwaters >500 m include the spring waters issuing from the exposed basement rocks in the mountain areas north and south of Osaka Prefecture. Part of the data is from Makino et al. (in press).

Na-HCO_3 type. Only two groundwaters from this area were studied, although similar groundwater chemistry is common in the Kawachi Basin (Nakaya et al. 2009), indicating stagnant condition of the aquifers. It is notable that the groundwaters at depths <100 m in the western basin have concentrated Na-Cl type chemistry (Fig. 3), suggesting the infiltration of seawater.

Dilute Na-HCO_3 chemistry is dominant among the groundwaters taken from the sedimentary formations >100 m depths. In Figure 4, the major chemistry of groundwaters used for hot spas is shown with the screen depth of the wells with the depth profile of the geology along the observation line A-B in Figure 1. It is notable that saline groundwaters are occasionally obtained from the base of the sedimentary strata and the underlying basement granitic rocks. Some of those, especially from the coastal area, would be derived seawater infiltration, while the others would be hydrothermal origin, as will be discussed later.

4.2 *Oxygen and hydrogen isotopes of water*

Figure 5 shows the O and H isotope ratios of studied groundwaters and river waters. Most of the groundwaters are on the local meteoric water line (MWL in Fig. 5), however, some of those are offset from the meteoric water line; the ones on the mixing line of seawater and local meteoric water, and the others show O isotope shift. The stable isotopes of river waters sampled on high tide are high, indicating mixing with upcoming seawater.

Shift of O isotope ratios is observed in some of the Na-Cl type groundwaters; it is prominent for the carbonated cold saline waters taken from the mountain area in the southern part of prefecture, while those are present at the bottom of the basin described above. The former has similar stable isotope ratios to those of hot brine found in Arima Hot Spring reported by Matsubaya et al. (1974). The Na-Cl type saline groundwaters at the bottom of the basin have temperatures >40 °C and O isotope shift but no CO_2. Fossil seawater has been known to occur in the deep aquifers up to a few hundred metres depths in the Kawachi Basin (e.g., Nakaya et al. 2009). However, the saline groundwaters at the bottom of the Osaka Basin are similar to those of the Arima-type brine but not fossil seawater found in the upper aquifers of the Kawachi Basin.

4.3 *Nitrogen and oxygen isotopes of nitrate in shallow groundwater*

In Katano city, located at the northeastern part of Osaka Prefecture (Box D in Fig. 1), and in the western Osaka Basin, N and O isotope ratios of nitrate were analyzed to document the source of N. Eutrophication due to anthropogenic S and N input is one of the serious problems of groundwater degradation in the world, although it is not a problem in the studied area. Concentration of nitrate-N was 2.0 mg/L on average and only two samples exceeded the WHO standard of 10 mg/L as N among 59 groundwaters, 48 from shallow wells <10 m and 11 from deep wells between 100 and 300 m depth. Stable isotope ratios of nitrate N and O were 20.8 and 6.5‰ relative to those of air, respectively. Because high nitrate contaminated wells are commonly found in cultivated areas, such heavy isotope enrichment must be caused by the application of organic fertilizers to the fields. Because the groundwaters in the lowland area of the Osaka Basin are anoxic, nitrate is rarely detected. Only group was obtained for the grounwater in the western Osaka Basin; 25.4 and 13.6‰ for N and O, respectively.

4.4 *VOCs in the groundwater*

In Japan, VOCs (volatile organic compounds) have been strictly prohibited to be discharged into waste water since 1989. Groundwaters from the western Osaka Basin are free from the VOCs (tetrachloroethylene (PCE), trichloroethylene (TCE) and dichloroethylene (DCE)) at present (Makino et al., in press). The groundwaters beneath Uemachi-daichi are also free of VOCs except one sample, in which there was detectable but not measurable DCE. In these areas, VOCs were removed in association with groundwater flow.

The VOCs are still present in groundwaters at >100 m depth in the northern and southern part of the Kawachi Basin according to the monitoring report of groundwater quality by the Osaka Prefectural Government (2008). The Kawachi Basin is in a rather stagnant condition regarding water flow, thus the VOCs would be expected there. The appearance of VOCs would be controlled by locations of the pollutant sources and underlying aquifer structures.

5 DISCUSSION

As described above, groundwaters are recharged in the terrace and hills at the center of and around the lowland in the Osaka Basin. Faults are recharging paths of confined deep aquifers beneath the plain. In Katano city, the groundwaters >200 m depth have the O and H isotope ratios similar to those of surface waters collected from the slope and foot of mountains, where active faults divides the basement rocks and aquifer sedimentary formations. At the centre of the basin, the faults that uplift basement rocks beneath the Uemachi-daichi must provide flow paths for groundwaters into the deep aquifers.

Present seawater, flowing upstream of the rivers, also recharges the groundwaters <100 m depths in the western Osaka Basin where salinization of the groundwater had been reported due to excess production until 1980s (Tsurumaki 1992). Altered seawater has been known to occur in the deep aquifers of the Kawachi Basin (e.g., Nakaya et al. 2009). However, seawater in the aquifers of the western Osaka Basin would not be older than 20 years, because the VOCs (TCE and its byproducts) were almost removed by groundwater flow.

Pore water would be squeezed from impermeable clay layers beneath the western Osaka Basin; groundwaters having Na-HCO_3 type major chemistry, containing less SiO_2 and lower stable isotope ratios than those of recharging surface waters were found from the aquifers at 100–300 m depths. It is occasionally observed that deep groundwaters used for hot spas change their major chemistry from saline Na-Cl type to dilute Na-HCO_3 type, implying that the squeezed groundwater from the less permeable layers compensates the lack of recharging water against the excess production of groundwater.

6 CONCLUSIONS

Although the groundwaters >100 m depth still contain anthropogenic pollutants discharged before 1990, the recharging water at present is not seriously polluted. Thus, proper use of groundwater is one of the mitigations to remove the pollutants from the aquifers. However, excess use of groundwaters from unconsolidated sedimentary formations, especially <300 m depths would cause subsidence, which has not occurred since middle 1970s after controlling the use of groundwater at these depths. Understanding the groundwater flow system in a whole basin is important to design a plan of groundwater use.

ACKNOWLEDGEMENTS

O and H isotope analyses were done at the Institute of Earth's Interior, Okayama University, and the Institute of Ecology, Kyoto University. VOCs were analyzed at the Faculty of Engineering, Osaka City University. Groundwater sampling and analyses using the monitoring data base were supported by the staff of the Osaka Prefectural Institute of Environment, Agriculture and Ecology. This work was financially supported by Nihon-Seimei Foundation. We thank all of them.

REFERENCES

Itihara, M. 1991. Quaternary geological map of Osaka and its surroundings—with special reference to Osaka Group. *Urban Kubota* 30 (in Japanese).

Makino, K., Masuda, H., Mitamura, M., Kanjo, Y. & Tayasu, I. (in press) Recharge sources of the groundwater in the Osaka City and the surroundings estimated from geochemical characteristics. *Journal of Groundwater Hydrology* 52 (in Japanese with English abstract).

Matsubaya, O., Sakai, H. & Tsurumaki, M. 1974. Hydrogen and oxygen isotopic ratios of thermal and mineral springs in Arima Spa. *Okayama Daigaku Onsen Kenkyusho Hokoku* 43: 15–28 (in Japanese).

Nakaya, S., Mitamura, M., Masuda, H., Uesugi, K., Motodate, Y., Kusakabe, M., IID T. & Muraoka, K. 2009. Recharge sources and flow system of groundwaters in Osaka Basin, estimated from environmental isotopes and water chemistry. *Journal of Groundwater Hydrology* 51(1): 15–41 (in Japanese with English abstract).

Tsurumaki, M. 1992. Salinitization of confined groundwater beneath Osaka Plain. *Chikasui Gijutsu (Groundwater Technology)* 34(10): 75–88 (in Japanese).

Uchiyama, E., Mitamura, M. & Yoshikawa, S. 2001. Displacement rate of the Uemachi Fault and basement block movement of the Osaka Plain. *Journal of Geological Society of Japan* 107(3): 228–236 (in Japanese with English abstract).

Water-Rock Interaction – Birkle & Torres-Alvarado (eds)
© 2010 Taylor & Francis Group, London, ISBN 978-0-415-60426-0

Stable carbon isotope ratios of natural gas seepage along normal faults in the southern Tokyo bay area (Mobara), Japan

E. Nakata, H. Suenaga, S. Tanaka & K. Nakagawa
Central Research Institute of Electric Power Industry, Japan

ABSTRACT: The stable carbon isotope ratio ($\delta^{13}C_{PDB}$) for natural gas, that is mainly composed of methane (C_1) and CO_2, was analysed between three normal faults within a total area 50 m wide, in the Pleistocene sedimentary rocks distributed around the Tokyo Bay area. Seepages were identified by borehole television survey (BTV) below a mudstone-rich layer in the upper part of a sandstone-rich sequence close to the faults. Gas samples were collected from the drill core and headspace used by the packer systems. $\delta^{13}C_{C1}$ in this area normally are −66‰. $\delta^{13}C_{C_1}$ of natural gas from the faults was about −50‰. The natural gas occurs as three types: methane, which is a light $\delta^{13}C_{C_1}$ gas distributed regionally; a biogenic gas derived by CO_2 reduction; and a heavy C_1, a thermogenic gas formed from organic matter at depth. The isotopically heavy C_1 rises along the faults.

1 INTRODUCTION

The Mobara gas field is part of the South Kanto natural gas field, which has accumulated in Pliocene–Pleistocene sedimentary rocks of the Kazusa Group, Japan. The South Kanto gas field produces 21% of the natural gas of Japan (Fig. 1). The Mobara gas field occupies 60% of the South Kanto gas field (Natori 1997).

The stable carbon isotope ratio $\delta^{13}C_{C1}$ is reported to be −66.4 ± 0.9‰, δD_{C1}–182.9 ± 4.6‰ for the natural gas produced from the Mobara area deposit (Kaneko et al. 2002). Our study was to survey the gas migration system in the shallow faults of the Mobara gas field using gas isotopes.

2 GEOLOGY OF THE GAS FIELD

The Kazusa Group, which reaches 2000–3000 m in thickness within a fore-arc basin, lies across a large area of the Tokyo bay region. The main gas-producing strata of the Kazusa Group are the Kiwada, Otadai and Umegase Formations. This formation, which represents turbidite deposition in a deep basin, consists of alternating sandstone and mudstone with thin tuff layers. The beds strike NE-SW and dip at 5–7° NW. The formation is cut by N-S trending normal faults with an easterly dip. The vertical displacements on these faults are normally 2–100 m, and the observed strike length of the faults is proportional to the displacement.

The high gas/water ratio gas produced from 200 to 600 m depth is consistent with the submarine fan sediment components of the Umegase Formation.

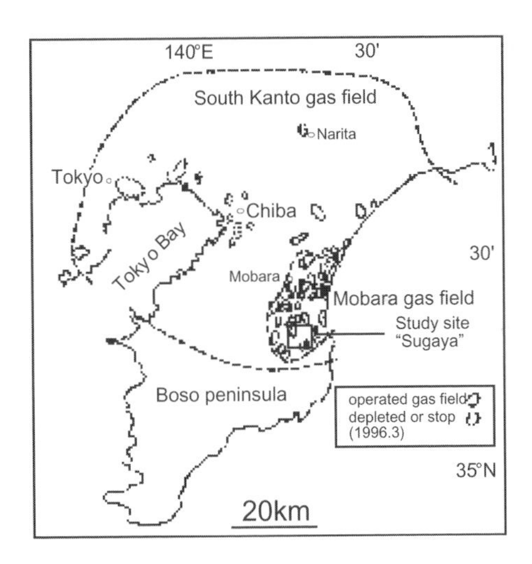

Figure 1. Location of the South Kanto gas field. Our study site centred in the Mobara gas field. Rearranged from Natori (1997).

The organic matter was supplied as a constituent of the turbidite. Gas concentrates to migrate through the sandstone layers, and this migration is restricted by the long normal faults (Kunisue et al. 2002).

3 SAMPLING

We located 12 vertical drill holes arranged 300 m apart to intersect the faults (Fig. 2). Samples having diameter of 70 mm and length of 100 mm were immediately taken from the core tube. Samples were collected in a He-filled glove bag and placed

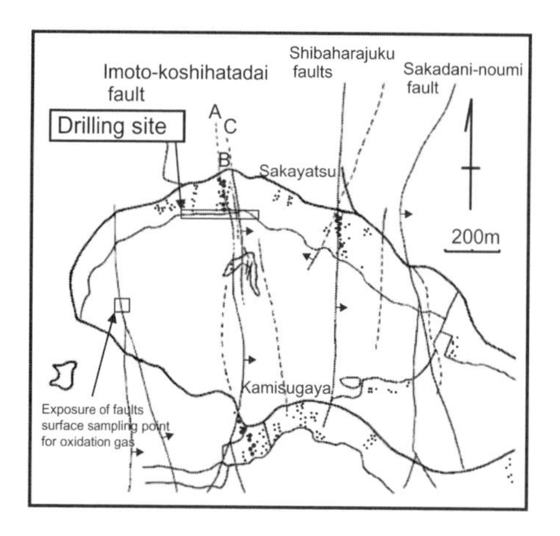

Figure 2. Distribution of gas seepage points. Dot; seepage, solid line; fault with direction of dips. A, B, C: The three target faults of this study.

in a stainless steel (SUS) container to avoid air contamination. This SUS container had septa to provide a gastight syringe. To prevent microbial reproduction, 70 cc of benzalkonium chloride 3% solution were introduced into the SUS container with each core sample. After enclosing, each sample was held for one month at room temperature to diffuse any natural gas in the He gas.

To collect the natural gas directly, we installed inflatable packers into the drill holes above the gas seepages. The seeping gas rises in a nylon tube. The gas was collected in glass containers with septa secured by water.

4 ANALYSIS

Gas analysis was performed using a GC/C/IRMS (GC: Agilent 6890 N, CT-PolaPLOT Q 30 m 30 μm, Combustion: GV instruments, Isoprime, CuO treatment at 1070°C to convert the CO_2 gas, IRMS) continuous flow type to analyse $\delta^{13}C$ (C_1; CH_4, CO_2, C_2; C_2H_6, C_3; C_3H_8, iC_4; iC_4H_{10}, and nC_4; nC_4H_{10}). The $\delta^{13}C$ error of this equipment was ±0.1‰ based on replicate analyses of our working standard.

5 RESULTS

5.1 *Geology of study site*

The Umegase Formation is composed of regularly alternating mudstone and sandstone beds of less than 1 m in thickness. Twelve volcanic ash layers 10–115 cm in thickness were found in the cores. These were labelled U6 A to U11. The mudstone-rich layer 30 m thick overlies a sandstone-rich layer that lies at GL (Ground level)–30 m depth. From the results of the core description and outcrop survey, we concluded that three normal faults were enclosed in a 50 m wide without the crush zone (Fig. 3). Displacements of the faults varied from 2 to 20 m.

5.2 *Pore size distribution and porosity*

The mudstone has a smaller pore size (0.02–0.6 μm diameter) in comparison with the sandstone (10–50 μm) and tuff (1–10 μm). The porosity of the sandstone and volcanic ash is about 30.8 ± 3.5 vol% and 32 vol%, respectively. That of the mudstone is 30.6–51.4 vol%. Mudstone is more porous than sandstone, notwithstanding that mudstone contains small pores. The permeability coefficient of the mudstone core is 1.0×10^{-9} m/s, less than that for sandstone and tuff which is 1.0×10^{-7} m/s.

5.3 *Seeping points*

Gas seepage was easily identified by bubbling locations in water stored in the rice paddy fields. Figure 2 shows the distribution of these points. The seepage is mainly distributed between the three normal faults. In the BTV survey, many seepage points were identified at shallow depth (20–40 m) within the upper part of the sandstone-rich layer (Fig. 3). Gas bubbles could not be distinctly traced beneath 40 m depth. Additionally, seepage points were distributed along faults in the mudstone layer and between the faults in the sandstone strata.

5.4 *Stable carbon isotope ratio and concentration*

Figure 3 summarizes the seepage points and faults with $\delta^{13}C_{C1}$ by cross section. The $\delta^{13}C_{C1}$ of core and headspace gas from the packer system is −74 to −45‰, ave. −65‰, $\delta^{13}C_{CO_2}$ is bimodal at −50‰ and −10‰, and $\delta^{13}C_{C_2}$ is −42 to −33‰. Other gases could not be detected by GC/C/IRMS. $\delta^{13}C_{C1}$ is divided into three types (Table 1). Type (I) gas is distributed in the shallow, upper parts to 50 m

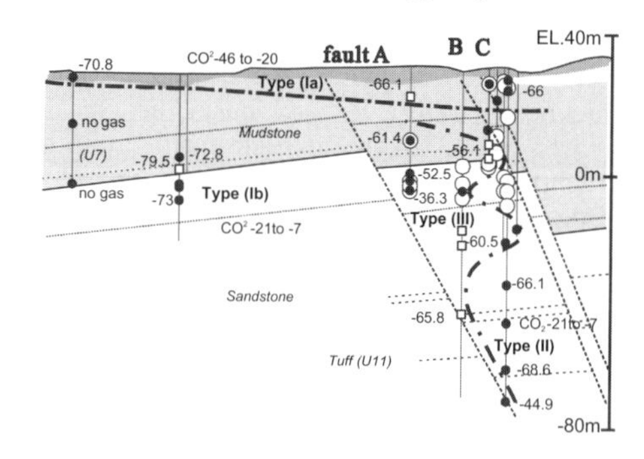

Figure 3. Cross section of geological information and distribution of the carbon isotope ratio of gases. Solid circle; core sampling point, Open square; headspace gas sampling point by packer system. Open circle; seeping point. Numeral; $\delta^{13}C_{C1}$.

Table 1. Estimation of gas types.

	$\delta^{13}C_{PDB}$(‰)		
	Methane	CO_2	Distribution
Type (I)			
Type (1a)	–70 to –53	–46 to –23	Surface
Type (1b)	–79 to –72		
Type (II)	–66 to –68	–23 to –7	Out of faults
Type (III)	–64 to –42		In faults

depth. Type (II) has normal Mobara $\delta^{13}C_{C1}$ from –66‰ to –68‰, and Type (III) has heavy $\delta^{13}C_{C1}$ of –60 to –42‰. Type (III) gas is mainly distributed between the faults. $\delta^{13}C_{CO_2}$ was divided into two sub-types as Type (I) and Type (Ia), –46 to –20% in the near surface area above 10 m depth and Type (Ib), –20 to –7% distributed beneath Type (Ia).

The composition of C_1 from headspace gas was 70 vol% between the faults. CO_2 and C_2 contents were 0.7–0.3 vol% and 0.01–0.002 vol% respectively. The only headspace gas sample from outside of the fault had C_1 of 1.3 vol% and CO_2 as 3.7 vol%. The content of CO_2 in the Type (Ia) zone determined from the core gas sample had a higher value of 2–38 vol%, in spite of its being diluted with He gas to avoid air contamination.

6 DISCUSSION

Type (Ia) gas only occurs in the thin, shallow, near-surface zone. The light $\delta^{13}C_{C1}$ gas (less than –68‰) was collected from another exposure of the faults (Fig.2), where the $\delta^{13}C_{CO_2}$ has a small value (–40 – –50‰) and a high CO_2 content, even diluted by He, compared with Type (II) and Type (III) gases. The seeping ratio SR ($mol_{gas}/m^3_{rock\ volume}$) from rock at 25°C, 1 atm is given by:

$$SR = A/2.24 \times (V/V_S - 1) \quad (1)$$

where A = gas content vol%, V = capacity of the SUS container in cm^3, and V_S = volume of rock. The SR_{CO_2} for Type (Ia) of about one is enriched by oxidation of organic matter; conversely, C_1 has low values in comparison with other types. It seems that the Type (Ia) gas has undergone considerable oxidation during its storage and migration to the surface.

Type (Ib) has a light C_1 value and a low SR_{CO_2}. These gases, which are produced by microbial reaction, are regionally distributed in the shallow area (to 100 m depth) away from the fault zone. Generally, the isotopically light $\delta^{13}C_{C1}$ is produced through biodegradation of organic matter (Whiticar et al. 1986) in marine sediments.

Type (II) gas, which has $\delta^{13}C_{C1}$ of –65‰, was mainly reported from 200–600 m deep production holes in the Mobara field (Kaneko et al. 2002). These values are commonly seen as the Mobara production gas formed by biodegradation from of organic matter (Igari and Sakata 1989). Some of this gas was detected between the faults in the study area. It seems that this gas rises from a deeper layer.

Type (III) gas consists of isotopically heavy C_1 (max. $\delta^{13}C_{C1}$ of –36.3‰) occurring along the western boundary fault (A, Fig.2). Type (III) gas apparently rises along fault A from a deeper layer without mixing with Type (II) gas dissolved in groundwater. Many gas bubbles seep from the sandstone underlain by the mudstone layer. These gas seepages are mainly traceable between the faults over a width of 50 m in the upper part of the sandstone layer. The SR_{C1} increases only between the faults. This shows that these gases rise along the fault to dissolve in groundwater prior to reaching a trapping depth between the faults. The sandstone-rich layer that is underlain by the mudstone-rich layer near the surface traps the Type (II) and Type (III) gases to produce a gas phase.

Figure 5 shows the origin of co-existing C_1 and CO_2 pairs as a biogenic gas formation. Type (Ia) plots on the oxidation trend with increasing $\delta^{13}C_{C1}$ and SR_{CO_2} (Whiticar 1999). Type (Ib) gas plots on the carbon isotope fractionation line at about ε = 60 indicating carbonate reduction in marine

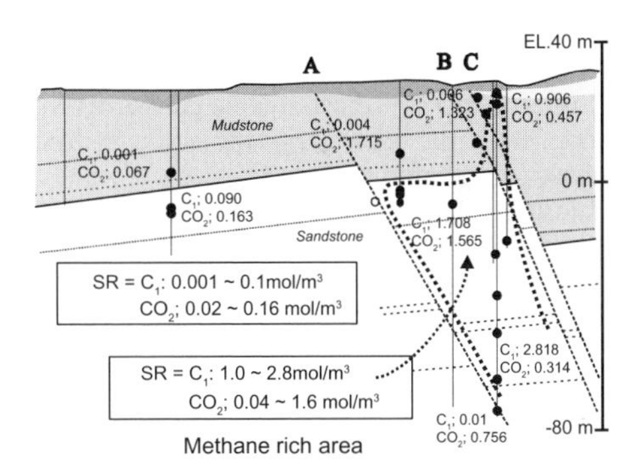

Figure 4. The distribution of gas seeping ratio (SR; C_1 (CO_2) mol/m³) from core. Methane mainly holds in fault zones. Solid circle; core sampling point.

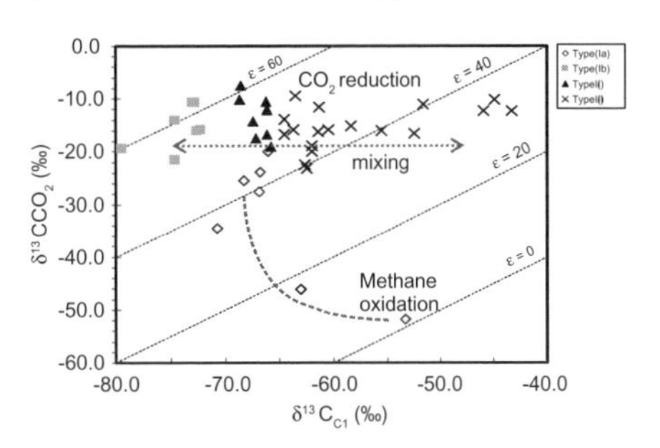

Figure 5. Combination plot of δC_{C1} and δC_{CO_2} with fractionation lines for the bacterial C1 formation process.

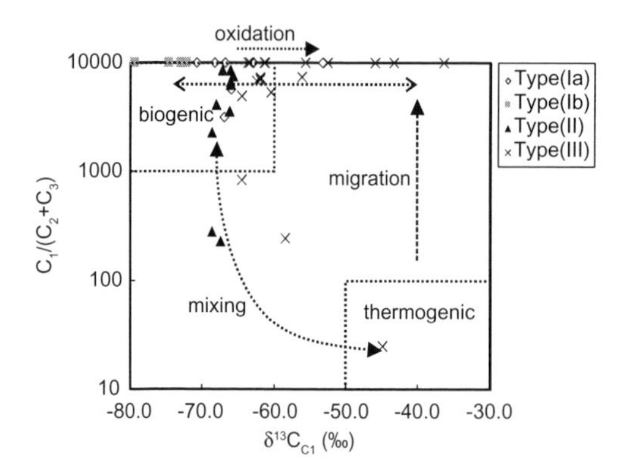

Figure 6. Relationship between the δC_{C_1} and $C_1/(C_2 + C_3)$ for interpreting the gas origin.

conditions. Type (III) gas of $\delta^{13}C_{CO_2}$ also has the same ratio as Type (Ib) gas. The $\delta^{13}C_{C_1}$ of Type (III) gas does not correspond to the fractionation line. This shows that Type (II) formed as a mix of Type (Ib) with Type (III) and is not a biogenic gas.

The origin of natural gas is inferred by the Bernard plot in Figure 6. $\delta^{13}C_{C_1}$ is distributed from −74 to −33‰ without other hydrocarbon gases lying in this area. Almost all the C_1 plots in the biogenic gas area. Except for the samples which have oxidized, isotopically heavy $\delta^{13}C_{C_1}$ between −50 and −33‰ in Type (III) without increasing CO_2 shows that the thermogenic gas has risen along the normal faults. Generally, thermogenic gas forms from, or co-exists with, other large molecule hydrocarbon gases, for example, C_2 to C_4.

Mudstone is porous compared with sandstone in spite of containing micropores. Furthermore, the surface area of mudstone is 12 m^2/g greater than sandstone, 3.2 m^2/g as determined by mercury intrusion porosimetry. We think that such gases produced by thermogenic reaction are trapped in a relatively deep sequence underlying the gas-producing layer in the Kanto area. The large molecule hydrocarbon gases are almost fixed (adsorbed) as a solid in this deep area.

We think that Type (II) gas was generated to mix with Type (Ib) and Type (III) gas in the deep area. Thermogenic gas including groundwater migrated upward from this deep area and along the normal faults. In this migration, C_2 to C_4 gases remained in the deep area. Groundwater was enriched in C_1 and CO_2 gases during upward migration. Type (III) gas was generated and mixed with Type (II) gas. Therefore, the $\delta^{13}C_{C_1}$ increases without variation of the $C_1/(C_2 + C_3)$ ratio. Apparently, the migration of thermogenic gas from the deep area leads to a depletion in C_2 to C_4 gases.

Seeping gas in this area has migrated from the deep area towards the surface and is stored under the impermeable small pore mudstone acting as a cap rock layer. After the migration, large molecule sized gases of up to C_3 are depleted and at shallow depth they can occur as bubbles on surface seepage. Thermogenic gas of is generated by organic matter in the deep layer, which produces petroleum and gas having $\delta^{13}C_{C_1}$ of up to −40‰ in the Japan oil field (Waseda and Omokawa, 1988). It seems that the Type (III) gas rises from a deeper layer than the Type (II) gas product using the three normal faults as the pathway. Biogenic gas consisting of C_1 and slight C_2 follows CO_2 reduction from the decomposition of organic matter at shallow depth.

We think that Type (II) gas having $\delta^{13}C_{C_1}$ of −66‰ is generated by mixing of Type (Ib) gas ($\delta^{13}C_{C_1}$ −74‰) with thermogenic gas ($\delta^{13}C_{C_1}$ −33‰, Type (III) gas), which was formed in the deep South Kanto gas field and is in the process of migration from this deep area.

7 CONCLUSIONS

Gas seepages on the surface are distributed between normal faults in this area. Seep points were identified in the upper part of the sandstone-rich layer underlain by the mudstone-rich layer. Isotopically heavy $\delta^{13}C_{C_1}$ was detected in core and headspace samples. It seems that gas having $\delta^{13}C_{C_1}$ of −66‰ mixes with biogenic gas having $\delta^{13}C_{C_1}$ of −74‰ and thermogenic gas having $\delta^{13}C_{C_1}$ of −33‰ from a deep zone. Thermogenic gas depleted in large molecule hydrocarbon gas rises along faults. Thermogenic gas generated in the deep zone was detected to have risen to surface in the 50 m wide normal fault zone.

REFERENCES

Igari, S. & Saklata, S. 1989. Origin of natural gas of dissolved-in-water type in Japan inferred from chemical and isotopic compositions: Occurrence of dissolved gas of thermogenic origin. *Geochem. J.* 23: 139–142.

Kaneko, N., Maekawa T. & Igari, S. 2002. Generation of archaeal methane and its accumulation mechanism into interstitial water. *J. Japanese Assoc. Petrol. Technol.* 67: 97–110.

Kunisue, S., Mita, I. & Waki, F. 2002. Relationship between subsurface geology and productivity of natural gas and iodine in the Mobara gas field, Boso Peninsula, central Japan. *J. Japanese Assoc. Petrol. Technol.* 67: 83–96.

Natori, H. 1997. Mobara type natural gas deposits seem to be methane hydrate origin. *Chishitsu News* 510: 59–66.

Waseda, A. & Omokawa, M. 1988. Geochemical study on origin of natural gases in Japanese oil and gas fields. *J. Japanese Assoc. Petrol. Technol.* 53: 213–222.

Whiticar, M.J. 1999. Carbon and hydrogen isotope systematics of bacterial formation and oxidation of methane. *Chem. Geol.* 161: 291–314.

Whiticar, M.J., Faber, E. & Schoell, M. 1986. Biogenic methane formation in marine and freshwater environments: CO_2 reduction vs. acetate fermentation–Isotope evidence. *Geochim. Cosmochim. Acta* 50: 693–709.

Water-Rock Interaction – Birkle & Torres-Alvarado (eds)
© 2010 Taylor & Francis Group, London, ISBN 978-0-415-60426-0

Interplay of bedrock and atmospheric sulfur sources for catchment runoff and biota

M. Novak, I. Jackova & E. Prechova
Czech Geological Survey, Prague, Czech Republic

ABSTRACT: Sulfur isotope compositions of spruce canopy throughfall, open-area precipitation and runoff were measured for a period of 12 months in two small catchments with contrasting $\delta^{34}S$ of bedrock, and contrasting pollution history. Sulfur in organic soil and above-ground biomass was isotopically lighter than S in bedrock and atmospheric input, indicating that organic S cycling fractionates isotopes. Sulfur in organic matter is not a simple mixture of bedrock S and atmospheric S.

1 INTRODUCTION

Catchments represent a dynamic system open for isotopes (Mayer et al. 1995). Abundance ratios of stable isotopes can be used to apportion sources of an element in individual compartments of a catchment, such as soil, water solutions or vegetation, if mixing end members are isotopically distinct, and if biogeochemical cycling does not redistribute isotopes (Alewell et al. 1999). In the case of sulfur, mixing end members are bedrock minerals (accessory sulfides or evaporites), and atmospheric deposition (water solutions and gases). Quantifying the contribution of bedrock (i.e., natural) and atmospheric (mainly pollutant) sulfur in soils, trees and stream water can be successful only if the direction and magnitude of within-site S isotope fractionations are known in addition to $\delta^{34}S$ values (Shanley et al. 2005, 2008). We present the results of a S isotope inventory of two contrasting sites in the Czech Republic, Central Europe. We compare new $\delta^{34}S$ data on two types of atmospheric deposition and runoff with known isotope signatures of catchment S pools. Our objective is to constrain the presence of bedrock S in runoff and organic matter in light of possible biotic S isotope fractionations.

2 METHODS

2.1 *Study sites*

Contours in Figure 1 show that, in the late 20th century, the north of the Czech Republic was industrially heavily polluted, whereas the south of the country was relatively unpolluted. JEZ, a spruce die-back affected catchment, is situated in the north, whereas LIZ, vegetated with healthy mature Norway spruce, is situated in the south. Sulfur pollution originated mainly from thermal power stations in the north emitting low-$\delta^{34}S$ sulfur (mean of 1.6 per mil; Mach et al. 1999), and has been sharply decreasing since the late 1980s. Site characteristics are given in Table 1.

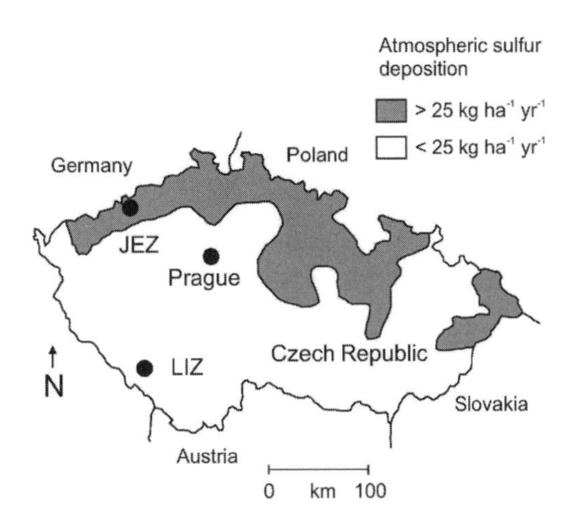

Figure 1. Location of the studied catchments.

Table 1. Study site characteristics.

Catchment	JEZ	LIZ
Area (ha)	261	99
Bedrock	Orthogneiss	Paragneiss
Soil	Dystric cambisol	Dystric cambisol
Elevation (m)	475–924	828–1024
Annual precipitation (mm)	930	905
Mean temperature (°C)	5.0	4.9

2.2 *Sampling*

Spruce canopy throughfall (9 collectors per site) and open-area precipitation (2 collectors per site) were sampled monthly between January and December 2002, stream water was collected at the end of each month. Details of the sampling protocol were given by Novak et al. (2007).

2.3 *Analytical procedure*

Sulfate was precipitated as $BaSO_4$, with yields of >0.5mg S (Chakrabarti 1978). Sulfate S was converted to SO_2 in a vacuum line (Yanagisawa & Sakai 1983) and $\delta^{34}S$ values determined on a Finnigan MAT 251 mass spectrometer with a reproducibility better than 0.3 per mil. Means and standard errors are given.

3 RESULTS

For all three sample types, spruce throughfall, open-area precipitation and runoff, $\delta^{34}S$ were mostly higher at LIZ compared to JEZ (Fig. 2). Mean $\delta^{34}S$ values were the highest in open-area deposition at both sites (5.2 ± 1.8 and 6.2 ± 1.4 per mil at JEZ and LIZ, respectively). Mean $\delta^{34}S$ values of runoff were higher at LIZ (4.5 ± 0.3 per mil) than at JEZ (3.8 ± 0.5 per mil). The lowest observed $\delta^{34}S$ values were those of spruce throughfall at JEZ (3.6 ± 0.5 per mil). Mean $\delta^{34}S$ value of spruce throughfall at LIZ was 5.3 ± 1.1 per mil. Temporal fluctuations in $\delta^{34}S$ values were higher for atmospheric input (Figs. 2a and 2b) than for runoff (Fig. 2c).

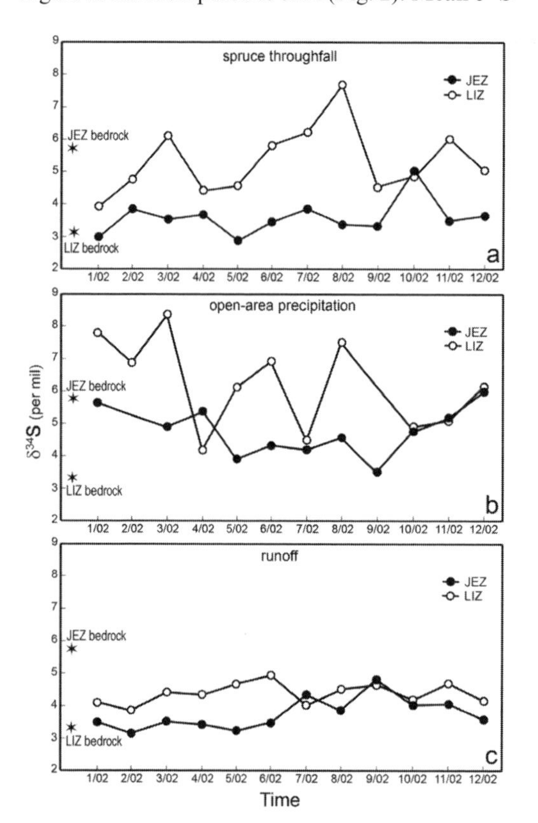

Figure 2. Time series of $\delta^{34}S$ values of input-output S fluxes. Asterisks denote $\delta^{34}S$ values of bedrock.

4 DISCUSSION

4.1 *Contribution of bedrock S to runoff*

We have previously published $\delta^{34}S$ values of whole-rock samples for both sites (Novak et al. 2000, 2005). Sulfur isotope signature of JEZ orthogneiss is higher, 5.8 per mil, that of LIZ paragneiss is lower, 3.2 per mil. If these values are compared with $\delta^{34}S$ of runoff, it appears that bedrock S does not dominate runoff S flux. JEZ, the site with isotopically heavier S in bedrock, exports isotopically lighter S via stream water. Conversely, LIZ, the site with isotopically lighter S in bedrock, exports isotopically heavier S via stream water. We note that weathering/dissolution of accessory bedrock minerals does not fractionate S isotopes, due to complete removal of surface layers of the weathered phase.

4.2 *Input $\delta^{34}S$ differences between the two sites*

Massive lignite burning near JEZ between ca. 1950 and 1990 injected isotopically relatively light sulfur ($\delta^{34}S$ of 1.6 per mil) to the atmosphere. $\delta^{34}S$ of the broader atmospheric background in mainland Europe was higher (5–6 per mil; Mayer et al. 1995, Novak et al. 2001). The pulse of low-$\delta^{34}S$ anthropogenic SO_2 into the atmosphere in the north explains lower $\delta^{34}S$ of atmospheric input at JEZ compared to LIZ, which is located 150 km from the power plants (Figs. 2a,b).

4.3 *$\delta^{34}S$ differences between spruce throughfall and open-area precipitation*

Because atmospheric oxidation of industrial SO_2 to sulfate represents a conversion of a gaseous phase into a liquid phase, the product becomes enriched in the heavier S isotope. The residual SO_2 with a lower $\delta^{34}S$ value is scavenged from the air by the large surfaces of spruce needles and becomes a major portion of the throughfall S flux. Open-area precipitation (wet deposition) is dominated by isotopically heavier sulfate S.

4.4 *Homogenization of $\delta^{34}S$ from input to output*

Lower range of $\delta^{34}S$ values of runoff compared to both types of atmospheric S input is a striking feature of Figure 2. Two factors can contribute to this phenomenon: i) long residence time for atmospheric S in the catchment, mainly in soil, prior to export, and ii) existence of S isotope fractionations. Long residence time of atmogenic S in the catchments removes monthly extremes in $\delta^{34}S$ of the input. Sulfur deposited at different times is mixed in the runoff and thus extreme input $\delta^{34}S$ values are not seen in runoff sulfate. Since inorganic adsorption and desorption of sulfate is not associated with isotope fractionations, organic S cycling is the only process that may redistribute S isotopes in the studied systems.

4.5 *Assimilatory S isotope fractionations*

Table 2 shows systematic differences between S fluxes and S pools. $\delta^{34}S$ values of soil are from Novak et al. (1996), $\delta^{34}S$ values of spruce xylem are from Novak (2009). Atmospheric input data are averages for 1997 (Novak et al. 2005) and 2001 (Fig. 2). At JEZ, S in both bedrock and atmospheric input is isotopically too heavy to explain $\delta^{34}S$ of organic soil horizons and of spruce xylem. These two catchment compartments may contain sulfur that has been fractionated upon assimilation. Sulfur assimilation appears to be associated with a negative $\delta^{34}S$ shift of as much as 4.4 per mil (Table 2). Sulfur in organic matter becomes isotopically lighter compared to the source reservoir (atmosphere, bedrock). At LIZ, both organic soil horizons and spruce xylem contain isotopically light sulfur compared to atmospheric input, similar to JEZ. In contrast to JEZ, however, xylem S has just slightly lower $\delta^{34}S$ value than bedrock, which happens to contain isotopically relatively light S. From S isotope systematics in Table 2 alone we cannot rule out a contribution of unfractionated bedrock S in spruce xylem. At the same time, however, we should keep in mind that we have previously concluded from Figure 2c that bedrock S contribution to runoff (which includes soil solutions) was probably low. Should sulfur in LIZ spruce xylem be mainly derived from atmospheric and not bedrock sources, the magnitude of the assimilation-related negative $\delta^{34}S$ shift is lower compared to JEZ (2.3 vs. 4.4 per mil, calculated as atmospheric input minus xylem values in Table 2). Organic soil horizons at LIZ exhibited lower $\delta^{34}S$ compared to both bedrock and atmospheric input (Table 2), again suggesting a S isotope fractionation upon organic cycling. Recently, several papers have suggested that dissimilatory bacterial sulfate reduction (BSR) may occur following precipitation events at anaerobic microsites even in normally well aerated upland forest soils (Giesler et al. 2009). Isotopically light sulfate S in runoff may contain some S resulting from reoxidation of sulfides, which had been formed by BSR. However, we have previously shown that in upland forests of the northern Czech Republic BSR results mainly in emissions of H_2S, not formation of solid sulfides (Novak et al. 2001). Sulfur isotope dynamics in forested catchments were studied by several other authors (e.g. Mayer et al. 1995, Alewell et al. 1999, Shanley et al. 2005, 2008). These authors did not find a systematic negative $\delta^{34}S$ shift upon sulfur assimilation.

Table 2. Sulfur isotope composition of catchment pools and fluxes (mean ± std. error).

	JEZ $\delta^{34}S$ (‰)	LIZ $\delta^{34}S$ (‰)
Bedrock	5.8	3.2
Soil B/C horizon	3.6 ± 1.6	2.8 ± 0.9
Soil A horizon	1.4 ± 0.4	2.4 ± 0.5
Spruce xylem	0.4 ± 0.6	3.0 ± 0.6
Runoff	3.6 ± 0.3	4.6 ± 0.3
Atmospheric input	4.8 ± 0.6	5.3 ± 1.1

4.6 *Runoff generation*

At JEZ, the $\delta^{34}S$ value of runoff (3.6 per mil) cannot be explained by simple mixing of bedrock S and atmospheric S, because both these end members contain S that is isotopically too heavy (Table 2). Low-$\delta^{34}S$ soil S, a result of organic S cycling, must have contributed to runoff S flux. In contrast, at LIZ, $\delta^{34}S$ value of runoff (4.6 per mil), if only data in Table 2 are considered, could result from simple mixing of lower $\delta^{34}S$ of bedrock and higher $\delta^{34}S$ of atmospheric input. Novak et al. (2001) have shown that decreasing atmospheric S inputs in the Czech Republic are not accompanied by an increasing $\delta^{34}S$ value. In fact, as S pollution eases, $\delta^{34}S$ values of atmospheric input still decrease. Therefore, we cannot invoke export of old pollutant low-$\delta^{34}S$ sulfur as an explanation of isotopically light S in present-day runoff.

4.7 *Usefulness of S source apportionment in catchments*

In recent past, motivation for catchment $\delta^{34}S$ studies stemmed from the need to track the rate of recovery of anthropogenic acidification. Sulfur isotopes showed that it may take many decades to reestablish pre-1950 conditions of soils and waters, and also indicated substantial organic S cycling. At present, more insights into the consequences of redox changes in catchment S reservoirs for S fluxes are needed in connection with climate change.

5 CONCLUSIONS

Input-output sulfur fluxes were studied at a previously heavily polluted site (JEZ) and a relatively less polluted site (LIZ). The site with isotopically heavier bedrock S exported isotopically lighter S via runoff, and vice versa. We conclude that unfractionated bedrock S does not dominate runoff S flux. Organic-rich S pools (organic soil horizons and above-ground biomass) contained isotopically lighter S compared to atmospheric input, indicating an isotope fractionation upon S assimilation. In the era of climatic warming, S isotopes may provide new insights into retention of natural and anthropogenic S in forested catchments.

ACKNOWLEDGEMENTS

FP7 Collaborative project of EC "SOIL TrEC" is thanked for funding.

REFERENCES

Alewell, C., Mitchell, M.J., Likens, G.E. & Krouse, H.R. 1999. Sources of stream sulfate at the Hubbard Brook Experimental Forest: Long-term analyses using stable isotopes. *Biogeochemistry* 44: 281–299.

Chakrabarti, J.N. 1978. Analytical procedures for sulfur in coal desulfurization products. *Analytical Methods for Coal and Coal Products*: 279–323.

Giesler, R., Bjorkvald, L., Laudon, H. & Morth, C.M. 2009. Spatial and seasonal variations in stream water delta S-34-dissolved organic matter in Northern Sweden. *Environmental Science and Technology* 43: 447–452.

Mach, K., Zak, K. & Jackova, I. 1999. Sulfur speciation and isotopic composition in a vertical profile of the main coal seam of the North Bohemian brown coal basin and their paleogeographic interpretation. *Bulletin of the Czech Geological Survey* 74: 51–66.

Mayer, B., Fritz, P., Prietzel, J. & Krouse, H.R. 1995. The use of stable sulfur and oxygen-isotope ratios for interpreting the mobility of sulfate in aerobic forest soils. *Applied Geochemistry* 10: 161–173.

Novak, M. 2009. Sulfur isotope composition of *Picea abies* growth rings. *Final Report, Czech Science Foundation*, Prague, Project No. 526/06/1589.

Novak, M., Bottrell, S.H., Fottova, D., Buzek, F., Groscheova, H. & Zak K. 1996. Sulfur isotope signals in forest soils of Central Europe along an air-pollution gradient. *Environmental Science and Technology* 30: 3473–3476.

Novak, M., Jackova, I. & Prechova, E. 2001. Temporal trends in the isotope signature of air-borne sulfur in Central Europe. *Environmental Science and Technology* 35: 255–260.

Novak, M., Kirchner, J.W., Fottova, D., Prechova, E., Jackova, I., Kram, P. & Hruska, J. 2005. Isotopic evidence for processes of sulfur retention/release in 13 Central European catchments spanning a strong pollution gradient. *Global Biogeochemical Cycles* 19, Art. No. GB4012.

Novak, M., Kirchner, J., Groscheova, H., Cerny, J., Havel, M., Krejci, R. & Buzek, F. 2000. Sulfur isotope dynamics in two Central European watersheds affected by high atmospheric deposition of SO_x. *Geochimica et Cosmochimica Acta* 64: 367–383.

Novak, M., Mitchell, M.J., Jackova, I., Buzek, F., Schweigstillova, J., Erbanova, L., Prikryl, R. & Fottova, D. 2007. Processes affecting oxygen isotope ratios of atmospheric and ecosystem sulfate in two contrasting forest catchments in Central Europe. *Environmental Science and Technology* 41: 703–709.

Shanley, J.B., Mayer, B., Mitchell, M.J. & Bailey, S.W. 2008. Seasonal and event variations in delta S-34 values of stream sulfate in a Vermont forested catchment: Implications for sulfur sources and cycling. *Science of the Total Environment* 404: 262–268.

Shanley, J.B., Mayer, B., Mitchell, M.J., Michel R.L., Bailey, S.W. & Kendall, C. 2005. Tracing sources of streamwater sulfate during snowmelt using S and O isotope ratios of sulfate and S-35 activity. *Biogeochemistry* 76: 161–185.

Yanagisawa, F. & Sakai, H. 1983. Precipitation of SO_2 for sulphur isotope ratio measurements by the thermal decomposition of $BaSO_4$-V_2O_5-SiO_2 mixtures. *Analytical Chemistry* 55: 985–987.

Water-Rock Interaction – Birkle & Torres-Alvarado (eds)
© 2010 Taylor & Francis Group, London, ISBN 978-0-415-60426-0

Groundwater geochemistry of a lacustrine zone in the Mexicali Valley, Northwest Mexico

E. Portugal & A.F. Hernández
Instituto de Investigaciones Eléctricas, Gerencia de Geotermia, Cuernavaca, México

J. Álvarez Rosales
Comisión Federal de Electricidad, Residencia General de Cerro Prieto, Mexicali, México

ABSTRACT: Major aqueous species and stable and radioactive environmental isotopes have been measured in ground waters in a lacustrine area of the Mexicali Valley. The groundwaters exhibit three chemical facies, of which, the Na-Cl facies predominates in the study area. The main mechanisms controlling sulfate concentrations are oxidation, dissolution, and bacterial reduction. The latter was identified in the lacustrine zone, where low salinity waters (TDS: 536–1,600 mg/L) reach the silt-clay area from the east, which is characterized by highly saline waters (TDS: 7,126–48,988 mg/L). The hydraulic, hydrochemical, and isotopic data reveal the existence of a phreatic, an intermediate, and a deep aquifer. Groundwater flows slowly in this area due to the low permeability of the host sediments, as evidenced by measured groundwater concentrations for ^3H and ^{14}C. The tritium concentration indicates an involvement of modern water only at the eastern section of the deep aquifer.

1 INTRODUCTION

In the Mexicali Valley, processes such as salinization caused by evaporation, dissolution of evaporites, and highly saline groundwater inflow from the USA have been recognized (Makdisi et al. 1982, Payne et al. 1979, Portugal et al. 2005). Its groundwater displays several chemical composition types (mainly sodium-chloride facies) and TDS values in a range from 536 to 48,988 mg/L (Portugal et al. 2005). The highest salinity waters are located in the lacustrine area where the Cerro Prieto geothermal field (CPGF) is situated (Fig. 1).

The spatial geochemical variation of groundwater in the valley has been described by Payne et al. (1979), Makdisi et al. (1982), and Portugal et al. (2005, 2006). Portugal et al. (2005) delineated a conceptual flow path model for the eastern aquifers, while Portugal et al. (2006) reported the hydrochemical characteristics for the central and western aquifers.

The present research is aimed to define the hydrogeological units of the lacustrine area and their vertical interaction, utilizing major elements, environmental isotopes ($\delta^{13}C_{DIC}$, $\delta^{34}S_{SO_4}$, $\delta^{18}O_{SO_4}$, $\delta^{18}O_{H_2O}$, δD_{H_2O}, T, and $^{14}C_{DIC}$), and potentiometric data. The lacustrine waters are hosted in a sedimentary unit, whose thickness varies from 900 to 2,500 m, and consists of deposits of silt, silt-clay-sand, and gravels of several sizes. The gravels are minor and are distributed throughout the non-

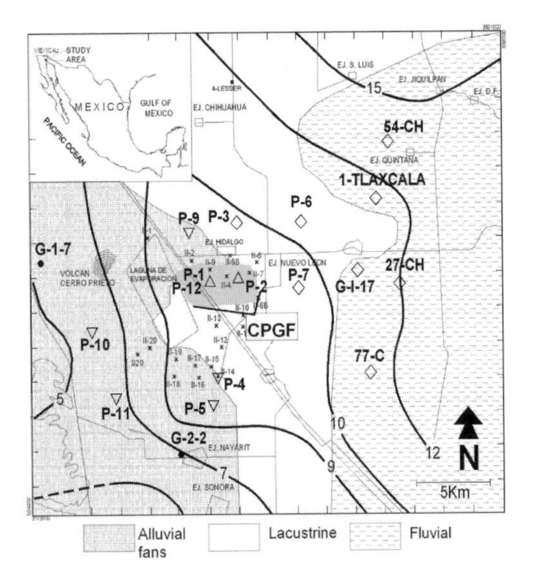

Figure 1. Hydrogeological map of the study area showing sampling sites (agricultural and observation wells), the potentiometric surface, and the main sedimentary units (after CICESE 1998).

consolidated unit. Currently, both the aquifer units and their water interaction are not well defined due to the native geological heterogeneity. However, from an environmental viewpoint, it is important to characterize them in spite of the high salinity characteristics and low commercial value of its hosted groundwater.

2 GEOLOGY AND HYDROGEOLOGY SETTING

The study area is located in the structural depression of the Salton Sea Basin, a transition area between the divergent boundary of the East Pacific Rise and the transform boundary of the San Andreas Fault system (Lippmann et al. 1991).

The geological formations of the Mexicali Valley are Palaeozoic to Quaternary and comprise; (1) Permian to Jurassic metamorphic complex with gneiss, schist, and shale. These units are intruded by pegmatitic dikes and granitic rocks; (2) Cretaceous-Jurassic igneous rocks composed of granite (tonalite and granodiorite); (3) Miocene igneous rocks, such as basalts, andesites, and rhyodacites. The formations (1) to (3) fill the valley basin and shape a consolidated unit. (4) Quaternary alluvium forming an unconsolidated unit. In the CPGF area, this unit increases in thickness from 350 m in the western zone towards 2000 m in the eastern section. Three sedimentary environments were identified in the area: fluvial to the east, with facies formed by clayey sands and gravel (fine, medium, and coarse grains); lagoonal in the center, with clays and silts as the main components along with sands; and alluvial from distal fans of the Sierra Cucapá to the west (not shown in Fig. 1), forming coarse sand, gravel in the deeper sections, and rounded granules (CICESE 1998).

3 RESULTS

3.1 *Geochemical characteristics*

Groundwater samples were collected and analyzed from 29 ("II", "P", and "G") observation boreholes and from 18 irrigation wells between 1997 and 2007. Figure 1 shows a simplified map of the study area, where well locations are indicated.

The groundwaters exhibit Na-Cl, Ca-Na-SO_4, and HCO_3 facies. Most of the groundwaters are characterized by the Na-Cl facies (Portugal et al. 2005), which is found throughout the study zone.

The deep waters coming from east are characterized by low salinities (TDS: 536–1,600 mg/L), whereas the waters are more saline in the central zone (lacustrine area) (TDS: 6,756–48,988 mg/L), particularly those hosted in shallower strata (TDS 6,945–66,636 mg/L). Finally, the western deep waters show less saline chemical characterisitcs (6,756–7,126 mg/L TDS).

3.2 *Potentiometric surface*

The regional hydraulic gradient of groundwater is from NE towards W and SW. Except in the northwestern zone (outside of the study area), where it flows towards the Salton depression (not shown in Fig. 1) with the lowest elevation in the area. In the CPGF area, the potentiometric data of observation boreholes reveal two aquifers whose hydraulic head differences of only 2.5 m indicate potential for downward leakage. The shallow aquifer (represented by the "II" wells) has a thickness between 50 and 80 m and overlies the confined aquifer (identified by the "P" wells), whose thickness is at least 150 m. No unit separating these aquifers has been identified. The deep aquifer is located within the depth range from which the eastern irrigation wells tap water. Only one observation well (P1), located in the central section, reaches more than 500 m of depth and its head is lower than those of the shallower aquifers. In this case, the hydraulic head difference (15 m) indicates also potential for downward leakage.

3.3 *Environmental isotopes*

In Figure 2, the $\delta^{18}O$ – δD values vary from −14.4 to 3.2‰ and from −111 to −64‰, respectively. The stable isotope values of the waters correlate with the hydrochemical data: The larger the mineralized water content, the higher the values of $\delta^{18}O$ and δD.

In the $\delta^{18}O$ – δD diagram, waters from the eastern zone, which circulate in a depth from 100 to150 m, plot near the Global Meteoric Water Line (GMWL). The other waters plot on a line that intercepts the GMWL with a depleted isotopic composition ($\delta^{18}O = -15.8‰$ and $\delta D = -116‰$), indicating that recharge occurs at a higher elevation.

The potential sources of dissolved sulfate in groundwater can be usually clarified by means of $\delta^{34}S$ values. The main contributions of sulfate to groundwater include atmospheric deposition (dust), sulfate minerals, microbial activity (oxidation of sulfide minerals), and mineralization of

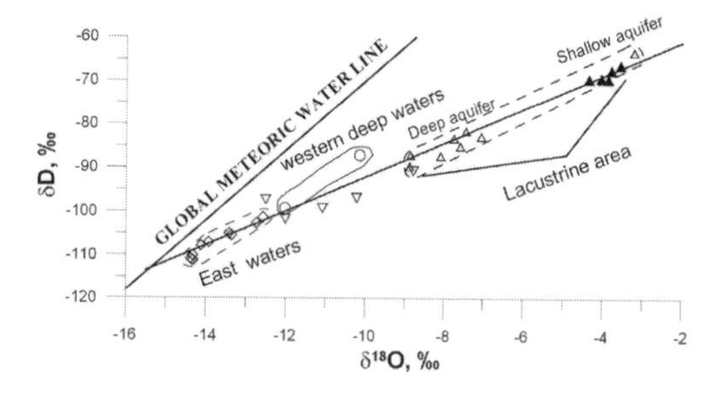

Figure 2. Stable composition of ground waters in the study area. The heavier isotopic composition correspond to the shallow waters in the lacustrine area (solid triangle).

S-bearing organic material. The $\delta^{34}S$ values of the waters ranged from +40.7 to +1.4‰. The aqueous sulfate concentration increased from 70 mg/L up to 1,750 mg/L with a corresponding decrease in $\delta^{34}S$ from +15‰ to + 5‰, suggesting the addition of ^{34}S-depleted sulfate.

This may be explained by the oxidation of a mineral such as pyrite or by dissolution of gypsum with $\delta^{34}S$ in the range from 0‰ to 15‰. Unfortunately, neither rock specimens nor $\delta^{18}O_{SO_4}$ measurements were available to test this hypothesis. Such type of data would have been very valuable for improving the interpretation. The waters east of the lacustrine zone showed highest $\delta^{34}S$ values, indicating an intensive microbial activity. $\delta^{34}S$ values for this area ranged from +25 to +40‰, which have a lower sulfate content (6 mg/L to 100 mg/L) than those of the lacustrine area. Sulfate concentrations in the waters of the western zone were extremely low and H_2S was not detected.

The tritium concentration in the eastern zone's groundwater was higher than 12 T.U. In contrast, the tritium in surface waters was 10 T.U. Makdisi et al. (1982) reported tritium values of up to 150 T.U. for some groundwater in the valley. This suggests infiltration of recent water, consistent with the presence of the bomb-period water from the Imperial Valley of the USA. No tritium was detected in shallow or deep aquifers at the geothermal field area, whereas tritium concentrations ranged between 0.7 and 2.4 T.U. in the surroundings of the CPGF.

Measured $\delta^{13}C$ values of dissolved inorganic carbon (DIC) varied from −19.7 to −2.1‰. The waters have a neutral pH indicating that HCO_3^- is the dominant species. The highly depleted ^{13}C value was measured in the deep aquifer (P1 site). For this investigation, ^{14}C activities were measured in four water samples from sites along the hydraulic gradient. These waters circulate at a depth between 100 and 150 m, except for water from the P1 site inside the CPGF, which was extracted from the deep aquifer ($\delta^{13}C = -19.7$ ‰ and $^{14}C_{Am} = 44.5$ pmC). The $\delta^{13}C$ and $^{14}C_{Am}$ values measured in the eastern zone were of −13.8 ‰ and 92.3 pmC, respectively and for the western zone were of −2.1 ‰ and 1.7 pmC, respectively. The water in the central zone near the P1 site resulted in $\delta^{13}C = -12.1$ ‰ and $^{14}C_{Am} = 19.3$ pmC.

4 DISCUSSION

The potentiometric surface shows that groundwater flows from northeast to southwest. Nevertheless, as groundwater approaches the silt area, the velocity decreases because of the low permeability of the deposits. The scarce annual precipitation and the high evaporation cause an increase in salinity, especially in the phreatic aquifer waters. Besides, the infiltration of surface water used in agriculture increases also salinity, especially in the eastern zone. The waters coming from the east reach the lacustrine area forming a mixing interface around the CPGF area, as shown by the chemical, stable-isotope, and tritium data. The sulfate concentration increases and the decrease of $\delta^{34}S$ to values >5 ‰ in the groundwater along the hydraulic gradient suggest an addition of $\delta^{34}S$ depleted sulfate. This might be derived from the oxidation of sulfide or from gypsum dissolution. The high variability in the concentrations of sulfate and the enrichment in $\delta^{34}S$ of the mixing zone (mixing interface) suggest that local mechanisms control the $\delta^{34}S$ composition of sulfate, though it is also suggested that bacterial sulfate reduction is the main mechanism in the area.

In the lacustrine area, characterized by silts, tritium was not detected neither in piezometers screened deeper than 100 m nor in shallow waters (10 to 25 m deep). This suggests that no contact occurs between waters hosted by the silt sediments and waters coming from either the eastern zone (whose age is recent, post-1952 age) or the atmosphere, at least over the last 50 years.

The low ^{14}C activity measured in the deep water site (P1 site) of this area supports this evidence. The radiocarbon method renders a residence time between 8 and 10 Ka, which suggests that the low permeability of the deposited material has isolated the aquifers. The mean residence time has been determined after correction of the initial activities for dead carbon from the matrix. The activity (Am) of ^{14}C shows a westward decrease along the hydraulic gradient, reaching a value of 1.7 pmC. Figure 3 presents a correlation between $\delta^{13}C$ and $^{14}C_{Am}$ where the data indicate a decrease of ^{14}C activity with a

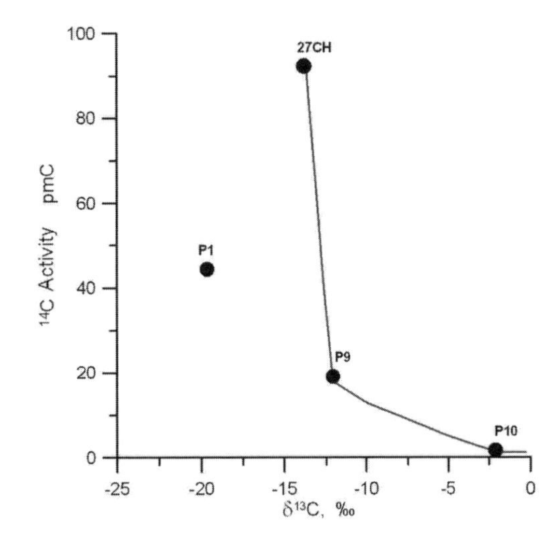

Figure 3. ^{14}C vs. $\delta^{13}C$ diagram for sampled water.

corresponding increase in $\delta^{13}C$ composition. This agrees with the existence of reactions with carbonate minerals in the matrix during water circulation in the subsurface. The P1 site, geographically near to the P9 site, plots in Figure 3 far off the solid ^{14}C—^{13}C line, suggesting that the P1 water resulted from a mixture with younger waters. The increase in the concentrations of chloride and the $\delta^{18}O$ content observed from 1998 to 2003 in the water of the P1 site can be also a result of this interaction. This interpretation is in agreement with the vertical downward flow from the phreatic aquifer towards the deeper aquifers, as revealed by the differences in hydraulic head.

5 CONCLUSIONS

The results are summarized as follows:

- The potentiometric data reveal the presence of at least three aquifers, although the deepest is evidenced by only one site.
- Vertical downward inter-aquifer flow in the CPGF area is evidenced by measured hydraulic head values as well as by chemical and isotopic data.

ACKNOWLEDGEMENTS

The authors wish to thank to the authorities of the Comisión Federal de Electricidad (CFE) for granting permission to publish this work.

REFERENCES

CICESE 1998. *Estudio hidrogeológico del campo geotérmico de Cerro Prieto*. Informe técnico RE 05/98.

Clark, I. & Fritz P. 1997. *Environmental isotopes in hydrogeology*. Lewis Publishers: 328 p.

Lippmann, M.J., Truesdell, A., Halfman-Dooley, S.E. & Mañón, M.A. 1991. A review of the hydrogeologic-geochemical model for Cerro Prieto. *Geothermics* 20: 39–52.

Makdisi, R.S., Truesdell A.H., Thompson, J.M., Coplen, T.B. & Sanchez, J. 1982. Chemical evolution of Mexicali Valley groundwater. *Proc. 4° Symp. Cerro Prieto Geothermal Field, CFE, Guadalajara, Jalisco, México*: 551–562.

Payne, B.R., Quijano, L. & La Torre, D.C. 1979. Environmental isotopes in a study of the origin of salinity of groundwater in Mexicali Valley. *Journal of Hydrology* 41: 201–215.

Portugal, E., Izquierdo, G., Truesdell, A. & Álvarez, J. 2005. The geochemistry and isotope hydrology of the southern Mexicali Valley in the area of the Cerro Prieto, Baja California (Mexico) geothermal field. *Journal of Hydrology* 313: 132–148.

Portugal, E., Álvarez, J. & Romero, B.I. 2006. Hydrochemical and isotopical tracer in the lacustrine aquifer of the Cerro Prieto area Baja California, México. *Journal of Geochemical Exploration* 88(1–3): 139–143.

Water-Rock Interaction – Birkle & Torres-Alvarado (eds)
ISBN 978-0-415-60426-0

Dissolved inorganic carbon isotopes in natural waters in Iceland

Á.E. Sveinbjörnsdóttir & S. Arnórsson
Institute of Earth Sciences University of Iceland, Iceland

J. Heinemeier
AMS ^{14}C Dating Center, Department of Physics and Astronomy, University of Aarhus, Denmark

H. Ármannsson
Icelandic Geosurvey, Iceland

H. Kristmannsdóttir
University of Akureyri, Borgir, Nordurslod, Iceland

ABSTRACT: In recent years numerous measurements on dissolved inorganic carbon (DIC) isotopes in Icelandic waters (surface, and cold and thermal groundwaters) have been performed. The dataset spans geographically all parts of Iceland and reflects wide range both in ^{14}C concentration (0 to 130 pMC) and $\delta^{13}C$ (+2 to −26‰). Despite the complexity of the DIC groundwater hydrology, the carbon isotopes display some systematic and distinctive groupings, especially reflected in the cold groundwater and the hot fluids circulating the five low geothermal systems within the Reykjavík region, SW Iceland.

1 INTRODUCTION

Isotopes of hydrogen and oxygen have been extensively used in geothermal studies in Iceland to define the source of groundwater, delineate groundwater flow and to study water-rock interaction. Since 1992 DIC isotopes have been measured in Icelandic surface- and groundwater in collaboration between the Institute of Earth Sciences, University of Iceland and the AMS ^{14}C center in Aarhus, Denmark.

Dissolved inorganic carbon is a major component of natural waters. Its isotopic composition is derived from atmospheric input, rock weathering, water-rock interaction and biological activities. Moreover, in volcanic areas there may be a supply of CO_2 from a magmatic source. Thus DIC carbon isotopes in groundwater can be derived from several different sources with different isotopic signals that can makes interpretation complex. However, Sveinbjörnsdóttir et al. (1998, 2001) demonstrated that for a well defined water shed within a homogeneous basalt bedrock DIC isotopes could be used to date the groundwater by using the boron concentration of the groundwater to correct for "dead" rock derived carbon. Earlier studies have also shown that for some low temperature geothermal systems the DIC fingerprints can be used to define the systems, detect changes with time and together with $\delta^{18}O$ and $\delta^{2}H$ trace the origin of thermal waters (Sveinbjörnsdóttir et al. 2005).

In the years 2002 to 2005 numerous water samples were collected in Iceland for geochemical and isotopic studies. The samples represent surface waters, non-thermal groundwaters and geothermal waters from natural springs and drillholes. The main objectives of the project were to map the chemical and isotopic evolution of the waters, to quantify the processes that regulate the chemical and isotopic composition of the waters and to determine the age and sources of cold and thermal groundwaters. A total of 322 samples were collected and analyzed for major and trace elements and the isotopes of oxygen and hydrogen. 228 of the samples were also analyzed for carbon isotopes, both for $\delta^{13}C$ and ^{14}C.

This paper will focus on the isotopic results of the samples that were analyzed for oxygen, hydrogen and DIC isotopes. Figure 1 shows the sampling sites for the samples studied.

2 ANALYTICAL METHODS

The stable isotopes of oxygen, hydrogen and carbon were analyzed on the Finnegan MAT 251 mass-spectrometer of the Institute of Earth Sciences, University of Iceland. The results are defined in the conventional δ-notation in ‰, relative to the standards VSMOW for oxygen and hydrogen and VPDB for carbon. The accuracy of the measurements is

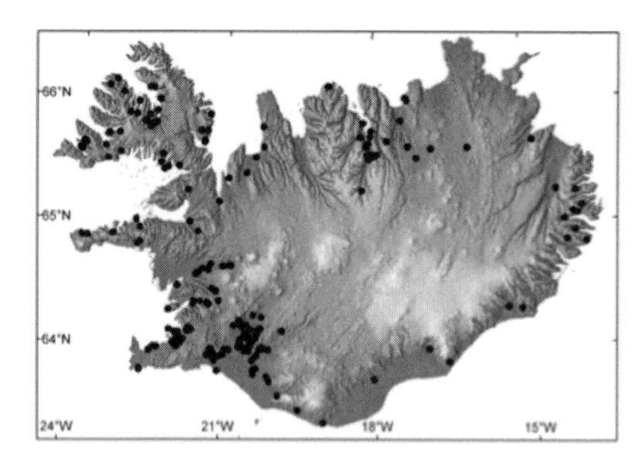

Figure 1. Location of sample sites.

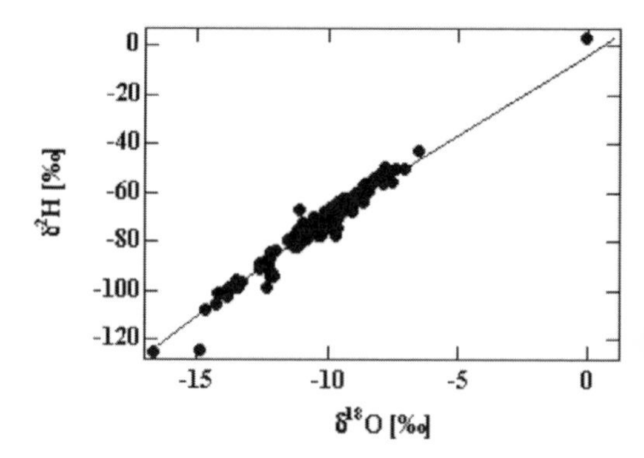

Figure 2. The relation between δ^2H and δ^{18}O for the whole dataset.

better than 0.05‰, 0.7‰ and 0.1‰ for oxygen, hydrogen and carbon, respectively.

Oxygen was extracted from the water samples by the method of Epstein and Mayeda (1953) and Hydrogen isotope analysis were based on the H_2-water equilibration method using a Pt-catalyst (Horita 1988). The sample preparation for carbon isotopes was carried out in accordance with McNichol et al. (1994), i.e. the water samples were acidified in a vacuum system and CO_2 extracted directly by nitrogen flow through the water sample. The collected CO_2 was then partly used for δ^{13}C measurements and partly converted to graphite for AMS ^{14}C measurements performed at the AMS Centre at Aarhus University, Denmark. (Sveinbjörnsdóttir et al. 1992). ^{14}C results are normalised to a δ^{13}C value of −25‰ PDB and expressed in percent modern carbon (pMC) relative to 0.95 times the ^{14}C concentration of the NBS oxalic acid standard (HoxI).

3 RESULTS

3.1 *Oxygen and hydrogen isotopes*

The oxygen and hydrogen isotope composition of the water samples range from about 0 to −17‰ and from +3.5 to −125‰, respectively and follow closely the meteoric water lines defined for Iceland (Sveinbjörnsdóttir et al. 1995) as demonstrated in Figure 2. Some of the samples show an oxygen shift up to ca. 3.5‰ due to water-rock interaction at elevated temperatures and two samples show an unusually high deuterium excess. The high deuterium excess (22‰) of the more ^{2}H depleted sample (δ^{18}O = −11‰) is explained by CO_2 exolution (Kristmannsdóttir & Sveinbjörnsdóttir 2010), whereas the more-heavy sample (δ^{18}O = −0.01‰) is reacted seawater.

The temperature of the water samples ranges from 3°C to 130°C and shows a slight correlation with δ^2H, as the isotopic values get slightly more negative with increasing temperature.

3.2 *Carbon isotopes*

Figure 3a shows the relation between water temperature and δ^{13}C. The δ^{13}C values of the cold water samples (< 10°C) is very variable and ranges from +2‰ to −22‰ as shown in the Figure. At water temperature of 20°C the range in δ^{13}C is considerably narrower or from −11‰ to −24‰. This is presumably due to the influence of soil organics on the water. Consequently most of the groundwater (T > 10°C) seems to have seeped through the soil zone and got ^{13}C depleted due to soil respiration. As it moves further into the ground it heats up and due to silicate weathering and water-rock interaction the δ^{13}C of the water becomes heavier in δ^{13}C (less negative number) as the temperature rises and water-rock interaction increases.

The ^{14}C of the cold water ranges from 10 to ca 130 pMC, as demonstrated in Figure 3b. Most of the cold water samples lie however between 50 and 130 pMC, and only three of the samples have ^{14}C < 40 pMC (Figure 3b). For the groundwater samples with temperature >10°C the ^{14}C concentration is also very variable from ca 80 to 0 pMC and hard to see any clear correlation with temoerature.

There is a slight trend between δ^2H and ^{14}C concentration as demonstrated in Figure 4. The samples are marked according to their geographical location (East-, West- South- North Iceland and Reykjavik). As expected the samples collected in northern Iceland are most depleted in ^{2}H whereas the samples from the southern lowlands are in general least depleted in ^{2}H. Several of the most depleted samples from the southern lowlands have been shown to have a pre-Holocene water component. This is also the case for three samples from western Iceland and two from the North (Sveinbjörnsdottir et al. 1998, 2005). Apart from these ^{2}H deleted pre-Holocene samples the correlation between δ^2H and ^{14}C concentration shown in Figure 4 is the same for samples from the different part of the country; a

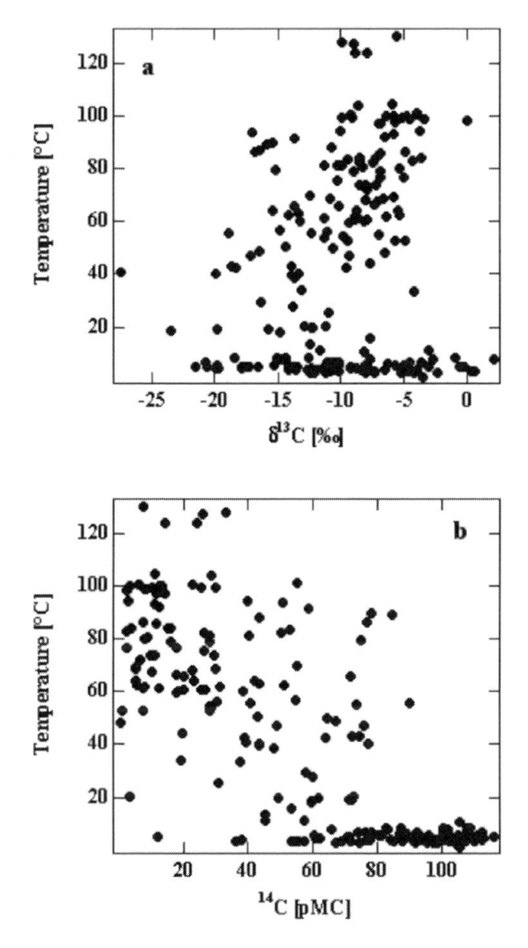

Figure 3. The relation between the water temperature and a) $\delta^{13}C$ and b) ^{14}C concentration in the water.

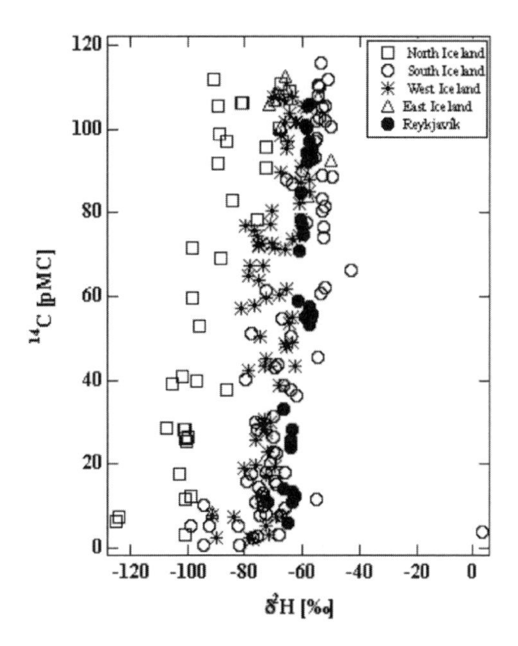

Figure 4. The water samples get more depleted in ^{2}H with decreasing ^{14}C concentration. The same trend is observed for all areas studied.

decrease in $\delta^{2}H$ with decreasing ^{14}C concentration in the water, reflecting increased water-rock interaction with increasing temperature.

The relation between $\delta^{13}C$ and ^{14}C concentration is shown in Figure 5. The figure demonstrates that $\delta^{13}C$ of the water samples that show the lowest ^{14}C concentration lie mostly in the range −2.5‰ to −6‰. A clear trend, with a few exceptions, is observed between $\delta^{13}C$ and ^{14}C below ca 40 pMC, where with decreasing ^{14}C pMC the water becomes less ^{13}C depleted. Some of the samples are more negative in $\delta^{13}C$ and can be explained by fast movement of deep groundwater through soil zone and thus the influence of very ^{13}C depleted organic carbon. At ^{14}C [pMC] > 40 the scatter becomes large. The samples that show the highest ^{14}C pMC [cold waters] range in $\delta^{13}C$ from +2‰ to −22‰ in accordance with observed values for surface rivers in Iceland. When this water seeps underground it

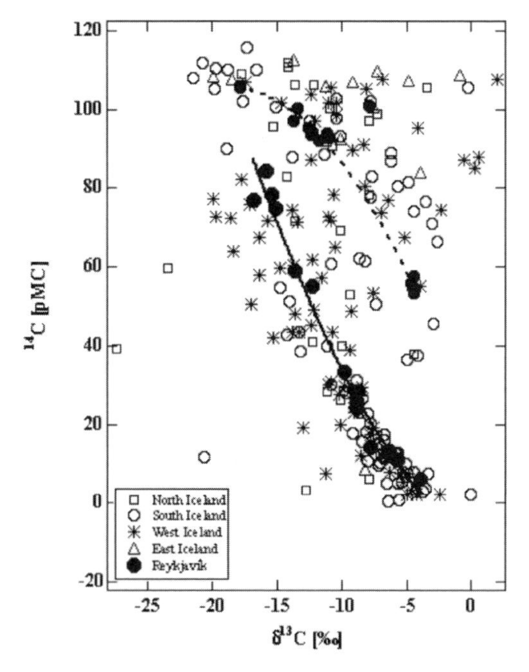

Figure 5. The relation between $\delta^{13}C$ and ^{14}C concentration of the water samples. The solid line represents the best fit between the low geothermal waters from the five thermal systems in the Reykjavik region. The dotted line represents the best fit between the cold water in the Reykjavik region.

starts to react with the basaltic bedrock (0 pMC and $\delta^{13}C$ ca –4‰) and subsequently the pMC of the water samples decreases both due to increasing age of the water and also because of dilution due to "dead" derived rock carbon. At the same time $\delta^{13}C$ of the water samples either increases (gets less negative) if the initial value of the water sample is more negative than the rocks or decreases if the initial value of the water sample is less negative than the rock. No difference can be seen with regards to geographical distribution of the samples.

Samples from five geothermal low temperature (70°C to 120°C) systems within the Reykjavík region are included in Figure 5. The systems are well defined on their $\delta^{18}O$ and δ^2H values. For these five systems the DIC carbon isotopes are very variable, with $\delta^{13}C$ ranging from –4‰ to –17‰ and ^{14}C ranging from 0 to 70 pMC. However, these samples fit remarkably well the solid curve shown in Figure 5.

The cold water ($T < 5°C$) from the same region are all collected from shallow drillholes and have similar $\delta^{18}O$ and δ^2H values as local precipitation. They have a very similar range in $\delta^{13}C$ as the geothermal waters, but range in ^{14}C from 106 to 53 pMC. The coldwater samples fit closely the dotted line on Figure 5.

4 DISCUSSION AND CONCLUDING REMARKS

The water samples are collected from various parts of Iceland. The geological bedrock is similar in all places (basaltic lava pile), except for the age of the bedrock. In general, the age of the bedrock decreases towards the middle of the country, with the highest bedrock age (ca 15 M years) in the NW peninsula and eastern Iceland.

Our results show no difference in the DIC isotopic behavior regarding geographical location, whereas variation of the vegetation cover of the sampling sites obviously affects the DIC isotopic composition of the waters.

From all parts of the country the cold and young (>100 pMC) ground- and surface waters show the same wide range in $\delta^{13}C$ values. Also identical to all geogapical locations is the relationship between ^{14}C concentration and $\delta^{13}C$ of groundwater with <40 pMC. The five low temperature systems in Reykjavík, identified on their distinct δ^2H and $\delta^{18}O$ fingerprints, follow the same curve even for the system that have the most ^{13}C depleted waters and highest ^{14}C concentration. For the rest of the dataset groundwaters with >40 pMC are scattered with relation to ^{14}C and $\delta^{13}C$. However there is an obvious trend from the cold surface waters towards deep ground water that can be explained by dilution of the ^{14}C concentration by increased water-rock interaction.

The cold water samples collected in the Reykjavík region are all young local precipitation by origin. The samples can be divided into 3 groups based on their $\delta^{13}C$ and ^{14}C concentrations, reflecting the depth of drillholes they were sampled from. The DIC isotopic values from the samples collected from the most shallow wells are characterized by least water-rock interaction with >100 pMC and the light $\delta^{13}C$ value of –18‰, samples from intermediate depth wells are characterized by pMC between 100 and 90 and $\delta^{13}C$ around –12‰, whereas the samples from the deepest wells reflect more intense water-rock interaction and show pMC values around 55 and $\delta^{13}C$ around –4.5‰.

ACKNOWLEDGEMENT

This study was financially supported by research grants from the Icelandic Research Council (now: the Icelandic Centre for Research) and the University of Iceland Research Fund.

REFERENCES

Epstein, S. & Mayeda, T.K. 1953. Variation in ^{18}O content of waters from natural sources. *Geochimica et Cosmochimica Acta* 4:213–224.

Horita, J. 1988. Hydrogen isotope analysis of natural waters using an H_2-water equilibration method: A special implication to brines. *Chemical Geology* (Isotope Geosciences Section) 72:89–94.

Kristmannsdottir, H. & Sveinbjörnsdottir, A.E. 2010. Geochemistry and origin of saline geothermal water at Hofstadir near Stykkilsholmur, Iceland. *Journal of Volcanology and Geothermal Research* (submitted).

McNichol, A.P., Jones, G.A., Hutton, D.L. & Gagnon, A.R. 1994. The rapid preparation of seawater CO_2 for radiocarbon analysis at the national ocean Sciences AMS facility. *Radiocarbon* 36:237–246.

Sveinbjörnsdóttir, Á.E., Arnórsson, S. & Heinemeier, J. 2001. Isotopic and chemical characteristics of old "ice age" groundwater, North Iceland. In R. Cidu (ed), *Water-Rock Interaction* 205–208.

Sveinbjörnsdóttir, Á.E., Arnórsson, S., Heinemeier, J. & Boaretto, E. 1998. Geochemistry of natural waters in Skagafjördur, N-Iceland. II Isotopes. In Arehart & Hulston (eds), *Water-Rock Interaction* 653–656.

Sveinbjörnsdóttir, Á.E., Heinemeier, J. & Arnórsson, S. 2005. Isotopic characteristics ($\delta^{18}O$, δD, $\delta^{13}C$, ^{14}C) of thermal waters in the Mosfellssveit and Reykjavík low-temperature areas, Iceland. *Proccedings World Geothermal Congress*, Anatyla, Turkey.

Sveinbjörnsdóttir, Á.E., Heinemeier, J., Rud, N. & Johnsen, S.J. 1992. ^{14}C anomalies observed for plants growing in Icelandic geothermal waters. *Radiocarbon* 34:696–703.

Sveinbjörnsdóttir, A.E., Johnsen, S.J. & Arnórsson, S. 1995. The use of stable isotopes of oxygen and hydrogen in geothermal studies in Iceland. *Proceedings of the World Geothermal Congress*, Florence, Italy, 1043–1048.

Water-Rock Interaction – Birkle & Torres-Alvarado (eds)
© 2010 Taylor & Francis Group, London, ISBN 978-0-415-60426-0

Occurrence of geopressured fluids with high δD and low Cl in northwestern Hokkaido, Japan

A. Ueda
Kyoto University, Katsura, Kyoto, Japan

K. Nagao
The University of Tokyo, Hongo, Tokyo, Japan

T. Shibata & T. Suzuki
Geological Survey of Hokkaido, Sapporo, Hokkaido, Japan

ABSTRACT: Stable (D, O) and noble gas isotopic compositions of 23 thermal and groundwaters along the west coast and in the inland region of Northwestern Hokkaido, Japan were analyzed to investigate their origin. The Cl concentrations vary from 45 to 19,300 mg/L and δD values are in the range from −90 to −8‰. The δD-δ^{18}O plots show a linear relationship which can be explained by a simple mixing of local meteoric water and altered sea water, with an oxygen isotopic shift of 5‰. However, the δD-Cl plot shows that there is also a contribution of an additional water component from different origin. This third component has a δD value of −20‰ and Cl concentration of 6,000 mg/L. This component is regarded as geopressured fluid which is widely distributed in northwestern Hokkaido.

1 INTRODUCTION

One of Professor Hitoshi Sakai's contributions to earth sciences was the identification of several groups of thermal waters in Japan, based on their isotopic and chemical compositions. He employed isotopic data to constrain the origin of Arima and coastal type waters (Sakai & Matsubaya 1974, 1976) and also applied these data to interpret the origin and evolution of Kuroko ore forming fluids (Hattori & Sakai 1979). In these papers, it was also recognized that the isotopic composition of fossil sea water differs from that of original sea water due to interaction with the surrounding rocks. In several areas, fossil sea water was found to have δ^{18}O values of +5 to +8‰ due to the oxygen isotopic shift caused by interaction, but to retain almost the same Cl concentration as fresh sea water (e.g. Sakai & Matsubaya 1974, Mizukami et al. 1977).

Recently, fluids that have oilfield-like characteristic, with unusually high δD (−20‰) and low Cl concentrations (ca. 6,000 mg/L) have been observed along faults in the Horonobe area (Ishii et al. 2006). These fluids are geopressured fluids and might be have originated as a mixture between sea water and local meteoric water and then undergone compositional changes during digenesis. However, the distribution of the geopressured fluid as well as detailed mechanism responsible for the apparent D enrichment and Cl depletion of the fluids are presently unclear.

The aim of the present study is to investigate the distribution of geopressured fluid with high δD and low Cl concentrations near the Horonobe area, Hokkaido and to compare the characteristics of these waters with rare gas isotopic compositions as well as other published data.

2 SAMPLES

Water samples from near the coast (No. 1 to 10) were obtained from wells with depths from 371 to 1807 m. This region is widely underlain by marine argillaceous sediments of Late Miocene age, which consist of diatomite, porcellanite and siliceous shale. Almost all the waters from the wells are derived from sedimentary rocks. The other water samples from inland were collected from hot springs, wells, and closed coal mines. In the area, pre-Tertiary sediments are present and locally contain coal-bearing strata.

In the present paper, water samples are grouped as thermal waters (>25 °C) and groundwaters (<25 °C) on the basis of the outlet temperature. The maximum temperature is 49.9 °C in Sample No. 2 (Tomamae). Water samples were collected in plastic bottles (1 L volume) from August 30 to September 2, 2004. The sampling points are shown in Figure 1. Sampling bottles used for the noble gas study were made of a low He-permeability glass containing Pb. Each bottle had an internal volume of about 50 cm^3 and

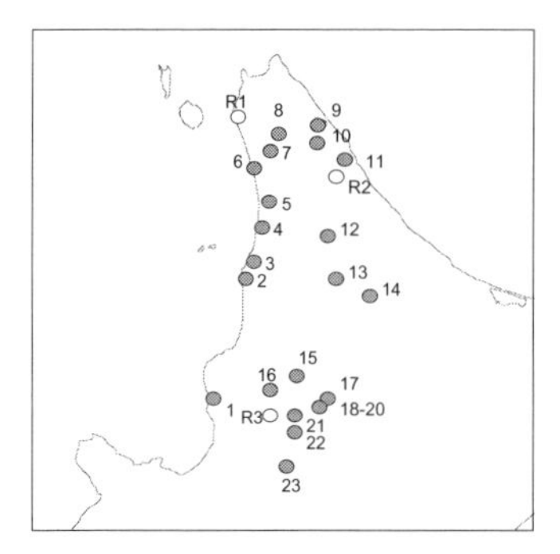

Figure 1. Localities of water and gas sampled.

high-vacuum stop-cocks were attached to both ends. Water was introduced directly into each glass bottle from the sampled water source through plastic tubing to avoid air contamination. After through rinsing with flowing water, the stop-cocks were closed.

3 RESULTS AND DISCUSSION

Studied groundwater and thermal water samples can be divided into 3 groups from their isotopic and chemical compositions; type 1 (samples No. 1, 10, 12 to 17, 19, 20, 22 and 23) mostly of meteoric origin, type 2 (No. 2, 3, 4, 6, 11 and 21) which is a mixture of local meteoric water and fossil sea water, and type 3 (No. 5, 7, 8, 9 and 18), which is a mixture of meteoric water, fossil sea water and deep groundwater (geopressured fluid). There is no good correlation between their types and the outlet temperature. In Figures 2 to 5, water samples from types 1, 2 and 3 are shown as cross, square, and circle, respectively. Geopressured (overpressured) fluid is defined as formation water with pressure between hydrostatic and lithostatic conditions, which is typically distributed below ca. 2 km depth in sedimentary rocks associated with oil and/or natural gas (e.g. Mayers 1968, Oki et al. 1999, Xu et al. 2006).

3.1 *δD and δ¹⁸O values*

The δD and $\delta^{18}O$ values of water samples are shown in Figure 2. Half of the samples plot along the meteoric water line and are of meteoric origin (as shown as cross symbols). In contrast, the others have a good correlation between δD and $\delta^{18}O$ values and may be a mixture of meteoric water and fossil sea water that is enriched in ^{18}O ($\delta^{18}O$ = ca. +6 ‰). Figure 3 shows the δD-Cl relationship which demonstrates that some of samples (square symbols) are plotted along a mixing line between meteoric and fresh sea water. In contrast, 5 samples (circle symbols; No. 5, 7, 8, 9 and 18) are distant from the mixing line. As discussed below, these samples may contain a third water component of different origin

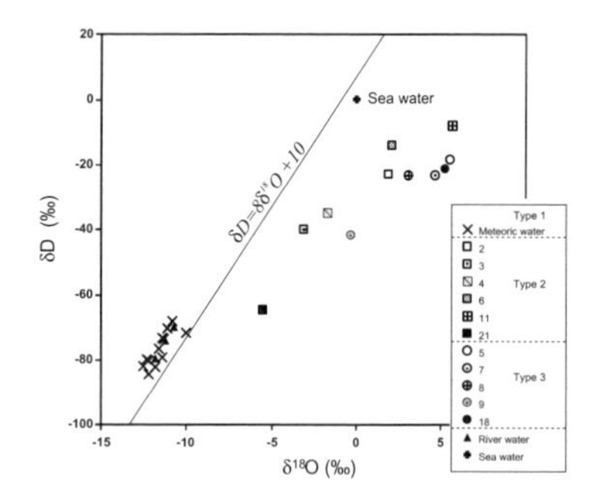

Figure 2. δD vs. $\delta^{18}O$ plots of water samples. A broken line in each figure is a meteoric water line.

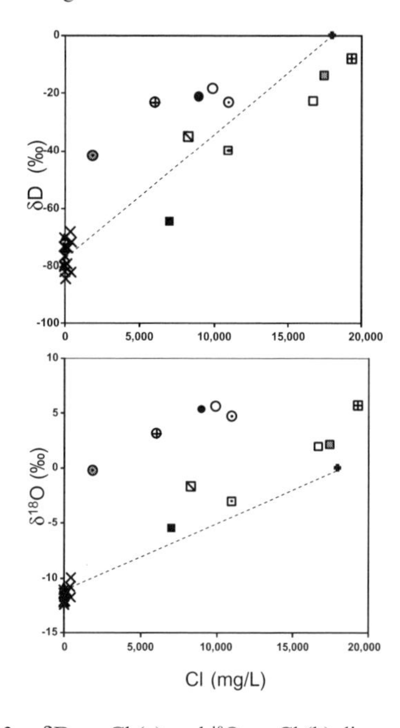

Figure 3. δD vs. Cl (a) and ^{18}O vs. Cl (b) diagrams. The symbols are the same as in Figure 1. The dashed line is a simple mixing line of sea water and local meteoric waters.

with a δD value of ca. −20 ‰ and Cl concentration of ca. 6,000 mg/L. Figure 3 also shows the $\delta^{18}O$-Cl relationship which demonstrates that type 3 waters have higher $\delta^{18}O$ values than type 2 waters, which results from water-rock interaction under long residence time and/or high rock/water ratio.

The sample No. 8 of this study, which is one of the fluids showing characteristics of the third component, was collected at Horonobe. More than 180 water samples from this site have been analyzed for their chemical and isotopic compositions as part of the Horonobe underground research laboratory (URL) project (e.g. Hama et al. 2007). Their results show that the groundwaters are divided into two types; Na-HCO_3 type water in the shallower part of the sedimentary formations and Na-Cl

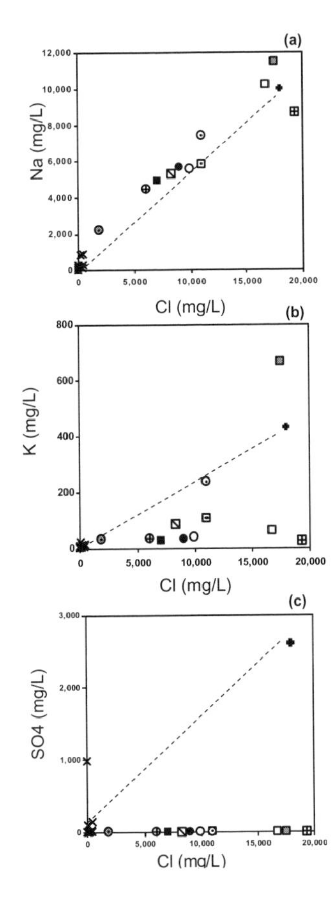

Figure 4. Correlation diagrams for some chemical constituents plotted against Cl. The symbols are the same as in Figure 1. The dashed line is a simple mixing line of sea water and local meteoric waters.

type saline water in the deeper part of the sedimentary formations. Water-rock interactions and mixing are considered to be the dominant evolution processes of the groundwater. However, the Na-Cl type waters have unusually high δD values and low Cl concentrations of –20 to –30 ‰ and 5,000 to 6,000 mg/L, respectively (Ishii et al. 2006), similar to those of sample No. 8 in the present study. These results indicate that groundwater with a high proportion of the third component is widely distributed in the deeper part of this area in the Wakkanai Formation. The enrichment factor (ΔD) is defined as the difference between δD that a sample would have if it originated by simple dilution of current sea water by local meteoric water (using the Cl content of the sample to estimate the degree of dilution), and the observed δD of the sample (Ishii et al. 2006). The ΔD increases up to 45 ‰ at depths below –500 m, where opal A changes to opal CT. At this depth, the H_2O^+ contents in the rocks decreases by 2 wt% (Ishii et al. 2006).

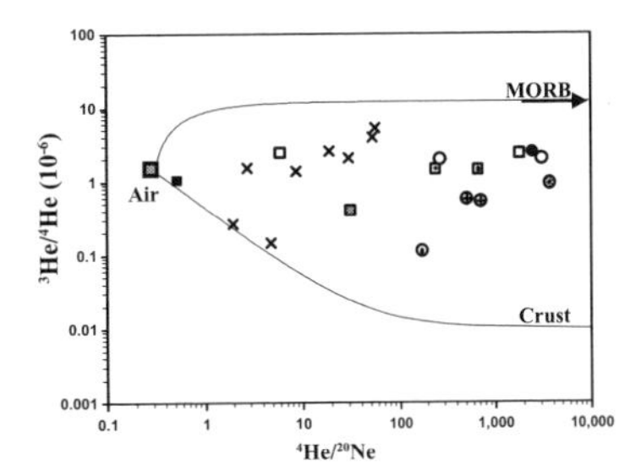

Figure 5. $^3He/^4He$ vs. $^3He/^{20}Ne$ plot of water samples. The symbols are the same as in Figure 1. The upper and lower solid lines are simple mixing lines of MORO and Air He and CRUST and AIR, respectively.

3.2 *Chemical composition*

Concentrations of some major chemical constituents of water samples are plotted against Cl concentrations in Figure 4. The broken line in the figure represents mixing between meteoric water and unmodified sea water. The Na concentration of the water samples plot along this mixing line, whereas K and SO_4 values are below the line. These characteristics are independent of the temperature and are similar to those that have been reported for groundwaters from various localities such as in the Seikan tunnel in Japan (e.g. Mizukami et al. 1977). Previous studies have attributed the observed chemical characteristics to water-rock interaction and mixing processes.

3.3 *Rare gas isotopes*

Figure 5 shows a plot for $^3He/^4He$ versus $^4He/^{20}Ne$ ratios. All gas and dissolved gas samples except for No. 21, 22 and 23 gave medium values between the MORB-AIR and CRUST-AIR mixing lines. There is no clear difference in $^3He/^4He$ ratios between all types. This means that these gases come from both, magmatic and radiogenic sources. The proportion of MORB component in the samples is calculated to be ca. 20% assuming that MORB gas has a $^3He/^4He$ ratio of 11×10^{-6} and CRUST gas of 1×10^{-8}. It is noted that the range of $^4He/^{20}Ne$ ratios for type 3 samples overlaps with that of type 2, whereas $^4He/^{20}Ne$ ratios are higher than for type 1, except for the samples 2, 6 and 21. Samples 21 to 23 plot along the CRUST-AIR mixing lines and are of radiogenic origin. From all samples, samples 1 and 16 from type 1 have the highest $^3He/^4He$ ratios due to a large contribution of magmatic gas. Groundwater samples from type 1 are of meteoric origin and have both, He and Ne from MORB and radiogenic components (Fig.5).

3.4 *Occurrence of geopressured fluid*

Recently, geopressured fluids with unusually high δD (–20‰) and low Cl concentrations (ca. 6,000 mg/L) have been observed along faults in several localities in Japan, e.g., Niigata (Oki et al. 1999, Xu et al. 2006), Horonobe (Ishii et al. 2006), Mobara gas field (Maekawa et al. 2006), and Miyazaki (Ohsawa personal comm.). Fluids in mud volcanoes in various part of the world also have such high δD (−20‰) and low Cl concentrations (e.g. Dia et al. 1999).

Ishii et al. (2006) examined the isotopic and chemical compositions of formation waters in the Horonobe area, including sample No. 8 of the present study (Toyotomi Spa, associated with CH_4). This sample is thought to be a geopressured fluid that has ascended through the Ohmagari fault. The marine nature of the geological sequences suggests that this fluid may have been formed by sea water undergoing a change in isotopic and chemical compositions during digenesis. Ishii et al. (2006) discussed the possibility that the isotopic change from 0 to −20‰ was caused by dehydration of opal A during the formation of opal C (2wt% water can be released). The D/H fractionation factor between opal (amorphous silica) and water is −145‰ and −108‰ at 25 °C and 50 °C, respectively (Kevin et al. 2000). This mineralogical transition is observed in the sedimentary rocks at about −500 m depth and the temperature is not clear but generally around 50 °C (Ishii et al. 2006). During the transition from opal A to C, the δD value and Cl concentration of connate water in sedimentary rocks might be decreased due to dilution of dehydrated water from opal A. However, the observed isotopic and chemical compositions cannot be explained by this model because the total amount of water dehydrated from rocks is too small compared to that of connate water due to the large porosity (35%). At greater depths, transition of smectite to illite is commonly observed in the world (e.g. Bruce 1984). During this transition, the connate water becomes a lower δD value and Cl concentration due to dilution by dehydrated water which has low δD values compared to smectite due to the D/H isotopic fractionation (e.g. −40 ‰ at 25 °C, Savin & Epstein 1970).

4 CONCLUSIONS

H, O and rare gas isotopic compositions were analyzed in groundwater and thermal waters (maximum temperature is 49.9 °C) in northwestern Hokkaido, Japan. The water samples can be divided into 3 types: type 1 is of meteoric origin, type 2 is a mixture of meteoric water and altered sea water (fossil sea water) and type 3 is a mixture of meteoric water, altered sea water and geopressured water. The geopressured fluids have a δD value of −20‰ and Cl concentrations of 6,000 mg/L and are widely distributed in northwestern Hokkaido. The $^3He/^4He$ ratios of 5.5×10^{-7} with a high $^3He/^{20}Ne$ ratio (~700) of type 3 samples indicate a crustal He input with a small amount of He derived from MORB source. In this paper, we could not examine the occurrence of the geopressured fluids and its geological sequences. From the detailed study at Horonobe (sample No. 8), the geopressured fluids might have migrated along fractures from deep crustal source.

REFERENCES

Bruce, C.H. 1984. Smectite dehydration—its relation to structural development and hydrocarbon accumulation in Northern Gulf of Mexico Basin. *AAPG Bulletin* 68: 673–683.

Dia, A.N., Castrec, M., Boulegue, J. & Boudou, J.P. 1995. Major-trace-element, and Sr isotope constraints on fluids circulations in the accretionary complex of Barbados. Part I: fluid origin. *Earth and Planetary Science Letters* 134: 69–85.

Hama, K., Kunimaru, T., Metcalfe, R. & Martin, A.J. 2007. The hydrogeochemistry of argillaceous rock formations at the Horonobe URL site, Japan. *Physics and Chemistry of the Earth* 32: 170–180.

Hattori, K. & Sakai, H. 1979. D/H ratios, origins, and evolution of the ore-forming fluids for the Neogene veins and Kuroko deposits of Japan. *Economic Geology* 74: 535–555.

Ishii, T., Haginuma, M., Suzuki, K., Hama, K., Kunimaru, T., K, Kobori, K., Shimoda, S., Ueda, A. & Sugiyama, K. 2006. Groundwater evolution processes in the sedimentary formation at the Horonobe, northern Hokkaido, Japan. *Geochimica et Cosmochimica Acta*, 70: Supplement 1, A280.

Kevin, F., Matsuhisa, Y. & Fujimoto, K. 2000. Water in amorphous silica and the hydrogen isotopic fractionation. *Rept. Geol. Surv. Japan* 51: 206 (in Japanese).

Maekawa, T., Igari, S. & Kaneko, N. 2006. Chemical and isotopic compositions of brines from dissolved-in-water type natural gas fields in Chiba, Japan. *Geochemical Journal* 40: 475–484.

Meyers, J.D. 1968. Differential pressures, a trapping mechanism in Gulf Coast oil and gas fields. *Gulf Coast Association of Geological Societies Transactions* 18: 56–80.

Mizukami, M., Sakai, H. & Matsubaya, O. 1977. Na-Ca-Cl-SO4-type submarine formation waters at seikan- undersea-tunnel, japan—chemical and isotopic documentation and its interpretation. *Geochimica et Cosmochimica Acta* 41: 1201–1212.

Oki, Y., Xu, H., Ishizaka, N. & Kawauchi, K. 1999. Geopressured hydrothermal system associated with active faults and historical destructive earthquakes Hydrothermal waters in the Niigata Basin. *Hot Spring Science* 48: 163–81 (in Japanese).

Sakai, H. & Matsubaya, O. 1974. Isotopic geochemistry of the thermal waters of Japan and its bearing on the Kuroko ore solution. *Economic Geology* 69: 974–991.

Sakai, H. & Matsubaya, O. 1976. Stable isotopic studies of Japanese thermal systems. *Geothermics* 5: 97–124.

Savin, S.M. & Epstein, S. 1970. The oxygen and hydrogen isotope geochemistry of clay minerals. *Geochimica et Cosmochimica Acta* 34: 25–42.

Xu, H.L., Shen, J.W. & Zhou, X.W. 2006. Geochemistry of geopressured hydrothermal waters in the Niigata Sedimentary Basin, Japan. *Island Arc* 15: 199–209.

Water-Rock Interaction – Birkle & Torres-Alvarado (eds)
© 2010 Taylor & Francis Group, London, ISBN 978-0-415-60426-0

Boron isotopes as a proxy for carbonate dissolution in groundwater—radiocarbon correction models

A. Vengosh
Duke University, Division of Earth & Ocean Sciences, Nicholas School of the Environment, Durham, North Carolina, USA

ABSTRACT: Radiocarbon dating of groundwater has been a major tool for assessing the residence time of groundwater. Several models have been proposed to quantify carbonate dissolution in groundwater system in attempts to correct for "dead carbon". Here I propose using boron isotopes as an additional indirect proxy for evaluating the input of carbonate dissolution. In coastal areas, meteoric boron has a high $\delta^{11}B$ signature (≥39‰) and thus recharge water would have a significant different $\delta^{11}B$ value relative to solution derived from carbonate dissolution. Preservation of high $\delta^{11}B$ rainwater composition in coastal groundwater infers lack of carbonate dissolution and thus an indirect proxy for ^{14}C correction. The model is applied for fossil groundwater from the Disi aquifer in Jordan where high $\delta^{11}B$ (25–48‰) and B/Cl ratio (>sea water) suggest that the recharge water originated from coastal rainwater of an early stage of air mass evolution with negligible water-rock interactions in the aquifer.

1 INTRODUCTION

The evolution of recharge water along groundwater flow paths and correction for water-rock interactions that introduce "dead carbon" are the basis of radiocarbon age dating, which has been applied in numerous studies in attempts to determine the residence time of fossil groundwater (e.g., Clark & Fritz 1997). Several models have been proposed to quantify the "dilution" of ^{14}C of the recharge water, including definition of the aquifer type (Vogel 1970), using the dissolved inorganic carbon (DIC) for mass-balance quantification of calcite dissolution (Tamers 1975), using major element mass-balance for carbonate dissolution (Fontes and Garnier 1979), applying $\delta^{13}C$ for correction of dissolved DIC from marine calcite (Pearson & Hanshaw 1970), and tracing with $^{87}Sr/^{86}Sr$ for carbonate dissolution (Bishop et al. 1994) (see reviews in Clark & Fritz 1997, Geyh 2005). In carbonate aquifers the correction for carbonate dissolution is more robust as the $\delta^{13}C$ composition of older marine calcite ($\delta^{13}C$~0‰) is significantly different from that of the DIC in groundwater ($\delta^{13}C$~−15‰). However, in sandstone aquifers, carbonate minerals are precipitated from the recharge water with a depleted $\delta^{13}C$ composition, and thus are isotopically indistinguishable from recharge DIC. Since many fossil groundwaters are found in non-marine sandstone aquifers (e.g., The Nubian Sandstone aquifer in North Africa and Middle East), the ability to provide adequate ^{14}C correction in these aquifers is limited.

This paper proposes using boron isotopes as an additional tool for ^{14}C correction. The method is based on the assumption that incorporation of boron into carbonate minerals involves isotope fractionation while carbonate dissolution does not, thus dissolution of carbonate would introduce boron with a different isotope composition relative to the recharge water. The proposed model is restricted to coastal areas where meteoric boron has a distinctive high $\delta^{11}B$ signature. The method is applied for low-saline groundwater from the Disi aquifer in southern Jordan (Vengosh et al. 2009).

2 THEORY

Aqueous boron occurs primarily as boric acid $B(OH)_3$ and borate ion $B(OH)_4^-$. The relative proportions of boron species in a solution depend on the pH, salinity, temperature, and pressure conditions (Dickson 1990, Millero 1995). The fractionation of boron isotopes relates to the vibrational and rotational energy differences between the two boron species; ^{11}B tends to incorporate preferentially into the trigonal species (boric acid) while ^{10}B is selectively fractionated into the tetrahedral species (borate ion) (Kakihana et al. 1977). The fractionation factor is defined as $\alpha_{B3-B4} = R_{B3}/R_{B4}$, where R is the isotope ratio $^{11}B/^{10}B$. The fractionation factor for the boric acid-borate ion system was originally determined theoretically by Kakihana et al. (1977) to be 1.0193, yielding an isotopic difference of 19.3‰ between boric acid and borate ion at 25°C. Recently, this fractionation factor has been reevaluated (Klochko et al. 2006), which suggests an isotopic difference of ~27‰ (α_{B3-B4} = 1.0272) at 25°C and seawater salinity. Since tetrahedral boron is assumed to incorporate preferentially into carbonate minerals (Vengosh et al. 1991, Hemming & Hanson 1992), the boron isotope composition of carbonate is expected to be depleted in ^{11}B relative to the water from which it precipitated. Modeling of the isotopic composition of calcite depends on the salinity, temperature, pressure, and pH, which control the boron species distribution in the solution and thus the δ^{11}B of borate ion. The expected δ^{11}B values of carbonate minerals under low salinity at 25°C (pk_B = 9.23) and different pH are presented in Figure 1. The model predicts an increase δ^{11}B of carbonate with pH (Fig. 1).

Boron in rainwater is derived from gaseous (90–95%) and particulate forms (Fogg and Duce, 1985). The gaseous boron is derived from evaporation of seawater and seawater aerosols and that is associated with isotope fractionation in which the vapour phase is 31‰ depleted in ^{11}B relative to the aquatic phase (Rose-Koga et al. 2006). During condensation and formation of rainwater, the gaseous-liquid fractionation induces ^{11}B-rich rainwater. Since boric acid from seawater is preferentially volatized to the atmosphere, the first

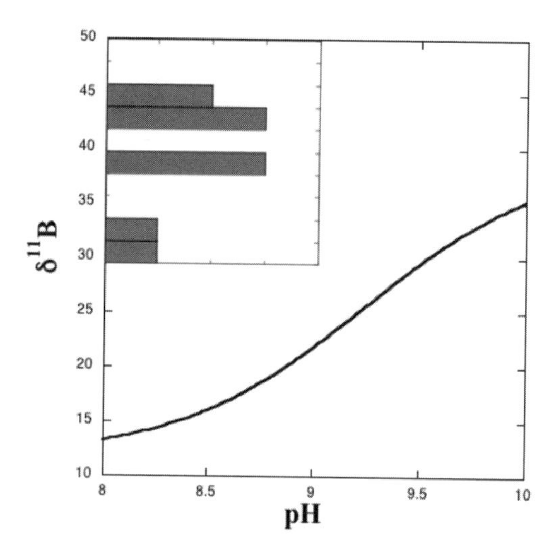

Figure 1. The predicted δ^{11}B of carbonate minerals originated from low-saline water at 25°C as compared to coastal rainwater composition (Chetelat et al. 2005). The model is based on initial water with δ^{11}B = 39‰, α_{B3-B4} = 1.0272, and boric acid dissociation constant (pkB) of 9.23 that corresponds to low saline water at 25°C. The histogram in the sub-plot represents the variations of δ^{11}B values in coastal rainwater from French Guiana (Chetelat et al. 2005).

rain that would condense from a marine air mass would have δ^{11}B > seawater (45–50‰) that corresponds to the δ^{11}B composition of boric acid in seawater. Further rainout process would cause progressive depletion of ^{11}B in the air mass, similar to the Rayleigh distillation process that characterizes ^{18}O and ^{2}H during rain evolution (Rose-Koga et al. 2006). This was demonstrated in rainwater that originated from the intertropical convergence zone (ITCZ) in French Guiana with δ^{11}B values ranging from 30‰ to 46‰ (Chetelat et al. 2005; Fig. 1) and chloride-rich rainwater from coastal California (up to 25‰; Rose-Koga et al. 2006).

The partition coefficient (Kd*) of boron in carbonate minerals is defined as $B_{carbonate}/B_{solution}$ and varies from 1.4 to 4 (Hemming et al. 1995, Sanyal et al. 2000), thus boron uptake into calcite would enrich the solid phase with boron relative to the initial solution. Dissolution of either marine carbonates (δ^{11}B = 10 to 20‰; Vengosh et al. 1991) or secondary carbonate originated from rainwater in sandstone aquifers (Fig. 1) would generate groundwater with a significantly lower δ^{11}B values relative to the recharge water in coastal areas. Since boron in groundwater can also be derived from desorption and/or weathering of silicate minerals (Lemarchand & Gaillardet 2006) that would also reduce the δ^{11}B of the recharge water (Vengosh & Spivack 2000), using the boron method exclusively without additional geochemical tools may be biased in an attempt to correct for carbonate dissolution. Nonetheless, preservation of the original high δ^{11}B rainwater in groundwater in coastal areas could indicate lack of carbonate dissolution.

3 METHODS

This study is part of investigation of the origin of fossil groundwater from the Disi aquifer system in southern Jordan (Vengosh et al. 2009; Fig. 2). The aquifer is part of a vast Cambro-Ordovician sandstone basin that extends to a larger section of the northern and central Saudi Arabia peninsula, and is composed mainly of highly conductive sandstone overlying the Precambrian basement. The aquifer system is generally divided into the Cambro-Ordovician Rum Group and the overlying Ordovician-Devonian Khreim Group (Lloyd & Pim 1990).

Boron isotopes were measured by a negative ion technique using a recently installed Thermo Scientific TRITON thermal ionization mass spectrometer at Duke University. Water samples were mixed with high-purity synthetic single-element standard solutions of Ca, Mg, Na, and K in proportions similar to those in seawater in a 5% HCl matrix salt solution with composition similar to boron-free seawater (Dywer & Vengosh 2008). The total loading blank

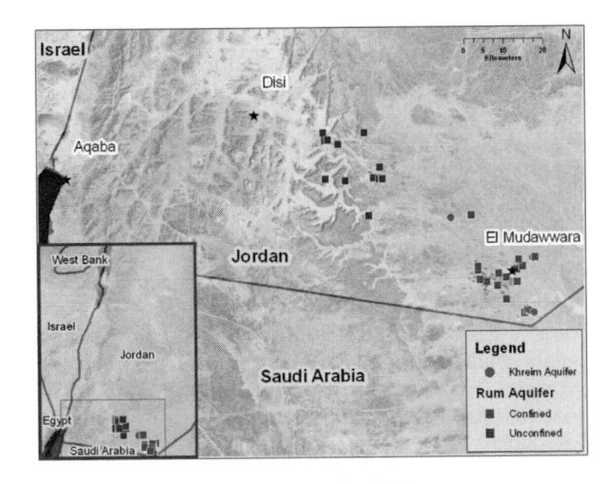

Figure 2. Map showing the distribution of wells from the unconfined and confined Rum Groups and the Khreim aquifer from the Disi aquifer system in Southern Jordan (from Vengosh et al. 2009).

is around 15 pg B as determined by isotope dilution mass spectrometry using ^{10}B enriched NIST952. Filaments were gradually heated to a temperature of 920°C. External replicate analyses of NIST SRM951 and modern seawater have yielded $^{11}B/^{10}B$ ratios of 4.0058 (± 0.0011, 0.3‰, n = 62) and 4.1630 (± 0.0025, n = 17; $\delta^{11}B$ = 39.2 ± 0.6‰), respectively. Isotopic fractionation during individual measurements (60–90 minutes) was typically less than 0.5‰.

4 RESULTS AND DISCUSSION

Groundwater from the Disi aquifer is characterized by low salinity (TDS of 250 mg/L) in the confined zone (Vengosh et al. 2009). Results from the investigated groundwater reveal high B/Cl (exceeding the marine ratio of $8x10^{-4}$) and $\delta^{11}B$ (range of 25–48‰; Fig. 3). Specific high $\delta^{11}B$ values were measured in groundwater from the confined Rum and Khreim aquifers (Fig. 3). These high $\delta^{11}B$ values suggest that the recharge water originated from rainwater at an early stage of air mass evolution with negligible water-rock interaction in the aquifer. This isotope composition is expected for an air mass that is derived from a marine moisture source, like the ITCZ coastal rain in French Guiana (Chetelat et al. 2005). It is therefore suggested that the recharge water of the Disi aquifer originated from coastal rains with minimum rainout evolution, thus preserving the original high $\delta^{11}B$ composition of the rainwater. The high $\delta^{11}B$ values are consistent also with relatively high $\delta^{18}O$ and $\delta^{2}H$ values, which confirm that the Disi groundwater was derived from rainwater with minimum rainout fractionation, particularly for groundwater from the confined Rum aquifer (Fig. 3). Consequently,

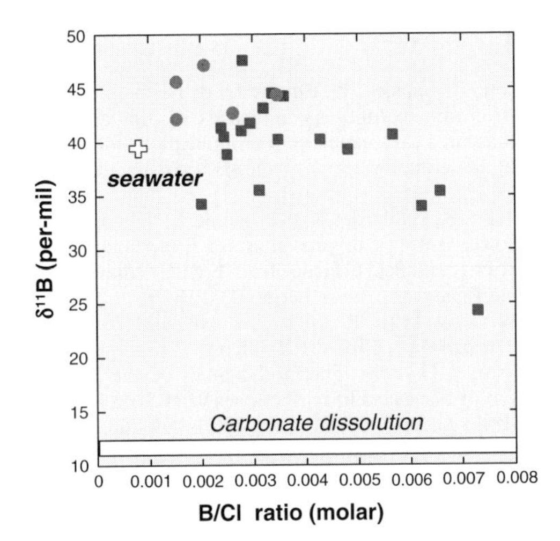

Figure 3. $\delta^{11}B$ versus B/Cl ratios of groundwater from the Disi aquifer in Southern Jordan. Groundwater symbols correspond to the legend in Figure 2. The higher $\delta^{11}B$ and B/Cl ratios of the groundwater relative to seawater (+) suggest that groundwater was originated from coastal rainwater. Carbonate dissolution refers to expected $\delta^{11}B$ range of carbonate minerals, calculated by assuming (1) $\delta^{11}B$ of initial water = seawater; (2) low salinity and pkB = 9.23; (3) temperature = 25°C; (4) fractionation of 27‰; and (5) pH range of 7 to 8.

the preservation of the original rainwater composition implies that carbonate dissolution in the aquifer is negligible, and therefore the ^{14}C activity in the Disi groundwater infers net decay of the original recharge water. The uncorrected ^{14}C data suggests three major replenishment events into the Disi aquifer (1) >30 ka BP to the Khrein aquifer; (2) 15–29 ka BP to the confined Rum aquifer; and (3) 9–12 ka BP to the unconfined Rum aquifer.

5 CONCLUSIONS

Quantification of secondary carbonate dissolution in sandstone aquifers is essential for ^{14}C correction models, yet a challenging task. This paper proposes using boron isotopes as an additional proxy for evaluating carbonate dissolution. The boron isotope methodology is indistinguishable for other water-rock interactions, and thus can be applied for only coastal areas where the recharge water has an elevated $\delta^{11}B$ that is different from the expected ^{11}B-depleted composition of carbonate dissolution. Results from the Disi aquifer in southern Jordan show high $\delta^{11}B$ values (>seawater), particularly for groundwater from the confined Rum aquifer, which infers preservation of original Late Pleistocene rainwater.

REFERENCES

Bishop, P., Smalley, P., Emery, D. & Dickson, J. 1994. Strontium isotopes as indicators of the dissolving phase in a carbonate aquifer: implications for ^{14}C dating of groundwater: *Journal of Hydrology* 154(1–4): 301–321.

Chetelat, B., Gaillardet, J., Freydier, R. & Négrel, P. 2005. Boron isotopes in precipitation: Experimental constraints and field evidence from French Guiana. *Earth and Planetary Science Letters* 235: 16–30.

Clark, I. & Fritz, P. 1997. *Environmental Isotopes in Hydrogeology*. CRC Press, 328 pp.

Dickson A.G. 1990. Thermodynamics of the dissociation of boric acid in synthetic seawater from 273.15 to 318.15 K. *Deep Sea Research* 37(5): 755–766.

Dwyer, G.S. & Vengosh, A. 2008. Alternative filament loading solution for accurate analysis of boron isotopes by negative thermal ionization mass spectrometry. *American Geophysical Union, Fall Meeting 2008, Abstract #H51C-0824, December 2008.*

Fogg, T.R. & Duce, R.A. 1985. Boron in the tr oposphere: distribution and fluxes. *Joural of Geophysical Res*earch 90 (D2): 3781–3796.

Fontes, J.-Ch. & Garnier, J.-M. 1979. Determination of the initial 14C activity of total dissolved carbon. A review of existing models and a new approach. *Water Resources Research* 15, 399–413.

Geyh, M.A. 2005. Dating of old groundwater- history, potential, limits and future. In Aggarwal, P.K., Gat, J.R. & Froehlich, K.F.O., (eds), *Isotopes in the Water Cycle: Past Present and Future of Developing Science*. Springer.

Hemming, N.G. & Hanson, G.N.,1992. Boron isotopic composition and concentration in modern marine carbonates. *Geochimica Cosmochimica Acta* 56: 537–543.

Hemming, N.G., Reeder, R.J. & Hanson, G.N. 1995. Mineral-fluid partitioning and isotopic fractionation of boron in synthetic calcium carbonate. *Geochimica Cosmochimica Acta* 59: 371–379.

Kakihana, H., Kotake, M., Satoh, S, Nomura, M. & Okamoto, M. 1977. Fundamental studies on the ion-exchange separation of boron isotopes. *Bulletin of the Chemical Society of Japan* 50: 158–163.

Klochko K., Alan J. Kaufman, A.J., Yao, W., Byrne, R.H. & Tossell, J.A. 2006. Experimental measurement of boron isotope fractionation in seawater. *Earth and Planetary Science Letters* 248: 261–270.

Lemarchand, D. & Gaillardet, J. 2006. Transient features of the erosion of shales in the Mackenzie basin (Canada), evidences from boron isotopes. *Earth and Planetary Science Letters* 245: 174–189.

Lloyd, J.W. & Pim, R.H. 1990. The hydrogeology and groundwater resources development of the Cambro-Ordician sandstone aquifer in Saudi Arabia and Jordan. *Journal of Hydrology* 121:1–20.

Millero, F.J. 1995. Thermodynamics of the carbon dioxide system in the oceans. *Geochimica et Cosmochimica Acta* 59: 661–667.

Pearson, F.J. & Hanshaw, B.B. 1970. Sources of dissolve carbonate species in groundwater and their effects on carbon-14 dating. In: *Isotope Hydrology 1970, IAEA Symposium 129, Vienna,* 271–286.

Rose-Koga, E.F., Sheppard, S.M.F., Chaussidon, M. & Carignan, J. 2006. Boron isotopic composition of atmospheric precipitations and liquid–vapour fractionations. *Geochimica et Cosmochimica Acta* 70:1603–1615.

Sanyal, A., Nugent, M., Reder, R.J.R. & Bijma, J. 2000. Seawater pH control on the boron isotopic composition of calcite: Evidence from inorganic calcite precipitation experiments. *Geochimica et Cosmochimica Acta* 64: 1551–1555.

Tamers, M.A. 1975. The validity of radiocarbon dates on groundwater. Geophysical Survey 2: 217–239.

Vengosh, A. & Spivack, A.J. 2000. Boron in Ground Water. *In:* P. Cook & A. Herczeg (eds), *Environmental Tracers in Groundwater Hydrology*: 479–485. Boston, Dorderecht, London: Kluwer Publisher.

Vengosh, A., Kolodny, Y., Starinsky, A., Chivas, A.R. & McCulloch, M.T. 1991. Coprecipitation and isotopic fractionation of boron in modern biogenic carbonates. *Geochimica et Cosmochimica Acta* 55: 2901–2910.

Vengosh, A., Hirschfeld, D., Vinson, D.S., Dwyer, G.S. Raanan, H., Rimawi, O., Al-Zoubi, A., Akkawi, E., Marie, A., Haquin, G., Zaarur, S. & Ganor, J. 2009. High Naturally Occurring Radioactivity in Fossil Groundwater in the Middle East. *Environmental Science and Technology* 43 (6), 1769–1775• DOI: 10.1021/es802969r.

Vogel, J.C. 1970. Carbon-14 dating of groundwater. In *Isotope Hydrology 1970, IAEA Symposium 129, Vienna:* 225–239.

Water-Rock Interaction – Birkle & Torres-Alvarado (eds)
 ISBN 978-0-415-60426-0

Sr isotopes, hydrogeologic setting, and water-rock interaction in the Mt. Simon sandstone (Minnesota, USA)

D.S. Vinson
Duke University, Division of Earth & Ocean Sciences, Durham, NC, USA

J.R. Lundy
Minnesota Department of Health, Source Water Protection, St. Paul, MN, USA

G.S. Dwyer & A. Vengosh
Duke University, Division of Earth & Ocean Sciences, Durham, NC, USA

ABSTRACT: We present new $^{87}Sr/^{86}Sr$ ratios in groundwater from Proterozoic and Cambrian sandstone aquifers of southeastern Minnesota (USA), emphasizing the Mt. Simon aquifer. In Mt. Simon waters, $^{87}Sr/^{86}Sr$ ranges from 0.7141 in the recharge area to 0.7085 at depth. Sampling of the overlying Franconia, Ironton, and Galesville aquifers similarly indicates $^{87}Sr/^{86}Sr$ from 0.7132 in shallow conditions to 0.7086 at depth. Proterozoic sandstone aquifers in shallow bedrock conditions exhibit $^{87}Sr/^{86}Sr$ of 0.7103–0.7119. Sr sources to the Mt. Simon include: (1) more radiogenic Sr derived from overlying Cretaceous rocks and/or Pleistocene glacial deposits that incorporate Precambrian crystalline rocks and Cretaceous sedimentary rocks; and (2) less radiogenic Sr associated with Na-bearing waters, possibly from plagioclase weathering. Overall, the observed $^{87}Sr/^{86}Sr$ and correlation with carbon-14 activity record the Sr system's sensitivity to groundwater evolution from the recharge area to deep, slow-circulating areas and contributions from multiple Sr sources associated with hydrogeologic setting.

1 INTRODUCTION

Isotopes of strontium (Sr) provide valuable constraints on weathering and other processes implied by alkaline earth metal concentrations, such as dilution of ^{14}C in dissolved inorganic carbon by $CaCO_3$ dissolution and cation exchange. In addition to tracing cation sources related to chemical processes, Sr isotopes can record conservative mixing of waters with different $^{87}Sr/^{86}Sr$. In regional sandstone aquifers in which groundwater residence time and major ion chemistry are observed to evolve down-gradient from the recharge area to slow-circulating zones, the relative importance of carbonate equilibrium, cation exchange, and other geochemical processes may vary within the aquifer.

In the Cambrian Mt. Simon sandstone aquifer (MTS) of southeastern Minnesota (USA), increasing Sr concentration and total dissolved solids coincide down-gradient with increasing groundwater residence time and a transition from Ca-Mg-bicarbonate composition to a more varied major ion chemistry. Because this transition also coincides with increased radium activity down-gradient (Lively et al. 1992), $^{87}Sr/^{86}Sr$ ratios have been determined as part of ongoing research on naturally-occurring radium in the Cambrian sandstone aquifers of Minnesota. The purpose of these isotopic measurements is to improve constraints on processes affecting alkaline earth metals (Mg, Ca, Sr, Ba, and Ra) and the apparent relationship between hydrogeologic setting, groundwater residence time, and Ra activity. In addition to Sr isotope ratios from the MTS, results are also presented from limited sampling of groundwater from the underlying Proterozoic sandstone aquifers, overlying Cambrian Franconia, Ironton, and Galesville sandstone aquifers, and Quaternary glacial drift aquifers.

2 METHODS

Water samples were collected from drinking water wells, mostly from high-capacity municipal water supply wells. Water was obtained from wellhead taps before chemical treatment. Stratigraphic and well construction data were obtained from the Minnesota Department of Health County Well Index. To supplement this data set, additional unfiltered, non-preserved water samples were analyzed for $^{87}Sr/^{86}Sr$. Sr concentration is not reported for the supplemental samples.

Sr isotope ratios were measured on a Thermo Scientific TRITON thermal ionization mass spectrometer after preconcentration of Sr using Eichrom SR-B50-S exchange resin. Replicate analyses of

NIST SRM987 over the course of this study yielded a mean $^{87}Sr/^{86}Sr$ ratio of 0.710245 ± 0.000010 (1σ, n = 109). Whole-process replicates of four samples differed by a median of 0.000008, which is consistent with precision for SRM987. Sr concentrations were determined on acidified water samples by direct-current plasma spectrometry.

3 RESULTS

The most radiogenic waters (>0.711) are observed in the portion of the MTS that occurs in relatively shallow settings beneath Pleistocene glacial deposits. $^{87}Sr/^{86}Sr$ ratios gradually decrease into the center of the Twin Cities basin, which is both a geologic basin affecting the Paleozoic formations and the site of a cone of depression in the Minneapolis-St. Paul metropolitan area. These less radiogenic values reach a minimum of 0.708472 near the eastern margin of the Twin Cities Basin. Samples in the western MTS exhibit $^{87}Sr/^{86}Sr$ in the narrow range of 0.7091–0.7096 (Figure 1).

These relationships are apparent vertically within the Mt. Simon aquifer results, as well as in limited supplemental data from the underlying Proterozoic sandstone aquifers and overlying Cambrian Franconia-Ironton-Galesville sandstone aquifers (Figure 2). Increasing depth of the aquifer below the bedrock surface (disregarding Quaternary deposits) is associated with progressively less radiogenic values. Although deep bedrock conditions were not sampled in the Proterozoic sandstone aquifers, the shallow conditions sampled in this study indicate relatively radiogenic $^{87}Sr/^{86}Sr$ > 0.710 (Figure 2).

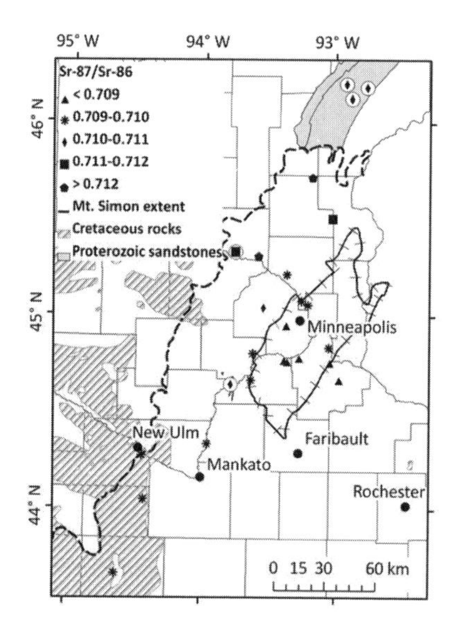

Figure 1. Map showing $^{87}Sr/^{86}Sr$ ratios for Mt. Simon sandstone aquifer wells and wells in primarily Proterozoic sandstone aquifers (circled points). Hachured line (Oronto Group extent from Minnesota Geological Survey shapefile) represents approximate extent of Twin Cities Basin. Other units shown are from Mossler (1992) and Morey & Meints (2000).

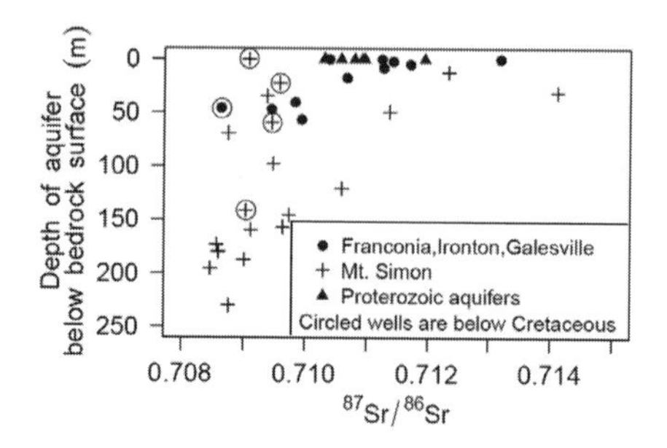

Figure 2. Variation of $^{87}Sr/^{86}Sr$ with depth below the bedrock surface to top of the sampled aquifer. Circled points are wells below or near observed Cretaceous rocks.

4 DISCUSSION

4.1 *Strontium source identification*

The plot of $^{87}Sr/^{86}Sr$ vs. 1/Sr (Figure 3) is consistent with Sr contributions from two or more sources: a more radiogenic source at low Sr concentration and a less radiogenic source at higher Sr concentration. Also, waters along the western margin of the MTS exhibit more radiogenic $^{87}Sr/^{86}Sr$ than samples in the vicinity of the Twin Cities Basin with comparable Sr concentrations (Figure 3).

The relatively radiogenic, Ca-Mg-HCO_3 waters generally exhibit molar (Ca+Mg)/HCO_3^- ratio of ~0.5 consistent with waters influenced by carbonate mineral dissolution. The more radiogenic $^{87}Sr/^{86}Sr$ values indicate that this cannot be the result of dissolution of marine carbonate (Figure 4). Although Sr sources depend on the specific formation, higher $^{87}Sr/^{86}Sr$ ratios have been documented in secondary relative to primary carbonate, due to secondary carbonate incorporating Sr derived from continental weathering of radiogenic minerals (Bishop et al. 1994, Frost & Toner 2004).

Analysis of five water samples from Quaternary drift aquifers in the Minneapolis-St. Paul metropolitan area indicates a wide range of $^{87}Sr/^{86}Sr$ (median 0.7108, range 0.7098–0.7131), most of which are more radiogenic than the bedrock aquifers. Thus, the diverse sediment sources of Des Moines Lobe tills present in the western part of the study area, which include material derived from Precambrian crystalline rocks and Cretaceous shales (Arneman & Wright 1959), may impart a large range of potential $^{87}Sr/^{86}Sr$ ratios on waters recharging through glacial deposits (e.g. Franklyn et al. 1991).

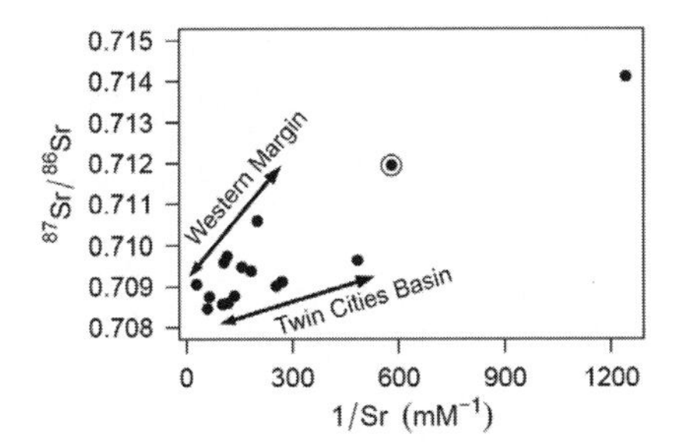

Figure 3. Relationship between Sr concentration and $^{87}Sr/^{86}Sr$ in MTS groundwater showing general geographic trends. Circled point represents a well mostly in Proterozoic sandstone.

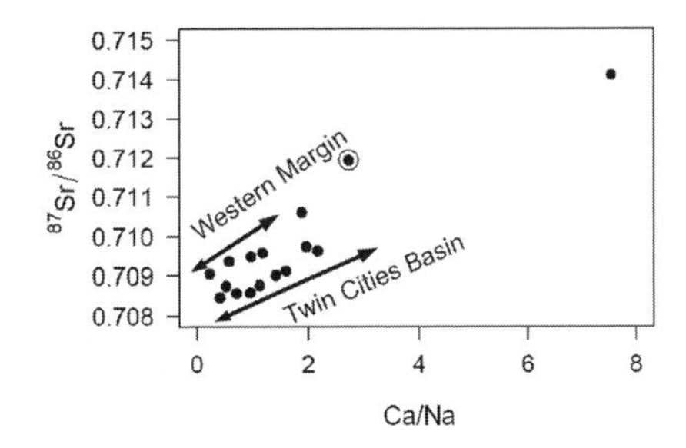

Figure 5. Relationship between $^{87}Sr/^{86}Sr$ and molar Ca/Na ratio in MTS groundwater showing general geographic trends. Circled point represents a well mostly in Proterozoic sandstone.

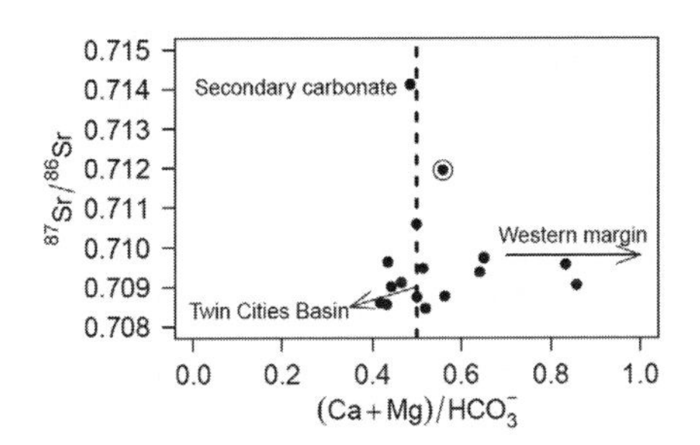

Figure 4. Relationship between (Ca+Mg)/HCO_3^- and $^{87}Sr/^{86}Sr$ in MTS groundwater. Circled point is a well mostly in Proterozoic sandstone; dashed line represents ratio of 0.5 (section 4.1).

In the Twin Cities Basin waters, the less radiogenic $^{87}Sr/^{86}Sr$ ratios of waters cannot be explained by dissolution of marine carbonate alone, as several samples exhibit ratios slightly lower than ~500 Ma (Guzhangian/Dresbachian) seawater (0.7088–0.7092; McArthur et al. 2001). Also, carbonate dissolution cannot explain the increasing proportion of Na in these waters as $^{87}Sr/^{86}Sr$ becomes less radiogenic (Figure 5). This relationship probably represents weathering of plagioclase which, with its high Sr concentration and low Rb/Sr ratio, can contribute large inputs of less radiogenic Sr to groundwater (e.g. Franklyn et al. 1991). This is also consistent with the observed feldspathic nature of the MTS (Mossler 1992).

Samples enriched in Ca (trend to right in Figure 4) coincide with wells in the southwestern portion of the MTS and exhibit a narrow range of $^{87}Sr/^{86}Sr$ (0.7091–0.7096). These Ca-Mg-SO_4-HCO_3 waters generally exhibit higher $^{87}Sr/^{86}Sr$ at comparable Ca/Na ratios than Twin Cities Basin waters (Figure 5). This group also exhibits higher sulfate concentrations (1.7–4.9 mM) that are probably derived from the overlying Cretaceous aquifer and/or Cretaceous-derived material in the Des Moines Lobe tills. $^{87}Sr/^{86}Sr$ in these areas (0.7090–0.7096) is more radiogenic than Sr contributed by Cretaceous seawater (<0.708; McArthur et al. 2001) and in that respect is somewhat consistent with $^{87}Sr/^{86}Sr$ of Ca-SO_4 groundwater in Cretaceous sandstone aquifers of eastern Nebraska (0.7100–0.7111) interpreted as water-rock interaction with the sandstone matrix (Gosselin et al. 2004). In the western MTS, $^{87}Sr/^{86}Sr$ could indicate a mixture of Cretaceous-derived Sr, other radiogenic sources in the Des Moines Lobe till, and/or Sr native to the MTS. Overall, the effect of Cretaceous-derived material is not well constrained, but the observed $^{87}Sr/^{86}Sr$ in the western MTS does not appear to result solely from weathering in the Mt. Simon sandstone because of the observed geographic variability with the Twin Cities Basin (Figure 3).

Another potential Sr source to the MTS could be mixing with a deep, saline contribution of unknown $^{87}Sr/^{86}Sr$. In the MTS, higher salinity water may occur locally (1) along basin-bounding faults (Lively et al. 1992) and (2) in the lower portion of the MTS in the Twin Cities Basin (Runkel et al. 2003). A fault-associated water has been documented with total dissolved solids of 7,500 mg L^{-1}, and Ca/Na (mol/mol) of 0.16 (Lively et al. 1992). This is slightly lower than the minimum Ca/Na of 0.23 observed in MTS waters in this study (Figure 5). However, the Na/Cl ratio of this saline water is 0.7, whereas the Na/Cl ratios of MTS waters in this study range from 0.6–10.8 (median 2.6), indicating that Na concentrations far exceed those expected from the conservative ion Cl. Also, vertical mixing within wells could provide a mechanism for intermediate $^{87}Sr/^{86}Sr$ ratios in wells with long open-hole intervals. However, 6 of the 7 wells with the longest open-hole intervals (65–88 m), mostly coinciding with the greatest MTS thickness in the Twin Cities basin, are associated with the lowest values of $^{87}Sr/^{86}Sr$ (<0.709) in the study area, which implies that these wells do not contain a mixture of documented more radiogenic and less radiogenic Sr. Mixing of these waters with undocumented Sr of low $^{87}Sr/^{86}Sr$ is possible but not

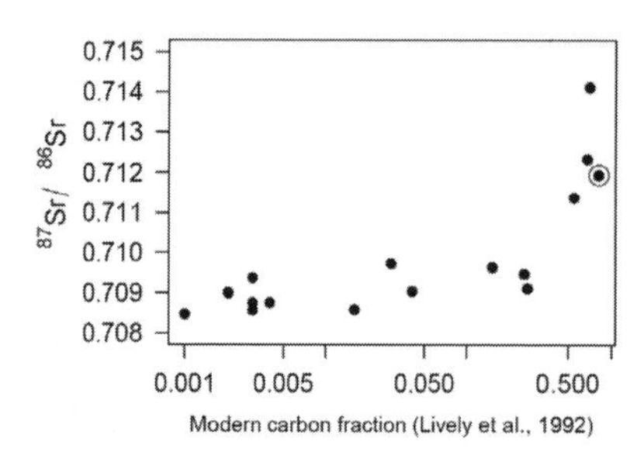

Figure 6. Relationship between $^{87}Sr/^{86}Sr$ and previously published carbon-14 activities (Lively et al. 1992). Circled point represents a well mostly in Proterozoic sandstone.

supported by ion ratios as discussed above. Thus, although deep saline contributions to $^{87}Sr/^{86}Sr$ cannot be ruled out, this appears to be a minor influence on waters sampled for this study.

4.2 *Relationship to groundwater residence time*

Previously published carbon-14 data (Lively et al. 1992) for these or nearby wells in the MTS permit examination of the relationship between Sr isotope ratios and groundwater residence time. Twelve of the 16 points in Figure 6 refer to wells sampled by Lively et al. (1992) that were resampled during 2007–2009 for this study; four points represent MTS wells sampled by Lively et al. (1992) nearby to 2007–2009 sample sites (within the same public water system). $^{87}Sr/^{86}Sr$ is less radiogenic in waters of long residence time indicated by low ^{14}C activity (Figure 6). The more radiogenic $^{87}Sr/^{86}Sr$ seen in waters with >50% modern carbon further implies that radiogenic Sr is obtained from a shallow rather than deep geologic source, probably as recently recharged water interacts with glacial deposits and/or secondary carbonate minerals in the subsurface. It should also be noted that $^{87}Sr/^{86}Sr$ can be used to constrain age models by identifying dilution of ^{14}C activity by dissolution of ^{14}C-free carbonate (e.g. Bishop et al. 1994).

5 CONCLUSIONS

The results of this study indicate several possible influences on observed $^{87}Sr/^{86}Sr$ in the Mt. Simon aquifer. In general, $^{87}Sr/^{86}Sr$ in the MTS is sensitive to hydrogeologic setting and groundwater residence time. The correlation of $^{87}Sr/^{86}Sr$ with Ca/Na in MTS further implies that more radiogenic Sr may be cycled in secondary carbonate, and less radiogenic Sr is not derived wholly from carbonate dissolution but perhaps from plagioclase weathering. In addition, Sr in MTS waters along the western margin of the study area is apparently derived from materials contained in Cretaceous rocks or glacial deposits containing Precambrian crystalline and/or Cretaceous sedimentary rocks; waters in the western area exhibit more radiogenic $^{87}Sr/^{86}Sr$ than waters in the Twin Cities Basin area at comparable Sr concentrations and Ca/Na ratios. Overall, interaction with near-surface sediments imparts more radiogenic $^{87}Sr/^{86}Sr$ on groundwater in the Mt. Simon aquifer, and the varied sediment components of glacial deposits may partially explain the complex pattern of $^{87}Sr/^{86}Sr$ observed in this aquifer system.

ACKNOWLEDGMENTS

We thank Bruce Olsen, Robert Tipping, Anthony Runkel, and E. Calvin Alexander for helpful discussions on the hydrogeology of southeastern Minnesota, and Nathaniel Warner for much assistance with Sr isotope analysis. We also thank Daniel Larsen and Thomas Bullen for their constructive reviews.

REFERENCES

Arneman, H.F. & Wright, H.E. 1959. Petrography of some Minnesota tills. *Journal of Sedimentary Petrology* 29: 540–554.

Bishop, P.K., Smalley, P.C., Emery, D. & Dickson, J.A.D. 1994. Strontium isotopes as indicators of the dissolving phase in a carbonate aquifer: implications for ^{14}C dating of groundwater. *Journal of Hydrology* 154: 301–321.

Franklyn, M.T., McNutt, R.H., Kamineni, D.C., Gascoyne, M. & Frape, S.K. 1991. Groundwater $^{87}Sr/^{86}Sr$ values in the Eye-Dashwa Lakes pluton, Canada: Evidence for plagioclase-water reaction. *Chemical Geology (Isotope Geoscience Section)* 86: 111–122.

Frost, C.D. & Toner, R.N. 2004. Strontium isotopic identification of water-rock interaction and ground water mixing. *Ground Water* 42: 418–432.

Gosselin, D.C., Harvey, E.F., Frost, C., Stotler, R. & Macfarlane, A.P. 2004. Strontium isotope geochemistry of groundwater in the central part of the Dakota (Great Plains) aquifer, USA. *Applied Geochemistry* 19: 359–377.

Lively, R.S., Jameson, R., Alexander, E.C. & Morey, G.B. 1992. Radium in the Mt. Simon-Hinckley aquifer, east-central and southeastern Minnesota. Minnesota Geological Survey, Information Circular 36.

McArthur, J.M., Howarth, R.J. & Bailey, T.R. 2001. Strontium isotope stratigraphy: LOWESS version 3: best fit to the marine Sr-isotope curve for 0–509 Ma and accompanying look-up table for deriving numerical age. *Journal of Geology* 109: 155–170.

Morey, G.B. & Meints, J. 2000. Geologic map of Minnesota, bedrock geology [1:1,000,000]. Minnesota Geological Survey, State Map Series S-20.

Mossler, J.H. 1992. Sedimentary rocks of Dresbachian Age (late Cambrian), Hollandale Embayment, southeastern Minnesota. Minnesota Geological Survey, Report of Investigations 40.

Runkel, A.C., Tipping, R.G., Alexander, E.C., Green, J.A., Mossler, J.H. & Alexander, S.C. 2003. Hydrogeology of the Paleozoic bedrock in southeastern Minnesota. Minnesota Geological Survey, Report of Investigations 61.

Water-Rock Interaction – Birkle & Torres-Alvarado (eds)
© 2010 Taylor & Francis Group, London, ISBN 978-0-415-60426-0

δD and $\delta^{18}O$ and chloride as indicators of groundwater recharge and discharge in Datong Basin, Northern China

X. Xie, Y. Wang, C. Su & M. Li
School of Environmental Studies & MOE Key Laboratory of Biogeology and Environmental Geology, China University of Geosciences, Wuhan, China

ABSTRACT: To better understand recharge and discharge processes of high arsenic and high fluoride groundwater at Datong Basin, environmental isotopes (δD and $\delta^{18}O$) and chloride concentration were analyzed for 29 groundwater samples. For the deep (depth >50 m) and shallow groundwater samples, the δD and $\delta^{18}O$ values range from −92 to −70‰ and from −12.5 to 8.3‰, and from −90 to −57‰ and from −12.3 to −6.7‰, respectively; and the Cl content from 6.59 to 1335 mg/L and from 12.03 to 385.6 mg/L respectively. High F groundwater occurring in northeastern part of the study area contains high Cl concentration and heavier isotope, which could be due to evaporation. High arsenic groundwater in the central of the basin with low Cl content and lighter isotopic composition could be related to irrigation using groundwater. Therefore, the spatial distribution of Cl and $\delta^{18}O$ can be correlated to the occurrence of high As/F groundwater at Datong.

1 INTRODUCTION

Datong Basin is a typical Cenozoic basin bounded by Pliocene to Pleistocene NE–SW trending normal faults that form part of the Shanxi Graben System (Wang & Shpeyzer 2000). It is located in northern part of the Shanxi province (Fig. 1) and is a typical arid and semi-arid region with a mean annual rainfall of 300–400 mm and a mean evaporation rate of about 2000 mm per year. About 80% of rainfall occurs from July to August (Wang et al. 2004). There are two seasonal rivers (Sanggan and Huanshui) in the Datong Basin. In this area, groundwater mainly occurs in two aquifers: shallow (5–50 m) and deep (>50 m). There are two general flow regimes for groundwater: (1) flow from marginal mountain front areas to the central of the basin, and (2) longitudinal flow within the basin along the direction of river flow from SW to NE. In this arid region, groundwater is the primary potable water source for drinking and irrigation purpose. Unfortunately, the extracted groundwater often contains high concentrations of arsenic and fluoride. The concentration of arsenic and fluoride is up to 1820 μg/L and 10.4 mg/L respectively (Xie et al. 2008, Wang et al. 2009). Although considerable studies have been conducted (Guo & Wang, 2003, Xie et al. 2008, Xie et al. 2009, Wang et al. 2009), the genesis of high arsenic and fluoride groundwater has not been fully understood. The recharge and discharge processes of groundwater

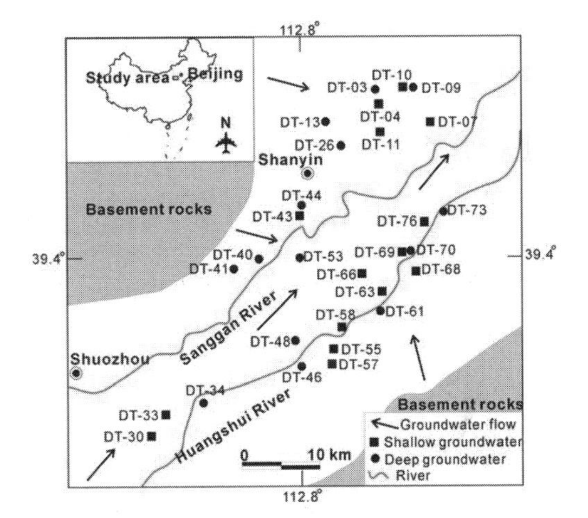

Figure 1. Location of study area and sampling sites.

are very important for understanding the occurrence of high arsenic and fluoride groundwater (Wang et al. 2009). In addition, understanding groundwater recharge and discharge in arid and semi-arid regions represents an essential component in the management of groundwater resource.

The present study was conducted in Datong Basin where the groundwater recharge and discharge processes still need more detailed investigation. In recent years, environmental isotopes and chloride content have been successfully employed as tools in understanding recharge and discharge patterns of aquifer systems (Salameh 2004, Zagana et al. 2007). The objectives of this study are therefore: (1) to examine the applicability of environmental isotopes and chloride concentration to recognize groundwater recharge/discharge relationships; and (2) to improve the understanding of recharge and discharge of high arsenic and fluoride groundwater in Datong.

2 MATERIAL AND METHODS

Twenty nine water samples were taken from Datong Basin for analysis of environmental isotopes Deuterium (D) and Oxygen (^{18}O) and chloride (Cl) concentration. Sampling sites covered the typical high arsenic and fluoride groundwater area at Datong. Water samples were filtered on site using 0.45 μm membrane filters. Samples for Cl analysis were collected in 125 ml polypropylene bottles. Another 125 ml sub-samples was collected for environmental isotope analysis. Chloride analysis was determined using ion chromatography (IC) (Metrohm 761 Compact IC). Isotope analysis for D and ^{18}O was performed by isotope ratio mass spectrometry (Thermo Finnigan MAT253). All measurements were conducted at the MOE Key Laboratory of Biogeology and Environmental Geology, China University of Geosciences in Wuhan.

Isotope ratios are reported in per mil (‰) as the conventional delta-notation (δ) δD and δ^{18}O is reported relative to the standard mean ocean water (SMOW). The reproducibility of the δD and δ^{18}O analysis of groundwater is better than 2‰ and 0.1‰, respectively. For chloride analysis, the reproducibility is better than 5%.

3 RESULTS

The 29 samples have a quite wide range of δD and δ^{18}O values (Figs. 2 and 4). For the deep groundwater (depth >50 m), the δD and δ^{18}O values range from −91.6 to −70.4‰ and from −12.5 to 8.3‰, respectively. As compared with the deep groundwater, shallow groundwater (depth < 50 m) samples had heavier isotopic composition, ranging between −90.1 and −56.9‰ for δD and between −12.4 and −6.7‰ for δ^{18}O (Fig. 4).

All of the water samples plot below the lower right side of the local meteoric water line (LMWL) and global meteoric water line (GMWL) (Fig. 2), which are defined as $\delta D = 7.6 \times \delta^{18}O + 9.3$ (Wang 1991) and $\delta D = 8 \times \delta^{18}O + 10$ (Craig 1961), respectively.

The concentration of Cl varied between 6.59 and 1335 mg/L and between 12.03 and 385.6 mg/L

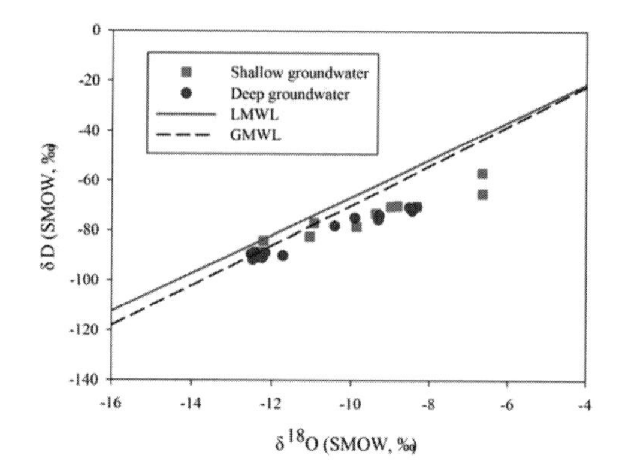

Figure 2. Relationship between δD and δ^{18}O for the groundwater samples from the Datong.

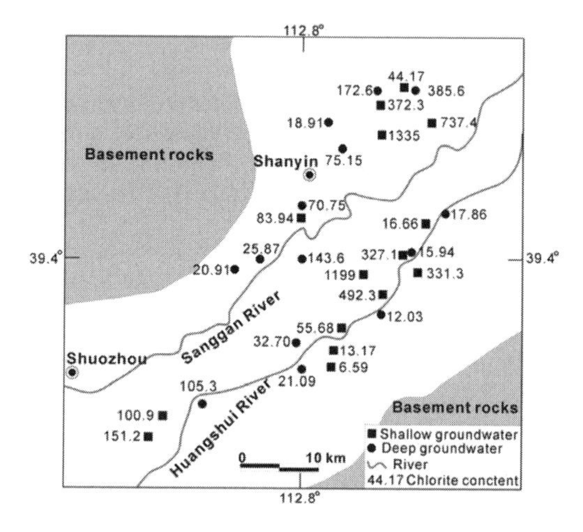

Figure 3. The spatial distribution of Chloride concentration in groundwater from Datong Basin.

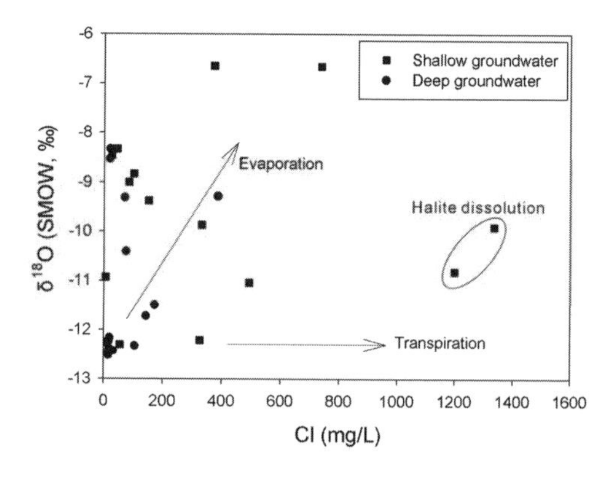

Figure 4. Relationship of Cl concentration and δ^{18}O values. The two highest Cl concentration samples could be due to the dissolution of halite.

for shallow and deep groundwater, respectively (Fig. 3). The spatial distribution of Cl in deep and shallow groundwater is shown in Figure 3. The lowest Cl concentration was observed in the southwestern part of the area for both shallow and deep groundwater. The highest Cl concentration (with 1199 and 1335 mg/L respectively) in the shallow groundwater is recorded in the northeastern part of the basin which is a local discharge zone of groundwater. And the highest Cl concentration (up to 385.6 mg/L) was also observed in the deep groundwater in the same area (Fig. 3).

4 DISCUSSION

As a conservative element, chloride concentration in groundwater can be elevated due to evaporation. The Cl concentration may increase along the flow path from recharge area to the discharge area and reach its maximum value in places where evaporation prevails. Evaporation can modify isotopic composition of water resulting in increasing $\delta^{18}O$ values with the increase of Cl concentration. In this study area, groundwater in the mountain front (recharge area) typically contained lower Cl concentration, whereas in the central of the basin (discharge area) higher Cl concentration (Fig. 3). Moreover, apart from two highest Cl concentrations of groundwater, positive correlation was observed between $\delta^{18}O$ value and Cl concentration for both shallow and deep groundwater (Fig. 4). Thus, in the northeastern part of the study area as a local discharge area, high Cl concentration and enriched isotopic composition could be due to the evaporation. However, in the central part of the basin, along the groundwater flow from SW to NE the isotopic composition became lighter with increase in Cl concentration (Figs. 3 and 5). The increase in Cl content can be attributed to the effect of evaporation, while the decrease in isotopic ratio could be related to the recharge of irrigation water which contains lighter isotopic compositions very similar to the local rainfall. Halite dissolution is more important Cl source than evaporation for two shallow groundwaters with highest Cl concentration (Fig. 4). The low Cl concentration and enrichment of lighter isotopic composition in deep groundwater can be related to regional recharge of infiltrating rainwater from the mountain front areas which contained low Cl concentration and lighter isotopic composition.

The relation between δD and burial depth of groundwater shows that δD values decreased with depth (Fig. 6), suggesting evaporation can strongly modify the isotopic composition of the shallow groundwater, but not that of deep groundwater. Therefore, the rainwater with lighter isotopic composition similar to the deep groundwaters is the primary recharge source for the groundwater in this area.

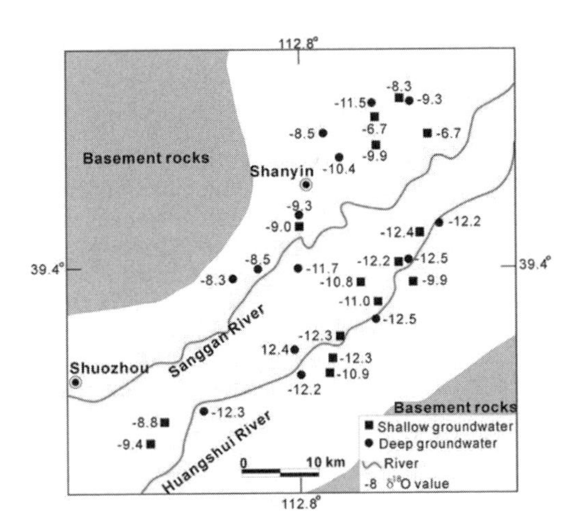

Figure 5. The spatial distribution of the $\delta^{18}O$ in the groundwater from Datong basin.

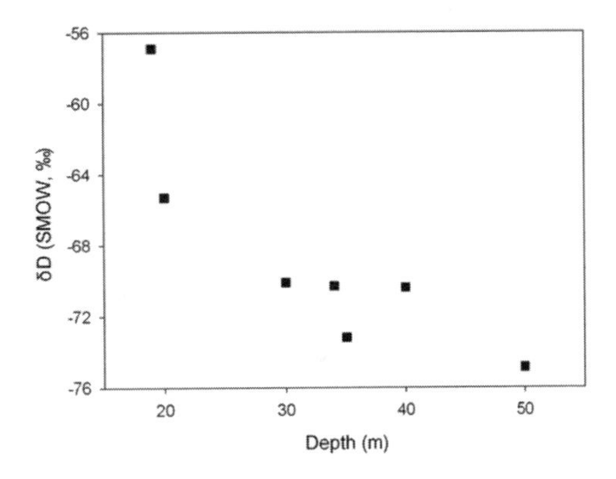

Figure 6. The relation between δD and depth of groundwater depth for samples collected from southeastern part of area.

In arid regions, evaporation can also greatly affect the isotopic composition of rain to make it isotopically enriched. Evaporation waters progressively evolve away from LMWL and GMWL to produce lower regression slope as evaporation process has a kinetic nature (Williams, 1997). The slopes of regression line of shallow and deep groundwater (4.7 and 5.1 respectively) clearly indicate the effect of evaporation on groundwater geochemistry. Most of the samples collected from southern part of the area along the Huangshui River shows lighter isotopic composition recorded (Fig.5). These samples are plotted very close to the LMWL or GMWL (Fig. 2), indicating their origin from local rainwater. However, for samples taken from the northern side of Sanggan River, they are shifted away from the LMWL and GMWL and contain heavier isotopic composition (Figs. 2 and 3), suggesting the effect of evaporation.

The general groundwater flow direction of the aquifer system is from the mountain front areas to the basin range and from SW to NE in the central part of the basin. The spatial distribution of Cl and $\delta^{18}O$ can be correlated to the occurrence of high fluoride and high arsenic groundwater, because differences in recharge and discharge processes result in variations of Cl content and isotopic composition in groundwater. High fluoride groundwater occurring in the northern part of the area contains higher Cl and heavier isotopic components indicating the effect of evaporation on fluoride enrichment. Evaporation process can result in accumulation of sodium.

Na-predominant groundwater is regarded as favorable for fluoride enrichment in Datong Basin (Wang et al. 2009). However, high arsenic groundwater in the southeastern part of the area has high Cl concentration but low heavier isotopic component. This characteristic indicates that the generation of high arsenic groundwater in this area can be related to the irrigation using arsenic-contained groundwater. Combining irrigation with evaporation can result in the enrichment of arsenic in shallow groundwater in this area. Therefore, the occurrence of high fluoride and arsenic groundwater could be related to the difference in recharge and discharge processes.

5 CONCLUSIONS

Environmental isotopes (δD and $\delta^{18}O$) and Cl concentration in high fluoride and arsenic groundwater from Datong Basin were studied. The spatial and vertical distribution characteristics of isotopic composition and Cl suggest that:

1. Rainwater recharge is significant to the groundwater in this area. In addition, in the southwestern part of the area, the recharge from the irrigation is important for both shallow and deep groundwater;
2. The spatial distribution of Cl and $\delta^{18}O$ can be correlated to the occurrence of high fluoride and high arsenic groundwater at Datong. More detailed work on groundwater hydraulics is needed to quantitatively characterize the regional groundwater flow and its relationship with the genesis of high As/F groundwater.

ACKNOWLEDGEMENT

The research work was supported by National Natural Science Foundation of China (40830748 & 40902071). The authors would like to thank Prof. Ian Cartwright and Prof. Paul Shand for constructive suggestions on this manuscript.

REFERENCES

Craig, H. 1961. Isotopic variation in meteoric waters. *Science* 133(3465): 1702–1703.

Guo H.M., Wang, Y.X. & Shpeizer, G.M. 2003. Natural occurrence of arsenic in shallow groundwater, Shanyin, Datong Basin, China. *Journal of Environmental Science and Health, Part A: Toxic/Hazardous Substances & Environmental Engineering* 38: 2565–80.

Salameh, E. 2004. Using environmental isotopes in the study of the recharge/discharge mechanisms of the Yarmouk catchment area in Jordan. *Hydrogeology Journal* 12: 451–463.

Wang H.C. 1991. *Isotopic hydrogeology.* Beijing: Geology Press.

Wang, Y.X. & Shpeyzer, G.M. 2000. *Hydrogeochemistry of Mineral Waters from Rift Systems on the East Asia Continent: Case Studies in Shanxi and Baikal.* Beijing: China Environmental Science Press.

Wang, Y.X., Shvartsev, S.L. & Su, C.L. 2009. Genesis of arsenic/fluoride-enriched soda water: A case study at Datong, northern China. *Applied Geochemistry* 24(4): 641–649.

Wang, Y.X. Guo, H.M. & Yan, S.L. 2004. *Geochemical evolution of shallow groundwater systems and their vulnerability to contaminants: A case study at Datong Basin, Shanxi province, China.* Beijing: Science Press.

Williams, A.E. 1997. Stable isotope tracers: natural and anthropogenic recharge, Orange County, California. *Journal of Hydrology* 201: 230–248.

Xie, X.X. Ellis, A. & Wang, Y.X. 2009. Geochemistry of redox-sensitive elements and sulfur isotopes in the high arsenic groundwater system of Datong Basin, China. *Science of the Total Environment* 407: 3823–3835.

Xie, X.X. Wang, Y.X. & Su, C.L. 2008. Arsenic mobilization in shallow aquifers of Datong Basin: Hydrochemical and mineralogical evidences. *Journal of Geochemical Exploration* 98: 107–115.

Zagana, E., Obeidat, M. & Kuells, Ch. 2007. Chloride, hydrochemical and isotope methods of groundwater recharge estimation in eastern Mediterranean areas: a case study in Jordan. *Hydrological Processes* 21(16): 2112–2123.

Water-Rock Interaction – Birkle & Torres-Alvarado (eds)
 ISBN 978-0-415-60426-0

Isotopic composition of some thermal springs in the South Rifian Rides, Morocco

A. Zian & L. Benaabidate
Laboratory of Georesources and Environment, Faculty of Sciences and Technology, Fez, Morocco

O. Sadki
National Office of Mining and Hydrocarbon, Rabat, Morocco

ABSTRACT: The abundances of 18O and deuterium were measured in water samples taken from thermal springs located in the southern Rif Rides, Morocco. The regional isotopic variations were evaluated based on hydrological, topographical and geochemical considerations. It was concluded that groundwater in the study area underwent various stages of evaporation before infiltration, and the high relief of the Rides receive rainfall from an atmosphere where the 18O content is more negative.

1 INTRODUCTION

Several thermal sources have been utilized in the southern Rif Rides, located in northern Morocco (Fig. 1), without any information on the hydrogeological flow paths, which may lead to poor operating conditions and eventually influence the quality and capacity of existing aquifers, especially in the presence of a remarkable hydrological drought experienced by the region (Zian 2005).

Taking into account the lithological diversity of aquifers in the Rides Basin, it is difficult to answer questions about the origin of water and its mineralization, the history of water until its emergence from the deeper parts where this water acquired most of its mineralization, and finally the length of the flow path.

The utility of isotopic techniques fits well into this kind of study and is often very successful.

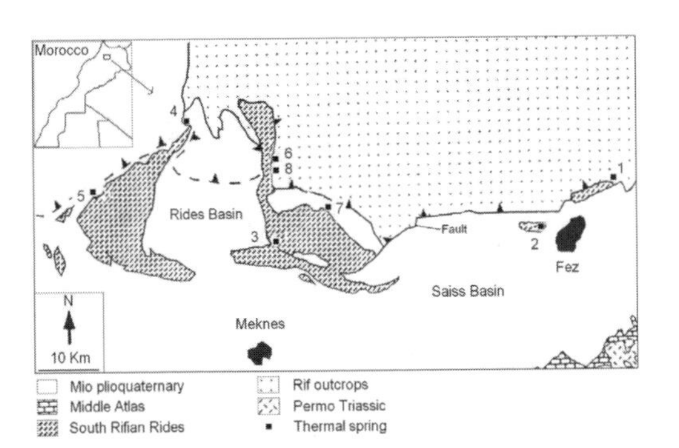

Figure 1. The South Rifian Rides (Faugères, 1978).

Indeed, these methods allow monitoring of water that is present in the environment or resulting from widespread anthropogenic action (environmental isotopes), or introduced intentionally and selectively, in an artificial way, into the hydrological system.

2 MATERIALS AND METHODS

In hydrology, studies focus particularly on three isotopes, 18O, D and T that are the original constituents of water molecules and therefore as such can be considered as ideal hydrogeological tracers (Etcheverry & Parriaux 1998).

The isotopic elements used in this study were determined on samples collected from different sources along the various thermal springs emerging in the south Rifain Rides.

The oxygen and hydrogen isotopic compositions (18O and D, in ‰, relative to V-SMOW standard) were determined by reaction of water with metallic Zn at 500°C (Kendall & Coplen 1985) in the laboratory of Geochemistry of the University of Kentucky at Lexington. The analysis results are expressed in the δ-notation in parts per thousand *vis-a-vis* the V-SMOW (Vienna—Standard Mean Ocean Water) for 18O and D.

3 RESULTS AND DISCUSSIONS

3.1 *Stable isotope contents in the Rides Basin*

The results for 18O and D in the Rides Basin waters are given in Table 1. These waters have values of

Table 1. Isotopic contents of thermal springs in the South Rifian Rides.

No	Springs	T (°C)	18O (%)	D (%)
1	Zalagh	34	99.65	0.35
2	Traht	37	99.37	0.63
3	Moulay Driss	31	96.79	3.16
4	Tiouka	26	99.78	0.22
5	Outita	41	97.59	2.40
6	Es-Skhoun	25	99.87	0.13
7	Anseur	24	99.91	0.09
8	Beida	22	99.89	0.11

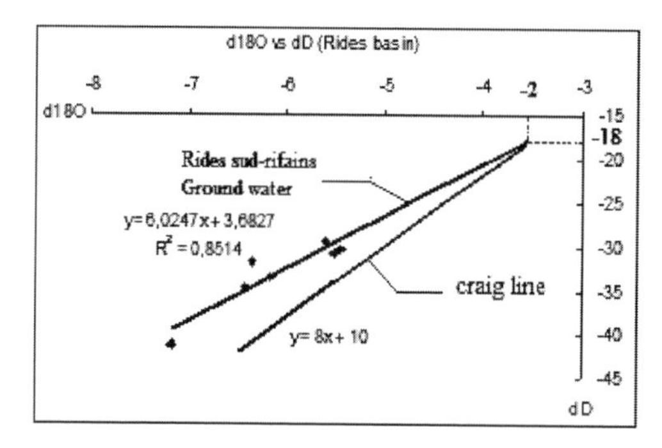

Figure 2. Relation of oxygen-18 and Deuterium.

stable isotopes ranging from –4 to –7 % vs. SMOW for 18O with an average of about –5.80 vs. SMOW and –29.5 to –41.5 % in D with an average of about –33.62 % vs. V-SMOW for deuterium.

The plot of stable isotopes results (18O, D) on the δ^2H *vs* δ^{18}O classical diagram (Fig. 2) shows that the line representing the groundwater in the Rides Basin lies above the average meteoric water line (Craig 1961). The observed lower slope shows, in fact, that the waters have suffered different degrees of evaporation due to isotopic exchange with the local atmosphere. Indeed, this differentiation is governed by the change in altitude of the South Rifain Rides. Infiltration altitudes increase from the west towards East and correspond to the mean altitude of the main tectonic front in the Couloir Pre-Rif and the Middle Atlas, respectively. The thermal waters from Morocco are of Meteoric origin (Cidu & Bahaj 2000).

However, we should not overlook the phenomenon which results from exchange of oxygen-18 with the deep reservoirs matrix, especially in the case of some of the thermal water sources (Es-Skhoun, Moulay Driss and Outita). The silicates and carbonates are generally richer in 18O than waters. 18O content of the geothermal water in contact to the minerals will increase towards the rock values. The intersection of this line with the Craig (1961) line provides the isotope content of water before exchange. They are respectively –2 % for deuterium and –18 % for oxygen-18. Furthermore, waters that have undergone evaporation are easily recognizable by their isotopic composition. They are plotted, in a δ^2H *vs* δ^{18}O diagram, under the line of precipitation (Craig & Gordon 1965, Fontes & Gonfiantini 1967).

The intercept of the equation ($D \approx 3.6$) shows that the excess of deuterium is low, and that the vapour from the ocean was not driven significantly by continental evaporation. It has not come from enclosed seas such as the Mediterranean Sea, which shows again, by inference, that these precipitations are associated with air masses coming from the Atlantic Ocean.

3.2 *Bivariate isotopic analyses*

3.2.1 *18O vs altitude*

Altitude is an important variable in the oxygen-18 content of precipitation. In the case of the South Rifian Rides, there is a clear decrease in the 18O content with altitude (Fig. 3). The mountain outcrops of the South Rifain Rides form a barrier in the path of prevailing ocean winds and are the site of significant precipitation. Consequently, the continental sites behind these leeward uplands receive rainfall from air masses with more negative 18O.

However, it has been noted that the sources of Beida and Anseur show different isotope contents despite the fact that they are located in the same range in altitude. The plot in Figure 3 indicates that water infiltration for the Anseur groundwater occurred at an altitude (about 400 m) higher than Beida (<300 m).

3.2.2 *Relation 18O–Cl⁻*

The diagram illustrating the relationship between 18O and Cl^- (Fig. 4) shows that the contents of 18O vary with the values of chloride recorded at each of the studied sources.

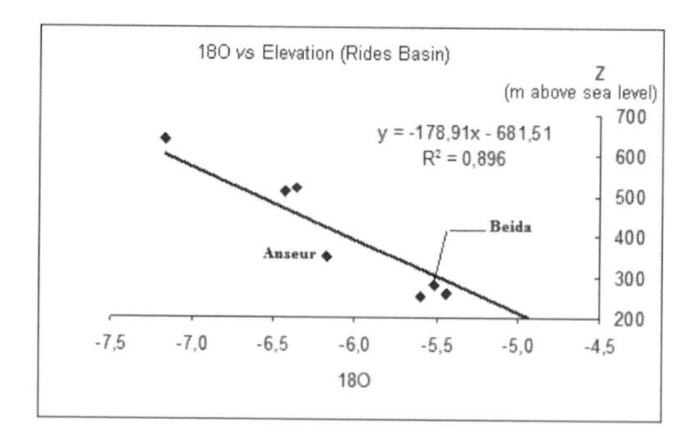

Figure 3. 18O content *vs* altitude.

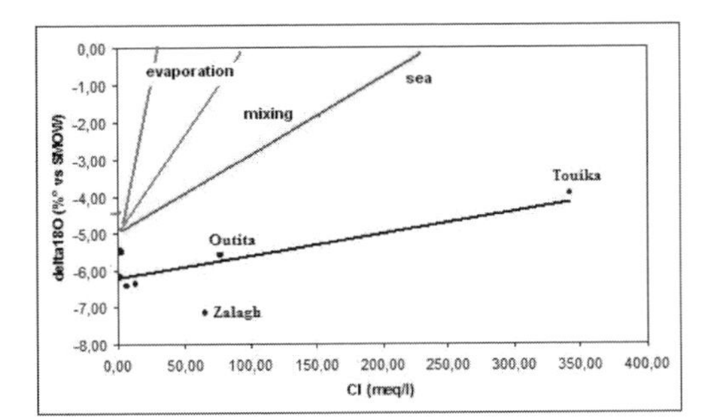

Figure 4. Chloride and Oxygen-18.

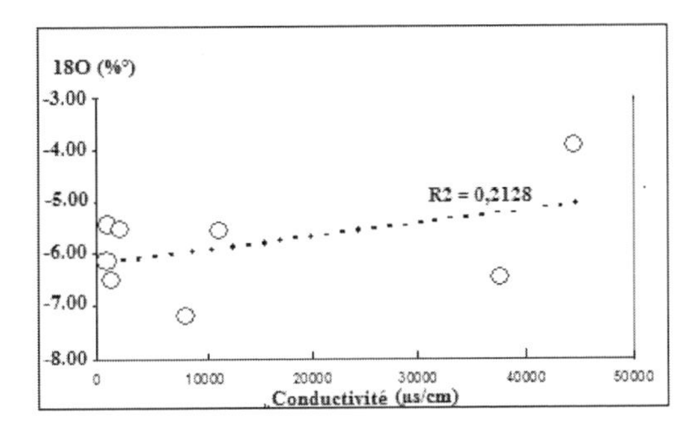

Figure 5. Oxygen-18 *vs* conductivity.

This suggests that the saline load of the waters is essentially local in origin. The dissolution of Cl^- bearing mineral species can be considered, from salt dissolution of deeper rocks, especially in Outita, Zalagh and Tiouka.

3.2.3 *Relation 18O—conductivity*

To determine the origin of the mineralization (salinity), a correlation between the values of conductivity and 18O content has been established (Fig. 5).

This suggests that the elements responsible for the salinity, mainly Na^+ and Cl^-, were introduced into the solution by a simple leaching from rocks present along the flow path, and thus the post-meteoric evaporation and pre-infiltration is very small.

4 CONCLUSIONS

The thermal springs in the South Rifian Rides have undergoing various stages of evaporation before infiltration; this is due to isotopic exchange with the local atmospheric moisture that is created as a result of altitude.

The strong correlation between the isotopic variations in 18O and altitude (r = 0.9), shows that high relief of the South Rifian Rides receive rainfall from an atmosphere with more negative 18O. The arrangement of slopes under the meteoric winds is another topographic factor in the case of sources of Anseur and Beida.

The relation of oxygen-18 and the chlorinated compounds do not show the phenomenon of mixing between freshwater and seawater, which verifies the results obtained from the classical 18O/D diagram. The saline load of the waters comes from the alteration of the surrounding rocks, rich in salts, since the evaporation is excluded before and during infiltration.

The low correlation between the isotopic values of 18O and those of the conductivity (r = 0.2) is explained by the intensive leaching of rocks in elements responsible for salinity.

REFERENCES

Cidu, R. & Bahaj, S. 2000. Geochemistry of thermal waters from Morocco. *Geothermics* 29: 407–430.

Craig, H. 1961. Isotopic variation in natural waters. *Science* 133: 1702–1703.

Craig, H. & Gordon, L.I. 1965. Deuterium and oxygen 18 variations in the ocean and marine atmosphere. In E. Tongiogi (ed.), *Proc. Stable Isotopes in Oceanographic Studies and Paleotemperature., Spoleto, Italy*: 9–130.

Etcheverry, D. & Parriaux, A. 1998. Isotope methods in hydrogeology. *GWA Gas, Wasser, Abwasser* 78: 10–17.

Faugères, J.C. 1978. Les Rides sud rifaines. Evolution sédimentaire d'un basin atlantico – mesogéen de la marge africaine. *Thèse doct. Es – sc.*, Univ. Bordeaux I, n°590.

Fontes, J.C. & Gonfiantini, R. 1967. Comportement isotopique au cours de l'évaporation de deux bassins sahariens. *Earth and Planetary Science Letters* 3: 258–266.

Kendall, C. & Coplen, T.B. 1985. Multi-sample conversion of water to hydrogen by zinc for stable isotope determination. *Analytical Chemistry* 57: 1437–1440.

Zian, A. 2005. Caractérisation climatique et hydraulique du bassin versant de l'Oued R'dom. *Mémoire de DESA*, Université Ibn Tofail. Fac. Sc. Kénitra, Maroc. 87p.

Water-rock interactions in geothermal systems

Water-Rock Interaction – Birkle & Torres-Alvarado (eds)
© 2010 Taylor & Francis Group, London, ISBN 978-0-415-60426-0

Geochemical processes camouflaging original signatures in the thermal waters of south-central Chile

M.A. Alam, P. Sánchez & M.A. Parada
Departamento de Geología, Universidad de Chile, Santiago, Chile

ABSTRACT: Several geothermal systems have been delineated and characterized along the Liquiñe-Ofqui Fault Zone, in the South-Central Volcanic Zone of Chile, from the geochemical signatures of the thermal discharges and structural analysis of the lineaments. Based on the ways of heating up of meteoric water, which is the feeder to these geothermal systems, two distinct domains of thermal discharges have been identified – (i) structural (or non-volcanic) and (ii) volcanic, the latter being directly associated with the regional volcanic centres. The process of heating is through deep circulation of meteoric water in the case of former; and absorption of heat and condensation of steam and gases by meteoric water during lateral circulation. However, these thermal discharges do not exhibit the typical signatures of steam heated waters, which are camouflaged or subdued by the means of water-rock interaction and other near surface processes.

1 INTRODUCTION

In this paper, we discuss the structurally controlled geothermal systems along the Liquiñe-Ofqui Fault Zone (LOFZ) in Villarrica-Chihuio area (39°15′–40°15′S, 71°40′–72°10′W) of South-Central Volcanic Zone (SCVZ) of Chile, with a note on the processes camouflaging the original geochemical signatures of the thermal discharges. LOFZ is a major intra-arc transpressional dextral strike-slip fault running for over 1200 km between 38°S and 47°S, trending NNE-SSW (Lara & Cembrano 2009, Lange et al. 2008, Cembrano et al. 2007). The study area has several stratovolcanoes, viz. Villarrica, Quetrupillán and Mocho-Choshuenco and minor (mainly monogenetic) volcanic centres of SCVZ, whose location is apparently controlled by LOFZ (Stern 2004, and references therein). The conceptual model for these geothermal systems presented here is based on the geochemical and structural characterization of the geothermal manifestations of the area.

A simplified geologic map of the study area (Fig. 1), based on previous works (Lara & Moreno 2004, and references therein) and inputs from the field observations during this study, shows two distinct lithological units - an impermeable basement (comprising crystalline rocks, please see the caption of Figure 1 for details) and a relatively permeable cover comprising pre- and post-glacial volcanic and sedimentary deposits associated with the erosion of the volcanic centers.

2 GEOCHEMICAL CHARACTERISTICS

Here we outline only those aspects of the geochemistry of thermal waters (Table 1), which are of immediate significance for the present discussion. Details on the geochemistry will be reported in a following paper.

i. Thermal discharges are Na-SO_4 and Na-SO_4–HCO_3 types.
ii. Low concentration range (~10–80 ppm) of chloride (Cl^-) largely indicates that these discharges are outflows, and not upflows of the system.
iii. Waters are relatively low in chloride (Cl^-, up to 80 ppm) and high in sulphate (SO_4^{2-}, up to 421 ppm), which is typical for steam-heated geothermal waters.
iv. All the sampled waters are slightly alkaline, which is unusual for steam-heated waters in the absence of additional neutralization processes.
v. Based on the concentration of conservative elements (viz. B, Cl^-), as well as on their ratio (B/Cl^-), two distinct domains (volcanic and structural domains) of thermal waters can be identified (Fig. 2), which is consistent with the spatial distribution of the thermal manifestations with respect to the Liquiñe-Ofqui fault zone and the volcanic centres in the area (Fig. 1).
vi. Variations of saturation indices of the mineral phases, viz. calcite, chalcedony and quartz, with temperature also support the

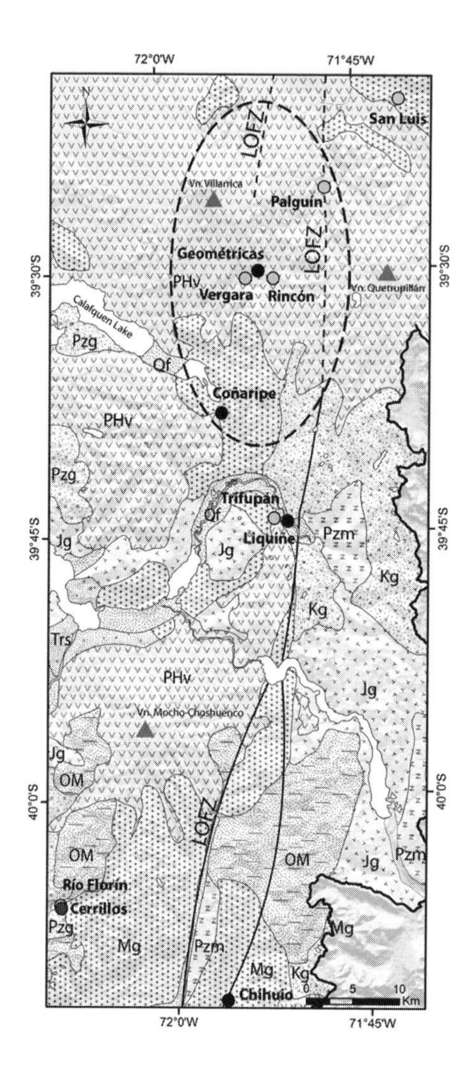

Figure 1. Geological map of the study area, with the locations of (i) geothermal manifestations (circles; darker circles represent higher temperatures of discharges), (ii) LOFZ, (iii) stratovolcanoes (solid triangles), and (iv) lithological units (Pzm: Palaeozoic Metamorphic Complexes; Pzg: Paleozoic Granitoids; Trs: Triassic Sedimentary Rocks; Jg: Jurassic Granitoids; Kg: Cretaceous Granitoids; OM: Oligocene Sedimentary Rocks; Mg: Miocene Granitoids; PHv: Pleistocene to Holocene Volcanic Deposits; Qf: Unconsolidated Quaternary Sediments). Geothermal manifestations in the encircled area belong to the volcanic domain; while the rest of them belong to the structural domain (please see Section 2).

idea of two different domains, as mentioned earlier.

vii. Base temperatures calculated by conventional water geothermometers (Giggenbach 1991, and references therein) also indicate the presence of two distinct domains, with base temperatures in the range 130–150°C for the volcanic domain and 100–120°C for the structural domain.

viii. Water-rock interaction is indicated by partial equilibrium (Fig. 3) for some mineral phases, viz. calcite, chalcedony and quartz.

ix. The values of SO_4^{2-}/Cl^- and SO_4^{2-}/HCO_3^- ratios decrease with increasing distance from the volcanic centres for discharges of the volcanic domain. This is particularly evident in the case of Coñaripe, which has the highest Cl^- content of all thermal discharges of the area.

x. Oxygen and hydrogen isotopes (not reported here) attest conclusively the meteoric origin of the thermal waters. The absence of a considerable shift in the $\delta^{18}O$ values from the local meteoric line suggests limited water-rock interaction.

3 STRUCTURAL ANALYSIS

From the structural analysis of the thermal areas, relation between the geothermal systems and fracture density (FD) is quite evident. FD correlates with the existence and location of the surface geothermal manifestations, as well as with recharge areas of meteoric water of these superficial geothermal systems. This association is particularly pronounced particularly in areas with crystalline rocks (granitic batholiths). To be consistent with this observation, the conceptual model must consider a considerable increase in the (secondary) permeability in the uppermost 200–300 m, in the areas of relatively high values of FD.

Although the lineaments scatter in a wide range, it is evident the absence of lineaments between N 60°E and N100°E (Fig. 4), which is consistent with displacement and stress data (Cembrano et al. 2007, Lavenu & Cembrano 1999) of LOFZ segment in the study area. This indicates that such lineaments, which represent fractures and faults, are the result of recent deformation, causing a secondary permeability that facilitates the subsurface flow particularly in NW-SE and N-S directions. This association will be discussed in detail elsewhere.

4 DISCUSSION

The process of heating of the meteoric water is (i) through their deep circulation for discharges of the structural domain and/or (ii) through absorption of heat and condensation of steam and gases during their lateral circulation, for discharges of the volcanic domain (steam-heated waters). Steam,

Table 1. Physical parameters and chemical composition of thermal water discharges.

Sample	Location	T °C	pH	TDS ppm	SiO_2	Na	K	Mg	Ca	B	Br^-	F^-	HCO_3^-	CO_3^{2-}	SO_4^{2-}	Cl^-	NO_3^-
1	San Luis	39	9.4	170	49	55	1.0	0.2	5.2	0.1	0	1.8	29	12	77	8	1.4
2	Palguin	35	8.7	110	47	42	1.9	1.3	4.2	0.4	1.3	1	67	0	32	10	1.4
3	Palguin	36	8.7	110	52	53	2.5	1.5	4.2	0.7	1.3	1.5	81	0	38	12	1.5
4	Geométricas	72	8.4	540	83	160	9.6	0.1	47	5.0	1.3	1.2	29	0	421	49	1.3
5	Rincón	36	8.0	250	69	62	4	2	10	1.4	1.3	0.9	52	0	103	17	1.5
6	Vergara	41	7.8	240	68	72	5.0	2.6	17	1.9	1.3	0.5	48	0	151	18	1.6
7	Coñaripe	55	7.9	230	55	82	2.7	0.9	7.2	4.7	1.4	0.9	78	0	61	50	1.6
8	Coñaripe	68	8.6	350	78	136	3.6	0.3	5.6	7.3	1.4	1.5	94	0.6	103	81	1.4
9	Coñaripe	60	8.3	330	61	99	3.2	0.2	6.9	5.4	1.4	1.1	83	0	78	62	1.3
10	Trifupán	37	8.9	160	42	60	1.4	0.9	9.8	0.5	1.3	0.7	42	0	77	23	1.7
11	Liquiñe	71	9.4	210	83	73	2.2	0	3.7	0.2	1.3	1.3	24	22	78	16	0
12	Liquiñe	71	9.5	260	87	71	2.2	0	3.7	0.2	1.3	1.5	67	0.9	81	16	1.3
13	Liquiñe	70	8.8	130	41	26	1.1	0.7	5.2	0	0	0.3	35	0	30	7.1	1.4
14	Liquiñe	70	9.4	210	84	69	2.2	0.1	3.8	0.2	1.3	1.4	40	1.2	80	16	1.3
15	Rio Florín	54	9.7	180	57	60	1.0	0	7.7	0.5	1.3	0.5	26	0	78	26	1.3
16	Cerrillos	41	9.4	330	54	55	0.8	0.1	8.3	0.4	1.3	0.4	29	8	75	21	1.5
17	Cerrillos	32	8.0	330	42	42	0.7	0.7	12	0.2	1.3	0.3	47	0	58	15	1.4
18	Chihuio	82	9.4	450	96	107	4.2	0.1	9.9	0	1.3	0.9	25	1.1	213	14	1.5
19	Chihuio	82	9.4	450	97	106	4	0	9.9	0	1.3	0.9	24	2.1	213	13	1.5

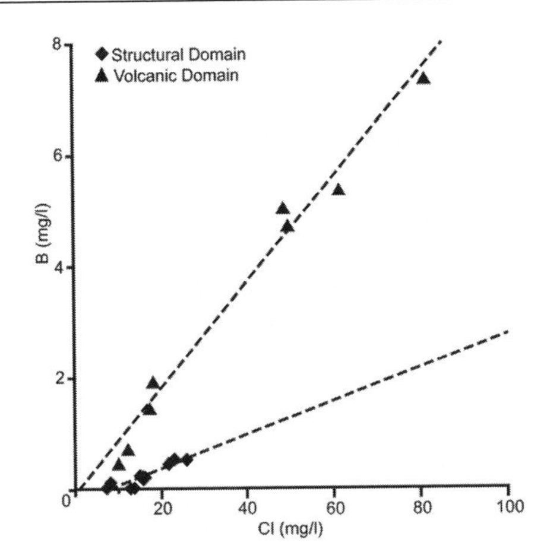

Figure 2. Cl-B relationship shows two distinct domains.

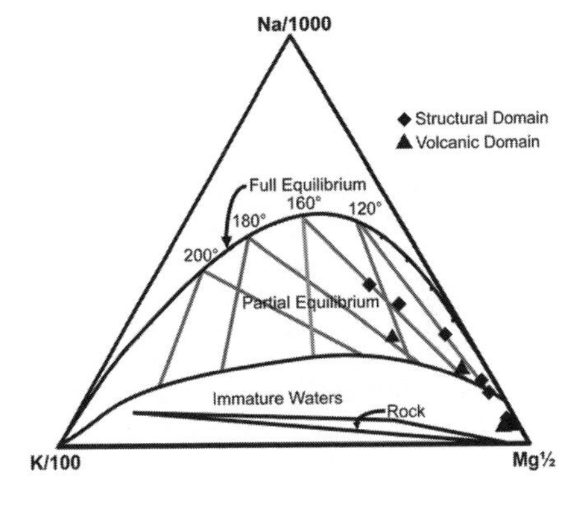

Figure 3. Na-K-Mg diagram (Giggenbach 1988) shows partial equilibrium in most of the cases.

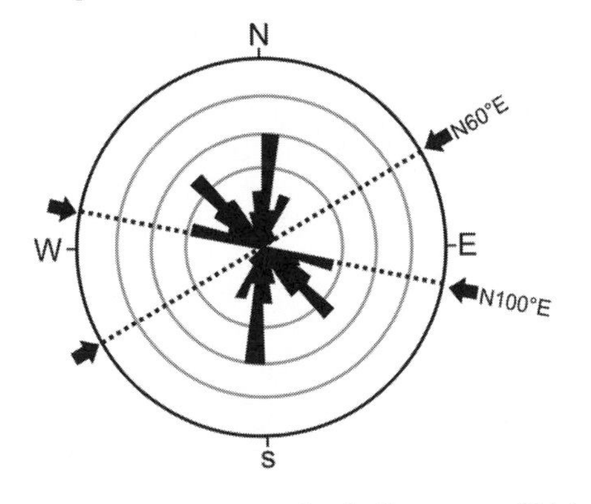

Figure 4. Rosette diagram for the lineaments of higher order representing major structural discontinuities.

rich in CO_2 and H_2S but low in Cl^-, generates bicarbonate-sulphate type waters. The acidity of such water is neutralized and further modified to alkaline by water-rock interaction.

In the case of discharges of the structural domain, infiltrated and percolated meteoric water get heated by convection as well as conduction, after reaching considerable depth. The waters thus heated are separated in two phases through adiabatic decompression, allowing the ascent of a gaseous phase. Chemical speciation in such thermal waters is controlled by the leaching of the host

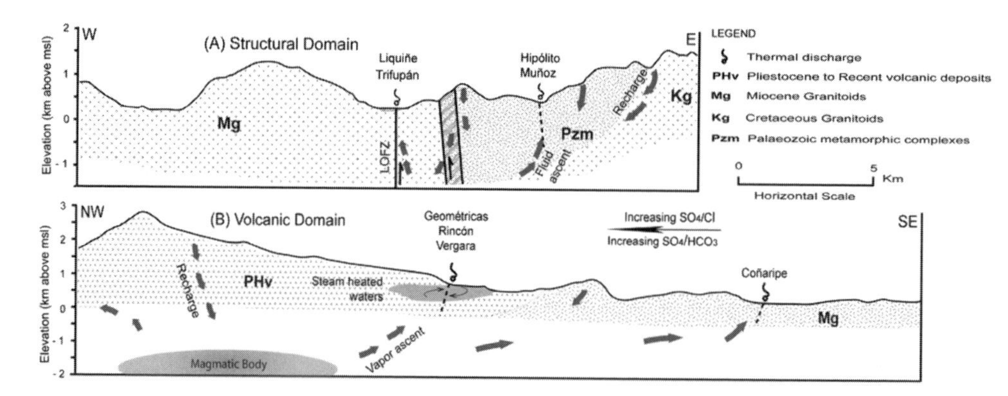

Figure 5. Conceptual model of the geothermal systems for the (a) structural and (b) volcanic domains.

rock, without any contribution of magmatic fluids, as supported by low Cl^- concentrations.

On the other hand, the geochemical signature of discharges of the volcanic domain is modified by limited interaction with magmatic fluids, as indicated by ongoing investigations on gas geochemistry.

In both the cases, steam absorption and condensation occurs in a permeable zone, with constant circulation of thermal water, to form convective cells of heated up water. In the structural domain, this zone is likely to have a dominantly vertical extension, with reservoir(s) restricted to sectors with high fracture density (Fig. 5). In the volcanic domain, on the other hand, due to the primary permeability of the volcano-sedimentary sequences, it is likely to have a more pronounced horizontal extension (Fig. 5), with predominant lateral flow.

In the uppermost zones of these geothermal systems, heated meteoric water is mixed with superficial waters, causing further dilution and cooling. This is particularly evident for the discharges in Vergara, Rincón and Geométricas.

5 CONCLUSIONS

Based on the geochemical signature of the thermal discharges and on the structural analysis of the lineaments, two distinct ways of heating of meteoric water, and thus two domains of thermal waters, have been identified: a structural (or non-volcanic) and a volcanic domain. It can be concluded that the thermal discharges in the studied area are largely superficial phenomena and do not represent high-enthalpy system(s) expected in the study area. However, they do indicate the presence of deep seated (blind) high-enthalpy geothermal system(s) that contributes to the superficial geothermal systems in the study area.

ACKNOWLEDGEMENTS

Authors would like to thank *Universidad de Chile* for the necessary facilities to carry out this work and acknowledge the financial support through PBCT PDA-07 project of CONICYT. The comments of the reviewers (Dr. Zsolt Berner and Dr. François Risacher) helped us to improve our paper. Thanks to Dr. Peter Birkle for efficient editorial handling and some invaluable suggestions.

REFERENCES

Cembrano, J. & Lara, L. 2009. The link between volcanism and tectonics in the southern volcanic zone of the Chilean Andes: A review. *Tectonophysics* 471: 96–113.

Cembrano, J., Lavenu, A., Yañez, G. (coordinators), Riquelme, R., Garcia, M., Gonzalez, G. & Herail, G. 2007. Neotectonics. In T. Moreno & W. Gibbons (eds), *The Geology of Chile*: 147–178. London: The Geological Society.

Giggenbach, W.F. 1988. Geothermal solute equilibria. Derivation of Na-K-Mg-Ca geoindicators. *Geochimica et Cosmochimica Acta* 52: 2749–2765.

Giggenbach, W.F. 1991. Chemical techniques in geothermal exploration. In F. D'Amore (ed.), *Application of Geochemistry in Geothermal Reservoir Development*: 119–142. Rome: UNITAR/UNDP.

Lange, D., Cembrano, J., Rietbrock, A., Haberland, C., Dahm, T. & Bataille, K. 2008. First seismic record for intra-arc strike-slip tectonics along the Liquine-Ofqui fault zone at the obliquely convergent plate margin of the southern Andes. *Tectonophysics* 455: 14–24.

Lavenu, A. & Cembrano, J. 1999. Compressional and transpressional-stress pattern for Pliocene and Quaternary brittle deformation in forearc and intra-arc zones (Andes of Central and Southern Chile). *Journal of Structural Geology* 21: 1669–1691.

Stern, C. 2004. Active Andean volcanism: its geologic and tectonic setting. *Revista Geológica de Chile* 31: 161–206.

Water-Rock Interaction – Birkle & Torres-Alvarado (eds)
© 2010 Taylor & Francis Group, London, ISBN 978-0-415-60426-0

Implications of rock composition and *PT* conditions in deep geothermal reservoirs for chemical processes during fluid circulation

P. Alt-Epping, L.W. Diamond & H.N. Waber
Rock-Water Interaction Group, Institute of Geological Sciences, University of Bern, Bern, Switzerland

ABSTRACT: We use reactive-transport simulations of the geothermal system at Bad Blumau, Austria, and of the planned geothermal system at Basel, Switzerland, to examine chemical processes during fluid circulation. In carbonate-dominated systems (Bad Blumau), separation of CO_2 gas from the production fluid leads to increases in *pH* and to carbonate precipitation. The *pH* change triggers disequilibrium reactions at the base of the injection well, altering the injectivity of the aquifer and the composition of its fluid. In silica-dominated systems (Basel), the cool injection fluid dissolves aluminosilicates in the reservoir rock, leading to silica scaling in the production well. A simple model suggests that incipient corrosion of the casing by SO_4^{-2}-rich reservoir fluids may be detected from an increase in the fraction of sulphide minerals in scales, whereas for SO_4^{-2}-depleted fluids, corrosion may be indicated by increased concentrations of CH_4 and/or H_2 and by solid Fe-oxides/hydroxides in scales.

1 INTRODUCTION

The conventional approach to simulating deep geothermal reservoirs involves modeling of coupled fluid flow and heat transfer. In recent years there has been increasing interest in considering fluid-rock interaction processes as well. Fluid-rock interaction and associated mineral dissolution and precipitation could have a major impact on the long-term performance of deep geothermal reservoirs. For instance, mineral scaling or corrosion caused by the circulating fluid may compromise the efficiency or integrity of the system. Similarly, precipitation of minerals at the base of the injection well could reduce the permeability of the reservoir rock and hence the injectivity of the system.

Here we present a numerical study of two examples. The first is the deep geothermal system at Bad Blumau, Austria which is fed by a carbonate-dominated reservoir at a depth of 2800 m. The second example is hypothetical; it is based on a borehole at Basel, Switzerland, which was drilled to 5000 m depth to induce a geothermal flow path to extract heat from granitic basement rocks. Initial injection tests caused seismicity which lead to termination of the project.

We first describe our approach to constructing a model that integrates and couples the relevant physical and chemical processes acting in the system and that reproduces key observations and data. Then we present results from these simulations and highlight some of the important differences between systems that are fed by carbonate-dominated and aluminosilicate-dominated reservoirs. The emphasis of the discussion will be on 1) constraining the geochemical conditions in the undisturbed reservoir, 2) the chemical and hydrological implications of reinjecting the used fluid into the reservoir, 3) mineral scaling and 4) borehole corrosion and geochemical fingerprints indicating incipient corrosion.

2 NUMERICAL SIMULATIONS

2.1 *Modelling approach*

Simulations were carried out with a modified version of the reactive transport code FLOTRAN (Lichtner 2007). FLOTRAN uses a kinetic formulation for mineral dissolution and precipitation reactions based on transition state theory (e.g. Lasaga 1998). This implies that the local mass transfer between the liquid and the solid phases is determined by the ratio of the local flow velocity to the local reaction rate. To make the simulations computationally efficient we simplify the geothermal circulation systems to 1D flow-through models (e.g. Alt-Epping & Diamond 2008). This simplification primarily concerns the physical aspects of the system while the chemical aspects, which are the main focus of this study, can be represented in greater detail. Perhaps the most important simplification concerning the physical aspects is the assumption that there is no coupling between fluid flow and the temperature distribution. The flow rate through the modeled system is constant and corresponds to that of the real system. Pressure and temperature (*PT*) are assumed to change linearly during ascent and descent of the fluid. Furthermore, the 1D representation of the flowpath implies that any effects of lateral dispersion or lateral flow and chemical and thermal mixing at the base of the injection well cannot be incorporated into the model. However, with appropriate scaling of the local flow velocity and/or local mineral reaction rates we can compute the correct local mass transfer between liquid and solid throughout most of the system and hence reduce some of the artifacts inherent to the simplified model. Details of the conceptual model underlying the numerical model and important parameters are summarized in Figure 1.

2.2 *Modelling results: The Bad Blumau geothermal system*

The geothermal system at Bad Blumau has been in operation for more than a decade. A peculiarity of the scheme is that as well as using the thermal fluid to generate heat and electricity, it is also used for recreational purposes in the form of a large thermal spa. Furthermore, because the geothermal system is fed by a carbonate-dominated reservoir, CO_2 gas exsolves from the fluid near the surface, where it is captured and used for industrial purposes. The system has been monitored throughout its operational lifespan and important information regarding the physical and chemical conditions is available to constrain the model. Information on composition, temperature and pressure are obtained only from discrete sampling points at the surface of the system. It is known from observations that CO_2 separates from the ascending liquid at about 350 m depth. Geological evidence suggests that the reservoir rock is composed primarily of calcite and dolomite, but a detailed modal analysis of the rock is not available.

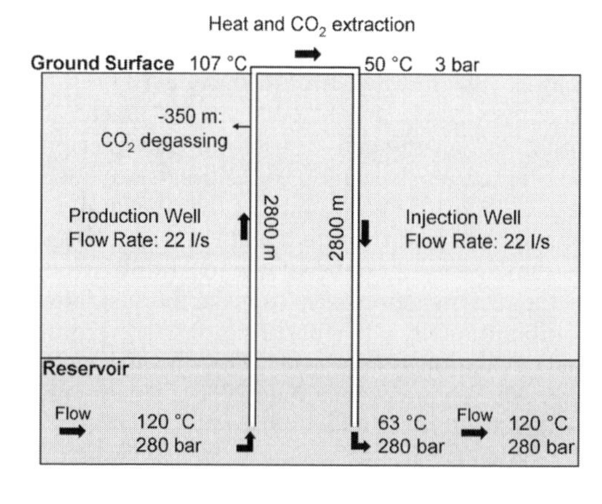

Figure 1. Conceptual model of the Bad Blumau geothermal system, including physical constraints and estimated pressure and temperature (*PT*) conditions in the reservoir. *P* and *T* change linearly between discrete points in the flowpath.

From the fluid composition data and the amount of gaseous CO_2 extracted at the surface we were able to estimate the CO_2 content in the reservoir fluid. Based on this, on the assumed *PT* distribution in the production well, and on CO_2 solubility data, we predicted the depth of CO_2 exsolution to be about 400 m. Considering the uncertainty in the CO_2 content in the reservoir fluid and the simplified representation of the *PT* conditions, the computed result agrees reasonably well with observations. The difference (50 m) may well be due to sluggish gas exsolution.

In addition to the constraints set by the behavior of CO_2, the model had to reproduce the fluid composition data at the surface to a reasonable degree of accuracy. Figure 2 shows computed profiles of selected species and the *pH* and the measured data at the surface. The dominant cations Ca^{+2} and Mg^{+2}, Fe^{+2} and the *pH* show excellent agreement between computed and measured values.

The model that best reproduced the observational constraints was subsequently used for predictive modeling. It was assumed that the fluid in the undisturbed reservoir is in equilibrium with calcite, dolomite, siderite, pyrite and quartz. It has a total aqueous CO_2 concentration of about 0.4 mol/kg and the *pH* is about 5.7. The reservoir temperature is around 120°C if we assume a reservoir pressure of 280 bar (i.e. "cold" hydrostatic conditions).

The release of CO_2 gas from the ascending fluid has two important chemical implications. The first is an increase in the *pH* (Fig. 2), the second is the precipitation of carbonate minerals. The second implication poses a substantial risk for the system at Bad Blumau, considering that the ascending fluid is close to calcite and dolomite saturation. Without preventive measures, the production well could be blocked by precipitated carbonate minerals within a few days. At Bad Blumau, the precipitation of carbonates is prevented by injecting chemical inhibitors into the production well at a depth well below the level of CO_2 degassing (350 m). Unfortunately, because it is propriety knowledge of the operator, the type of inhibitor(s) used to prevent carbonate precipitation was not revealed to us and therefore details about the effect of the inhibitor are not known.

Although the *pH* increase following the release of CO_2 gas may not appear significant (Fig. 2) (which is perhaps why it has not received much attention), it does induce a chemical disequilibrium between the reinjected fluid and the reservoir.

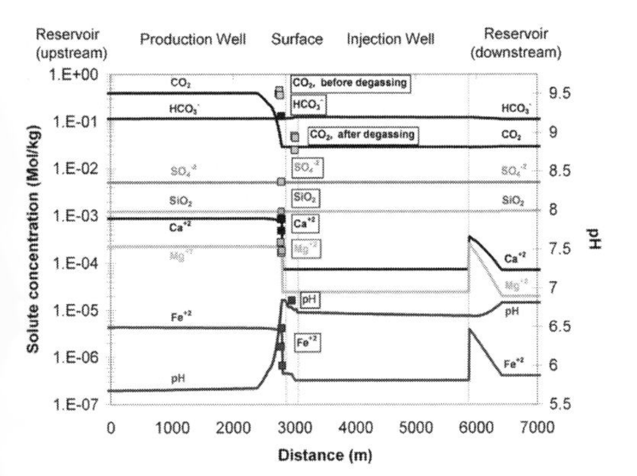

Figure 2. Computed profiles of selected species along the flowpath through the geothermal system at Bad Blumau. Shaded squares are concentrations measured in the fluid at the the surface. This simulation assumes that no inhibitor is injected. Consequently, calcite and dolomite precipitate, decreasing aqueous Ca^{+2} and Mg^{+2} concentrations.

The consequence of this disequilibrium is the dissolution of carbonate minerals at the base of the injection well. The fluid is relatively cool when it reaches the bottom of the injection well and only acquires reservoir temperatures as it moves away from that point. Carbonate minerals that dissolve at the base of the injection well therefore start to re-precipitate at some distance from the well owing to their retrograde solubility behavior. The effect is a carbonate dissolution/(re)precipitation front that over time shifts away from the injection well, as primary carbonate minerals nearest to the well become exhausted. There are two implications of this process. The first is that the front is associated with increases in porosity and permeability where carbonate minerals dissolve and decreases where precipitation occurs. Hence, the re-injection of a chemically modified fluid may affect the hydraulic conditions in the reservoir and, in the worst case, compromise the injectivity of the system. The second implication is that, if potentially hazardous constituents (e.g. organics or trace metals such as Cd or Pb) are present in the aquifer, they may be released to the fluid, possibly rendering it unfit for use further downstream.

Changes in pressure and temperature during ascent and descent of the fluid induce changes in mineral solubilities. Which minerals become oversaturated and form scales within the borehole is primarily dictated by the composition of the reservoir fluid and by the pressure/temperature dependence of the mineral solubility functions. Thus, mineral assemblages that precipitate in the production well may be different from those in the injection well. For instance, if the fluid in the reservoir is saturated in silica, then the prograde solubility behaviour of silica favours its precipitation in the production well. Conversely, carbonate minerals tend to precipitate in the injection well owing to their retrograde solubility behaviour. Also, if in carbonate-dominated systems the fluid undergoes only minor temperature but relatively large pressure changes during ascent in the production well, then carbonate minerals may precipitate owing to the relatively large sensitivity of their solubility to pressure.

Calculations suggest that pyrite precipitates within the production and the injection boreholes. The fluid is oversaturated with respect to quartz in the production well, but undersaturated with respect to amorphous silica, which is the silica phase most commonly found in geothermal systems. Carbonate minerals are all undersaturated within the production well.

2.3 *Modelling results: The Basel geothermal system*

The planned geothermal system beneath the city of Basel, Switzerland is quite different from that at Bad Blumau. The reservoir is deeper (5000 m), the fluid temperature is higher (reservoir temperature: ~200°C), the flow rate was intended to be higher, and the reservoir consists of dry gneisses and granites. The flow system at Basel was to be created by inducing hydrofractures via injection of pressurized riverwater. A second borehole was envisaged to permit closed-loop circulation and heat extraction.

Unfortunately, the geothermal project in Basel was abandoned when the first hydrofracturing tests triggered seismic activity. As a result, there are no observations of the operating system. While the thermal properties of the system are relatively well known, geochemical data are scarce. Thus, we only loosely pattern our model after the Basel case to explore some of the fundamental and generally valid differences between carbonate- and silicate-hosted geothermal systems.

We assume that the reservoir rock is a generic granite composed of quartz, feldspars and mica. The composition of the injected fluid is that computed for a fluid that has circulated through the system once. The re-injected fluid, after extraction of heat and electricity generation, has a temperature of 70°C, the reservoir temperature is 200°C and the fluid that reaches the surface has a temperature of 170°C. We assume a cold, hydrostatic pressure distribution with atmospheric pressure along the top boundary. Both pressure and temperature change linearly.

In contrast to carbonate-dominated reservoirs, the fluid in aluminosilicate-dominated systems carries relatively little dissolved CO_2 and SO_4^{-2}. The dominant anion in solution is Cl^-. The forced flow of a fluid through the granitic basement at temperatures of around 200°C alters the igneous feldspars to hydrated aluminosilicates such as chlorite, epidote/clinozoisite and muscovite (sericite). Whereas the *pH* (~7.0) and the cation concentrations in the geothermal fluid are controlled by these types of hydration reactions, the greatest impact on the system as a whole is imposed by silica.

Silica poses the greatest risk of mineral scaling and of potentially inhibiting fluid flow. It also exerts the greatest controls on dissolution/precipitation reactions at the base of the injection well. Silica typically precipitates as amorphous SiO_2, which has a higher solubility than quartz. For amorphous silica to precipitate the fluid has to undergo substantial cooling. Thus, whereas the risk of clogging by amorphous silica is relatively small within the production well, it is high in the heat exchanger. Any silica that precipitates along the geothermal loop will cause the reinjected fluid to be undersaturated with respect to aluminosilicate minerals, causing a halo of strong mineral alteration and associated changes in porosity and permeability around the base of the injection well. Primary aluminosilicates also dissolve as the cool injected fluid moves away from the injection well and is heated up to reservoir temperature. In all the simulations of the Basel system, porosity is generated at the base of the injection well. This is likely to enhance injectivity. However, a quantitative evaluation of the porosity change is difficult because it involves the kinetic description of multiple simultaneous dissolution-precipitation reactions.

2.4 *Corrosion*

One concern during geothermal energy production is that of chemical corrosion of the borehole casing. We explore the effect of corrosion on the fluid composition and on mineral precipitation to identify chemical fingerprints that could be used as corrosion indicators. Once suitable indicators are identified, incipient corrosion could be detected early on during regular chemical monitoring.

Corrosion of the casing is typically associated with the release of Fe and H_2 into the circulating fluid. This process is implemented into the simulation by a simplified corrosion model in which we assume that the borehole casing is composed of Fe^0. The corrosion rate can then be described by the dissolution rate of Fe^0 in the circulating fluid.

The effects of corrosion show dramatic differences between the carbonate-dominated system at Bad Blumau and a silicate-dominated system such as that in Basel. In Bad Blumau, the ascending fluid is rich in SO_4^{-2} and therefore has a strong redox buffering capacity. Any H_2, aq that is released into the fluid from the corrosion of the casing, reacts with SO_4^{-2} in the fluid and reduces it to HS^-. Thus, even at high corrosion rates there is no noticeable increase in H_2,aq. However, the release of Fe^{+2} from the corroding casing and the reduction of sulphate increase the amount of pyrite in mineral scales. Accordingly, when checking for signs of corrosion it is advisable to monitor the composition of the fluid as well as that of precipitated solid phases.

The effect of the corrosion process is much more pronounced in systems that do not have a strong redox buffering capacity, such as in many SO_4^{-2}-depleted, silicate-dominated systems like the planned situation in Basel. With very low SO_4^{-2} in the fluid, most of the H_2 released during the corrosion process reacts with CO_2 in the fluid to produce CH_4. When all CO_2 has been reduced, the concentration of H_2,aq in the fluid increases. The amount of HS^- in the fluid is not sufficient to cause pyrite precipitation. Instead, mineral scales are composed of Fe-oxides and hydroxides such as magnetite or goethite. Thus, in systems that are SO_4^{-2}-depleted, incipient corrosion leaves its fingerprint in the composition of the fluid as in the composition of mineral scales. Both of these should be detectable by regular monitoring.

3 CONCLUSION

We used numerical simulations to gain a theoretical understanding of chemical processes in geothermal systems. We demonstrated that these processes are strongly dependent on the rock composition and *PT* conditions in the reservoir. This theoretical understanding can be used to make predictions about processes in different parts of the system, potential risks of failure, monitoring strategies and the effectiveness of remediation measures if failure occurs. Therefore, this type of theoretical analysis should become part of regular system monitoring to ensure safe and sustained operation and energy production.

REFERENCES

Alt-Epping, P. & L.W. Diamond. 2008. Reactive transport and numerical modeling of seafloor hydrothermal systems: a review. In R.P. Lowell, J.S. Seewald, A. Metaxas & M.R. Perfit (eds), *Modeling Hydrothermal Processes at Oceanic Spreading Centers: Magma to Microbe*. Washington, DC: AGU Monograph.

Lasaga, A.C. 1998. Kinetic Theory in the Earth Sciences, Princeton Univ. Press, Princeton, NJ.

Lichtner, P.C. 2007. *FLOTRAN Users Manual: Two-phase non-isothermal coupled thermal-hydrologic-chemical (THC) reactive flow and transport code, Version 2.* Los Alamos National Laboratory, Los Alamos, New Mexico.

Water-Rock Interaction – Birkle & Torres-Alvarado (eds)
 ISBN 978-0-415-60426-0

Geochemical patterns of scale deposition in saline high temperature geothermal systems

H. Ármannsson & V. Hardardóttir
Iceland GeoSurvey, Reykjavík, Iceland

ABSTRACT: Studies of scales in the Asal, Djibouti and Reykjanes, Iceland geothermal systems are described and the results compared to those for the Milos, Greece and Salton Sea, California geothermal systems. At >16 bar a sulphides, such as galena, sphalerite, wurtzite, troilite, pyrite, chalcocite, chalcopyrite, and bornite, are the predominant deposits, but below that and down to amorphous silica saturation pressure iron silicates are most abundant after which amorphous silica predominates. The sulphides may be chemically inhibited and so can the iron silicates although their deposition can be prevented by keeping the wellhead pressure well above 16 bar a, and the amorphous silica deposition by keeping the separator pressure above the saturation pressure of amorphous silica.

1 INTRODUCTION

From October 1989 to April 1990 Virkir-Orkint, carried out a comprehensive scaling/corrosion study in Asal, Djibouti (Virkir-Orkint 1990). Since 2000 Orkustofnun and ÍSOR have carried out scaling studies in Reykjanes, Iceland (Hardardóttir et al. 2005). Results from these two studies constitute the backbone of this paper although references will be made to similar studies in other areas. The objective is to find out whether there is a certain pattern of scaling in saline high-temperature geothermal systems with reference to temperature, pressure and salinity.

2 METHODS

2.1 *Test apparatus*

A layout of the equipment used for the Djibouti study is shown in Figure 1, and a schematic drawing of the experimental manifold used in Reykjanes in Figure 2. In Djibouti the "Flow line"was used for enthalpy and flow measurements and for drawing fluid samples at the beginning and end of the test but remained closed for the rest of the time. Coupons were placed in the "Separator line" and the "Brine pipe", at various spots in the "Ageing tank" and the three pipes of the "Inhibitor line." Flow was determined by the method of James (1962),

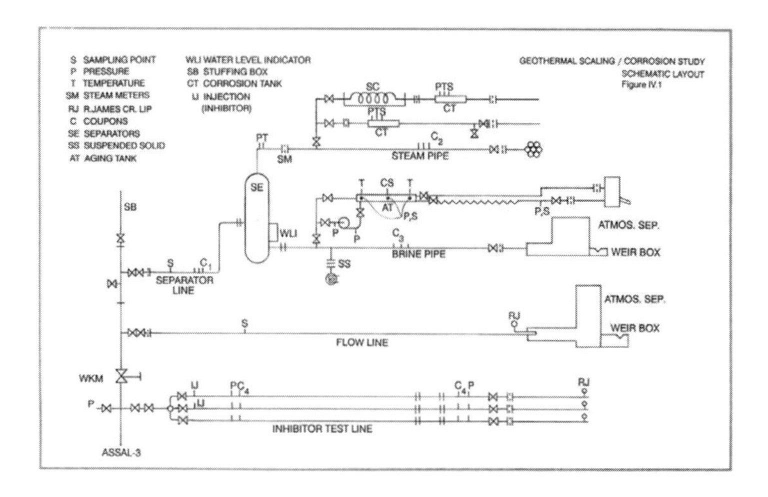

Figure 1. Layout of equipment used for the scaling/corrosion test on well Asal-3, Djibouti.

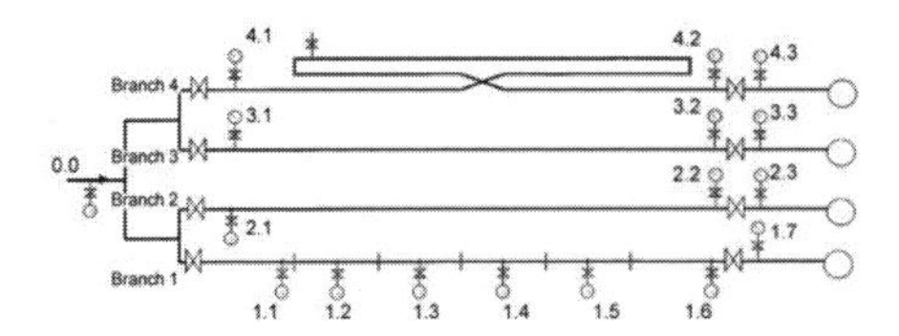

Figure 2. Schematic drawing of the experimental manifold at Reykjanes, Iceland. Individual branches are labeled 1, 2, 3, and 4 and the locations of individual experimental stations are indicated by location numbers (i.e. 0.0, 1.1, 1.2 etc.) (Hardardóttir et al. 2005).

and the flow in the "Inhibitor line" pipes adjusted using orifices and valve adjustment.

2.2 *Sampling and analysis*

Two phase fluid samples were separated using a mini Webre separator. Brine samples were partitioned into raw, filtered, untreated, acidified, precipitated and extracted according to the constituent to be determined. Steam samples were collected into NaOH for the determination of acid gases by titration and the head space gas was determined by gas chromatography (See e.g. Ármannsson and Ólafsson 2006).

In the Djibouti brine samples Na, K, Li, Ca, Mg, Sr, Fe, Zn, Cd, Hg, Cu, Ni, Pb and Ag were determined by AAS, Al and Ba by ICP/AES, CO_2, H_2S and Cl by titration, NH_3, SiO_2, B, Fe and Mn by UV/Vis Spectrophotometry, Cl, Br, F and SO_4 by I,C, Al by Fluorimetry and pH and conductivity by Electrometry. The same methods were used for the fluid at Reykjanes except that Li, Sr, Fe, Mn, Zn, Cd, Hg, Cu, Ni and Pb were determined by ICP/AES or ICP/MS and Cd, Hg and Ag were not reported. General precision and detection limits for AAS are found in http://www.cdc.gov/NCHS/data/nhanes/frequency/lab06_met_lead_and_cadmium.pdf, but for a specific extraction technique for Zn, Cd, Cu, Ni, Pb in Ármannsson (1979), Ag in Ármannsson and Ovenden (1980), and a specific technique for Hg in Ólafsson (1974).

Scales from Djibouti and Reykjanes were characterized by microscopic examination, X-ray diffraction and Scanning Electron Microscopy (SEM), those from Djibouti analysed by Energy Dispersive Spectrometry (EDS) (D.l. 1000–3000 ppm; precision ± 0.2%) but those from Reykjanes by EDS and wet chemical analysis. Limits for ICP are found in http://www.analytica.se/hem2005/eng/miljo/vatten_naturliga.asp, for titration, IC, electrometry, fluorimetry and spectrophotometry (for B and SiO_2) techniques in Pang & Ármannsson (2006). Limits for spectrophotometric techniques for NH_3, Fe and Mn are shown in Grasshoff et al. (1983).

Metal coupons from three types of steel, i.e. Carbon steel 37, and stainless steels AISI 304 and 316 measuring 76 × 13 × 3.5 mm, as well as a few ORMAT coupons, 70 × 16 × 1 mm were fixed on coupon holders in the Djibouti tests but mild steel coupons were used for the Reykjanes experiments.

3 CHEMICAL COMPOSITION OF THE BRINES

3.1 *Analytical results*

The results of chemical analysis of fluids from well Asal-3, Djibouti and well Reykjanes-11, Iceland are reported in Ta ble 1 as total composition calculated from vapour and steam analyses, and compared to the composition of sea water (Turekian 1969).

3.2 *Comparison of fluids*

The salinity of the Reykjanes brine is similar to that of sea water but the Djibouti brine is evaporated and is about 3.5 times more saline than sea water. Both fluids show obvious alteration relative to sea water due to high temperature, i.e relatively high SiO_2, K, Ca, and Mn concentrations, but low Mg and SO_4 concentrations. The Ca and Mn concentrations are however considerably higher than would be expected from the evaporative effect alone. Al and Fe concentrations were significantly increased over those of sea water in the Reykjanes brine but again much higher in the Djibouti brine than would be expected from evaporation alone.

It has been shown that in the downhole brine in Reykjanes Zn, Cu and Pb concentrations in the brine are orders of magnitude higher demonstrating

Table 1. Some chemicals in total fluid in wells Asal-3, and Reykjanes 11, compared to sea water (35%) (Turekian 1969).

Constituent mg/kg	Geothermal brine		Sea water 35%
	Asal 3	Reykjanes 11	
°C	260	295	
Enthalpy kJ/kg	1133	1317	
P_0 bar	20.4	42	
pH[1)]	4.1	4.7	
SiO_2	460	731	6.4
Na	26471	9291	10800
K	4451	1358	392
Cl	70979	18034	19800
Fe	32.3	1.00	0.003
Zn	37.0	0.02	0.005
Pb	2.6	<0.00006	
Cu	0.27	<0.0005	0.0009
Ni	<0.1	0.0007	

Calculated using WATCH (Arnórsson et al. 1982, Bjarnason 1994).

that the low wellhead concentrations are due to deposition during the ascent of the fluid (Hardardóttir et al. 2009). A large increase in Zn over seawater was observed in the Asal discharge water along with a significant concentration of Pb and a moderate concentration of Cu.

4 COMPOSITION OF SCALES

Certain patterns can be discerned in distance from wellhead, deposition at different pressures and possibly the environment of deposition. In Table 2 the analysis from different sampling locations in Asal-3, is reported normalized to a sum of 100%

The composition differs greatly according to the distance from the wellhead. Sulphides, mostly galena is more prominent close to the wellhead but silicates and silica further away from the wellhead. At Reykjanes sulphide, wurtzite at the highest pressures, sphalerite at lower pressures, were the most prominent sulphides with traces of galena observed at medium pressures. In Asal-3 chalcopyrite was also observed in the P_c line. In Reykjanes chalcopyrite and traces of bornite were observed. In Asal-3 scales significant concentrations of carbonate (0.5–2.2% as CO_2), characterized as siderite were also found.

The distance from the wellhead does not tell the whole story. A sample from an orifice at the opening of the separator line in Asal-3 contained nearly exclusively galena and an orifice at high pressure in Reykjanes contained nearly pure wurtzite. The thick- ness of the scale is also pressure dependent and an experiment on the separator line in Asal-3, Djibouti, shows a significant increase in scaling rate between 17.7 bar and 16.2 bar (Fig. 3) the increase

Table 2. Composition of scales from the well Asal-3, Djibouti.

Constituent%	WH	OR[1)]	TP	SP[1)]	BP	SS[1)]	WB
P_0 bar	20.0	17.7	17.7	17.7	17.7	0	0
SiO_2	19.6	0	6.7	40.7	30.5	56.4	72.9
Al_2O_3	3.7	0	1.0	4.3	3.4	8.6	2.7
Fe_2O_3	22.5	0	6.7	31.8	25.8	14.8	2.7
MnO	2.3	0	0.9	5.8	3.7	0.7	0.2
MgO	1.6	0	0.1	0.7	1.1	0.2	0.2
CaO	0.6	0	0.6	1.6	1.4	8.4	12.8
Na_2O	4.4	0	0.3	1.4	1.7	8.1	0.8
K_2O	0.1	0	0	0.7	0.4	1.9	2.9
S	13.7	14.9	18.3	4.0	8.0	0.2	0.4
Cu	0.4	0	0	0.1	0	0.1	0.1
Pb	22.3	85.1	65.4	7.2	23.3	0.2	0.4
Zn	8.8	0	0	1.7	1.0	0.4	0.1

WH: Wellhead; OR: 90 mm orifice to separator line; TP: Two phase pipe on separator line SP: Separator; BP: Brine pipe SS: Single drum silencer; WB: Weir box. [1)]Some inhomogeniety (>±10%).

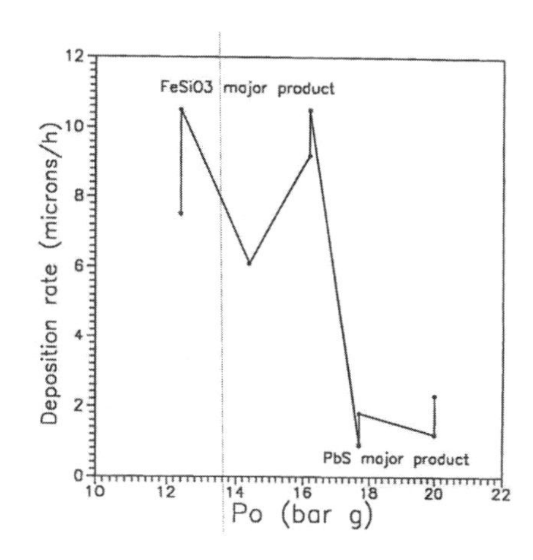

Figure 3. Asal-3, Djibouti. Thickness of scales on coupons at different pressures.

Table 3. Fe/Si mole ratio for some samples from well Asal-3, Djibouti and well 11, Reykjanes in which scales had formed at different temperatures.

Well	Temperature °C	Fe/Si mole ratio
Reykjanes 11	253	1.7
	184	0.32
	161	0.13
Asal-3	215	0.70
	209	0.60
	200	0.45
	100	0.07

being concomitant with a large increase in iron silicate deposition. The Fe/Si mole ratio of the scales (Table 3) varies similarly to that of deposits from Salton Sea which varied linearly from 0.1 at 100°C to 1 at 220°C (Gallup 1989) suggesting that iron silicates predominate at high temperatures but silica at lower ones. The characterization of the scales shows that at relatively high temperatures a compound with a composition close to minnesotaite is observed but at lower temperatures silica and iron oxides exist as separate compounds. Silica absorption on iron oxides (Swedlund and Webster 1999) may further contribute to the variation in the Fe:Si ratio.

5 DISCUSSION

5.1 *Scale formation in saline geothermal systems*

Deposits of heavy metal sulphides and iron silicate-have been reported in the Salton Sea, California (Cl: 155000 ppm, Gallup et al. 1990) and Milos, Greece (Cl: 65400 ppm, Karabelas et al. 1989) geothermal systems. In both cases galena is the most prominent sulphide scale followed by sphalerite.

In the Salton Sea system troilite is also prominent with smaller amounts of chalcopyrite, bornite and pyrite. Chalcopyrite and pyrite are also observed in the Milos system. In the more saline systems (Salton Sea, Asal, Milos) galena is apparently the major deposit as might be expected from solubility considerations (Helgeson 1969), but in the Reykjanes system the zinc sulphides (wurtzite, sphalerite).

In all systems iron silicates start precipitating at temperatures below about 200°C (15.5 bar a) but at lower temperatures (< 150°C) amorphous silica precipitation becomes prominent. Bench scale studies suggested that silica colloidal formation was not important at >100°C but but became fast at a lower temperature and a higher pH.

5.2 *Scale prevention*

Injection of air and chemical oxidizing agents to reduce sulphide concentrations has been suggested to reduce the extent of sulphide scaling but suffers from possible corrosion effects and/or formation of secondary precipitates (Jost 1980). In Asal two types of inhibitors from the Nadar Chemical Company, Italy were tested for sulphide inhibition, a sequestration agent for heavy metals (Nadar 4093) and a sequestering and dispersing against calcium and magnesium salts and silica (Nadar 1008). Both inhibited metal sulphide formation. The former was deemed unsuitable due to iron silicate formation and but the latter caused the formation of calcium chloride, avoided by acidification but corrosion problems still have to be resolved. Gallup (1993) has used reducing agents, namely sodium formate, to control ferric silicate deposition, and found that the formate also mitigates against acid corrosion.

As has been observed the extent of iron silicate scaling in Asal is small above 16 bar a and the recommendation is to keep the wellhead pressure well above that. The same is not true in Reykjanes Amorphous silica scales similarly are best avoided by keeping the separator pressure of the power plant above that of amorphous silica saturation.

6 CONCLUSIONS

In the Asal geothermal system the pattern of deposition is such that at pressures >16 bar a the major scales are sulphides, but at lower pressures iron silicate scales predominate down to amorphous silica saturation. Such sulphide scales, mostly lead sulphide (galena) and zinc sulphides (wurtzite, sphalerite) with iron and copper sulphides (troilite, pyrite, chalcopyrite, bornite, chalcocite) are also prominent in other saline geothermal systems as are the iron silicate scales which may also form at higher pressures. The sulphide scales may be dealt with by inhibition, the iron silicate scales by inhibition but the amorphous silica deposits by pressure (temperature) control.

REFERENCES

Ármannsson, H. 1979. Dithizone extraction and flame atomic absorption spectrometry for the determination of cadmium, zinc, lead, copper, nickel, cobalt and silver in sea water and biological tissues. *Analytica Chimica Acta* 110: 21–28.

Ármannsson, H. & Ólafsson, M. 2006. *Collection of geothermal fluids for chemical analysis.* ÍSOR report. ÍSOR-2006/016, 17 pp.

Ármannsson, H. & Ovenden, P.J. 1980. The use of dithizone extraction and atomic absorption spectrometry for the determination of silver and bismuth in rocks and sediments and of a demountable hollow cathode lamp for the determination of bismuth and indium. *Journal of Environmental Analytical Chemistry* 8: 127–136.

Arnórsson, S., Sigurdsson, S. & Svavarsson, H. 1982. The chemistry of geothermal waters in Iceland I. Calculation of aqueous speciation from 0°C to 370°C. *Geochimica et Cosmochimica Acta* 46: 1513–1532.

Bjarnason, J.Ö. 1994. *The speciation program WATCH, version 2.1.* Orkustofnun, Reykjavík, 7 pp.

Gallup, D.L. 1993. The use of reducing agents for control of ferric silicate scale deposition. *Geothermics* 22(1): 39–48.

Gallup D.L., Andersen, G.R. & Holligan, D. 1990. Heavy metal sulfide scaling in a production well at the Salton Sea geothermal field. *Geothermal Resources Council Transactions* 14: 1583–1590.

Hardardóttir, V., Ármannsson, H. & Thórhallsson, S. 2005. Characterization of sulfide-rich scales in brines at Reykjanes. *Proeedings of the World Geothermal Congress 2005, Antalya, Turkey, 24–29 April 2005*, 8 pp.

Hardardóttir, V., Brown, K.L., Fridriksson, Th., Hedenquist, J.W., Hannington, M.D. & Thórhallsson, S. 2009. Metals in deep liquid of the Reykjanes geothermal system, southwest Iceland: Implications for the composition orf seafloor black smoker fluids. *Geology* 37: 1103–1106.

Helgeson, H.C. 1969. Thermodynamics of hydro- thermal systems at elevated temperatures and pressures: *American Journal of Science* 267: 2441–2453.

James, R. 1962. Steam water critical flow through pipes. *Proceedings of the Institute of Mechanical Engineers* 176: 741–745.

Jost, 1980. *US Patent 4,224,151.*

Karabelas, A.J., Andritsos, N., Mouza, A., Mitrakas, M., Vrouzi, F. & Christanis, K. 1989. *Geothermics* 18: 169–174.

Ólafsson, J. 1974. Determination of nanogram quantities of mercury in seawater. *Analytica Chimica Acta* 68: 207–211.

Pang, Z. & Ármannsson, H. (eds.) 2006. *Analytical procedures and quality assurance for geothermal water chemistry.* United Nations University Geothermal Training Programme 2006-Report 1, 172 pp.

Swedlund, P.J. & Webster, J.G. 1999: Adsorption and polymerisation of silicic acid on ferrihydrite, and its effect on arsenic adsorption. *Water Resources* 33: 3413–3422.

Turekian, K.K. 1969. The oceans, streams and atmosphere. In K.H. Wedepohl (ed) *Handbook of Geochemistry,* vol. 1, ch. 10: 297–323, Berlin: Springer-Verlag.

Virkir-Orkint 1990. Djibouti. *Geothermal scaling and corrosion study. Final report.* Electricité de Djibouti, Virkir-Orkint, Reykjavík.

Water-Rock Interaction – Birkle & Torres-Alvarado (eds)
 ISBN 978-0-415-60426-0

Geochemical data analysis (2009) of Los Azufres geothermal fluids (Mexico)

R.M. Barragán, V.M. Arellano, A. Aragón & J.I. Martínez
Instituto de Investigaciones Eléctricas, Cuernavaca, Mexico

A. Mendoza & L. Reyes
Comisión Federal de Electricidad, Residencia Los Azufres, Mexico

ABSTRACT: Analysis of 2009 geochemical data of the Los Azufres geothermal fluids demonstrated reservoir processes related to exploitation and fluid reinjection. Minimum reservoir temperature and reservoir liquid CO_2 but maximum chlorides, N_2 and enriched δ^2H, $\delta^{18}O$ values are seen at the southwest part of the field where injection wells are found. In contrast, maximum reservoir temperatures (>300°C) and reservoir liquid CO_2 (>5‰ vol.) but low N_2 (~10 mmol/kg) areas were correlated to natural up-flows.

1 INTRODUCTION

The Los Azufres geothermal field is located in the northern part of the Mexican Volcanic Belt. At present (2009), it has an installed power capacity of 188 MW (Gutiérrez-Negrín et al. 2010). The field consists of two well-defined areas of production, Maritaro in the north and Tejamaniles in the south, separated by a distance of several kilometers with no surface manifestations within the intervening area. The Comisión Federal de Electricidad (CFE) started field development in 1980 while commercial exploitation began in 1987. Injection in Los Azufres started in 1983 at an early stage of development of the field and has been beneficial to the reservoir (Torres-Rodríguez & Flores-Armenta 2000).

Since 2003, due to the 100 MW increase in power capacity installed at the north zone, concerns on reservoir performance because of higher rates in fluids extraction, have arisen. In order to estimate the occurrence of reservoir processes induced by exploitation (fluids extraction and reinjection), the CFE routinely performs the monitoring of chemical, isotopic (Barragán et al. 2003, 2005a, 2009a,b) and production data of wells (Arellano et al. 2003, 2005). The reinjection effects on the geochemical behavior of fluids produced during 2005–2007 has been discussed by Barragán et al. (2009a). In this work, the geochemical behavior of fluids was investigated through the analysis of 2009 data. Geochemical data for 50 wells provided by CFE were included (Fig. 1).

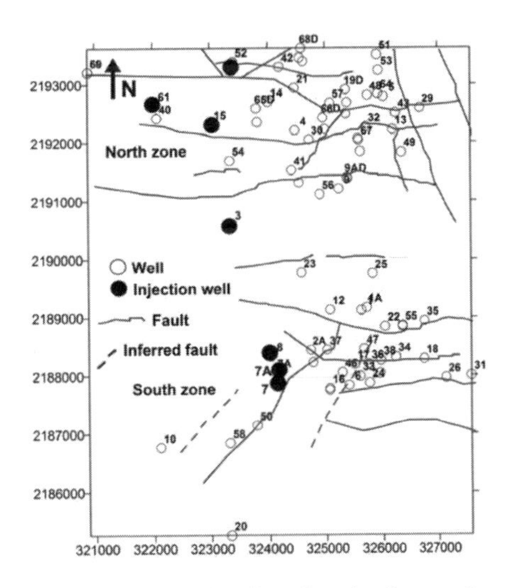

Figure 1. Location of wells at Los Azufres geothermal field.

2 METHODOLOGY

In 2009 samples for isotope monitoring from production and reinjection wells of Los Azufres geothermal field were collected (Barragán et al. 2009b). Isotopic data were used to calculate the total discharge composition of wells. The Na/K geothermometer (Nieva & Nieva 1987) was used to estimate reservoir temperatures of two-phase wells while temperatures of dry steam wells were obtained by the FT-HSH2 method (Barragán et al. 2005b). Reservoir liquid

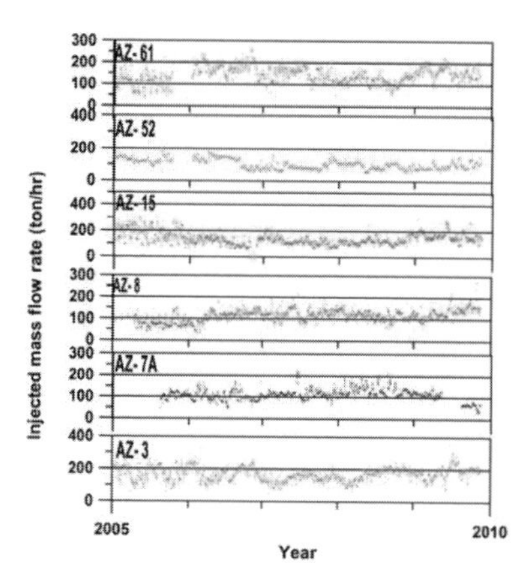

Figure 2. 2005–2009 injected mass flow rates.

CO_2 concentrations were calculated by using the SCEXVAP program (Nieva et al. 1987).

Average values of chemical compositions of fluids were used. The results were plotted as iso-lines on the map of the field in order to define the behavior of the geochemical indicators. For reference, mass flow rates injected (2005–2009) are given in Figure 2; the reinjection wells AZ-7A and AZ-8 are located in the southern zone while the wells AZ-3, AZ-15, AZ-52 and AZ-61 in the north. Reinjection fluids consist of a mixture of separated water, condensed steam and air, this mixture is highly evaporated at ambient conditions before injection.

3 RESULTS

The total discharge chloride concentration and the Na/K reservoir temperature distributions are given in Figures 3 and 4, respectively. Both distributions show that the influence of reinjection is more important in the southern zone since maximum chlorides and minimum reservoir temperatures are found close to the injection wells. Chloride concentrations decrease but temperatures increase toward the northeast of the field, where no injection takes place.

However reinjection effects in the north zone are slightly seen in the west through inflections (toward the east) of the 1,500 mg/kg chloride contour and both the 270 and 280°C temperature contours, with respect to the general patterns.

Reinjection fluids are highly CO_2 depleted as compared to the reservoir fluids, thus the reinjection influence is noticed through minimum CO_2 concentrations in fluids. CO_2 concentrations (Fig. 5) increase from the west toward the east, with maximum CO_2 concentrations in the southeast (well AZ-34) but also in the northeast (well AZ-13), where important natural up-flows seem to occur.

Air is injected to the reservoir, thus N_2 distribution provides actual trajectories of volatile species.

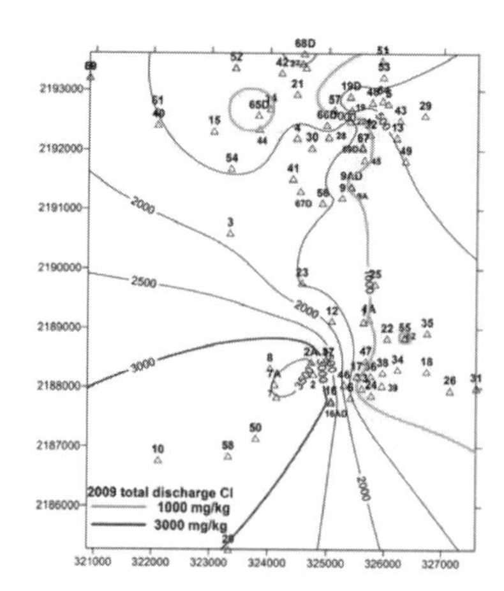

Figure 3. 2009 distribution of the total discharge chlorides.

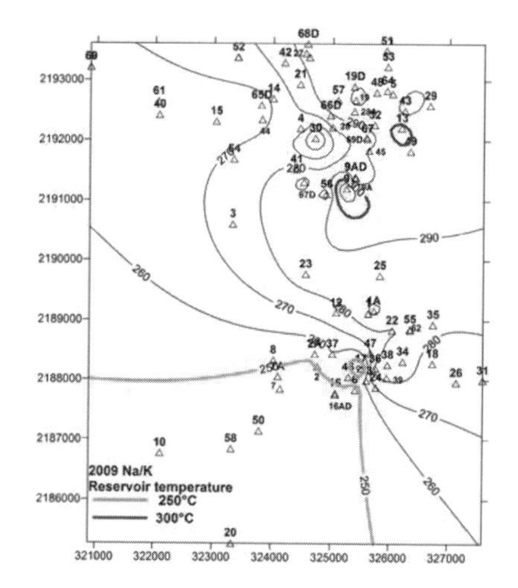

Figure 4. 2009 distribution of the Na/K reservoir temperatures.

Maximum N_2 concentrations (Fig. 6) are seen in the southwest of the field (where reinjection wells are located) with a decreasing trend toward the northeast. Important similarities in the shapes of both the temperature and the N_2 distributions are seen in Figures 4 and 6 as follows. In the south zone the inflections of both the 280°C and the 15 mmol/kg of N_2 iso-lines confirm the up-flow area (where maximum CO_2 occurs). In the northeast of the field maximum temperature and CO_2 and minimum N_2 indicate practically no influence of reinjection.

In Figures 7 and 8 the $\delta^{18}O$ and δ^2H distributions for October 2009 data are given. Both distributions show the more isotopically enriched fluids at the southwest of the field, close to the reinjection wells. The mass flow rates injected in 2009, as compared with 2008 trends (Fig. 2) show a decrease in well AZ-7A and a slight increase in well AZ-8.

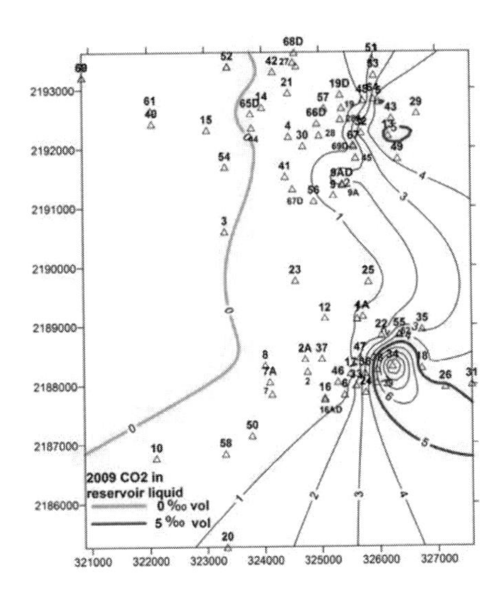

Figure 5. 2009 distribution of CO_2 in the reservoir liquid.

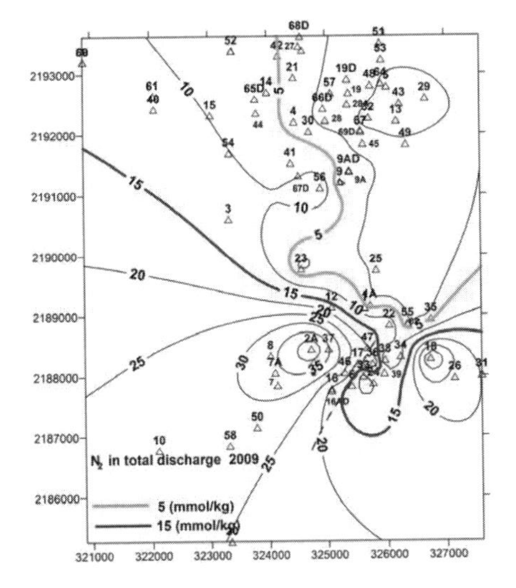

Figure 6. 2009 distribution of N_2 in the total discharge.

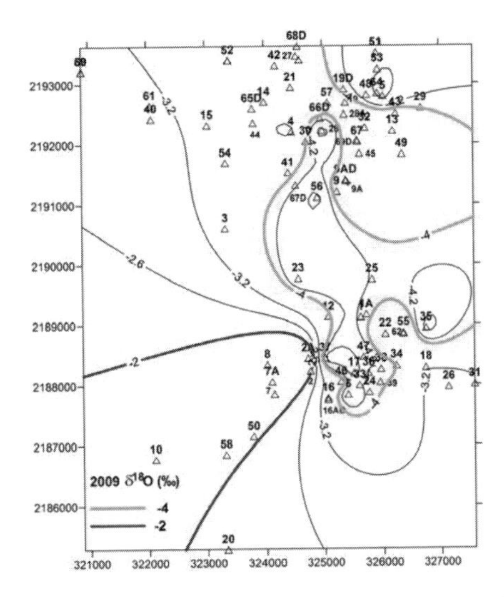

Figure 7. Distribution of $\delta^{18}O$ in total discharge fluids for October 2009 data.

The $\delta^{18}O$ distribution (Fig. 7) shows the more depleted values (–4.4‰) in steam-dominated zones as follows: (a) at the center of the southern zone, (b) toward the northeast of the southern zone (well AZ-35); (c) at the center of the field (wells AZ-56 and AZ-23); (e) at the center of the north zone (well AZ-28) where important boiling has developed (Barragán et al. 2009b) and (f) at the northeast of the field (wells AZ-48 and AZ-5). The boiling process occurring at the center of the north zone started after increased fluid extraction to meet steam demands of the new power plants. As a result, the well AZ-28 became a steam well (it was first a two-phase well). The same transition occurred by 1995 to wells AZ-5 and AZ-13 which were both two-phase wells but after years of production became steam wells.

$\delta^{18}O$ values increase toward the east of the south zone in the wells AZ-18 (–3.5‰) and AZ-26 (–3‰) (Fig. 7), where a natural up-flow was suggested.

Minimum $\delta^{18}O$ values (<–5‰) are seen at the center of the field (well AZ-25). Enriched $\delta^{18}O$ fluids (≥ –3.5‰) are found in the north zone (wells AZ-4, AZ-9, AZ-9A, AZ-9AD, AZ-13 and AZ-45). These wells produce more representative reservoir fluids since reinjection has less influence in the north as compared to the south zone (Arellano et al. 2005). However in 2009 some south zone wells that produced important fractions of reinjection returns (such as AZ-33 and AZ-46), when the original well AZ-7 was in operation (it was closed and a new well was drilled, the AZ-7A) now produce dry steam. Maximum $\delta^{18}O$ values in the south zone are seen in the well AZ-2A which is strongly influenced by reinjection in well AZ-7A.

The more enriched δ^2H values are seen in the southwest of the field in Figure 8, with a decreasing trend toward the north-central part of the field. A small inflection of the –62‰ contour toward the east is noticed in the north zone indicating some minor reinjection effect.

δ^2H vs $\delta^{18}O$ relationships in Figures 9 (A) and (B) indicate the mixing effects between reservoir fluids and reinjection returns in both zones of the field. The depleted end-members are the compositions of wells not affected (or little affected) by reinjection while the reinjection fluids constitute the enriched end-members. In Figures 9 (A) and (B) most of the steam wells are located above the δ^2H vs $\delta^{18}O$ linear trends. This is due to the convection processes in the reservoir at temperatures higher than

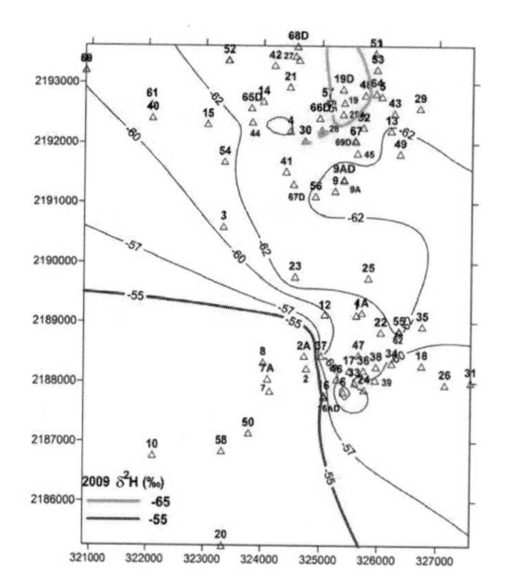

Figure 8. Distribution δ^2H in total discharge fluids.

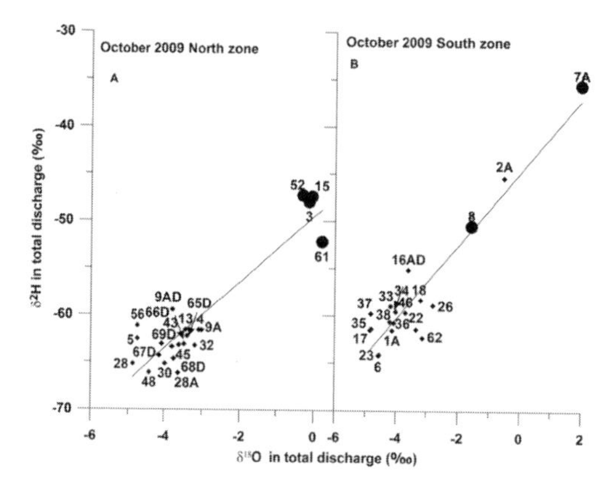

Figure 9. δ^2H vs $\delta^{18}O$ in total discharge fluids of (A) north zone and (B) south zone wells.

220°C. Above 220°C deuterium slightly partitions to the steam, leaving the condensate relatively δ^2H depleted (Truesdell et al. 1977).

In Figure 9 (B) it is seen that about 50% of the fluids produced by the well AZ-2A consist of reinjection returns. Most of the wells located below the linear trends in Figures 9 (A) and (B) are two-phase. These wells produce some condensate which downflows to the reservoir because of convection.

4 CONCLUSIONS

The analysis of 2009 chemical and isotopic ($\delta^{18}O$, δ^2H) data of Los Azufres geothermal fluids was useful to investigate changes occurred due to exploitation. Geochemical indicators included: total discharge $\delta^{18}O$, δ^2H, N_2 and chlorides; reservoir temperatures and reservoir liquid CO_2 concentrations. The main results for 2009 showed that minimum reservoir temperatures, maximum chlorides and isotopic enrichment occur at the west of the field, where reinjection wells are located. 2009 δ^2H vs $\delta^{18}O$ relationships showed that some wells produce different proportions of reinjection returns and that reinjection effects are more important in the south zone. It was also seen that minimum reservoir liquid CO_2 contour lines are good tracers of reinjection returns in the reservoir.

REFERENCES

Arellano, V.M., Torres, M.A., Barragán, R.M., Sandoval, F. & Lozada, R. 2003. Chemical, isotopic and production well data analysis for the Los Azufres (Mexico) geothermal field. *Geothermal Resources Council Transactions* 27: 275–279.

Arellano, V.M., Torres, M.A. & Barragán, R.M. 2005b. Thermodynamic evolution (1982–2002) of the Los Azufres (Mexico) geothermal reservoir. *Geothermics* 34(5): 592–616.

Barragán, R.M., Arellano, V.M., Portugal, E., Sandoval, F., Gonzalez, R., Hernandez, J. & Martínez, J. 2003. Chemical and isotopic ($\delta^{18}O$, δD) behavior of the Los Azufres (Mexico) geothermal fluids related to exploitation. *Geothermal Resources Council Transactions* 27: 281–285.

Barragán, R.M., Arellano, V.M., Portugal, E. & Sandoval, F. 2005a. Isotopic ($\delta^{18}O$, δD) patterns in Los Azufres (Mexico) geothermal fluids related to reservoir exploitation. *Geothermics* 34(4): 527–547.

Barragán, R.M., Arellano, V.M., Portugal, E., Sandoval, F. & Segovia, N. 2005b. Gas geochemistry for the Los Azufres (Michoacán) geothermal reservoir, México. *Annals of Geophysics* 48: 145–157.

Barragán, R.M., Arellano, V.M., Martínez, I., Aragón, A. & González, R. 2009a. Patrones de comportamiento de especies químicas e isotópicas (2006–2007) en el campo geotérmico de Los Azufres, Mich., en respuesta a la reinyección. *Geotermia* 22(2): 19–27.

Barragán, R.M., Arellano, V.M., Aragón, A. & Martínez, I. 2009b. *Monitoreo isotópico de fluidos de pozos productores y de reinyección del campo geotérmico de Los Azufres*. Report IIE/11/13768/I 01/F Instituto de Investigaciones Eléctricas, Cuernavaca, México: 75 p.

Gutiérrez-Negrín, L.C.A., Maya, R. & Quijano J.L. 2010. In *Current status of geothermics in Mexico, Proc. World Geothermal Congress 2010, Bali, Indonesia*: 11 p. (in press).

Nieva, D. & Nieva, R. 1987. Developments in geothermal energy in Mexico-Part Twelve. A cationic geothermometer for prospecting of geothermal resources. *Heat Recovery Systems & CHP* 7: 243–258.

Nieva, D., Verma, M., Santoyo, E., Barragán, R.M., Portugal, E., Ortíz, J. & Quijano, L. 1987. Chemical and isotopic evidence of steam upflow and partial condensation in Los Azufres reservoir; *Proc. 12th Workshop on Geothermal Reservoir Engineering, Stanford University*: 253–259.

Torres-Rodríguez, M.A. & Flores-Armenta, M. 2000. Reservoir behaviour of the Los Azufres geothermal field, after 16 years of exploitation; *Proc. World Geothermal Congress 2000, Kyushu-Tohoku, Japan:* 2269–2275.

Truesdell, A.H., Nathenson, M. & Rye, R.O. 1977. The effects of subsurface boiling and dilution on the isotopic compositions of Yellowstone thermal waters. *Journal of Geophysical Research* 82(26): 3694–3704.

Water-Rock Interaction – Birkle & Torres-Alvarado (eds)
© 2010 Taylor & Francis Group, London, ISBN 978-0-415-60426-0

Determination of flowing pressure gradients in producing geothermal wells by using artificial neural networks

A. Bassam, A. Álvarez del Castillo, O. García-Valladares & E. Santoyo
Departamento de Sistemas Energéticos, Centro de Investigación en Energía, Universidad Nacional Autónoma de México (UNAM), Temixco, Morelos, Mexico

ABSTRACT: A predictive model based on an application of the artificial neural networks (ANN) for obtained the flowing pressure gradients in geothermal wells was applied. The ANN model uses geometrical and physical data commonly measured in producing wells. The prediction of pressure gradients was successfully achieved using the following input data: wellbore geometry (i.e., wellbore depth, inclination angle of pipe, and wellbore diameter), mass flow rate and bottomhole temperature measurements collected from a world database of geothermal wellbores. For the ANN, several computational architectures based on the Levenberg-Marquardt optimization algorithm, the hyperbolic tangent sigmoid transfer-function, and the linear transfer-function were effectively used. The best fitting training data set was obtained with an ANN architecture of 15 neurons in the hidden layer, which made possible to predict the flowing pressure gradients with a satisfactory efficiency ($R^2 = 0.9910$). The results provided by the ANN model between measured and simulated data were in good agreement.

1 INTRODUCTION

The evaluation of wellbore production data are normally required for several engineering tasks, such as, the optimum design of the wellbore geometry, the identification of permeable zones inside the wellbore, the design of surface equipment, the wellbore deliverability studies, among other applications (Tian & Finger 2000). Wellbore production data are commonly obtained from stable flowing measurements performed in geothermal fields. Fluid and heat flows inside wellbores play an important role for the better understanding of the geothermal wellbore production (Hadgu et al. 1995). Numerical simulators have been widely accepted as the most effective and cheaper tools used to analyze the fluid and heat flows, in terms of the calculation of flowing pressure and temperature gradients. These tools have the advantage to minimize the number of field measurements in geothermal wells, which are normally difficult, costly or sometimes unfeasible to be performed.

Numerical simulations have been sometimes criticized due to significant differences found between simulated and measured field data. These differences are mainly caused by the use of non-suitable empirical equations for describing the fluid and heat flow process in the wellbore model, as well as by the consideration of assumptions that do not describe the actual physical phenomena in the wells.

The aim of this work is to propose the use of ANN as an alternative tool for determining flowing pressure profiles or gradients. The ANN has been suggested as a suitable computational tool to model, to control and to optimize the actual non-linear phenomena involved with the flow mechanisms of geothermal wells.

2 THEORETICAL BACKGROUND

2.1 *Pressure gradient*

In the geothermal literature there is a large effort that has been carried out for determining flowing pressure and temperature gradients in geothermal wells. Several wellbore numerical simulators have been recently developed (e.g., García-Valladares et al. 2006; Álvarez del Castillo et al. 2010). Most of these simulators calculate such pressure and temperature gradients using the mass, momentum and energy governing equations subject to a set of initial and boundary conditions under several assumptions. The total pressure gradient under one-dimensional conditions in a well is usually estimated by the following equation:

$$\left(\frac{dP}{dz}\right) = \left[\left(\frac{dP}{dz}\right)_f + \left(\frac{dP}{dz}\right)_a + \left(\frac{dP}{dz}\right)_g\right] \tag{1}$$

where: z is the vertical coordinate; and P is the flowing pressure. The first term in square brackets (Eq. 1) represents the pressure gradient due to friction (f), the second term denotes the pressure gradient due to acceleration (a), and the last term is the gravitational pressure gradient (g). Pressure gradients are normally computed using the following equations:

$$\left(\frac{dP}{dz}\right)_f = f_M \rho \frac{V^2}{2D} \tag{2}$$

$$\left(\frac{dP}{dz}\right)_a = \rho V \frac{dV}{dz} \tag{3}$$

$$\left(\frac{dP}{dz}\right)_g = \rho g sin\theta \tag{4}$$

where V is the fluid velocity; D is the inner pipe diameter; is fluid density; θ is the wellbore inclination angle; and f_M is the Moody friction factor.

Evidently, the solution of these equations require the accurate and reliable knowledge of several data sources, such as, the two-phase flow parameters (the liquid hold-up or void fraction, which are calculated from empirical correlations), the thermodynamic and transport properties of geothermal fluids (which are rarely reported with accuracy), as well as complex numerical schemes for obtaining reliable predictions of the flowing pressure gradients.

The knowledge of some of these fundamental parameters (or input data) is sometimes unreliable and difficult to obtain under actual flow conditions. These input data requirements have pushed the searching for other practical or analytical methods to compute the pressure and temperature gradients for the geothermal reservoir engineering. Within this context, in the present work, we are proposing a new practical method based on the well-known ANN to predict flowing pressure gradients of producing geothermal wells.

2.2 *Artificial neural networks (ANN)*

A neural network is a massively parallel distributed processor, which has a natural propensity for storing and learning an experimental knowledge, and a later use of it in a wide variety of engineering applications (Haykin 1999). ANN can be trained to solve particular problems that are difficult to find out solutions with conventional methods.

ANN are composed of simple elements operating in parallel (Demuth & Beale 2007). These elements are inspired by biological nervous systems. As in nature, the network function is largely determined by connections between elements or neurons. The training of an ANN to perform any particular function is usually carried out by adjusting the connection values (weights) among elements.

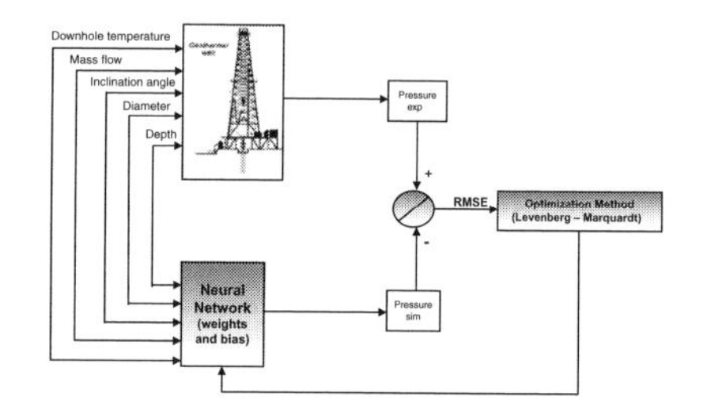

Figure 1. Numerical procedure used for the ANN learning process, and the iterative architecture used by the model to predict the flowing pressure gradients of geothermal wells.

ANN are commonly adjusted or trained for leading a particular input to a specific output (target). Such a situation is schematically shown in Figure 1.

Here, the network is adjusted based on a statistical comparison between the ANN simulated output and the experimental target of the flowing pressure gradients.

3 ARTIFICIAL NEURAL NETWORKS MODEL

A world geothermal database containing 1218 production data from 71 producing wells was created. These data was used for the network training process. The production data include input variables, such as the wellbore depth, the inclination angle of the wellbore in the producing zone, the wellbore diameter, the mass flow rate, and the down-hole temperatures.

These data were employed to carry out the ANN simulation, and for comparing the simulated results against the actual pressure gradients measured in the geothermal wells.

To test the ANN model, the experimental database was divided into a learning database (with 70% of experimental data set) and a testing database (with 30% of experimental data set) in order to obtain a good representation of the data distribution, as well as to avoid bias. Thus, we have used a learning database for calculating optimal weights and biases, as well as to have a testing database for validating the ANN model (Rumelhart 1986). The error of the learning process decrease when the number of iterations increases. This calculation was considered as a criterion for model adequacy (see Fig. 1). The number of neurons in the input and output layers was given by the number of input and output variables in the process, respectively.

In this work, we have used a neural network model with 15 neurons in the hidden layer (Fig. 2). Such architecture was found to be efficient for predicting most of the pressure gradients of the geothermal wells. The input data used by this ANN are summarized in Table 1.

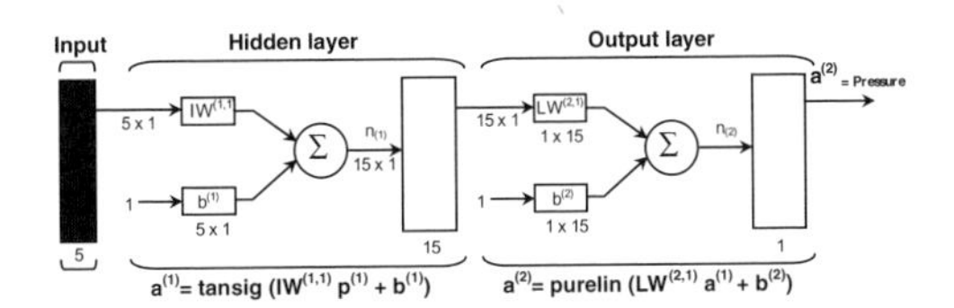

Figure 2. Computing architecture of the ANN model developed.

Table 1. Geometrical and physical parameters used as input data for the ANN model training.

Parameters	Values range	Units
Input parameters:		
Wellbore depth	[0–2600]	[m]
Inclination angle	[0–32.1]	[°]
Wellbore diameter	[0.102–0.385]	[m]
Mass flow rate	[1.1–202]	[kg/s]
Temperature	[104.4–364]	[°C]
Output parameters:		
Pressure	[2–195]	[bar]

For normalizing the ANN, the input and output variables was fixed to work in the close computing domain [–1, 1] (i.e., zero mean and the standard deviation depends on the input variable distribution), which constitutes the classical method proposed by many authors (e.g., Demuth & Beale 2007).

4 RESULTS AND DISCUSSION

4.1 *Evaluation model*

The simulated pressure results obtained with the ANN model were compared with actual data measured in field experimental tests (Fig. 3). An acceptable agreement between simulated and measured data was observed.

The ANN model was well fitted to the behavior of the learning database (R2 > 0.99). For the evaluation of the ANN model, we used data from two different geothermal wells (KW-2 and KE-1), which were reported by Garg et al. (2004). These data were included in the ANN training process.

The response of the ANN model was compared with the results provided by the GEOWELLS wellbore simulator (García-Valladares et al. 2006), which uses different empirical correlations, first, to estimate the void fraction parameter (Duns & Ros 1963; Orkiszewski 1967), and then to calculate the pressure gradients in the geothermal wells.

Figures 4 and 5 present the flowing pressure gradients calculated for the KW2 and KE1 wells with three different methods: the ANN predictions (red diamond symbols), and the simulated results reported by the GEOWELLS simulator using two different correlations (Duns-Ros: represented by blue diamonds, and Orkiszewski: blue circles) to compute the void fractions, respectively.

It was also included the measured field data reported for the KW-2 and KE-1 wells (black cross symbols).

As can be observed, ANN provides the better results for the geothermal well KE-1, whereas for the KW-2 shows some significant differences of about ±3% were found.

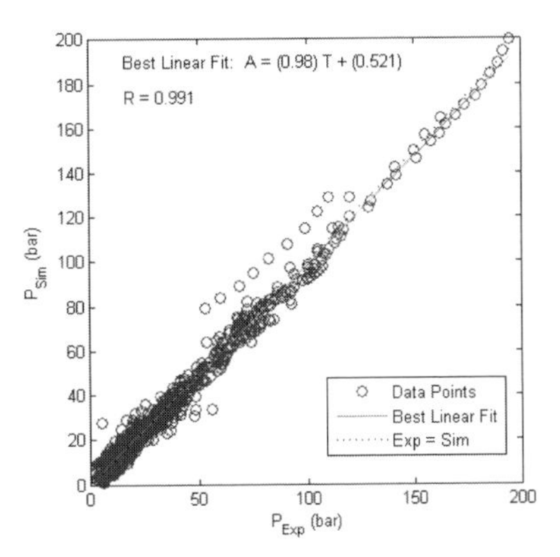

Figure 3. Statistical comparison of flowing pressure gradients between experimental and simulated (obtained during the learning database).

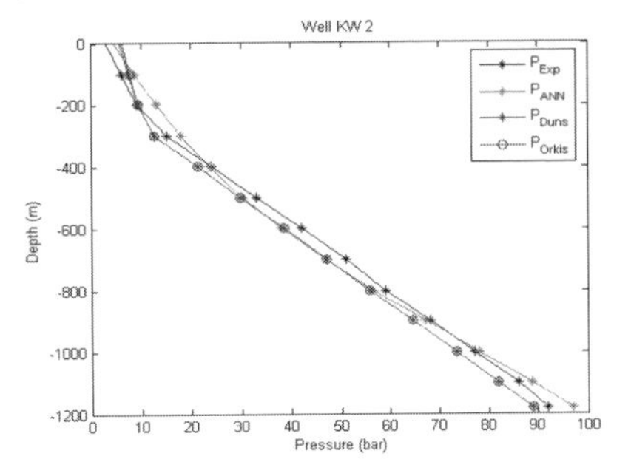

Figure 4. Flowing pressure gradients calculated for the geothermal well KW-2 using the ANN model and the GEOWELLS simulator.

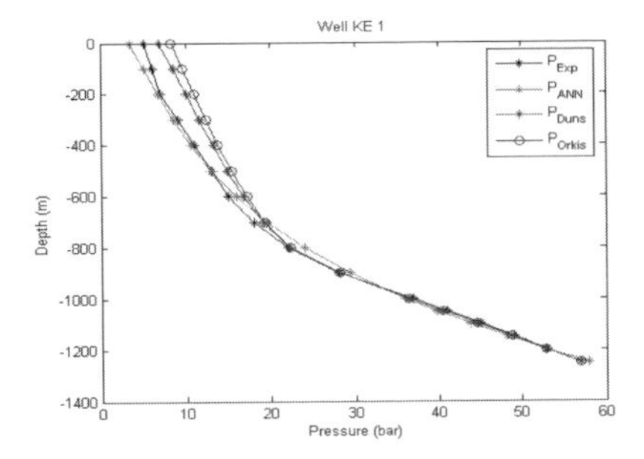

Figure 5. Flowing pressure gradients calculated for the well KE-1 using the ANN model and the GEOWELLS simulator.

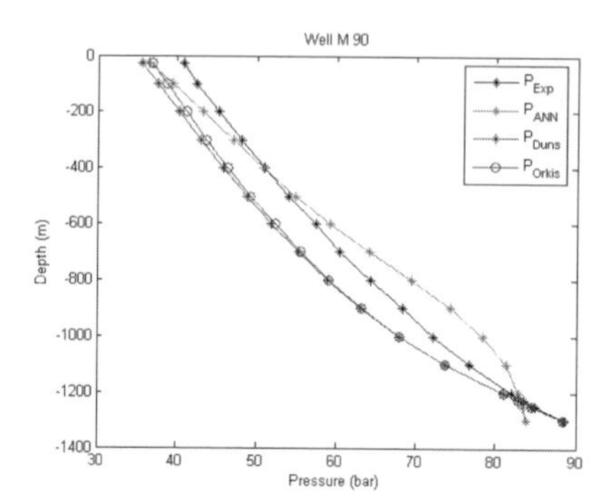

Figure 6. Flowing pressure gradients calculated for the well M-90 using the ANN model and the GEOWELLS simulator.

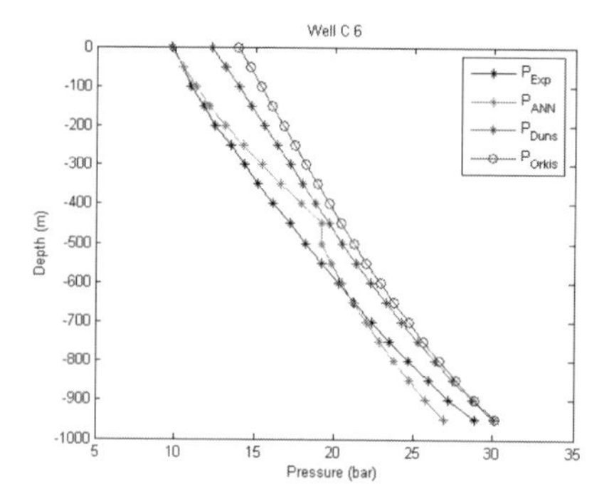

Figure 7. Flowing pressure gradients calculated for the well C-6 using the ANN model and the GEOWELLS simulator.

4.2 *Validation of model*

The validation of the ANN model was carried out using well-known cases of producing wells (M-90: Cerro Prieto, México, and C-6: Japan), which were reported by Ambastha & Gudmundsson (1986) and Garg et al. (2004), respectively. These data were not included in the training database. The capability of the ANN model to predict pressure gradients of the geothermal wells M-90 and C-6 is presented in Figures 6 and 7. As can be observed, ANN provides the better results for the C-6 well, whereas for the M-90 some small differences (<10%) were found. The residuals of the testing database were small and their distribution was well-balanced (data not shown). The global regression coefficient of evaluated model, $R^2 = 0.9910$, which confirms the flexibility of the ANN model to predict the pressure gradients of producing geothermal wells.

5 CONCLUSIONS

The validity of the pressure gradients computed by the ANN was confirmed by a comparison between simulated and measured field data. An acceptable agreement was obtained, which suggests that the ANN model can be used for a reliable determination of flowing pressure gradients in producing geothermal wells.

The model developed may therefore enable the implementation of smart sensors for on-line determination and control of pressure gradients in geothermal boreholes.

ACKNOWLEDGEMENTS

The first and second authors wish to thank to the PhD Graduate Program of Engineering (UNAM) and CONACyT for the scholarships and facilities provided to carry out the present work.

REFERENCES

Álvarez del Castillo, A., Santoyo, E., García-Valladares, O., & Sánchez-Upton, P. 2010. Evaluación estadística de correlaciones de fracción volumétrica de vapor para la modelación numérica de flujo bifásico en pozos geotérmicos. *Revista Mexicana de Ingeniería Química*, in Press.

Ambastha, A.K. & Gudmundsson, J.S. 1986. Pressure gradients in two-phase geothermal wells: comparison of field data and model calculations. Proc. *11th Workshop on Geothermal Reservoir Engineering*, Stanford, California, USA, 6p.

Demuth, H. & Beale, M. 2007. Neural network toolbox for Matlab, User's guide version 3. The MathWorks.

Duns, H. & Ros, N.C.J. 1963. Vertical flow of gas and liquid mixtures in wells. Proc. *6th World Petroleum Congress*, Frankfurt, Germany, Section 11, Paper 22, PD 6, 451–465.

García-Valladares, O., Sánchez-Upton, P. and Santoyo, E. 2006. Numerical modeling of flow processes inside geothermal wells: An approach for predicting production characteristics with uncertainties. *Energy Conversion and Management*, 47, 1621–1643.

Garg, S.K., Pritchett, J.W. & Alexander, J.H. 2004. Development of new geothermal wellbore holdup correlations using flowing well data. *Idaho National Engineering and Environmental*, Project Report No. INEEL/EXT-04-01760.

Hadgu, T., Zimmerman, R.W. & Bodvarsson, G.S. 1995. Coupled reservoir-wellbore simulation of geothermal reservoir behavior, *Geothermics*, 24, 145–166.

Haykin, S. 1999. Neural networks. Second Ed., Prentice Hall.

Orkiszewski, J. 1967. Predicting two-phase pressure drop in vertical pipes", *Journal of Petroleum Technology*, 19, 829–838.

Rumelhart, D.E., Hinton, G.E. & Williams, R.J. 1986. Learning internal representations by error propagation, *Parallel Data Processing*, 1, 318–362.

Tian, S.F. & Finger, J.T. 2000. Advanced geothermal wellbore hydraulics model. *Journal of Energy Resources & Technology*, ASME Transactions, 122, 142–146.

Water-Rock Interaction – Birkle & Torres-Alvarado (eds)
 ISBN 978-0-415-60426-0

Geochemistry of boron in fluids of Los Humeros and Los Azufres hydrothermal systems, Mexico

R.A. Bernard-Romero & Y.A. Taran
Instituto de Geofísica, Universidad Nacional Autónoma de México, México

M. Pennisi
Istituto di Geoscienze e Georisorse, CNR, Pisa, Italy

ABSTRACT: Separated water from geothermal wells of high-temperature geothermal fields Los Humeros and Los Azufres in Mexico are unusually rich in boron. According to the isotopic composition, boron in both fields has magmatic origin. Since the Los Humeros fluids, in contrast to Los Azufres, do not show any correlation between B and Cl and have Cl/B < 1, a mechanism is proposed based on the B and Cl partitioning between a deep acidic brine and the upper steam reservoir.

1 INTRODUCTION

High-temperature hydrothermal systems Los Humeros and Los Azufres located within Trans-Mexican Volcanic belt (Fig. 1) are characterized by very high concentrations of boron in thermal fluids. Geothermal wells from the water-dominated Los Azufres system discharge fluids with up to 400 mg/kg of boron in the total fluid, but with a nearly constant Cl/B weight ratio ~10 (Arellano et al. 2005). In contrast, the predominantly vapor-dominated Los Humeros system is characterized by the irregular Cl/B weight ratio from 0.02 to 0.5 (B > Cl). Separated water from Los Humeros wells sometimes contains up to 5000 mg/kg of boron (Arellano et al. 2003, this work) and Cl < 100 mg/kg.

It is generally accepted that boron in hydrothermal fluids is originated mostly from two sources: from magmatic fluids feeding hydrothermal aquifers with heat, volatile components, and boron extracted from local reservoir rocks. High-temperature volcanic gases are characterized by constant Cl/B supporting the idea about a common magmatic source for Cl and B in volcanic and hydrothermal gases. Therefore, boron behavior in Los Humeros fluids should be explained whether by a specific mechanism of the steam-water partitioning at depth and/or by an additional independent source of B.

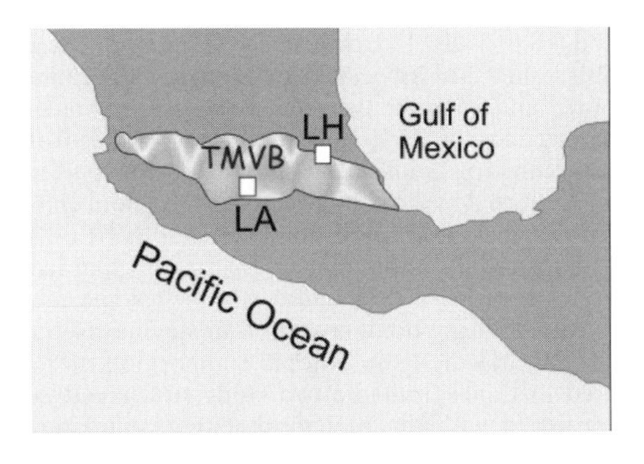

Figure 1. Trans-Mexican Volcanic Belt (TMVB) and location of Los Humeros (LH) and Los Azufres (LA) geothermal fields.

Here, we present data on boron and chloride concentrations and boron isotopes in fluids from Los Azufres and Los Humeros geothermal wells and propose a possible elucidation for the lack of correlation between Cl and B in Los Humeros fluids.

2 CL/B IN VOLCANIC AND HYDROTHERMAL FLUIDS

High-temperature volcanic gases from andesitic and more evolved arc volcanoes usually have Cl/B ratio >500 (Fig. 2). Gases from volcanoes with magmas contaminated by crustal material like Italian volcanoes are characterized by a higher B content and lower Cl/B ratios (Vulcano, Fig. 2). It follows from data reported by Goff and McMurtry (2000) that Cl/B in volcanic gases correlates well with Cl/B in associated rocks but usually ((Cl/B) fluid > (Cl/B) rock)).

Deep hot hydrothermal fluids from geothermal wells are characterized by a wide range of Cl/B ratios, but they are usually, between 10 and 100. Generally, B and Cl are well correlated, and all Na-Cl waters discharging within a geothermal field

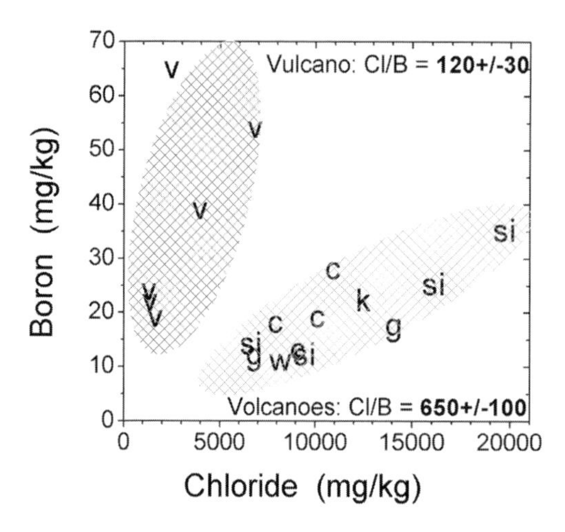

Figure 2. Correlation between Cl and B in high-temperature volcanic gases. V—Vulcano; C—Colima; G—Galeras; SI—Satsuma Iwojima; K—Kudryavy; W—White Island. From Taran et al. (1995) and Goff and McMurtry (2000).

including hot springs and geothermal wells show a single linear trend indicating a common source of boron and chloride. The highest boron content in the total fluid ~1000 mg/kg and the lowest Cl/B ~1 was reported for Ngawha geothermal system in New Zealand (Ellis & Mahon 1977).

3 BORON AND CHLORIDE IN LOS AZUFRES AND LOS HUMEROS FLUIDS

Relationships between B and Cl in separated water from geothermal wells of Los Azufres and Los Humeros are shown in Figure 3. It should be noted that the concentration of dissolved species in the total fluid can be estimated as $C_s(1-Y_s)$ where C_s is the concentration of a specie in separated water and Y_s is the weight fraction of the separated steam. Using data on the fluid enthalpies (Arellano et al. 2005) one can estimate that the deep water of the Los Azufres liquid-dominated aquifer contains about 1700 mg/kg of Cl and ~160 mg/kg of B. The B-Cl correlation for Los Azufres shown in Figure 3 is simply an evidence for fluids of different enthalpies (steam fractions). The Los Humeros case is more complicated. Very high concentrations of B in separated water of Los Humeros fluids are usually associated with very low water fraction and hence, low concentration of B in the total fluid. This concentration could be estimated as 100–200 mg/kg, similar to Los Azufres fluids, and only 15–50 mg/kg in the steam phase at 300°C. These values can be estimated using data on enthalpies and the temperature dependence of the distribution coefficient of B between steam and water. However, the concentrations of B in Los Humeros, in contrast to Los Azufres, are accompanied with low and irregular Cl concentrations from 10 to 100 mg/kg.

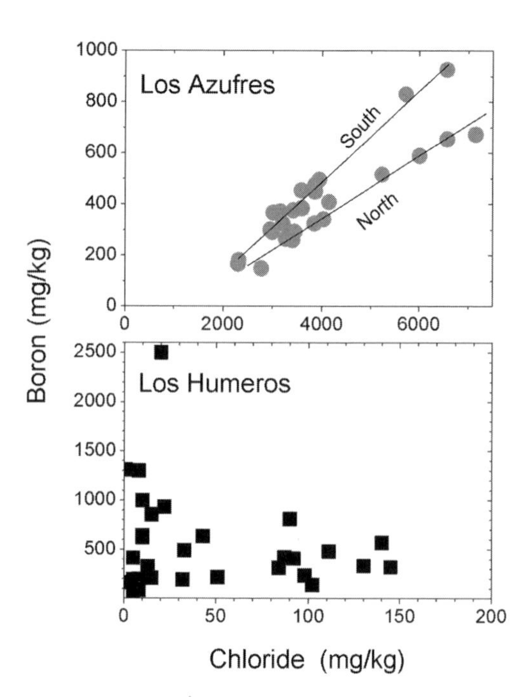

Figure 3. Cl-B relationships in separated waters from geothermal wells of Los Azufres and Los Humeros. Two trends for Los Azufres are observed for the Northern and Southern sections of the field.

4 ORIGIN OF BORON

The isotopic composition of boron in the fluids from Los Humeros and Los Azufres is –0.8‰ (±1) and –8.5‰ (±0.1), respectively (Table 1). These $\delta^{11}B$ values are associated to average Cl/B ratios of 0.2 and 10 in the two sites. According to values reported for the Pacific arc lavas, the average $\delta^{11}B$ of Los Humeros fluids falls within the range reported for Central America volcanic rocks (Doglioni et al. 2009), while the lighter isotopic signature of Los.

Azufres falls in the range –7‰ to –9‰ that has been measured for Cascades and Andes volcanic rocks. There are no data on $\delta^{11}B$ in magmas of the TMVB, however, the isotopic composition measured in fluids from the two study sites could be consistent with the geological setting, suggesting a different contribution of crustal material and hydrous fluids from the subducting slab to the TMVB magmas (Doglioni et al. 2009).

Table 1. Isotopic composition of boron ($\delta^{11}B$ ‰ relative to NBS standard) and Cl and B concentrations (mg/kg) in separated waters from Los Humeros and Los Azufres geothermal wells.

Well n°	$\delta^{11}B$	B	Cl
Los Humeros			
H-1	−0.2 ± 0.4	179	98
H-2	0.3 ± 0.5	295	111
H-3	−1.7 ± 0.4	725	27
H-6	−1.6 ± 0.4	1105	1.0
Los Azufres			
Az-1a	−8.6 ± 0.5	816	7860
Az-13	−8.5 ± 0.5	666	5420
Az-22	−8.1 ± 0.5	333	3450
Az-62	−8.8 ± 0.5	278	3062

5 POSSIBLE MECHANISMS OF CL/B VARIATIONS IN FLUIDS OF LOS HUMEROS

The behavior of Cl/B in the water-dominated Los Azufres system is similar to many other geothermal systems in the world and is the result of a common source for Cl and B in hydrothermal fluids associated with magmatic plutons feeding hydrothermal systems with heat and volatiles. Los Humeros system is not a water-dominated system, but it also differs from "classic" vapor-dominated systems like Larderello (Italy) and The Geysers (USA) by the heterogeneity of phase conditions at depth. On the other hand, Los Humeros is similar to Larderello in some important geological and water-rock interaction aspects. Both systems are associated with limestones. Scarn-type contacts between limestone and magmatic rocks were reported for deep levels of the aquifer of Los Humeros (Martinez-Serrano 2002). Another similarity is the presence of acidic HCl-bearing fluids at both Larderello and Los Humeros fields (Truesdell et al. 1988). Truesdell et al. (1988) reported the composition of the acidic steam condensates from Larderello with Cl from 20 to 120 mg/kg and B from 120 to 500 mg/kg and no correlation between Cl and B.

It can be suggested that the acidity of hydrothermal fluids formed due to the incomplete neutralization of host rocks by a mixture of the condensed magmatic fluid and meteoric water is the main reason for the variation of Cl/B in hydrothermal steam. The apparent Cl distribution between liquid and vapor is a strong function of pH (Simonson & Palmer 1993), whereas the distribution of B is practically pH independent. On the other hand, both steam-water partition coefficients (for HCl and H_3BO_3) are temperature-dependent (Siminson & Palmer 1993, Leeman et al. 1992). The deepest parts of the Los Humeros and Larderello aquifers are thought to be partially filled with the boiling acidic brine. The heterogeneity in pH of the brine (due to different reasons including water/rock ratio, sulfur precipitate, degassing) should cause a variable transfer of HCl in the steam phase, but the partition of boric acid depends only on temperature. At 300–350°C the steam-water partition coefficient for H_3BO_3 is close to 0.2 (Leeman et al. 1992), which indicates that at 15–50 mg/kg of B in the deep steam phase, concentration of B in the brine should be 100–250 mg/kg.

Another reason for lacking correlation between Cl and B in Los Humeros fluids would be the existence of an additional source of boron in B-bearing minerals associated with the scarn sequence found at depth. These minerals can be considered as primary traps for magmatic boron and then as a secondary source of boron in hydrothermal fluid.

6 CONCLUDING REMARKS

Relationships between $\delta^{11}B$ and Cl/B ratio (Fig. 4) for a number of hydrothermal systems over the world in general demonstrate a mixing trend from magmatic values to seawater (sedimentary) values. The opposite trend directions for Iceland and Larderello, Italy (Aggarwal et al. 1992, 2000; Pennisi et al. 2001) are most probably related to different endmembers. In the case of water-dominated systems of Iceland, relatively Cl-poor magmatic-meteoric mixture mixes with Cl-rich, B-poor

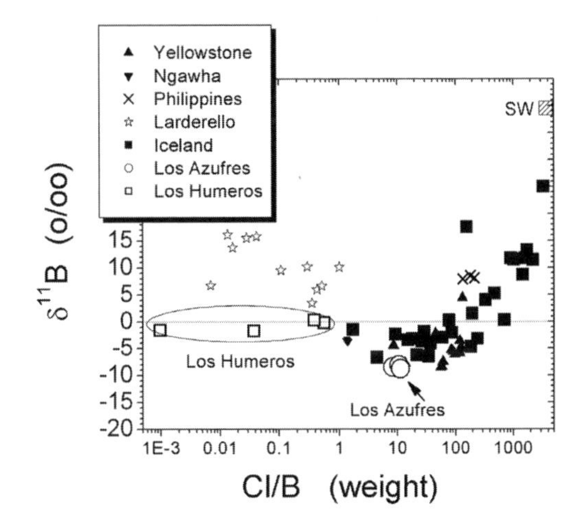

Figure 4. Boron isotopic composition vs. Cl/B (weight ratio) in fluids of high-temperature hydrothermal systems including Los Humeros and Los Azufres. SW—seawater value. Data by Palmer and Sturchio (1990); Aggarwal et al. (1992, 2000, 2003); Pennisi et al. (2001); Bayong et al. (2008).

seawater. In the case of the vapor-dominated system of Larderello, Cl- and B-rich magmatic fluid is mixed with a Cl-free, B-rich vapor derived from a sedimentary source.

Data for Los Azufres are plotted on the main trend for fluids from water-dominated systems with Cl/B > 1. Data for Los Humeros do not show any dependence on the Cl/B ratio, supporting the idea about fractionation of Cl and boric acid at boiling of a brine of variable acidity.

More work is needed to resolve the behavior of boron in high-temperature vapor-dominated aquifers.

ACKNOWLEDGEMENTS

This work was supported by grant from DGAPA-UNAM # IN100709. Authors thank the personal of the Los Humeros campus for the assistance in sampling geothermal wells.

REFERENCES

Aggarwal, J.K., Palmer, M.R. & Ragnarsdottir, K.V. 1992. Boron isotope composition of Iceland Hydrothermal systems. In Y.K. Kharaka, & A.S. Maest (eds.), *Proc. Symposium on Water-rock Interaction 7, Park City, UT, 13–18 July 1992.* Rotterdam: Balkema.

Aggarwal, J.K., Palmer, M.R., Bullen, D., Arnorsson, S. & Ragnarsdottir V. 2000. The boron isotope systematics of Icelandic geothermal waters: 1.Meteoric water charged systems. *Geochimica et Cosmochimica Acta* 64: 579–585.

Aggarwal, J.K., Sheppard, D., Mezger, K., & Pernicka, E. 2003. Precise and accurate determination of boron isotope ratios by multiple collerctor ICP-MS: origin of boron in the Ngawha geothermal system, New Zealand. *Chemical Geology* 199: 331–342.

Arellano, V.M., Garcia, A., Barragan, R.M., Izquierdo, G., Aragon, A. & Nieva, D. 2003. An updated conceptual model of Los Humeros geopthermal reservoir (Mexico). *Journal of Volcanology and Geothermal Research* 124: 67–88.

Arellano, V.M., Torres, M.A. & Barragan, R.M. 2005. Thermodynamic evolution of the Los Azufres, Mexico, geothermal reservoir from 1982 to 2002. *Geothermics* 34: 592–616.

Bayong, F.E.B., See, F.S., Magro, G. & Pennisi, M. 2008. Noble gases and boron isotopic signatures of the Bacon-Manito geothermal fluid, Philippines. *Geofluids* 8: 230–238.

Doglioni, C., Tonarini, S. & Innocenti, F. 2009. Mantle wedge asymmetries and geochemical signatures along W-and E-NE directed subduction zones. *Lithos* 113: 179–189.

Ellis, A.J. & Mahon, W.A.J. 1977. *Chemistry and Geothermal Systems.* New York: Academic Press.

Goff, F. & McMurtry, C. 2000. Tritium and stable isotopes of magmatic waters. *Journal of Volcanology and Geothermal Research* 97: 347–396.

Leeman, W.P., Vocke, R.D. & McKibben, M.A. 1992. Boron isotopic fractionation between coexisting vapor and liquid in natural geothermal system. In Y.K. Kharaka, & A.S. Maest (eds.), *Proc. Symposium on Water-rock Interaction 7, Park City, UT, 13–18 July 1992.* Rotterdam: Balkema.

Martínez-Serrano, R.G. 2002. Chemical variations in hidrotermal minerals of the Los Humeros geothermal system. *Geothermics* 31: 579–612.

Palmer, M.R. & Sturchio, N.C., 1990, The boron isotope systematic of the Yellowstone National park *Geochim. Cosmochim. Acta* 54: 2811–2815.

Pennisi, M., Magro, G., Adorni-Braccesi, A. & Scandiffio G. 2001. Boron and Helium isotopes in geothermal fluids from Larderello (Italy). *Proceedings 10° International Symposium on Water Rock Interaction,* Villasimius, Italy.

Simonson, J.M. & Palmer, D.A. 1993. Liquid-vapor partitioning of HCl(aq) to 350°C. *Geochimica et Cosmochimica Acta* 57: 1–7.

Taran, Y.A., Hedenquist, J.W., Korzhinsky, M.A., Tkachenko S.I. & Shmulovich, K.I. 1995 Geochemistry of magmatic gases from Kudryavy volcano, Iturup, Kuril Islands. *Geochimica et Cosmochimca Acta* 59: 1749–1761.

Truesdell, A.H., Haizlip, J.R., Armansson, H. & D'Amore, F. 1988. Origin and transport of chloride in superheated geothermal steam. *Proc. Deposition of Solids in Geothermal Systems, Reykjavik*, 1988.

Water-Rock Interaction – Birkle & Torres-Alvarado (eds)
© 2010 Taylor & Francis Group, London, ISBN 978-0-415-60426-0

Geochemistry and origin model for thermal springs from Kamchatka and Kuril Islands

E.P. Bessonova
Institute of Geology and Mineralogy SB RAS, Novosibirsk, Russia

S.B. Bortnikova
Trofimuk Institute of Petroleum Geology and Geophysics SB RAS, Novosibirsk, Russia

ABSTRACT: This study deals with the genesis of contrasting thermal water types from active volcanoes. The objects of our study are volcanic springs from the Kamchatka and Kuril islands (Russia). The results of hydrochemistry, geophysical study and physicochemical modeling show that contrasting types of thermal waters can be originated from a single magmatogene fluid. The geometrical zoning of the thermal subsurface zone has been obtained and phase barriers were located using frequency electric sounding. We developed a new software package which allows us to take into account the dynamics of thermo-physical and chemical parameters. The obtained results show that the occurrence of phase barriers on the pathways of ascending fluids represents the decisive factor controlling variations in thermal spring's composition.

1 INTRODUCTION

The hydrogeochemical features of thermal springs that discharge on the flanks and in craters of active volcanoes depend on a number of exogenous and endogenous factors, which are affecting magmatic fluids, separated from a melt. The degree of dilution of the original fluid with groundwater (the fluid/meteoric water ratio) is the leading factor controlling the chemical element concentrations, as well as the physicochemical characteristics of the thermal water. This ratio depends on meteorological and hydrological conditions of the volcano, as well as on geo-structural features of fluid migration paths. Another important factor influencing the formation of volcanic thermal waters is the composition of host volcanogenic rocks through which the waters circulate leaching a number of rock-forming and impurity elements. However, the occurrence of geochemical barriers can significantly change the composition of the surface discharge. The division of a mixed solution into condensates and phase separation by secondary boiling complicates the process for the compositional formation of solutions. It can radically change the hydrogeochemical parameters of thermal springs. Recent data about a wide range of trace elements and their specific associations in active volcano thermal waters (Giammanco et al. 1998, Aiuppa et al. 2000, Bortnikova et al. 2008, 2010 and others) show that these elements allow recognizing and quantifying the sources (endogenous and exogenous) and the source mixing proportion in water samples.

In this paper, we present a detailed analysis of the hydrogeochemical features of various thermal springs (water and mud pots, geyers), and their internal structures of the subsurface (based on geophysical studies). Physicochemical modeling allowed us to understand the processes and mechanisms during the extraction of chemical elements and their deposition during fluid migration, depending on the structure of the volcano-hydrothermal system.

2 OBJECTIVES AND METHODS

Investigations of the gas-hydrothermal springs of various active volcanoes, which are located on the South Kamchatka (Mutnovsky volcano, Karymsky volcano, Uzon caldera, North-Mutnovskoe fumarolic field) and Paramushir Islands (Ebeko volcano) (Fig. 1) have shown considerable differences in the composition and physicochemical parameters of thermal waters.

A detailed sampling of the thermal springs was conducted during 2001–2008. Water samples were collected in plastic containers, filtered and then acidified with ultra-pure HNO_3 (except acid and ultra-acid solutions). pH, Eh and the anions Cl^-, F^- were measured in situ. In addition to the water samples, the tests of the wet substance, which were collected on the thermal fields of the Mutnovsky and Ebeko volcanoes

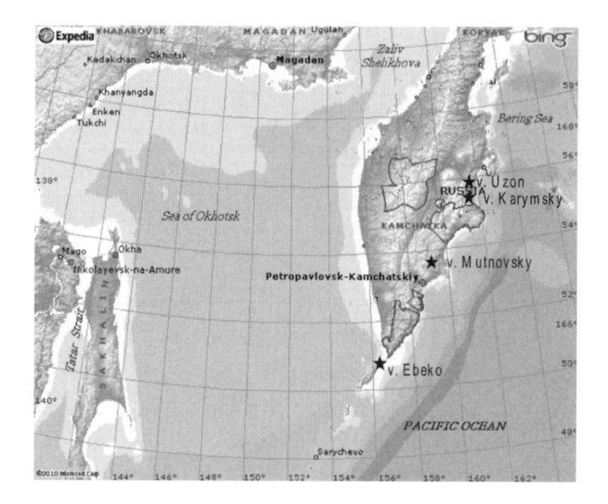

Figure 1. Map of the studied volcanoes.

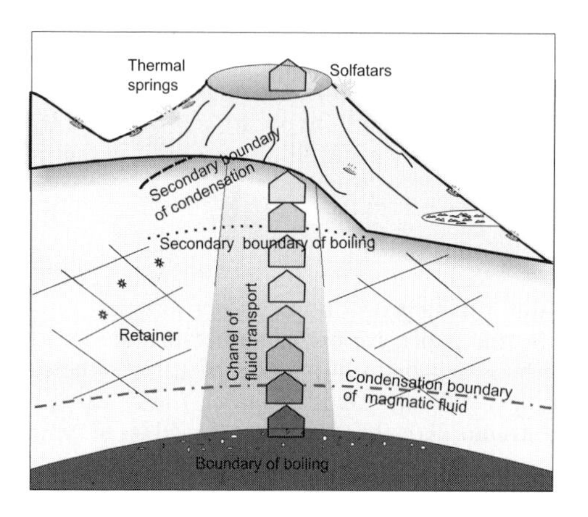

Figure 2. Conceptual model of volcano-hydrothermal system.

(North-Mutnovskoe field) were pressed out in the laboratory to recover pore solutions. All samples were analyzed by ICP-AES, AAS. REE and PGE contents for some samples were determined by ICP-MS.

Electromagnetic frequency sounding was performed using NEMFIS technique during 2007–2009. NEMFIS was developed by IPGG SB RAS. The soundings were performed at 14 frequencies in the range between 2.5 and 250 kHz. The obtained data were processed by ISystem software, which forms part of NEMFIS. The software package combines the SP Selector (Karpov et al. 2002) and SP Fluid (Sharapov et al. 2008) to model hydrothermal process taking into account thermo-physical, hydrodynamic and physic-chemical parameters. This SP is based on a numerical model of quasi-two-dimensional descriptions of non-isothermic physicochemical dynamics of fluid-rock interaction in the presence of changing phase states in the flow of magmatic fluid (Fig. 2).

The numerical model is based on the combination of the Stefan approach (to describe the appearance and disappearance of phase boundaries in the fluid flow within porous fractured media) and computing of equilibrium physicochemical dynamics (using SP Selector). First of all, thermo-physical and hydrodynamic parameters are calculated in all tanks at each point of the timeline. These parameters are further used in determining the physicochemical equilibrium in the direction of the multi-fluid "wave" propagation through the multiple-tank column. This approach first allowed to pass from conditional time used in similar SP (accounting as the water/rock (W/R) ratio), to real-time from the beginning of life of the magma chamber to produce heat and magmatic fluid.

3 GEOCHEMISTRY OF THERMAL SPRINGS

The studied thermal waters represent the discharge of distinct solution types contrasting in physicochemical parameters and composition. The values for pH – Eh vary from alkaline, weakly reducing conditions (Hot stream in the New Year peninsula, Karymskoe lake) to ultra-acidic, strongly oxidized ones (mud pots in the Mutnovsky volcano, Fig. 3). Eh and pH of mud pots from the Uzon caldera and the North Mutnovskoe fumarolic fields are located on the pH-Eh diagram between both extreme types.

The TDS of the solutions is associated with their acidity by a weak negative correlation (Fig. 4). However, there is dissimilarity in the different source types. The relation is most clearly expressed for thermal waters from the North-Mutnovskoe field, whereas it is less expressed for solutions from the Mutnovsky volcano. In the Karymskoe springs, TDS are almost unrelated to the solution's acidity. Sulfates and chlorides show different proportions and represent the major anions of all sampled

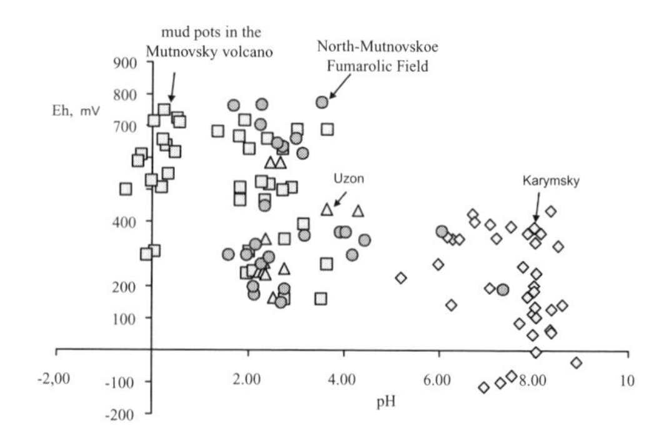

Figure 3. Physicochemical features of studied thermal springs.

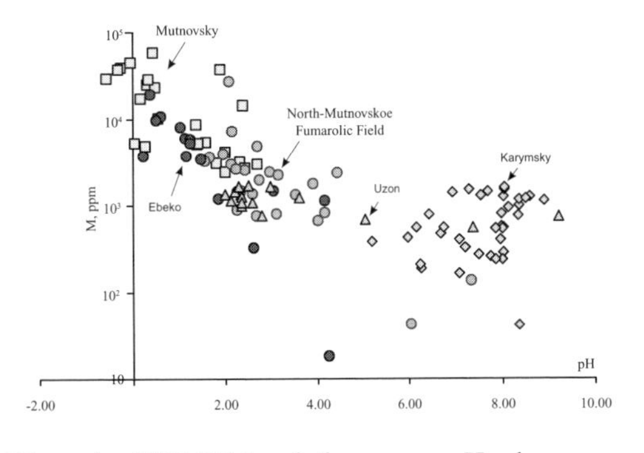

Figure 4. TDS (M) in solutions versus pH values.

thermal waters. HCO_3^- is the major anion of only a few Karymskoe springs. The chloride and sulfate concentrations vary within 5 to 6 orders of magnitude. In the ultra-acidic and acidic solutions of the Mutnovsky volcano and North-Mutnovskoe field, the chloride concentrations increase with decreasing pH value, while no clear relationship between the concentrations of chloride and pH values exists for solutions from Karymskoe springs.

Sulfate concentrations increase with decreasing pH in North-Mutnovsky acidic waters. In Mutnovsky mud pots as well as in the Karymskoe springs, sulfate concentrations are irregularly changing. Na is the predominant cation in the solutions from Karymsky volcano, whereas Fe and Al are predominant (in different proportions) in the solutions from the Mutnovsky volcano and North Mutnovsky. Each group contains springs which differ in composition. For example, Ca is a major cation for several Karymskoe springs along the lake shore, whereas for Bears Sources K is a major cation. In one of the mud pots of the North-Mutnovskoe field, NH_4^+ is a major cation. It can be concluded from the short review about the concentrations of major ions in sampled waters, that their composition is highly variable even within individual groups and apparently depends on many factors.

The trace element concentrations also vary by a wide range. In this case, we define trace elements those below 1 mg, eq % in the solutions. Therefore, elements such as Fe and Al are major cations in the solutions of the Mutnovsky volcano and North Mutnovsky springs, but trace elements in the Karymsky geysers. The elements pointed in the thermal waters can be divided into the several associations, based primarily on similarities of their geochemical properties, their mineral types in the rocks (basalts) and correlations dependences: 1) Cr+Ni+Co+Ti+V; 2) Zn+Cu+Pb+Cd+Ag; 3) As+Sb+Se+Bi. The total concentration of each group elements (as well as major anions concentrations) range 5 to 6 orders of magnitude and can reach up elevated values. The specific associations of elements are isolated for different groups of the springs: Ti-V-Co-Cr-Ni and Zn-Cu (Pb-Cd-Ag have a subordinate importance in this association) associations for the Mutnovsky volcano thermal waters, and As-Sb and Li-Rb are the associations for the springs of Uzon caldera. With the exception of the anion forming association, the total concentration of the elements increases with increasing TDS for the thermal waters from Mutnovsky volcano and North Mutnovsky springs. In the Karymskoe springs, such dependence is not observed. The problem lies in the explanation of differences for thermal waters and in the construction of a most adequate model for the path of chemical elements towards the surface, discharging into hydrothermal systems.

4 GEOPHYSICAL INVESTIGATION

Geophysical prospecting (geoelectric tomography) revealed the inner ground structure in the study regions down to a depth of 40 m (Fig. 5), giving additional information for the discussion. The resistive (blue color) and conductive (red color) zones can be seen. It is likely that low resistivity (ρ) is related to the saturation of the ground with highly mineralized thermal waters, while high resistivity is characteristic of ground washed by surface water or host rocks. The zones describes spring

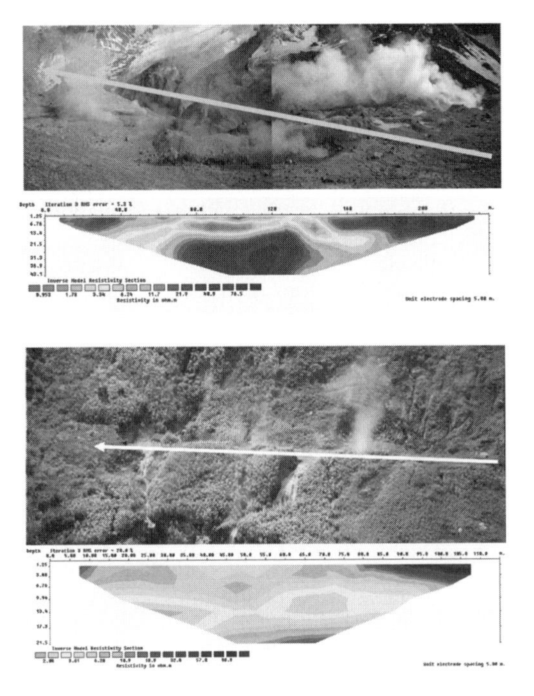

Figure 5. Vertical cross-sections of subsurface space at the Donnoe field (upper photo + geoelectric cross-section) and for Karymsky geysers (lower photo).

channels and phase boundaries structure. There is an evident vertical channel, which supplied the thermal mud pots of the Donnoe field, whereas a subhorizontal underground reservoir of thermal water was revealed for the Karymsky geysers (springs of the Academy of Science).

Moreover, the specific resistivity of subsurface space at these two areas is different. Its minimal values in subsurface space at the Donnoe field are about 0.5–0.9 Ohm × m, but at Karymsky geysers –2 Ohm × m. This indicates a more elevated proportion of high mineralized magmatic fluids in the composition of surface discharges at the Donnoe mud pots. In contrast, the source of Karymsky geysers is a stable underground reservoir, replenished by meteoric water and heated inductively.

5 PHYSICOCHEMICAL MODELING

Differences in formation processes are related to the fluid phase composition and, as a result, the fluid ability to transport components. If only one parameter (heart-emission from the fluid channel wall, α_2) varies, the location and the number of phase boundaries within the section change significantly (Fig. 6). In a system with a low heat-transfer coefficient from the wall, heating will end quickly. Up to 50 years at a depth of 1500 m, solutions dominated in the fluids. After 130 years, vapor escaped to the surface. Migration of phase boundaries was discontinued after 400 years, and a transition zone (gas/solution) remained at a depth of 400 m. At this stage, "aggressive", weakly reduced, acidic (pH = 2, Eh = –0.2 mV), hydrothermal solutions formed above the magmatic gas condensation boundary. At the high heat-transfer coefficient from the side wall, heating with the formation of stable phase boundaries ended after 300 years. An area with clear gas predominance is absent and several stable phase transitions were formed.

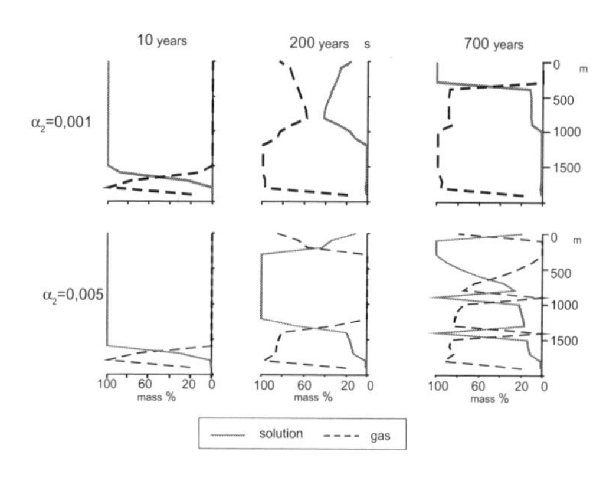

Figure 6. Distribution of fluid phases in cross section depending on age and α_2.

Physicochemical modeling was carried out to determine the genesis of different types of thermal waters. It was shown that subalkaline, Na-Cl sources (Karymsky type) can be formed by magmatic fluids rising to the surface without significant host rock interaction and, consequently, without leaching of elements. Acidic and ultra-acidic waters of the mud pots from Mutnovsky volcano are the result of multiple transformations of the original fluids by reaction with the volcanic reservoir rocks during their pathway. The deepest part of the sequences of volcanic structures may be the source of metals. At earlier stages of metasomatism, Co, Cr, Ni, and V are probably deposited and further involved in the hydrothermal process. As concluded from the results of numerical modeling, the dynamics during the formation of mineral zoning in fluid-magmatic systems are generally defined during the non-stationary stage of the temperature-related field evolution.

ACKNOWLEDGMENTS

This research was carried out with the financial support of Government of Russian Federation (MK-167.2010.5) the Russian Fund of Basic Researches (RFBR, grant N 10-05-00588) and Integration Project SB-FEB RAS number 96.

REFERENCES

Auippa, A., Dongara, G., Capasso, G. & Allard, P. 2002. Trace elements in the thermal groundwaters of Vulcano Island (Sicily). *Journal of Volcanology and Geothermal Research* 98: 189–207.

Bortnikova, S.B., Bessonova, E.P., Bessonov, D.Yu., Kolmogorov, Yu.P, Lapuchov, A.S., Palchik, N.A., Prisekina, N.A. & Kotenko T.A. 2010. Trace Elements in Native Sulfur as Indicator of Substance Sources in Fumaroles of Active Volcanic Regions (Ebeco Volcano, Paramushir Island). *Proc. World Geothermal Congress 2010, Bali, Indonesia, April 26–30, 2010.*

Bortnikova, S.B., Bessonova, E.P., Gavrilenko, G.M., Vernikovskaya, I.V., Bortnikova, S.P. & Palchik N.A. 2008. Hydrogeochemistry of thermal sources, Mutnovsky Volcano, South Kamchatka (Russia). *Proc. Thirty-Third Workshop on Geothermal Reservoir Engineering Stanford University,* Stanford, California, SGP-TR-185.

Giammanco, S., Ottavian, M., Valenza, M. et al. 1998. Major and trace elements geochemistry in the ground waters of a volcanic area: Mount Etna (Sicily, Italy). *Water Res*earch 32(1): 19–30.

Karpov, I.K., Chudnenko, K.V., Kulik, D.A. & Bychinskii, V.A. 2002. The convex programming minimization of five thermodynamic potentials other than Gibbs energy in geochemical modeling *American Journal of Science* 302: 281–311.

Sharapov, V.N., Cherepanov, A.N., Cherepanova, V.K. & Bessonova, E.P. 2008 Dynamics of phase fronts in ore-forming fluid systems of volcanic arcs. *Russian Geology and Geophysics* 49(11): 827–835.

Water-Rock Interaction – Birkle & Torres-Alvarado (eds)
© 2010 Taylor & Francis Group, London, ISBN 978-0-415-60426-0

Geochemistry of the thermal waters of the Far East Russia and Siberia

O.V. Chudaev, V.A. Chudaeva, I.V. Bragin & E.V. Elovskii
Far East Geological Institute FEB RAS, Russia

V.V. Kulakov
Institute of Water Problem and Ecology FEB RAS, Russia

A.M. Plysnin
Institute of Geology SB RAS, Russia

ABSTRACT: New geochemical data are reported on the low temperature thermal waters in Siberia and the Far East Russia. The studied alkaline waters belong to the HCO_3–Na type with significant trace element variations. The oxygen and hydrogen isotopic data of thermal waters suggest their meteoric origin. Dissolved gas is mainly nitrogen of atmospheric origin. The chemical composition of these results from water–rock interaction and strongly depends strongly on residence time and water/rock.

1 INTRODUCTION

The alkaline thermal groundwater of areas where absent recent volcanic activities are typically confined to fault zones, which facilitate penetration of infiltration waters in the high temperature horizons of the Earth. Examples of such water in Siberia and the Far East Russia are found in Belokuriha (Altay ridge), Baikal rift, Primorye (Sikhote-Alin ridge) and the Okhotsk region (Fig. 1).

The formation of thermal waters, especially in the Russian Far East, has not been studied very thoroughly. The available geochemical database on thermal waters is out of date, and contains only the major ions and some specific ions assumed to have medical effects (B, Br, and others). The first reliable trace element data on the Primorye thermal springs are reported by (Chudaeva et al. 1995) and those for the Okhotsk region reported by (Bragin et al. 2007 and Chudaev et al. 2008). However, the data are fragmentary and do not allow detailed interpretation of the regional geochemical characteristics for the formation of this type of thermal waters. In this paper, we report new geochemical data, obtained in 2006–2007 on the alkaline thermal waters of the Belokuriha springs, Baikal rift (North Baikal, Angarskii, Zipa-Bauntovskii springs) the Okhotsk region (Kuldyr, Annenskii, Tumninskii, Turma, and Lazarevskii springs) and the Primorye region (Chistovodnoe and Amgu springs).

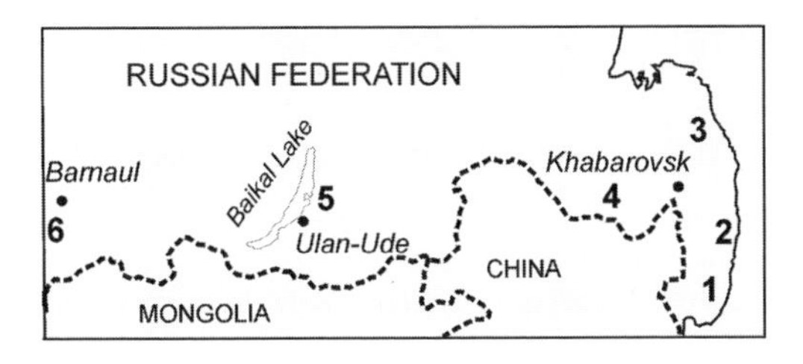

Figure 1. Location of studied hot springs. *Primorye region:* 1-Chistovodnoe, 2- Amgu. Okhotsk region: 3-Annenskii, Tumninskii, Turma & Lazarevskii; 4-Kuldur; Baikal rift: 5- North Baikal, Angarskii, Zipa-Bauntovskii; Al-tay region: 6-Belokuriha

2 METHODS

The unstable water parameters such as the pH, Eh, SEC (electroconductivity), DO (dissolved oxygen), and bicarbonate were measured *in situ*. The major cations and trace elements were analyzed using ICP-AES and ICP-MS on Thermo scientific iCAP 6000 and Agilent 7500c devices. The anion concentrations were measured using a LC-10 Avp (Shimadzu) liquid ion chromatograph. The oxygen and hydrogen isotopes were determined using a Finnigan-MAT 252 mass spectrometer. All these analyses were carried out at the Far East Geological Institute. The tritium concentrations were determined by liquid–scintillation counting using a Quantulus 1220 low-noise [alfa]–[beta] spectrometer at the Center of Isotope Research (VSEGEI). The computer modeling and calculation of the equilibrium reactions were done using the SOLMINEQ software package (Kharaka et al. 1988).

3 RESULTS AND DISCUSSION

The geological and hydrogeological characteristics of the nitric thermal waters of the Okhotsk and Primorye regions were reported in (Kiryukhin & Reznikov 1962, Bogatkov & Kulakov 1966, Chudaeva et al. 1999, Chudaev 2003, Chudaev et al. 2008). Most of the studied springs are located in the granite and/or their contact zones. The age of the granite intrusions varies widely (PR-K). The location of the intrusions are tectonically controlled and the waters penetrates along the fractures and faults. Maximum depth of water circulation is about 400 m. Measured temperature of the waters on the surface <100°C, and TDS is less than 500 mg/L.

3.1 *Major ions*

The studied thermal waters are characterized by pH >9 with Na as the predominant cation (Fig. 2). The concentration of sodium in this type of water as rule increases with increasing temperature of the water. According to the data of Ryzhenko et al. (1999, 2000), HCO_3–Na waters are formed in the granite/water system at the initial stages with high water/rock ratio. The sodium concentration in the studied thermal waters varies within 19–153 mg/L. The lowest contents were found in the Promorye region and the highest ones in Baikal rift springs.

It is lower than in the main European springs of this type and several times lower than in the similar water of the Republic of Korean (Michard 1990, Yum 1995). According to the thermodynamic calculations, sodium mainly occurs in the ionic form, and is an order of magnitude lower in the complexes of $NaCO_3^-$-$NaSO_4^{2-}$ than ionic. The potassium contents show no significant variations (1.2–0.33 mg/L) and practically do not differ from those in the water of shallow circulation. The thermal waters also have calcium contents within 2.0–5.4 mg/L. It is close to calcium variations in the thermal waters of the granite massifs of Europe and corresponds to the similar springs in Korea (Michard 1990, Yum 1995). Magnesium occurs in small amount in the alkaline thermal waters containing less than 0.03 mg/L. Similar thermal waters of Europe and Korea also have a low Mg content. The silica content in the thermal waters is 14.8–72 mg/L. The minimum concentration was found in Primorye, maximum in the Baikal rift. The contents of Si depends on the temperature water. High content of Si in Baikal rift (72 mg/L) corresponds to the highest calculated (quartz geothermometer) water temperature (116°C). The concentration (14.8 mg/L) of Si in Primorye is low as compared to the water temperature (60.8°C). Bicarbonate content in studied springs varies in widely range from 63.4 mg/L in Primorye to 207 mg/L (Baikal rift). The thermal

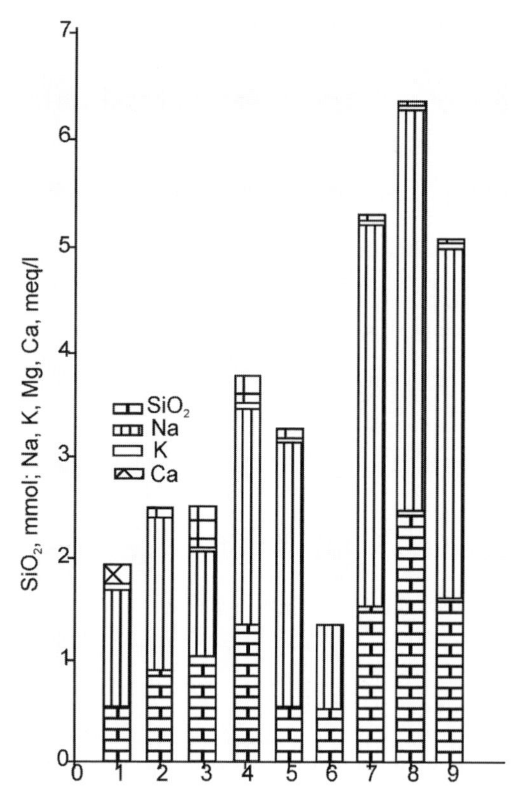

Figure 2. Proportion of the major cations and SiO_2 in thermal waters. 1–Chistovodnoe, 2–Amgu, 3–Tumninskii, 4–Annenskii, 5–Turma, 6–Lazarevskii, 7–Kuldur, 8–North Baikal region (Angarskii, Zipa-Bauntovskii), 9–Belokurikha.

waters contain low and weakly variable Cl contents, which correspond to those in surface waters: 1.4–23 mg/L. Sulfate ion behaves as a mobile complex. No saturation in calcium sulfate (gypsum or anhydrite) is attained in the alkaline waters. Its lowest contents were found in the springs of the Chistovodnoe springs (5.7 mg/l), and the highest contents, in the Belokuriha waters (22.6 mg/l). The TDS is mainly Composed of by sodium and bicarbonate, and partly, by silicon, whose content depends on the water temperatures.

3.2 *Trace elements*

Trace elements are subdivided into three main groups: siderophile, chalcophile, and lithophile. Rare earth elements (REE) will be described separately due to their particular geochemical features, though they belong to lithophile elements. Among the obtained trace element data (>50 elements), we deal with only the elements that emphasize the geochemical features of each of the selected water groups. Among the siderophile elements of most interest are iron, manganese, molybdenum, cobalt, and nickel. The thermal waters have a low Fe content within 0.013–0.005 mg/L, which somewhat increases only in the surrounding cold ground waters reaching 0.04 mg/L. The manganese concentration in the thermal waters, as that of iron, is extremely low, mainly below 0.1 μg/L. All the waters have high molybdenum contents, which are the highest in the Belokuriha springs (68.7 μg/L) and the lowest (6.2 μg/L) in the Annenskii springs. As is known, the molybdenum content in an alkaline environment is significantly higher than that in an acid environment. Cobalt varies within the limits 0.005–0.009 μg/L. The highest content was found in the Annenskii springs and the lowest, in the Chistovodnoe springs. The fresh river waters of the surrounding springs have an order of magnitude higher Co content. The nickel content increases northward from 0.033 μg/L in the Chistovodnoe springs to 0.43 μg/L in the Annenskii springs. The surrounding fresh waters contain 0.1–0.2 μg/L nickel. Among the chalcophile elements, we will consider copper and arsenic. The copper content in the studied thermal waters is low. In Belokuriha springs is about 8 μg/L; Baikal rift –1.2 μg/L. Lowest contents is 0.09 μg/L in Chistovodnoe springs. The lowest arsenic content was obtained in the Turma and Lazarevskii springs (2.41 μg/L) and the highest content, in the Kuldur springs (86.02 μg/L). Among the lithophile elements, we will consider Al, F, Li, The Al content in the studied waters varies within 3.6–19.5 μg/L. The highest contents were found in the Annenskii springs (19.5 μg/L). and the lowest contents, in the Chistovodnoe springs (3.6 μg/L). These values are insignificantly higher than those in the fresh waters. The F content in the studied thermal waters is within 0.8–24.7 mg/L with the lowest contents found in the Tumninskii springs (0.8 mg/L) and Amgu (0.9 mg/L) and the highest contents, in the Baikal rift. The Li content varies within 6.26–270 μg/L. The highest content was obtained in the waters of the Kuldur and the lowest content, in the Annenskii springs. The alkaline waters typically have low REE contents often within the detection limit of ICP-MS. An Agilent 7500s ICP MS equipped with a nebulaser allowed the direct determination of the REE contents in the studied thermal waters. Figure 3 demonstrates the REE patterns for springs of Primorye normalized to the North American Shale Composite.

The REE level in the thermal waters is close to that in the atmospheric precipitation (Chudaev 2003). All the studied waters show some HREE enrichment and Eu anomaly. Eu is supplied in the thermal waters from plagioclase during its interaction with the water.

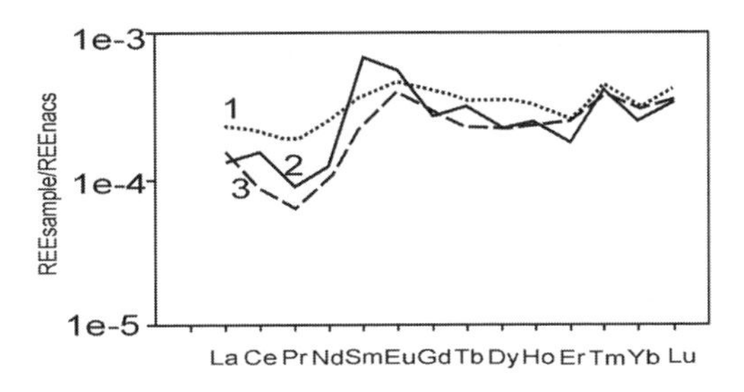

Figure 3. REE distribution pattern in studied springs of Primorye. 1–Amgu, 2–Chistovodnoe, 3–Tumninskii.

3.3 *Gas composition*

No significant deep supply of volatile components along the faults is observed at the discharge sites of the thermal waters. The absence of the influence of juvenile gases is confirmed by the low He isotopic ratio $^3He/^4He$ $(0.1–0.24)*10^{-6}$ for the thermal waters of the Chistovodnoe (Bogolyubov et al. 1984). Data of Chudaev et al. (2008) indicate that the dissolved gases of the thermal waters of Primorye, as other thermal gases of the Far East, are dominated by nitrogen (up to 99%). The gas component presumably had an atmospheric origin.

3.4 *Isotope composition*

The oxygen and hydrogen isotopic compositions signify the meteoric origin of the water component in springs (Fig. 4).

The $^{87}Sr/^{86}Sr$ ratios are about 0.70638 in the Chistovodnoe group of thermal waters and 0.71027 in the fresh ground waters of this area. This ratio in the thermal springs of the Amgu group is within 0.70458–0.70483. In our opinion, the difference in the Sr isotopic ratios between the northern and southern groups of the thermal waters of Primorye is caused by the different water exchange rate. In particular, the tritium data on all the studied thermal waters showed that the tritium activity is less than the MDC, i.e., <0.97 Bq/l. This may indicate an older age than 50 years. Our data indicate that the cations in the atmospheric precipitation of the Far East Russia are dominated by sodium, whereas the ground waters have a mixed cation composition (Chudaev 2003). This can be explained by the fact that the HCO_3–Cl–Na rain water in the zone of the groundwater formation accumulates mainly calcium and carbonate ions owing to the decomposition of the soil organics, which leads to the formation of HCO_3–Ca–Na–Mg fresh ground waters. Further subsidence and heating of the waters is accompanied by the predominant accumulation of sodium via mainly plagioclase (albite) decomposition. This results in the formation of HCO_3–Na water. The results of Ryzhenko et al. (1999) of the modeling of the granite–water system showed that,

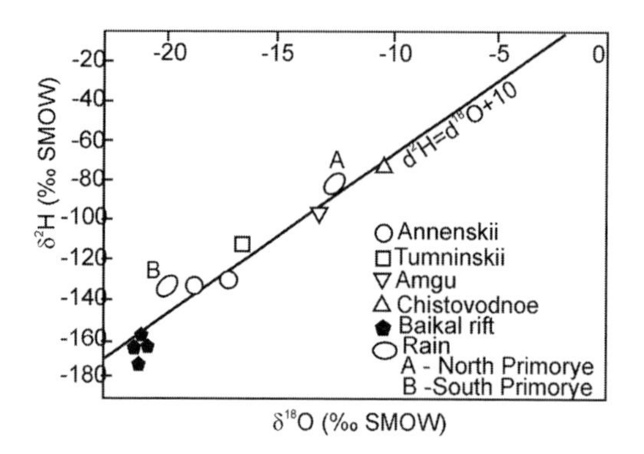

Figure 4. Variations of hydrogen and oxygen isotopes in thermal waters and atmospheric precipitation. The Craig (1961) world meteoric line is shown for comparison.

at the initial stage of the interaction (the water/rock ratio >> 1), weakly mineralized HCO_3–Na waters are formed. According to the calculations of Ryzhenko, the liquid phase in the water–granite system at a temperature >25°C has an alkaline reaction. The thermodynamic calculations showed that the studied thermal waters are supersaturated with respect to clay minerals (smectite, illite, and kaolinite), the group of low-temperature zeolites and albite. The calculation temperatures were close to 61°C in the Chistovodnoe springs, 81°C in the Amgu, 86°C in the Tumninskii, and 99°C in the Annenskii, 103°C in Kuldur and Belokuriha springs, 116°C in Baikal rift.

4 CONCLUSIONS

1. The considered thermal waters studied are low-salinity waters of the HCO_3–Na type. The level of most of the elements is lower than in similar waters of Europe and Korea. The content and behavior of the chemical elements in the waters is controlled by the water exchange rate and the formation of secondary equilibrium minerals.
2. The first obtained REE data showed their low contents close to those in the atmospheric precipitation. The REE distribution patterns indicate that their main source in the water were from plagioclases. The low salinity of the waters and the REE distribution pattern manifest to the rapid water circulation in the rock sequence.
3. The oxygen and hydrogen isotopic data indicate that the main water component is of meteoric origin.

ACKNOWLEDGMENTS

This work was supported by the RFBR (project no. 10-05-91158), FEB RAS (09-2-CO-08-005; 09-III-A-08-414), and by the RF Program Scientific Schools (NSh-3561.2008.5).

REFERENCES

Bogatkov, N.M. & Kulakov V.V. 1966. The Annenskii Thermal Springs. *Sovetskaya Geology* 5: 153–155.

Bogolyubov, A.N., Korplyakov, O.P., Benkevich, L.G. & Yudenich V.S. 1984. Helium Isotopes in the Groundwaters of Primorye. *Geokhimiya* 22(8): 1241–1244.

Bragin, I.V., Chelnokov, G.A., Chudaev O.V. & Chudaeva, V.A. 2007. Low-Temperature Geothermal Waters of Continental Margin of the Russian Far East. In T.D. Bullen & Y. Wang (eds). *Proceedings of International Symposium on Water-Rock Interaction* 1: 481–484. Rotterdam: Balkema.

Chudaev, O.V. 2003. *Composition and Formation Conditions of the Modern Hydrothermal Systems of the Russian Far East*. Vladivostok: Dal'nauka.

Chudaev, O.V., Chudaeva,V.A. & Bragin, I.V. 2008. Geochemistry of the thermal waters of Sikhote-Alin. *Russian Journal of Pacific Geology* 2(6): 528–534.

Chudaeva, V.A., Chudaev, O.V., Chelnokov, A.N., Edmunds M. & Shand, P. 1999. *Mineral Waters of Primorye (Chemical Aspects)*. Vladivostok: Dal'nauka.

Chudaeva, V.A., Lutsenko, T. & Chudaev, O.V. 1995. Thermal Waters of the Primorye Region. Eastern Russia. In Y. Kharaja & O. Chudaev (eds.), *Proceedings of International Symposium on Water-Rock Interaction*: 375–378. Rotterdam: Balkema.

Craig, H. 1961. Isotopic variations in meteoric waters. *Science* 133: 1702–1703.

Kharaka, Y.K., Gunter, W.D. & Aggarwal, P.K. 1988. *SOLMINEQ.88: Computer Program for Geochemical Modeling of Water-Rock Interactions. U.S. Geol. Survey Water—Resources Investigations*. Report 88–4227.

Kiryukhin, V.A. & Reznikov, A.A. 1962. New Data on Chemical Composition of Nitric Springs of the South Far East. In: *Problems of Special Hydrogeology of Siberia and Far East*. Irkutsk: 71–83.

Michard, G. 1990. Behaviour or Major Elements and Some Trace Elements (Li, Rb, Cs, Sr, Fe, Mn, W, F) in Deep Hot Waters from Granitic Areas. *Chemical Geoogy* 89: 117–134.

Ryzhenko, B.N. & Krainov S.R. 2000. On the Influence of Mass Proportion of Reacting Rock and Water on the Chemical Composition of Natural Aqueous Solutions in systems Open to CO_2. *Geokhimiya*. 38(8): 803–815.

Ryzhenko, B.N., Barsukov, V. & Knyazeva,S.N. 1999. Chemical Characteristics (Composition, pH, and Eh) of a Rock–Water System: 1. Granitoids–Water System. *Geokhimiya* 34(5): 436–454.

Yum, B.W. 1995. Movement and Hydrochemistry of Thermal Waters in Granite at Cosung, Republic of Korea. In: Y. Kharaka & O. Chudaev (eds.), *Proceedings of the 8th International Symposium on Water-Rock Interaction*: 401–404. Rotterdam: Balkema.

Water-Rock Interaction – Birkle & Torres-Alvarado (eds)
© 2010 Taylor & Francis Group, London, ISBN 978-0-415-60426-0

Geochemical characterization of thermal waters in the Borateras geothermal zone, Peru

V. Cruz & V. Vargas
Instituto Geológico Minero y Metalúrgico-INGEMMET, Lima, Perú

K. Matsuda
West Japan Engineering Consultants, INC, Fukuoka, Japan

ABSTRACT: The Borateras Geothermal Zone (BGZ) is located in the Western Cordillera of the Andes in southern Peru. In Borateras volcanic chains which are aligned from NW to SE, volcanic rocks have been deposited over sedimentary Cretaceous basement. The geochemical interpretation of the samples using Langelier and Piper diagrams indicates a alkaline-chloride type for the thermal waters. Most of the geothermal waters plot close to the Cl corner in the triangular diagram, which is typical for mature geothermal deep fluids. High B concentrations lead to a relatively high B/Cl ratio as shown on a B-Cl binary diagram, which suggests the reaction of fluids with sedimentary marine rocks at deep levels. The $\delta^{18}O$ *vs* δD diagram suggests a relatively high temperature and the mixing of magmatic fluids with meteoric water. The application of chemical geothermometers allowed to estimating a temperature as high as 200°C for the geothermal resources.

1 INTRODUCTION

Borateras is located inside the Maure basin close to the Maure river in the Western Cordillera of the Andes in southern Peru, at an altitude of 4300 m asl in the Tacna Region (Fig. 1). The weather is cold and dry, with strong sunshine during the day and low temperatures at night. There is sparse vegetation, and arid and stony soils can be observed around the entire zone. The most representative species are *Polypelis* and *Azorella compacta*, commonly called Queñuales and Yareta, respectively.

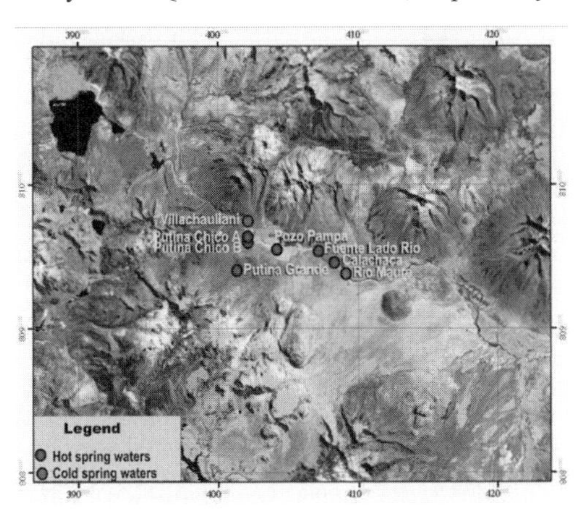

Figure 1. Localization map of the thermal waters in Borateras.

The zone is characterized by the presence of a considerable number of geothermal manifestations such as hot springs, boiling springs and mud pools. In October 2007, we carried out a study of the geothermal manifestations with geochemical methods to interpret their chemical and isotopic characteristics. This work was carried out for West Japan Engineering Consultants INC (WEST JEC) and INGEMMET.

2 GEOLOGICAL SETTING

Borateras is surrounded by a chain of volcanic centers with important structural systems. NW-SE aligned volcanic chains are found towards the southwest of the region. The following volcanic centers are recognized: Jaruma, Coverane and Purupuruni, of which the last one is formed by a group of dacitic domes with a diameter of approximately 850 m. These volcanic rocks were deposited over the sedimentary Cretaceous basement composed mainly by intercalating sandstones, limestones and shales. Towards the southern part of the field, the main lithology consists of andesitic lava (with crystals of plagioclases, olivines and pyroxenes), interlayered by some porphyritic lavas and sequences of pyroclastic flows. These volcanic deposits are from the upper Miocene to Pliocene time period and cover almost 90% of the entire zone. There are some Holocene deposit as alluvials and colluvials, mainly near the river. Some moraines cover the slopes of the Purupuruni domes.

In Borateras, there are also some important tectonic structures. Regional and local faults are aligned NW-SE and NE-SW, respectively, whereby the latter ones form the main structure that controls the flow of geothermal water.

3 CHEMICAL COMPOSITION OF THE GEOTHERMAL FLUIDS

During the field work carried out in October 2007, we collected six samples of geothermal water in the temperature range from 43.5 to 83°C and a pH from 6.9 to 7.6. We have also collected two cold water samples in the lower part of the Maure River, as well as from a cold source. The isotope ratios for deuterium (δD) and oxygen-18 ($\delta^{18}O$) and the chemical composition of water samples were determined by WEST JEC in Japan.

The 5 thermal springs (54111–001, 54111–009, 54111–017, 54111–015, 54111–016) are directly associated with the BGZ, and showed concentrations of Na and K from 253 to 1310 mg/L and from 29.5 to 96.3 mg/L, respectively. We have observed elevated Cl concentrations, ranging from 350 to 2150 mg/L, and B contents from 16.5 to 94.5 mg/L.

The Putina Grande spring (54111–035) shows low Cl concentrations, 0.38 mg/L, and SO_4 is the predominant anion with 47.6 mg/L.

The majority of these thermal springs are bubbling which indicates the presence of CO_2. The HCO_3 concentrations range from 21 to 224 mg/L and SO_4 concentrations from 47.6 to 73.3 mg/L. The concentrations of Ca and Mg are much lower than those of the other major anions. Cl is the principal ion in these waters, followed by Na. Waters have mainly a neutral pH and are of Na-Cl type, typical of mature deep geothermal fluids in high-temperature systems.

The chemical results of the thermal waters were plotted in the Langelier diagram (Fig. 2) (Fyticas et al. 1989). All thermal water samples (54111–001, 54111–009, 54111–015, 54111–016) are located specifically in the Na-Cl sector. This indicates that the waters are sodium-chloride or alkaline water type, characteristic for geothermal waters. The Calachaca (54111–017) spring is different, which is due to the mixing with meteoric waters.

According to the Cl-SO_4-HCO_3 ternary diagram (Giggenbach 1988) (Fig. 3), the thermal springs are within the chloride waters group, which is typical for mature deep geothermal fluids in high-temperature systems. Areas which contain hot, large-flow springs with highest Cl concentration are fed directly from the deep reservoir, and can be used to identify permeable zones within the field. This means that the waters in the BGZ come from deep level reservoirs (Nicholson 1993). In Putina Grande, the high SO_4 and HCO_3 concentrations may indicate mixing of magmatic fluids with shallow waters.

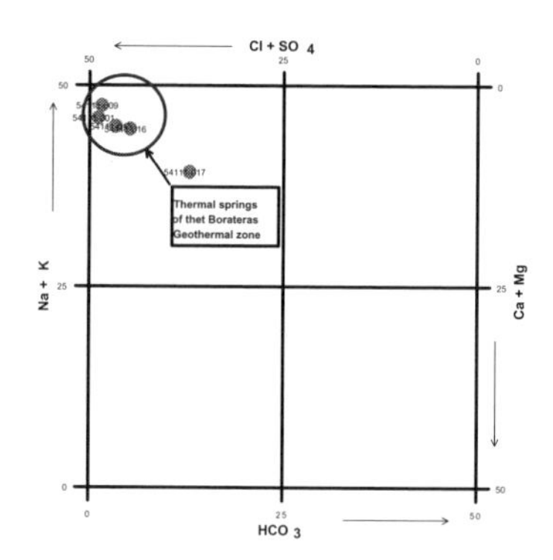

Figure 2. Langelier diagram.

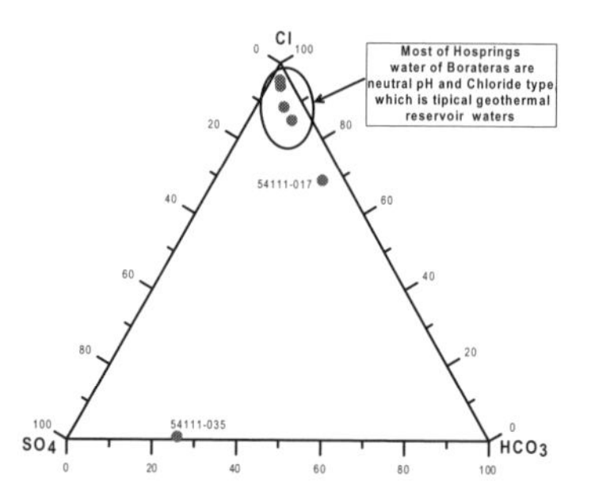

Figure 3. Cl-SO_4-HCO_3 ternary diagram (Giggenbach, 1988).

Table 1. Chemical composition (in mg/L) of the thermal waters in the Borateras geothermal zones.

Samples	Na	K	Ca	Mg	Cl	SO_4	HCO_3	B
54111–035					0.38	47.6	21.0	
54111–001	1310	91.0	68.0	0.48	2150	71.9	86.0	95.4
54111–009	1020	96.3	44.4	2.84	1630	70.3	93.0	72.7
54111–017	253	29.5	36.8	17.0	350	37.1	224	16.5
54111–015	643	77.2	40.1	9.17	981	73.3	128	43.4
54111–016	592	68.1	45.2	13.0	874	66.3	184	39.6
54113–001					0.88			
54114–001					510			

As shown by the B-Cl diagram (Fig. 4), high B concentrations were detected for geothermal waters in Borateras. This may occur when volcanic rocks are affected by hydrothermal activity (Risacher 1984) or it may be related to leaching by meteoric water and/or hydrothermalization of rocks, enriched in B (Murray 1996).

The Cl/B molar ratio for geothermal waters in Borateras is high, as typical for neutral Na-Cl type geothermal water of low salinity, reflecting the geologic structures and the volcano distributions in these areas (Shigeno & Abe 1983). The main geothermal reservoir developed in marine sedimentary rocks, probably with a relatively high porosity and permeability with abundant fractures (Shigeno 1993). In the case of BGZ, high B concentrations and B/Cl ratios of the geothermal waters could imply reaction with marine sedimentary rocks at deep levels.

Two samples fall on the equilibrium line in the Na-K-Mg ternary diagram (Giggenbach et al. 1983) (Fig. 5), which is characteristic for mature or geothermal waters, suggesting that their cation ratios are controlled by mineral-solution equilibria. The spring samples 54111–015, 54111–016 and 54111–017 are located in the partial equilibrium field, probably due to dilution by surface waters.

4 ISOTOPIC COMPOSITION

The isotopic ratios for $\delta^{34}S(SO_4)$, $\delta^{18}O(SO_4)$, $\delta^{18}O$ and δD were determined. The $\delta^{18}O$ values are more negative for cold waters in comparison to geothermal waters (Table 2). The latter ones slightly deviate from the local meteoric water line, indicating that the geothermal waters are related to the mixing of magmatic fluids with meteoric water (Fig. 6) (Craig et al. 1956, Craig 1963). Based on the entalphy-chloride model, a Cl content of 2300–2500 mg/L is estimated for the main reservoir, and the mixing ratio of magmatic fluid in the Borateras geothermal zone is estimated to be around 25%.

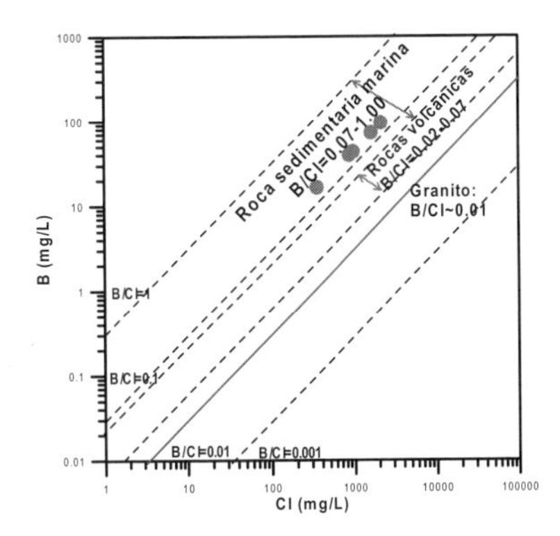

Figure 4. Boron vs. Chloride diagram (Shigeno 1993).

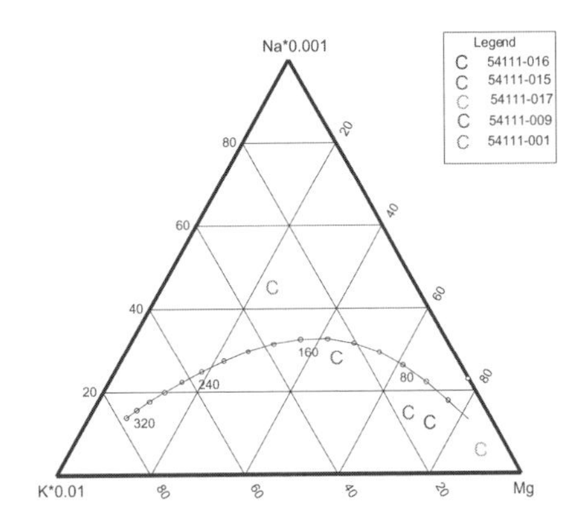

Figure 5. Na-K-Mg ternary diagram (Giggenbach et al. 1983).

Table 2. Isotope results.

	Samples			
Isotopes	54111–001	54111–009	54111–017	54111–015
$\delta D(H_2O)$	–106	–109	–119	–114
$\delta 18O(H_2O)$	–11.1	–12.3	–15.8	–13.9
$\delta 34S(SO_4)$	9		4.5	
$\delta 18O(SO_4)$	–5.7		–1	

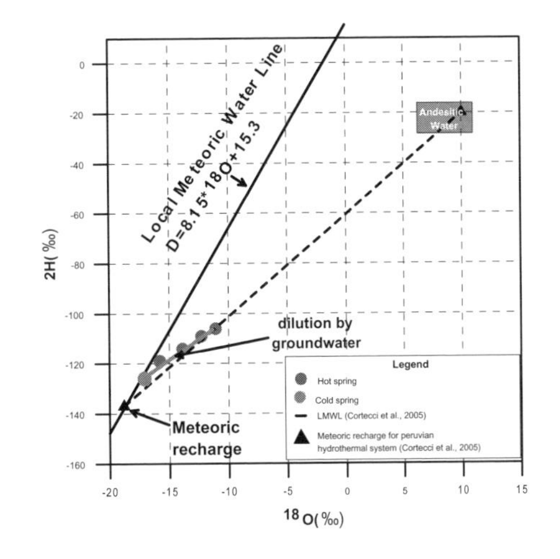

Figure 6. δD vs $\delta^{18}O$ in the Borateras geothermal zone.

5 APPLICATION OF GEOTHERMOMETERS

Chemical geothermometer results for geothermal water samples are presented in Table 3, with an estimation of the liquid phase temperature at depth in the geothermal reservoir of Borateras. The temperature of the present geothermal resource may be as high as 200°C (T-Na/K, Truesdell 1976) (Table 3 and Fig. 7).

Table 3. Chemical geothermometer results (in °C).

Geotermometer	Sample					
	54111–035	54111–001	54111–009	54111–017	54111–015	54111–016
T-Qu	151	197	191	154	171	169
T-Calc.	126	179	171	129	149	146
T-α Crist.	101	148	141	103	121	119
T-β Crist.	51	98	91	54	72	69
T-Na/K		152	182	205	208	203
T-NaKCa		206	222	140	203	189
T- NaKCa-Mg		183	181	140	100	68
T-K/Mg		181	149	87	123	114
T-Na/Li		253			216	
T –δ 18O(H_2O-SO_4)		276		116		

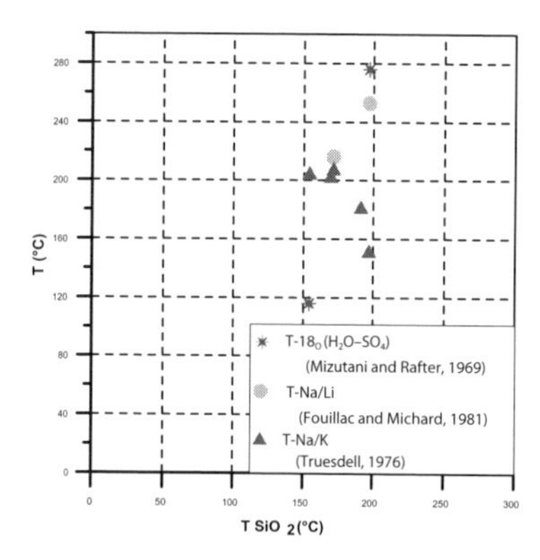

Figure 7. T-Na/Li, T-^{18}O(H_2O-SO_4, T-Na/K) vs. T-SiO_2.

6 CONCLUSIONS

- The geochemical characterization of the thermal springs in Borateras resulted in an alkaline-chloride (Na-Cl) water type.
- The stable isotope relationship between $\delta^{18}O$ and δD suggests that the geothermal waters are a mixtue of meteoric and magmatic waters.
- The application of different types of geothermometers resulted in a broad range of estimated reservoir temperatures, but conditions around 200°C seem plausible.
- According to the extent of geothermal manifestations and estimated subsurface temperature by geothermometers, the geothermal resources in this zone seem to be promising.

REFERENCES

Cortecci, G., Boschetti, T., Mussi, M., Lameli, C.H., Mucchino, C. & Barbieri, M. 2005. New chemical and original isotopic data on waters from El Tatio geothermal field, nothern Chile. *Geochemical Journal* 39: 547–571.

Craig, H. 1963. Isotopic geochemistry of water and carbon in geothermal Areas. In Tongiorgi, E. (ed), *Nuclear Geology in Geothermal Areas*; Consiglio Nazionale delle Ricerche, Laboratorio di Geologia Nucleare, Pias, Spoleto: 17–53.

Craig, H., Boato, G. & White, D.E. 1956. Isotopic geochemistry of thermal Waters. *National Acad. Sci. National Research Council Publication* (400): 29–38.

Fyticas, M., Kavouridis, T., Leonis, C. & Marini, L. 1989. Geochemical exploration of the three most significant geothermal areas of Lesbos Island, Greece. *Geothermics* 18: 465–475.

Fouillac, C. & Michard, G. 1981. Sodium/Lithium ratio in water applied to geothermometry of geothermal reservoirs. *Geothermics* 10: 55–70.

Giggenbach, W.F. 1988. Geothermal solute equilibria; derivation of Na-K-Ma-Ca geoindicators. *Geochimica et Cosmochimica Acta* 52: 2749–2765.

Giggenbach, W.F., Gonfiantini, R., Jangi, B.L. & Truesdell, A.H. 1983. Isotopic and chemical composition of Parbati valley geothermal discharges NW-Himalaya, India. *Geothermics* 12: 199–222.

Mizutani, Y. & Rafter, T.A. 1969. Oxygen isotopic composition of sulphates: Isotopic composition of sulphate in rain water, Gracefield, New Zealand. *New Zealand Journal of Sciences* 12, 69–80.

Murray, K.S. 1996. Hydrology and Geochemistry of Thermal Waters in the Upper Napa Valley, Califomía. *Ground Water* 34(6): 1115–1124.

Nicholson, K. 1993. *Geothermal Fluids, Chemistry and Exploration Tecniques*. Berlin: Springer-Verlag.

Risacher, F. 1984. Origine des concentrations extremes en bore et lithium dans saumeres de L´Altiplano Bolivien. *C.R. Acad. Sci. Paris* 299 (II): 701–70.

Shigeno, H. 1993. Reservoir environment of the Onuma geotermal power plant, northeast Japan, estimated by forward analysis of long-term artificial tracer concentration change, using singe-box-model simulator. *Workshop on Geothermal Reservoir Engineering. California: Stanford University.*

Shigeno, H. & Abe, K. 1983. B-Cl geochemistry applied to geothermal fluids in Japan, especially as an indicator for deep-rooted hydrothermal systems. *Extended Abstr. 4th Internat. Symp. On Water–Rock Interaction, Misasa*, 437–440.

Truesdell, A.H., 1976, Geochemical techniques in exploration (Summary of Section III). In *Proceedings of the 2nd United Nations Symposium on the Development and Use of Geothermal Resources: San Francisco, CA, USA* 1: 53–79.

Water-Rock Interaction – Birkle & Torres-Alvarado (eds)
© 2010 Taylor & Francis Group, London, ISBN 978-0-415-60426-0

Laboratory work on the response of reservoir rocks from the Los Humeros geothermal field, Puebla, Mexico, with acid solutions

G. Izquierdo
Instituto de Investigaciones Eléctricas, Gerencia de Geotermia, Cuernavaca, Mexico

M. Flores & M. Ramírez
Residencia de Proyectos Geotermoeléctricos, Comisión Federal de Electricidad, Mexico

ABSTRACT: Rock matrix stimulation is a methodology used to enhance well productivity in oil systems. Some years ago this methodology began to be applied in geothermal systems. In order to investigate the behavior of altered volcanic rocks interacting with an acid solution, experiments were carried out using rock samples from Los Humeros geothermal reservoir. Two acid solutions, 10% HCl and a mixture of 10% HCL–5% HF, were used to investigate rock dissolution at atmospheric pressure and 150°C. Chemistry, mineralogy and permeability of rocks were determined before and after the reaction with acid solutions. Mineral dissolution depends on the mineral reactivity with acids. Calcite in veins and micro fractures reacts immediately with the acid solution. Calc silicates react by dissolution of minerals in contact while leaving the rock matrix untreated. Preliminary experimental data indicate that both solutions are effective dissolution agents. Dissolution capacity increases with time. No precipitation of secondary minerals was detected.

1 INTRODUCTION

1.1 *Stimulation methods*

Secondary mineralogy is the result of water-rock interaction in hydrothermal systems. Minerals may be formed commonly by isochemical rearrangement or by deposition in pores and fractures. Deposition in pores and fractures may reduce porosity and productivity. Another common problem in geothermal wells is the scale formation which in many cases also contributes to reduce productivity.

To minimize or to eliminate the effects of scale deposition as well as restore or improve permeability, several methodologies have been used in geothermal fields. Among others: matrix acidicing, hydraulic fracturing, thermal fracturing and chemical stimulation.

Hydraulic fracturing is commonly used although not many successful cases are known; it is considered as an option to improve wells with poor reservoir connectivity (Flores et al. 2005). Thermal fracturing produces thermal shock by injection of cool water. It is a well documented method but it is not suitable to eliminate scales. Matrix stimulation is an old methodology used to enhance well productivity in oil systems. It has been extended successfully to the geothermal industry in wells that have shown reduced productivity either by clogged pores and fractures or scale formation. Hydrochloric acid is known to dissolve scales such as calcium carbonate and is used extensively in oil field operations throughout the world.

HCl and mixtures of HF and HCl have been used in acidizing operations. HCl is selected to treat limestone and calcite in veins, pores and scales. A mixture of HF and HCl is used to dissolve silicates and silica. A mixture of 12% HCl–3% HF (called regular mud acid) is used (Portier et al. 2009; Malate et al. 1998).

Chemical stimulation using chelating agents such as ethylenediaminetetraacetic acid (EDTA) or nitrilotriacetic acid (NTA) have been proposed as an alternative treatment. Such agents have the ability to chelate, or bond, metals such as calcium. This procedure has been studied in the laboratory as a method to dissolve calcite in geothermal reservoirs (Mella et al. 2006). They found that the rate of calcite dissolution is not as fast as using strong mineral acids.

Acid stimulation techniques have been successfully used in many geothermal systems. In Mexican geothermal fields acid stimulation of wells has been carried out in Los Azufres, Mich. and Las Tres Virgenes B.C. Soon, a well from the Los Humeros geothermal field will be acid stimulated.

In this work mineralogical, chemical and petrophysical results before and after acid treatment are given for rocks from the Los Humeros geothermal field. The objective of this study is to evaluate the response of the reservoir rocks to acid stimulation.

2 THE LOS HUMEROS GEOTHERMAL FIELD

2.1 *Geologic setting*

The Los Humeros geothermal field is located at central-eastern Mexico, Figure 1 1a shows a panoramic view of the geothermal field (Xalapazco-Maztaloya), 1b shows the location of the geothermal field and 1c illustrates the location and topograpy of the geothermal field. The field is inside the Los Humeros volcanic caldera, which lies at the eastern end of the Mexican Volcanic Belt. Los Humeros is one of the four geothermal fields currently operating in Mexico, with an installed capacity of 40 MW.

In relation to the Los Humeros caldera, a series of geologic events have occurred (Gutiérrez-Negrín 1982b; Viggiano & Robles 1988). A simplified description of the subsurface lithology is:

Unit 1. Post-caldera volcanism. Quaternary (<100,000 years). It is composed of andesites, basalts, dacites, rhyolites, flow and ash tuffs, pumices, ashes and materials from phreatic eruptions. The unit contains shallow aquifers.

Unit 2. Caldera volcanism. Quaternary (510,000–100,000 years). This unit is mainly composed of lithic and vitreous ignimbrites from the two collapses (Los Humeros and Los Potreros). It also includes rhyolites, pumices, tuffs and some andesitic lava flows, as well as the peripheral rhyolitic domes. This unit acts as an aquitard (Cedillo 2000).

Unit 3. Pre-caldera volcanism. Tertiary (Miocene-Pliocene, 10–1.9 Ma). It is composed of thick andesitic lava flows, with some intercalations of horizons of tuffs. The characteristic accessory mineral of the upper andesites is augite and the lower andesites contain mainly hornblende. Both packages include minor and local flows of basalts, dacites and eventually rhyolites. This unit contains the geothermal fluids.

Unit 4. Basement. Mesozoic-Tertiary (Jurassic-Oligocene, 140–31 Ma). This basement unit is composed of limestones and subordinated shales and flint. Therefore, this unit includes also intrusive rocks (granite, granodiorite and tonalite) and metamorphic (marble, skarn, hornfels), and eventually some more recent (Miocene?) diabasic to andesitic dikes.

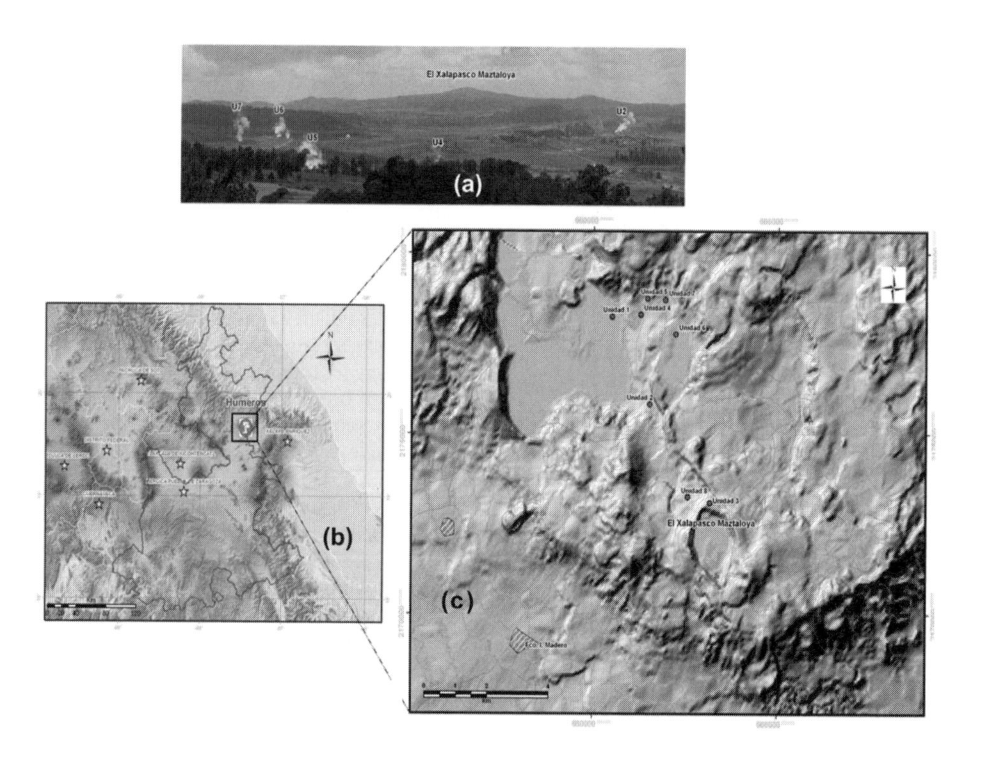

Figure 1. Los Humeros geothermal field and well location.

3 EXPERIMENTS

3.1 *Rock samples*

Matrix acidicing experimental work was carried out on selected samples coming from the Unit 3, which is considered the production zone of the reservoir.

As Unit 3 contains the geothermal fluids, well H-13 was chosen as representative of the upper andesite or augite-andesite and a core sample of well H-40 was chosen as representative of the lower andesite or hornblende-andesite. The core fragment from well H-13 comes from a depth between 1200 and 1203 m (Fig. 2). It is classified as andesite with augite as characteristic accessory mineral. It is almost 60% hydrothermally altered. Minerals formiing this rock (identified by optical microscopy and X-ray diffraction) are: plagioclase, quartz, augite (altered to chlorite), calcite, epidote, hematite and scarce zeolites.

Another rock fragment selected for this experiment is from well H-40; it is an andesite (with hornblende as characteristic accessory mineral) from a depth between 1300 and 1300 m (Fig. 3). It shows a micro crystalline texture. It is almost 80% altered to chlorite, epidote, calcite, quartz and mica (illite).

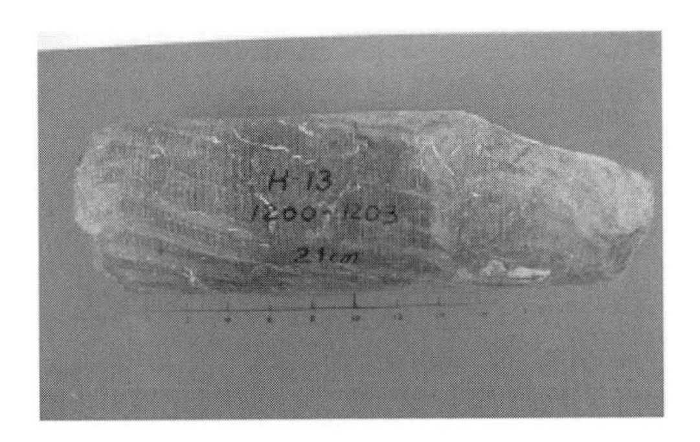

Figure 2. Fragment of core from well H-13.

Figure 3. Fragment of core from well H-40.

3.2 *Experimental conditions*

Experiments were carried out in acid and temperature resistant vessels. Pre-weigh core cutting sample was placed in each acid solution (10% HCl and 10% HCL + 5% HF).

Experiments were conducted at atmospheric pressure in a controlled temperature oil bath at 100°C for 5 hours using different specimen each time.

After treatments, samples were recovered from the acid solution, immersed in distilled water and dried at room temperature. Small chips finely powdered were analyzed for mineral identification by X-ray diffraction using a diffractometer model APDO 2000 from ITAL STRUCTURES with filtered CuKα radiation. Chemical characterization of rocks was carried out by ICP OS (Thermo scientific, iCAP 6300).

Klinkernberg permeability was determined in the same briquette before and after the acid treatment by measuring the absolute permeability by the stable state technique at room temperature and constant pressure using nitrogen as work fluid (Contreras & García, pers. comm.).

4 RESULTS

X-ray diffraction of the chemically treated samples shows the same mineralogy as the original rock samples. That means that acid react by dissolving first contact minerals leaving much of the rock matrix unaffected. Samples of H-13 show empty micro fractures indicating complete calcite dissolution, also is noted that acid penetrates the rock matrix. Figures 4a and 4b show the same piece of rock before and after HCL treatment.

This is not the case of samples of well H-40, where the rocks show that acid reacted only with first contact minerals. So mineral bulk composition does not show changes respect to the non treated samples.

Chemical analyses of treated samples indicated lower concentration of major elements compared to the original/untreated rock. Changes show relation to the original rock composition, time of

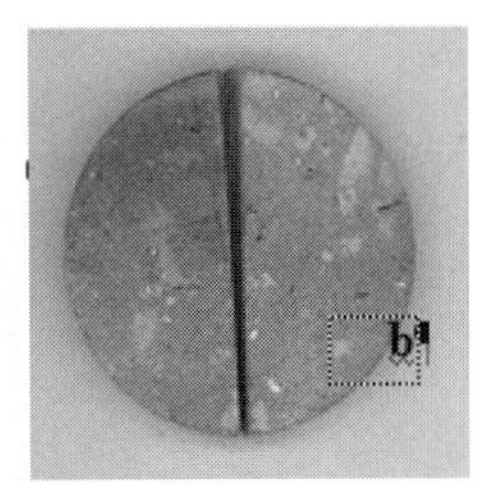

Figure 4. A fragment of core from well H-13. (a) before HCl treatment, (b) after HCl treatment.

Table 1. Klinkenberg permeability before and after acid reaction.

Well	Klinkenberg permeability (milidarcys)	
	Before treatment	After treatment
H-13	0.004	6.66
H-40	0.035	6.52

reaction and the type of acid solution. In well H-13 Ca shows important variation due to calcite dissolution. Calc-silicates react with the mixture of HCl-HF showing decrease in Si. The major weight loss is observed in well H-13 due to the important dissolution of calcite.

The Klinkernberg permeability increases after the reaction with 10% HCl–5% HF solution (Table 1); particularly in sample of well H-13, where channels were open by the action of the acid solution.

Solid residues in the acid solution were filtered, dried and analyzed by X-ray diffraction no new crystalline phases were identified.

5 CONCLUSIONS

Laboratory studies show the effectiveness of acid treatment of rocks.

The results depend on the type of alteration minerals of rocks. Calcite reacts rapidly with both acid solutions leaving open pores and fractures while calc silicates react only superficially leaving much of the rock matrix unreacted.

After treatment changes in permeability are considerable for the purposes of acid stimulation of geothermal wells. Further work has to be done controlling flowing of acid fluids thought the rock matrix and mineralogical characterization and acid treatment of scales.

No new phases are formed by the interaction between the rocks and acid solutions.

ACKNOWLEDGEMENTS

The authors want to express their gratitude to the authorities of the Gerencia de Proyectos Geotermoelectricos from the Comisión Federal de Electricidad of Mexico (CFE) for permission to present information relative to contract number 9400046929 between Instituto de Investigaciones Electricas and the Comisión Federal de Electricidad. Results are part of the research that the Comisión Federal de Electricidad carries out in Mexican geothermal fields. Also we want to express our gratitude to Dr. P. Kailasa and to Dr. Z. Pang for their suggestions to improve the manuscript.

REFERENCES

Cedillo, R.F. 2000. Hydrogeologic model of the geothermal reservoirs from Los Humeros, Puebla, México. *Proc. World Geothermal Congress 2000.* Kyushu-Tohoku, Japan: 1639–1644.

Flores M., Davies D., Couples G. & Palsson B. 2005. Stimulation of geothermal wells, can afford it? *Proc. World Geothermal Congress 2005*. Antalya, Turkey.

Gutiérrez-Negrín, L.C.A. 1982. Litología y zoneamiento hidrotermal de los pozos H-1 y H-2 del campo geotérmico de Los Humeros, Pue. CFE, Internal report number 23–82. Unpublished.

Malate, R.C.E., Austria, J.J.C., Sarmiento, Z.F., Di Lullo, G., Sookprason, P.A. & Francis, P. 1998. Matrix stimulation treatment of geothermal wells using sandstone acid. *Proc. Twenty-Third Workshop on Geothermal Reservoir Engineering Stanford University.* Stanford, California.

Mella, M., Kovac, K., Xu, T., Rose, P. & McCulloch, J. 2006. Calcite dissolution in geothermal reservoirs using chelants.

Sandrine, P., Francois-David, V., Patrick, B.N.S. & André G. 2009. Chemical stimulation techniques for geothermal wells: experiments on the three-well EGS system at Soultz-sous-Forêts, France. *Geothermics* 38: 349–359.

Viggiano, J.C. & Robles, J. 1988. Mineralogía hidrotermal en el campo geotérmico de Los Humeros, Pue. I: Sus usos como indicadora de temperatura y del régimen hidrológico. *Geotermia* 4(1): 15–28.

Water-Rock Interaction – Birkle & Torres-Alvarado (eds)
 ISBN 978-0-415-60426-0

Assessment of two low-temperature geothermal resources in North Portugal

J.M. Marques & H.G.M. Eggenkamp
Instituto Superior Técnico (CEPGIST), Technical University of Lisbon, Portugal

P.M. Carreira
Instituto Tecnológico e Nuclear, Sacavém, Portugal

J. Espinha Marques
Departamento e Centro de Geologia, Faculdade de Ciências da Universidade do Porto, Portugal

H.I. Chaminé & J. Teixeira
Instituto Superior de Engenharia do Porto (LABCARGA|ISEP/Centro GeoBioTec|UA), Portugal

ABSTRACT: This work outlines some results of coupled geologic, hydrogeologic, tectonic, geochemical and isotopic techniques applied to low-temperature (40°C to 76°C) geothermal resources investigations. Two case studies of N-Portugal are presented and discussed. This multidisciplinary approach has been used to build up a "picture" of the geothermal systems. The results obtained provided key information on the origin and "age" of the geofluids, underground flow patterns, and water-rock interaction occurring at depth. The studied low-temperature geothermal systems present similar hydrogeological conceptual models but rather different geochemical signatures (e.g., HCO_3-Na/CO_2-rich with pH ≈ 7 and HCO_3-Na with pH ≈ 9, type waters), due to water-rock interaction within different granitic environments at depth. In such cases, the local/regional high altitude sites associated with high fractured rocks play an important role in conducting the infiltrated meteoric waters towards the discharge zones near the Spas.

1 INTRODUCTION

The investigation of geothermal systems takes place in progressive stages that frequently overlap. These studies embrace many activities, including research and exploration, and must be carried out within the framework of a conceptual model of the system, which is improved as more and more information is collected.

The aim of this paper is to present the results of the assessment of low-temperature geothermal resources (discharge temperatures between 40°C and 76°C) occurring in the Portuguese mainland. For this purpose, a multidisciplinary approach, including geologic, tectonic, geochemical and isotopic (δ^2H, $\delta^{18}O$, $\delta^{13}C$, ^3H and ^{14}C) techniques, was applied in order to update local and/or regional conceptual circulation models. Two case studies of Northern Portugal are presented and discussed.

2 GEOLOGICAL AND TECTONIC SETTING

Case study 1: The Chaves low-temperature geothermal system (Fig. 1), associated with the hottest

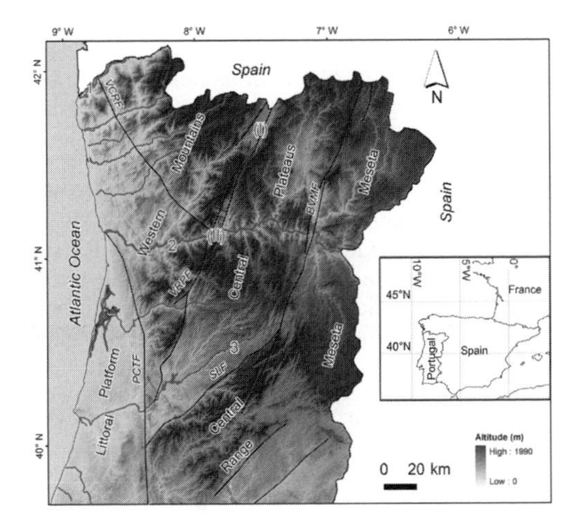

Figure 1. Regional morphotectonic features of northern Portugal. The studied low-temperature geothermal regions: (I) Chaves and (II) Caldas do Moledo. *Explanation:* 1) Minho River, 2) Douro River, 3) Mondego River. Major regional fault zones: PCTF–Porto–Coimbra–Tomar shear zone; VCRF–Vigo–V. N. Cerveira–Régua shear zone; VRPF–Verin–Régua–Penacova fault; BVMF–Bragança–Vilariça–Manteigas fault; SLF–Seia–Lousã fault.

(76°C) springs in the Portuguese mainland, is situated in the N part of the country along one of the most important NNE trending faults, the so-called Verin-Régua-Penacova fault zone (Cabral 1989, Ribeiro et al. 2007). Along that tectonic lineament lie not only the Chaves hot CO_2-rich geothermal waters but also several saline cold (17°C) CO_2-rich mineral waters (e.g. Vilarelho da Raia, Vidago, and Pedras Salgadas) that are used in local spas.

The Chaves CO_2-rich geothermal waters flow from natural springs and boreholes (150 m depth). Nowadays they are used in balneotherapy (at the local spa) and for space heating (a hotel and a municipal swimming pool).

The morphotectonics is dominated by the "Chaves Depression" which is a graben whose axis is oriented NNE-SSW (Fig. 1).

The geology of Chaves region has been described by Portugal Ferreira et al. (1992) and Baptista et al. (1993). The region is situated in the pre-Mesozoic Iberian Massif that consists mainly of Variscan granites and Paleozoic metasediments (e.g. phyllites, quartzites and schists). The most recent formations are Miocene-Pleistocene sedimentary cover deposits with variable thickness showing their maximum development along the central axis of the Chaves graben. The Alpine Orogeny developed extensive tectonic features responsible for emplacement of several hydrothermal ore deposits.

Case study 2: At Caldas do Moledo region (Fig. 1), the low-temperature (40°C $< T <$ 46°C) geothermal waters from boreholes are used in the local spa for diversified treatments.

The main geomorphologic feature of the Caldas do Moledo region is the Douro River valley, bordered to the N by the Marão mountain ridge (c. 1415 m a.s.l.). This region is located in the Central Iberian Zone of the Iberian Massif (Ribeiro et al. 2007).

Previous studies (e.g. Espinha Marques et al. 2001) pointed out that the most important tectonic structures in the region are the NNE–SSW Verin-Régua-Penacova fault zone and the WNW–ESE to NW–SE Vigo-Vila Nova de Cerveira-Régua shear zone (Fig. 1). Caldas do Moledo region is dominated by metasedimentary rocks of the so-called Schist-Greywacke Complex (Douro Group) of lower Cambrian age cut by aplite and pegmatite veins (Teixeira et al. 1967, Bernardo de Sousa & Sequeira 1989). Granitic rocks were newly mapped in this region (Espinha Marques et al. 2001).

3 CONCEPTUAL MODELS

Exploration and development must be carried out within the framework of a conceptual hydrogeological model of a given geothermal system, which is improved as more and more information is collected.

Geologic, tectonic, hydrogeologic, geochemical and isotopic methodologies are used to build a "picture" of the geothermal systems. Used in parallel, they have provided key information on origin and "age" of the geofluids, underground flow paths and water-rock interactions.

Case study 1: The Chaves low-temperature geothermal waters belong to the HCO_3-Na/CO_2-rich type (with pH $\approx$ 7). Total dissolved solids (TDS) range between 1600 mg/L and 1850 mg/L. One can also observe a range between 350 mg/L and 500 mg/L of free CO_2 (Aires-Barros et al. 1998).

The application of SiO_2 and K^2/Mg geothermometers (Aires-Barros et al. 1998) indicates equilibrium temperatures around 120°C, which agree with the discharge temperature (76°C). Considering the mean geothermal gradient of 30°C/km (Duque et al. 1998), the Chaves system circulates at a maximum depth of about 3.5 km. This value is obtained from:

$$\text{depth} = (T_r - T_a)/gg \quad (1)$$

where T_r is the reservoir temperature (120°C), T_a the mean annual air temperature (15°C) and gg the local geothermal gradient.

The $\delta^{18}O$ and δ^2H values of Chaves low-temperature CO_2-rich geothermal waters lie on the Global Meteoric Water Line—GMWL ($\delta^2H = 8\delta^{18}O + 10$) defined by Craig (1961), indicating that they are meteoric waters that have been recharged without evaporation (Fig. 2).

As described by Aires-Barros et al. (1998) and Andrade (2003), the more depleted cold dilute groundwaters (plotted along the Local Meteoric Water Line–LMWL) are those related to sampling sites (springs) located at higher altitudes, showing $\delta^{18}O$ and δ^2H values close to those of Chaves low-temperature CO_2-rich geothermal waters.

The isotopic gradient obtained for ^{18}O (–0.22‰ per 100 m of altitude) are in good agreement with the values found in Mediterranean regions (IAEA 1981).

The low $\delta^{18}O$ values of Chaves low-temperature CO_2-rich geothermal waters require that these waters were derived from precipitation at more than 1150 m

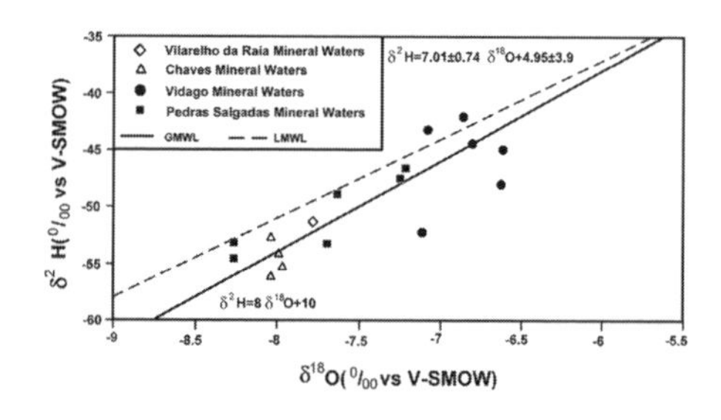

Figure 2. δ^2H vs. $\delta^{18}O$ relationship of Chaves low-temperature CO_2-rich geothermal waters. The isotopic values of the cold CO_2-rich mineral waters from Vilarelho da Raia, Vidago and Pedras Salgadas areas are also plotted (after Andrade 2003).

a.s.l. These elevations are obtained in the Padrela Mountain (E block of Chaves graben), which presumably feeds the local infiltration (Fig. 3).

The $\delta^{13}C$ determinations carried out on total dissolved inorganic carbon (TDIC) of the CO_2-rich saline waters gave values lying in the range of −6‰ to −1‰ (Marques et al. 1998). A deep-seated (upper mantle) origin for the CO_2 should be considered as the most plausible hypothesis, considering the tectonic / fracture scenario. Such assumption is supported by the $^3He/^4He$ ratios, measured in the gas phase, between 0.89 and 2.68 times the atmospheric ratio (Ra) and higher than that those expected for a pure crustal origin ≈ 0.02 Ra (Carreira et al. 2010).

The CO_2 seems to be transported from its mantle source to the surface by migration as a separate gas phase, being incorporated in the infiltrated meteoric waters, at considerable depth in the case of the Chaves low-temperature CO_2-rich geothermal waters.

Carreira et al. (2008) stated that the input of carbon-14 free gas to the CO_2-rich thermomineral water systems must induce changes in the groundwater "*age*" determinations. In fact, the radiocarbon content (^{14}C activity from 4.3 pmC up to 9.9 pmC) determined in some of the cold CO_2-rich mineral waters from Vidago and Pedras Salgadas region is incompatible with the systematic presence of ^3H (from 1.7 to 7.9 TU).

Case study 2: At Caldas do Moledo region, several low-temperature geothermal spring and borehole waters emerge with different temperatures (between 40°C and 46°C). Their chemical signatures are quite similar, being characterized by the following main features: i) relatively high pH values (between 8.0 and 9.0), ii) TDS values in the range of 200 mg/L to 350 mg/L, iii) HCO_3 as the dominant anion, iv) Na as the dominant cation, v) the presence of reduced species of sulphur ($HS^- \approx$ 2.5 mg/L), vi) high SiO_2 values, representing more than 15% of total mineralization, and vii) high fluoride concentrations (up to 10 mg/L), indicating that the reservoir rock should be mainly the granitoids.

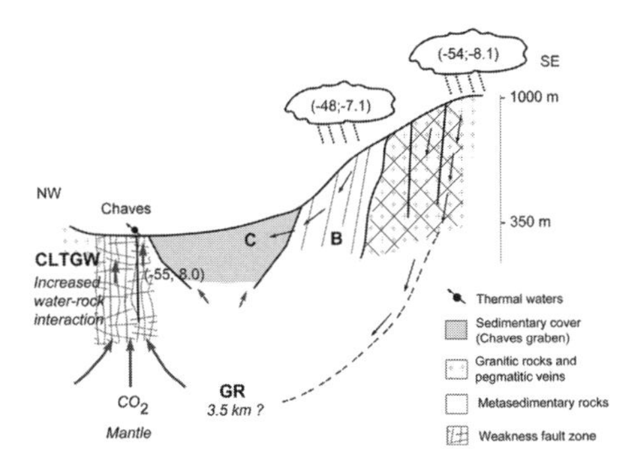

Figure 3. Hydrogeological conceptual model of Chaves geothermal system. B—granitic and metasedimentary rocks; C—cover deposits; CLTGW—Chaves low-temperature geothermal waters; GR—geothermal reservoir; (δ^2H; $\delta^{18}O$)—isotopic composition of the waters.

The results of the application of chemical geothermometers to the Caldas do Moledo low-temperature geothermal waters were discussed in detail by Marques et al. (2003). The mean Na-K-Ca and K^2/Mg temperatures indicated 71 ± 14°C and 71 ± 6°C, respectively, as the most feasible reservoir temperatures.

In the case of Caldas do Moledo region, the low-temperature geothermal waters and the local shallow groundwaters (group-a) lie on or close to the Global Meteoric Water Line–GMWL indicating that they are of meteoric origin (Fig. 4). No oxygen-isotope shift due to water-rock interaction at high temperatures was found (Marques et al. 2003).

Isotope data from Douro River water samples (group-b) indicate depletion in heavy isotopes (Fig. 4). Most of the Douro River basin upstream of Caldas do Moledo is characterized by elevations higher than 500 m a.s.l. The river source is located in the Urbion Mountains (Spain) at 1700 m a.s.l. Thus, the contribution of Douro River waters to the recharge of the geothermal system should be nonexistent.

The ^{14}C content determined in the TDIC of borehole AC1 waters was 14.0 ± 1.3 pmC. Carbon-14 age calculations, using the closed system model proposed by Salem et al. (1980), indicated an apparent ^{14}C age of 15.66 ± 2.86 ka BP for Caldas do Moledo low-temperature geothermal borehole waters (Marques et al. 2003).

The recharge altitudes were determined using the isotopic signatures of Caldas do Moledo borehole waters ($\delta^{18}O_{mean}$ = −6.75 ± 0.25‰ vs V-SMOW) and the isotopic gradient for $\delta^{18}O$ in Caldas do Moledo area (−0.12‰/100 m of altitude) (Marques et al. 2003). The recharge elevations (≈ 1000 m a.s.l.) occur at a granitic outcrop (NW of the spa area), suggesting that the NW-SE Vigo–V. N. Cerveira–Régua shear zone plays an important role in groundwater recharge and deep circulation towards the NNE–SSW Verin–Régua–Penacova fault zone, which seems to act as the most likely channel way for the ascent of geothermal waters (Fig. 5).

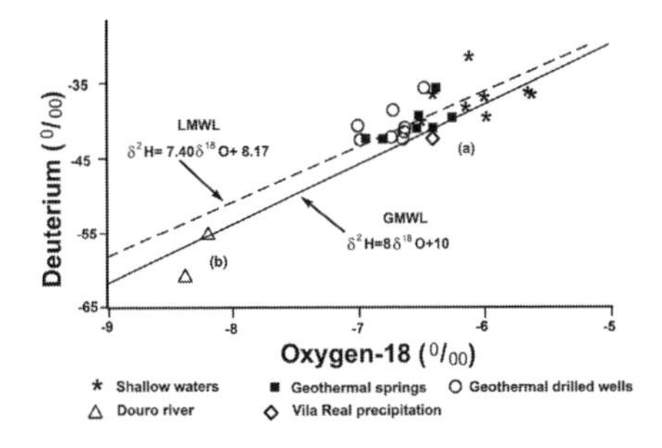

Figure 4. δ^2H vs. $\delta^{18}O$ plot for water samples from Caldas do Moledo region. Adapted from Marques et al. (2003).

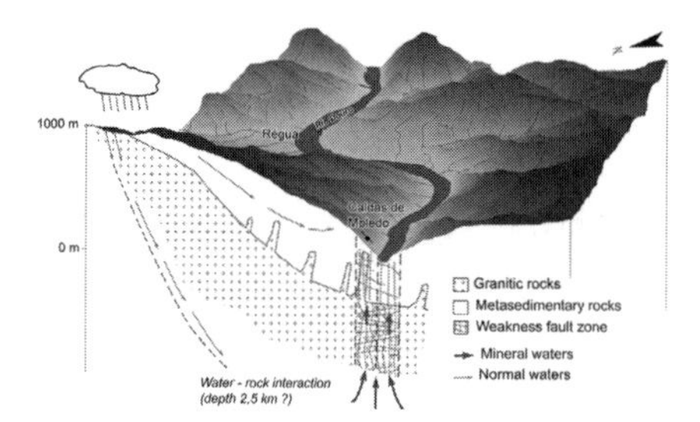

Figure 5. Schematic cross-section of the possible regional/local flow paths at the Caldas do Moledo low-temperature geothermal system (adapted from Espinha Marques et al. 2001).

4 CONCLUDING REMARKS

This paper illustrates two case studies ascribed to low-temperature geothermal systems, presenting similar hydrogeological conceptual models but different geochemical signatures. The local/regional high altitude sites associated with highly fractured rocks play an important role in conducting the infiltrated meteoric waters towards the discharge zones near the spas. The discharge zones are related to the intersection of the major regional deep fault structures (namely highly sheared tectonic fractures), responsible for the geothermal fluids ascent. This work shows the important role of geosciences (e.g. geology, geomorphology, tectonics, hydrogeology, geochemistry and isotope hydrology) in the establishment of robust hydrogeologic conceptual models describing low-temperature geothermal systems.

ACKNOWLEDGEMENTS

This paper was performed under the framework of several R&D Projects granted by the FCT and EU FEDER funds. This manuscript was critically read by Fraser Goff and Yoseph Yechieli, and we gratefully acknowledge their contributions.

REFERENCES

Aires-Barros, L., Marques, J.M., Graça, R.C., Matias, M.J., Weijden, C.H. van der, Kreulen, R. & Eggenkamp, H.G.M. 1998. Hot and cold CO_2-rich mineral waters in Chaves geothermal area (Northern Portugal). *Geothermics* 27(1): 89–107.

Andrade, M.P.L. 2003. *Isotopic geochemistry and thermomineral waters. Contribution of Sr ($^{87}Sr/^{86}Sr$) and Cl ($^{37}Cl/^{35}Cl$) isotopes to the elaboration of circulation models. The case of some CO_2-rich waters from N Portugal.* MSc Thesis, Technical University of Lisbon (IST): 104 p. (in Portuguese).

Baptista, J., Coke, C., Dias, R. & Ribeiro, A. 1993. Tectonics and geomorphology of Pedras Salgadas region and associated mineral springs. In A. Chambel (ed.), *Comunicações da XII Reunião de Geologia do Oeste Peninsular* 1: 125–139. Évora University.

Bernardo de Sousa, M. & Sequeira, A.J.D. 1989. Geological Report on the Alijó Sheet No. 10-D (1:50,000). *Portuguese Geological Survey*, Lisbon: 59 p. (in Portuguese).

Cabral, J. 1989. An example of intraplate neotectonic activity Vilariça Basin, Northeast Portugal. *Tectonics* 8: 285–303.

Carreira, P.M., Marques, J.M., Graça, R.C. & Aires-Barros, L. 2008. Radiocarbon application in dating "complex" hot and cold CO_2-rich mineral water systems: a review of case studies ascribed to the northern Portugal. *Applied Geochemistry* 23: 2817–2828.

Carreira, P.M., Marques, J.M., Rosário Carvalho, M., Giorgio C. & Fausto, G. 2010. Mantle-derived carbon in Hercynian granites. Stable isotopes signatures and C/He associations in the thermomineral waters, N-Portugal. *Journal of Volcanology and Geothermal Research* 189: 49–56.

Craig, H. 1961. Isotope variations in meteoric waters. *Science* 133: 1702–1703.

Duque, R., Monteiro Santos, F.A. & Mendes-Victor, L.A. 1998. Heat flow and deep temperatures in the Chaves Geothermal system, northern Portugal. *Geothermics* 27(1): 75–87.

Espinha Marques, J., Marques, J.M., Chaminé, H.I., Graça, R.C., Carvalho, J.M., Aires-Barros, L. & Borges, F.S. 2001. The newly described 'Poço Quente' thermal spring (Granjão–Caldas do Moledo sector, N Portugal): hydrogeological and tectonic implications. *Geociências, Revista da Universidade de Aveiro* 15: 19–35.

IAEA (ed) 1981. *Stable Isotope Hydrology. Deuterium and Oxygen-18 in the Water Cycle.* Technical Reports Series 210. Vienna: IAEA.

Marques, J.M., Carreira, P.M., Aires-Barros, L. & Graça, R.C. 1998. About the origin of CO_2 in some HCO_3/Na/CO_2-rich Portuguese mineral waters. *Geothermal Resources Council Transactions* 22: 113–117.

Marques, J.M., Espinha Marques, J., Carreira, P.M., Graça, R.C., Aires-Barros, L., Carvalho, J.M., Chaminé, H.I. & Borges, F.S. 2003. Geothermal fluids circulation at Caldas do Moledo area, Northern Portugal: geochemical and isotopic signatures. *Geofluids* 3: 189–201.

Portugal Ferreira, M., Sousa Oliveira, A. & Trota A.N. 1992. Chaves geothermal pole. Geological Survey, I and II. Joule I Program, DGXII, CEE. *University of Trás-os-Montes and Alto Douro*, Portugal, Internal Report: 44 p.

Ribeiro, A., Munhá, J., Dias, R., Mateus, A., Pereira, E., Ribeiro, L., Fonseca, P.E., Araújo, A., Oliveira, J.T., Romão, J., Chaminé, H.I., Coke, C. & Pedro, J. 2007. Geodynamic evolution of the SW Europe Variscides. *Tectonics* 26, TC6009. Doi: 10.1029/2006TC002058.

Salem, O., Visser, J.M., Deay, M. & Gonfiantini, R. 1980. Groundwater flow patterns in western Lybian Arab Jamahitiya evaluated from isotope data. In IAEA (ed.), *Arid zone hydrology: investigations with isotope techniques*: 165–179. Vienna: IAEA.

Teixeira, C., Fernandes, A.P. & Peres, A. 1967. Geological Report on the Peso da Régua Sheet No. 10-C (1:50,000). *Portuguese Geological Survey*, Lisbon: 60 p. (in Portuguese).

Water-Rock Interaction – Birkle & Torres-Alvarado (eds)
© 2010 Taylor & Francis Group, London, ISBN 978-0-415-60426-0

Hydrochemical and isotopic interpretation of thermal waters from the Felgueira area (Central Portugal)

Manuel Morais
Centre for Geophysics of the University of Coimbra, Portugal

ABSTRACT: The hydrothermal system of Felgueira has long been appreciated for the recreational and therapeutic benefits of the thermal baths. In order to protect and maintain these thermal resources it is necessary to understand the characteristics of these waters and determine its origin and that of its solutes. In this paper we use chemical and isotopic data as tools for addressing some of those questions. Early Holocene meteoric waters reacted, at low reservoir temperatures, with minerals from a granite aquifer, in an environment with available deep CO_2, thus acquiring a sodium bicarbonate-chloride composition, with alkaline and strongly reducing characteristics. Biogenic sulphate reduction results in a wide variation of the S isotopic composition of both oxidized and reduced sulphur species. The drawdown imposed by exploitation wells results in some degree of vulnerability of the hydrothermal system.

1 INTRODUCTION

Caldas da Felgueira is an active thermal spa whose waters have been used for medicinal purposes since ancient times. Located at Central Portugal (Fig. 1), it has a mild climate, with an average rainfall of 1100 mm and a mean annual air temperature of 15°C.

There are two natural spring discharges, one with a water temperature of 32.5°C, which serves as Buvette of the spa, and a cold spring with a temperature of 18°C. For balneological activities, 32.0°C water was obtained from a 16 m deep excavated well until the 90′s of the last century. Since then, 3 wells have been drilled in order to satisfy the development of the spa, with increased consumption of thermal water.

A considerable amount of chemical data is available from 1991 until present, derived from monitoring programs of those wells. Additionally, two sampling campaigns for isotopic determinations were made in 1996 and 2008, as reported on Table 1.

The overall goal is to obtain a geochemical model of the hydrothermal system based on chemical and isotopic data.

2 HYDROGEOLOGICAL SETTING

Caldas da Felgueira region is located onto fractured granitic rocks intruding a metamorphic complex of low Palaeozoic age, belonging to the Central Iberian geotectonic zone of the Hesperian Massif. The area was cut by large, deep faults during the Hercynian orogeny, some of which were reactivated during Alpine times, allowing for extensive deep circulation of meteoric waters which resulted in several hydrothermal fields, including Caldas da Felgueira.

The locally significant features of the hydrogeologic setting are shown in Figure 1 (map after Rodrigues et al. 2000). The rock exposed in the neighbourhood of the thermal springs is porphyritic medium to fine-grained two-mica granite, gradually passing into a coarser grain size. These granites belong to the calc-alkaline series and are associated to a post-tectonic Hercynian pluton. A dense fracture network affects the crystalline basement, with the main fault system -trending ENE-WSW- controlling the spring occurrences. It was along this structurally controlled corridor that 3 wells were drilled; namely AC1, a 64 m deep flowing well with a water temperature of 35.8°C; AC2 (250 m; 26,7°C) and the deepest AC3 well (306 m) with water at 33.2°C. All the wells yield warm water whose chemical and physical characteristics are similar to those of the spring waters, but with a minor mixing with surficial waters. This suggests that the waters emerge from the same aquifer system and follow identical flow paths, and are subjected to similar geochemical processes during its trajectory in the hydraulic circuit.

3 ORIGIN AND "AGE" (MEAN RESIDENCE TIME) OF THERMAL WATERS

The meteoric origin of these waters has been established by comparison of $\delta^{18}O$ and δD values (Fig. 2) with those of the World Meteoric Water Line (WMWL) for rainfall as defined by Craig (1961). The lack of significant deviation of the

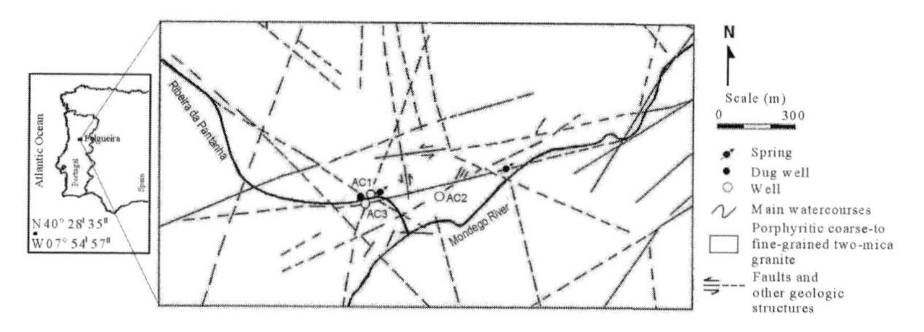

Figure 1. Location and geologic map of the Felgueira area (modified from Rodrigues et al. 2000).

Table 1. Chemical and isotopic data and saturation indices for Felgueira thermal waters.

Sampling Point Date	AC1 96/09	AC1 08/07	AC2 96/09	AC3 96/09	AC3 08/07
Field Parameters					
Temp., °C	35.5	35.1	23.3	33.5	34.1
Cond., S cm^{-1}	680	528	580	690	529
pH	8.10	8.00	7.85	8.05	8.17
Constituents, mg/L					
Na^+	109	107	104	108	108
K^+	1.6	2.4	2.2	2.6	2.4
Ca^{++}	5.3	5.1	7.2	6.2	5.0
Mg^{++}	0.11	0.11	0.38	0.2	0.13
Li^+	1.06	0.99	1.06	1.12	0.99
NH_4^+	0.12	0.10	0.07	0.17	0.08
Cl^-	51.8	50.7	49.0	51.8	48.6
HCO_3^-	143	148	138	145	149
$SO_4^=$	26.0	16.8	36.9	27.6	20.3
F^-	15.1	15.7	13.0	14.7	15.5
HS^-	0.6	0.7	0.3	0.5	0.3
NO_3^-	0.24	<0.10	0.27	0.33	0.10
SiO_2	50.5	52.2	50.4	52.2	56.2
Calculated Parameters*					
Eh, mV	−303	−297	−258	−295	−303
TDS, mg/L	404.4	399.9	402.8	410.4	406.6
Isotopes					
Stable**					
$\delta^{18}O(H_2O)$	−5.0	−5.4	−5.3	−5.1	−5.0
$\delta^{18}O(SO_4)$	+10.9	–	+7.3	+10.3	–
$\delta^2H(H_2O)$		−34.3			−34.8
$\delta^{13}C(DIC)$	−14.6	−13.8	−14.5	−13.6	−14.0
$\delta^{34}S(SO_4)$	+6.4	+9.0	−1.3	+6.0	+10.6
$\delta^{34}S(H_2S)$	−25.1	−23.5	−29.9	−25.2	−25.3
Radioactive					
T., TU	<d.l.	n.m.	0.82	<d.l	n.m.
^{14}C, pmC	17.9	n.m.	31.2	18.5	n.m.
Saturation Indices					
Calcite	−0.33	−0.43	−0.62	−0.33	0.28
Fluorite	0.17	0.20	0.31	0.23	0.18
Chalcedony	0.35	0.37	0.49	0.39	0.41
Quartz	0.74	0.77	0.93	0.79	0.81

*Calculated with Wateq-4F version 2.63 (2004), (Ball &Nordstrom, 2001). Eh calculated with the pair HS/SO_4.

** $\delta^{18}O$ ‰ *vs* VSMOW; δ^2H ‰ *vs* VSMOW; $\delta^{13}C$ ‰ *vs* PDB; $\delta^{34}S$ ‰ *vs* CDT; n.m.—not measured; <d.l. below detection limit of 0.7 TU.

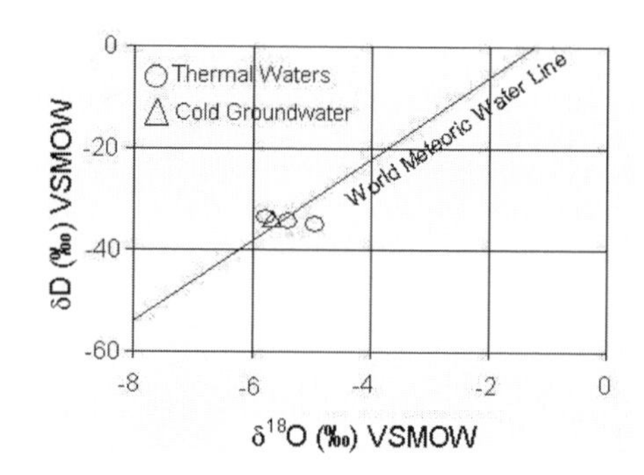

Figure 2. The relationship between $\delta^{18}O$ and δD values in the thermal and cold water samples from the study hydrothermal system.

$\delta^{18}O$ values of the thermal waters from WMWL ("oxygen isotope shift") limits the reservoir temperatures to a maximum of 150°C (Ellis & Mahon 1977), suggesting a low enthalpy hydrothermal system corresponding to the deep circulation of meteoric water in a fractured granite of Hercynian age. Tritium is absent from wells AC1 and AC3, while the small con-tent measured at well AC2 gives an indication of a major vulnerability to phreatic waters (Table 1). That hint is corroborated by ^{14}C content: AC1 and AC3 have the same content of ^{14}C, and are the less hybrid fluids of the system, while ^{14}C contents of well AC2 are significantly larger. The value of the initial activity of ^{14}C was calculated with the isotopic mixing model of Pearson (Pearson & Hanshaw 1970). Applying the decay equation to the measured ^{14}C activities of the waters, an adjusted radiocarbon age of 10.3 Ka is derived for AC1 and AC3 waters.

4 CLASSIFICATION OF THERMAL FLUIDS

Chemically, Felgueira thermal waters (Table 1) have low mineralization (total dissolved solids around 400 mg/L), pH in the alkaline range (around 8.10) and a strong negative calculated redox potential (always < −250 mV), which allow for measurable quantities of NH_4^+ and HS^- species, representing nitrogen and sulphur at lower oxidation states. Nitrate and magnesium are usually low. Additional main constituents are Cl^- (50 mg/L), F^- (15 mg/L) and Li^+ (1 mg/L). The mean values of sulphate concentration were ca. 25 mg/L. Na represent 90% of cations (rNa>>rCa>>rMg), but none of the anions clearly dominates ($rHCO_3$ > rCl > rSO_4), thus belonging to a sodium-bicarbonate-chloride type. The observed characteristics give a complex chemical structure, rendering these waters very susceptible to aging effects.

5 MINERAL-SOLUTION EQUILIBRIA AND GEOTHERMOMETRY

The composition in terms of cations (Na^+, K^+ and Mg^{++}) plotted on the triangular diagram of Giggenbach (1988) and Giggenbach & Corrales (1992) is used to classify the state of rock-water interaction of these waters as partially equilibrated at 70–140°C with hydrothermally altered rock in the geothermal reservoir (Fig. 3).

However, the most probable temperature range for the reservoir system is 70–80°C, as calculated with some degree of confidence by independent temperature estimates using chemical geothermometers (chalcedony: Fournier 1991, Na-K-Ca: Fournier & Truesdell 1973, *in* Fournier 1991, Na-K: Nieva & Nieva 1987, *in* Fournier, 1991, and K-Mg: Giggenbach 1988).

The saturation indices were calculated using the thermodynamic equilibrium model from the geochemical software Wateq-4F (Ball & Nordstrom 2001) for minerals assumed to be relevant for the aquifer. Waters are oversaturated with respect to quartz, chalcedony and fluorite, and undersaturated with respect to calcite. At the temperature of the reservoir these waters approach equilibrium with calcite and chalcedony; are slightly undersaturated in fluorite, and slightly oversaturated in quartz.

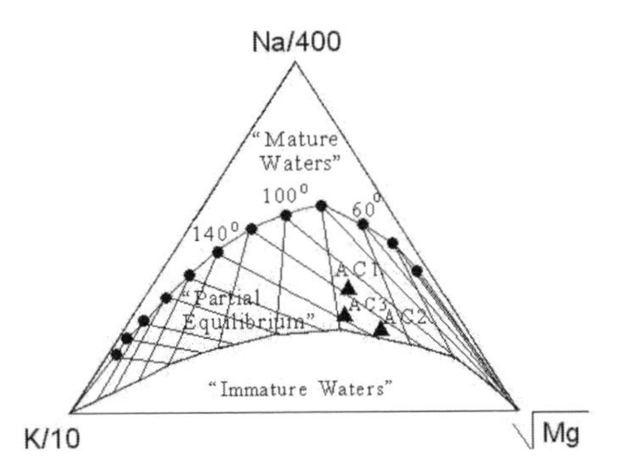

Figure 3. Triangular diagram of Na/400-K/10-√Mg for the thermal waters (modified from Giggenbach, & Corrales 1992).

The investigation of the fluid-mineral equilibrium for aluminosilicates employed activity diagrams constructed at reservoir temperatures, for primary minerals like microcline, albite and anorthite. According to these diagrams, thermal waters plot in the stability field of microcline, and are most likely to produce montmorillonites (Na and Ca) as alteration minerals.

6 ISOTOPIC STUDIES

If groundwater circulates in an aquifer with apparent absence of carbonate minerals, and in a system closed to CO_2, its C isotopic ratios do not evolve far beyond those of the soil CO_2 contribution, *i.e.* $\delta^{13}C$ around −23 to −25‰ for C3 type vegetation in temperate climate (Clark & Fritz 1997). However, dissolved inorganic carbon (DIC) $\delta^{13}C$ values at Caldas da Felgueira, approach ca. −14‰. This probably records a contribution of isotopically heavy carbon from a source external to the granite aquifer. We can only speculate about this source as being metamorphic carbon or mantle derived CO_2.

The $\delta^{34}S$ values of SO_4 from AC1 and AC3 waters are positive and between +6 and +10‰, while $\delta^{34}S$ in the associated dissolved sulphide is around −25‰. Keeping in mind the values reported in Table 1, the S isotope ratios for those waters could be explained by some bacterial reduction of $SO_4^=$ which results in an enrichment in ^{34}S of the remaining $SO_4^=$; Δ $SO_4^= - HS^- \approx +30$ to +35 ‰ (van Everdingen et al. 1982, Hoefs 2004). Moreover, the $\delta^{18}O_{(SO4)}$ values of +10 to +11‰ are enriched in ^{18}O relative to $\delta^{18}O_{(H2O)}$ (≈ -5‰), which is also to be expected in the reduction process of dissolved $SO_4^=$ (Fritz et al. 1989). The negative $\delta^{34}S$ value and increased content of $SO_4^=$ (oxidized S) in water from well AC2 possibly reflects a contribution from oxidation of dissolved sulphide derived by more significant mixing of the thermal waters with phreatic waters. The lower value of $\delta^{18}O_{(SO4)}$ relative to waters from wells AC1 and AC3 also supports this interpretation.

7 CHEMICAL MONITORING

Time series plots of the chemical data obtained from the monitoring programme exhibit a consistent increase in dissolved sulphate concentrations up to 1998, followed by a decrease later on. This variation is relative to "historical" mineral water composition, *e.g.* data from 1951 analysis (Fig. 4). Other parameters hardly changed, although small variations in concentrations with time of some elements have been carefully analysed for the period 1991–98 by Morais (2008). Such contamination by sulphates is attributed to the impact of the extraction and ore processing of uranium at Urgeiriça mine, located

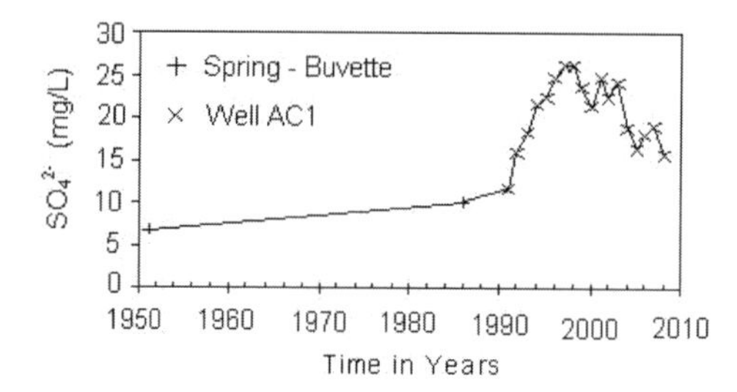

Figure 4. Evolution in sulphate concentrations in the thermal waters during the period 1951–2008.

4 kms to the NW of Caldas da Felgueira. The mining area is drained by the Ribeira da Pantanha watercourse that flows through the Caldas da Felgueira area, transporting wastewater from the treatment plant of Urgeiriça. This plant used components like sulphuric acid to benefit the ore. In a sample collected in September 1996 the content of sulphates in water from Ribeira da Pantanha were 968 mg/L, with a value of $\delta^{34}S$ of +4.57‰. The significant increase of sulphate levels in water after the construction of the wells is consistent with admixture with contaminated surficial waters, as a consequence of the new dynamic equilibrium in water pressures within the productive geological structure imposed by pumping the wells. Extraction and treatment of uranium began in 1951, declined towards 1991 and completely ended in 2004. Although anthropogenic impacts are now much diminished, initial sulphate values have not yet recovered (Fig. 4).

8 CONCLUSIONS

The Felgueira hydrothermal system flows through a fractured granite aquifer, heating early Holocene aged water to a maximum temperature of 70–80°C. Its chemistry is controlled by the interaction of meteoric water, enriched in deep CO_2, with the minerals of the granite aquifer. The stable carbon isotopic composition reflects both organic and inorganic carbon sources for DIC, in a system partially open to inorganic CO_2. Biogenic reduction of dissolved sulphate is the process which significantly determines the isotopic geochemistry of sulphur. The medicinal utilization of Felgueira water was hampered from 1951 onwards by the coexistence at its neighbourhood of environmentally aggressive uranium mining. Drawdown caused by wells pumping the thermal water resulted in surficial sulphate rich waters, derived from mining activities, entering the system, which result in a marked increase in sulphate contents of the water. The adverse effect of mining activities still continues, although environmental restoration began in 2006 by sealing of mill tailings.

ACKNOWLEDGMENT

We gratefully acknowledge "Companhia das Águas Medicinais da Felgueira, SA." for all the support given to this study. The author benefited from a graduate research fellowship from "Fundação para a Ciência e Tecnologia".

REFERENCES

Ball, J.W. & Nordstrom, D.K. 2001. User's manual for Wateq4F, with revised thermodynamic data base and test cases for calculating speciation of major, trace, and redox elements in natural waters. *U.S.G.S. Open-File Report*: 91–183.

Craig, H. 1961. Isotopic variations in meteoric waters. *Science* 133: 1702–1703.

Clark, I.D. & Fritz, P. 1997. *Environmental Isotopes in Hydrogeology*. New York: Lewis Publishers.

Ellis, A.J. & Mahon, W.A.J. 1977. Hydrothermal Solutions (ch.4). In *Chemistry and Geothermal Systems*: 117–161. New York: Academic Press.

Fournier, R.O. 1991. Water geothermometers applied to geothermal energy. In F. D'Amore (ed.), *Applications of geochemistry in geothermal reservoir development*: 37–69. United Nations Institute for Training and Research, USA.

Fritz, P. Basharmal, G.M. Drimmie, R.J. Ibsen, J. & Qureshi, R.M. 1989. Oxygen isotope exchange between sulphate and water during bacterial reduction of sulphate. *Chemical Geology* (Isot. Geosc. Sec.) 79: 99–105.

Giggenbach, W.F. 1988. Geothermal solute equilibria. Derivation of Na-K-Mg-Ca geoindicators. *Geochimica et Cosmochimica Acta* 52: 2749–2765.

Giggenbach, W.F. & Corrales, S.R. 1992. Isotopic and chemical composition of water and steam discharges from volcanic-magmatic-hydrothermal systems of the Guanacaste Geothermal Province, Costa Rica. *Applied Geochem*istry 7: 309–332.

Hoefs, J. 2004. *Stable Isotope Geochemistry*. Berlin: Springer.

Morais, M. 2008. Trend analysis using nonparametric statistical techniques for detection and evaluation of spatial and temporal chemical changes at a hydrothermal exploitation (Felgueira Spa—Central Portugal). In Bruthans, Kovar & Hrkal (eds.), *Proc. of the International Interdisciplinary Conference on Predictions for Hydrology, Ecology, and Water Resources Management*: 179–182. Prague.

Pearson, F.J. & Hanshaw, B.B. 1970. Sources of dissolved carbonate species in groundwater and their effects on carbon-14 dating. In *Isotope Hydrology*: 271–285. Viena: IAEA.

Rodrigues, N.V., Correia, C., Telles, M. & Dias, J.M. 2000. Ensaios de bombagem nas Caldas da Felgueira. *Proc. of the "Congresso da Água 2000"*. APRH.

van Everdingen, R.O., Shakur, M.A. & Krouse, H.R. 1982. Isotope geochemistry of dissolved, precipitated, airborne, and fallout sulphur species associated with springs near Paige Mountain, Normam Range, N.W.T. *Canadian Journal of Earth Sci*ences 19: 1395–1407.

Water-Rock Interaction – Birkle & Torres-Alvarado (eds)
© 2010 Taylor & Francis Group, London, ISBN 978-0-415-60426-0

Geochemical evidence for two hydrothermal aquifers at El Chichón

L. Peiffer & Y. Taran
Instituto de Geofísica, Universidad Nacional Autónoma de México, México D.F., México

E. Lounejeva & G. Solis-Pichardo
Instituto de Geología, Universidad Nacional Autónoma de México, México D.F., México

D. Rouwet
Istituto Nacionale di Geofisica e Vulcanologia, Palermo, Italy

ABSTRACT: Major, trace and rare earth element compositions and $^{87}Sr/^{86}Sr$ ratios of waters from the thermal springs and crater lake at El Chichón volcano suggest the existence of two distinct thermal aquifers (Aq. 1 and Aq. 2). Na/K and K/Mg ratios indicate that the Aq. 2 composition represents a deep mature hydrothermal fluid. The Ca/Sr ratios are unusually low for Aq. 2 (~17) compared to Aq. 1 (~130) and not obviously related to Ca/Sr in the wall rocks. On the other hand, based on the similarity of $^{87}Sr/^{86}Sr$ from waters and wall rocks, we suggest that Aq. 1 is shallow and composed of volcanic rocks, while Aq. 2 is deeper and composed of sedimentary rocks. The REE concentrations are strongly pH-dependent and for acidic waters match the pattern of El Chichón volcanic rocks.

1 INTRODUCTION

Hydrothermal waters and gases have been monitored at El Chichón volcano, Chiapas State, Mexico, for 15 years (Taran et al. 1998, Taran & Peiffer 2009 and references therein). However, the El Chichón hydrogeological and thermal structure is still not well understood. Here, we report new data on trace elements, REE and $^{87}Sr/^{86}Sr$ ratio in all types of thermal waters of the El Chichón volcano-hydrothermal system and discuss a hypothesis about two independent high-temperature aquifers beneath the volcano.

2 GENERAL SETTING

El Chichón is a small (1100 m asl) but complex volcanic structure of trachyandesitic composition covered by pyroclastic deposits. It represents the youngest volcano of the Chiapanecan Volcanic Arc (CVA), located between the Central America Volcanic Arc to the south and the Trans Mexican Volcanic Belt to the north. El Chichón basement consists of Paleogene-Neogene sandstones and siltstones underlain by Jurassic-Cretaceous dolomitic limestones and evaporites. Evaporites outcrop together with limestone 2 km south of the crater in the Agua Salada canyon. They were also identified at a depth of 2200 m bsl in a borehole, 7 km east of El Chichón. Four major structures can be distinguished: the 0.2 Ma outer 'Somma' crater rim, the 1 km wide-200 m deep 1982 crater and two extrusive domes on the NW and SW flanks (Duffield et al. 1984).

Thermal waters from El Chichón flanks (Fig. 1) can be divided into 3 groups according to their geochemical compositions. The first group consists of Agua Caliente—AC, Agua Caliente New—ACn, Agua Tibia 1—AT1, Agua Tibia 2—AT2, and Agua Tibia 1 new—AT1n. These hot springs have common characteristics: 50–79 °C, near-neutral pH and chlorine content between 1500–2200 mg/l. They all discharge

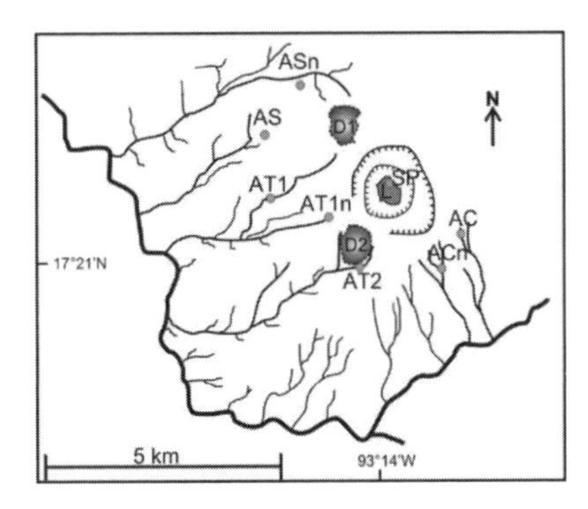

Figure 1. Map of thermal manifestations at El Chichón volcano. AC, ACn, AT1, AT2, AT1n, AS, ASn and SP are thermal springs. L: lake. D1 and D2: NW and SW dome. Indented lines: 1982 and 'Somma' crater rims (from inside to outside).

hot water from south-east to west flanks of the volcano (Fig.1) with the total outflow rate ~230 l/s.

The second group, known as Agua Salada (AS) springs, discharges acidic to neutral water on the north-west flank (pH from 2.2 to 7.8) with high Cl content (>10,000 mg/l) at an average flow rate of 10 l/s and temperature between 48 and 79 °C.

Crater manifestations including neutral saline 'Soap Pool' (SP) boiling spring and lake water belong to the first group. However, their composition can differ from this group by boiling, dilution and steam-heating processes.

3 METHODS

Temperature (±0.1 °C) and pH (±0.05 units) were measured in situ by an Orion multimeter. In the field, spring waters were filtered through 0.45 µm MILLIPORE filters into 250 ml plastic bottles for anions and into ultra-pure HDPE Nalgene flasks, with 5 ml of ultra-pure HNO_3, for cations and trace elements. Major species were analyzed by a Metrohm 761 ion chromatography, alkalinity by titration with 0.1 N HCl. Trace element analysis was performed by ICP-MS (Agilent 7500 CE). All determinations used the external standard calibration method, with Re and In as the internal standards. The accuracy of the results (±5%) was obtained by analyzing certified reference materials. Sr isotope ratios were determined using a Finnigan MAT262 thermal ionization mass spectrometer equipped with 8 Faraday collectors.

4 RESULTS AND DISCUSSION

During 2009, a new group of neutral pH hot springs named Agua Salada new—ASn was discovered (Fig.1). The existence of these springs was predicted by Taran & Peiffer (2009) who found anomalous Cl concentrations in a stream draining these springs (62 mg/l). The ASn springs discharge hot (72–75 °C) neutral, high salinity (5,000 < Cl < 10,000 mg/l) water with low Mg (<2 mg/l) and relatively low SO_4. The AS springs in the neighboring canyon, sampled in 2006–2009, have a similar composition but differ by being more acidic (pH: 2.7–5.4) with higher Mg and SO_4 contents. The acidity of the AS springs results from the shallow oxidation of H_2S. A thermal field with bubbling gas, steaming ground and steam-heated pools, was discovered in 2009 some 100 m up from the main acidic group of Agua Salada. This gas has a low $^3He/^4He$ ratio (2.3 Ra, He/Ne = 180) in contrast to very high $^3He/^4He$, up to 8 Ra, in steam of the crater fumaroles. Asn springs discovery constitute a major advance in El Chichón hydrothermal system understanding; their highly saline and neutral composition could represent the deep mature fluid.

Water compositions of springs collected before 2009 (Taran et al. 1998, Taran et al. 2008, Taran & Peiffer 2009) are also considered here in the mixing diagrams (Fig. 2). AR springs ('Agua roja') are superficial waters draining pyroclastic deposits of the 1982 eruption of El Chichón. They were also plotted for comparison with thermal springs.

4.1 *Mixing trends and evidence for two aquifers*

Mixing plots Cl vs. Mg, Ca and B indicate the existence of two distinct aquifers at El Chichón (Fig. 2). Flank springs AC, ACn, AT1, AT2, AT1n and the crater springs SP are fed from one aquifer (Aq. 1) with Cl content ~ 2000 mg/l, Mg up to 60 mg/l and Cl/B ~ 80.

All springs of the AS groups originate from a second aquifer 2 (Aq. 2). This aquifer is characterized by neutral pH and higher salinity with Cl > 10,000 mg/l, Mg < 2 mg/l and Cl/B ~ 250. Both end-members are Ca-rich. According to Giggenbach's (1988) classification, the Aq. 2 water is a mature hydrothermal solution, in equilibrium with altered rock at ~ 230 °C. It represents the deepest fluid composition. Water from Aq. 1 with high Mg is an

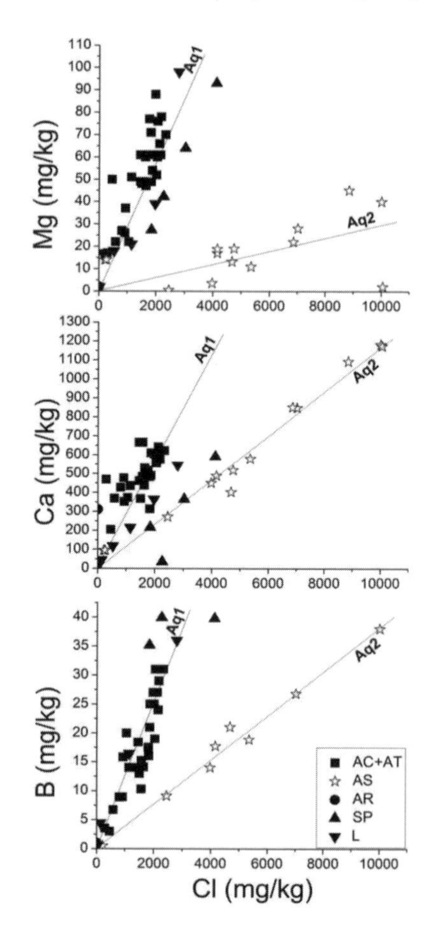

Figure 2. Mixing plots 'ion-Cl'. Aq1: aquifer 1. Aq2: aquifer 2.

immature hot water with Na/K and K/Mg model equilibrium temperatures of 250 °C and 100 °C, respectively, indicating that equilibrium has not been attained. This water is related, most probably, to a shallow level of the volcano-hydrothermal system and interaction of ground water with the magmatic gas condensate inside the volcano edifice.

Figure 3a shows the mixing relationship between Ca and Sr concentrations in thermal waters. This is further evidence for the existence of two aquifers with different host-rock compositions. The AS waters are characterized by unusually low Ca/Sr weight ratios of ~17, whereas all other springs on the volcano slopes, SP and lake water from the crater present Ca/Sr ~130. The latter value is two times greater than Ca/Sr ratios analyzed by Luhr et al. (1984) in magmatic anhydrites of the 1982 El Chichon eruption (~60) and five times the Ca/Sr ratio of El Chichon glass (~30, Jones et al. 2008). The AS ratio of ~17 is lower than the values of seawater (~50) and marine limestone (~150).

Figure 4 shows Na-normalized enrichment factors relative to the average El Chichón volcanic rock for trace elements from El Chichón waters. AR springs show enrichment in all trace elements in comparison to thermal springs, with the exception of Cl, As and alkali elements Li, Rb, and Cs. This can be explained by the fact that AR waters are very superficial; the compositions of AR waters are formed by leaching of the anhydrite-rich 1982 pyroclastic deposits (2 wt%, Luhr et al. 1984). Arsenic depletion is probably due to the adsorption of As on Fe-oxyhydroxides that precipitate from the AR

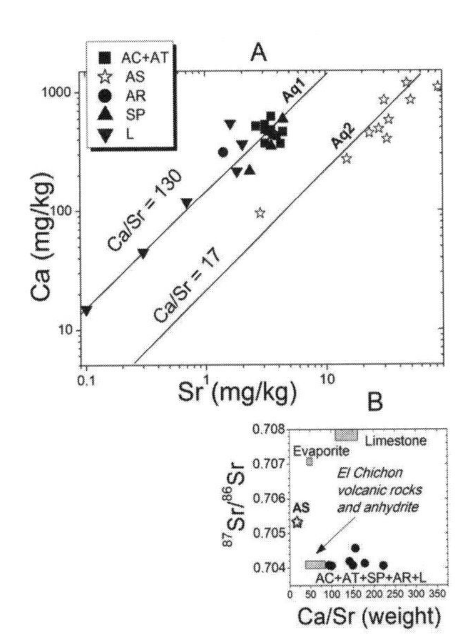

Figure 3. A. Mixing plot of 'Ca-Sr' ratios for waters from El Chichón. B. $^{87}Sr/^{86}Sr$ and Ca/Sr ratios for waters and rocks from El Chichón. Rock data taken from Luhr et al. (1984), Tepley et al. (2000), and Peterman et al. (1970).

waters. Unlike to previous plots, AS waters and other thermal springs display a remarkably similar enrichment pattern; however, it can be seen that Sr in AS springs is a half log unit more enriched while Mg is almost one order of magnitude less enriched.

4.2 *Concentration and isotopic composition of Sr*

The $^{87}Sr/^{86}Sr$ ratios of dissolved Sr in natural waters generally match host rock $^{87}Sr/^{86}Sr$ ratios, therefore waters originating from different aquifers are supposed to have distinct isotopic Sr ratios (Ishikawa et al. 2007).

Data on $^{87}Sr/^{86}Sr$ of El Chichón waters strongly support the existence of two thermal aquifers (Fig. 3b) and provide unambiguous evidence for the origin of wall rocks for both aquifers. Aq. 1 is represented by waters with lower Sr (0.2–4.4 mg/l) and $^{87}Sr/^{86}Sr$ between 0.7041–0.7042. Waters from the Aq. 2 show higher concentrations of Sr (14.8–83.4 mg/l), and $^{87}Sr/^{86}Sr$ around 0.7053 (Fig. 3b). The $^{87}Sr/^{86}Sr$ ratios of the Aq. 1 closely fit the El Chichón trachyandesite $^{87}Sr/^{86}Sr$ ratio which was estimated by Tepley et al. (2000) to be around 0.7041. Values for the Aq. 2 fall between volcanic values and value for the limestone basement (0.7085 ± 0.0015) and/or Late Jurassic evaporites (0.7068, Peterman et al. 1970).

Aq. 1 is thus thought to be shallow and composed of volcanic rocks while Aq. 2 is probably deeper, located in carbonate and evaporite rocks. When deep fluid rises to the surface, its $^{87}Sr/^{86}Sr$ ratio may be lowered due to Sr dissolution from volcanic rocks.

4.3 *Rare Earth Elements (REE)*

The REE patterns of all sampled waters show a significant depletion compared to chondrite (Fig. 5). Neutral waters have the least REE concentrations. This is consistent with the well-known relationship between REE mobility and pH; low pH values enhance mineral solubility and reduce adsorption phenomena (Wood, 2006).

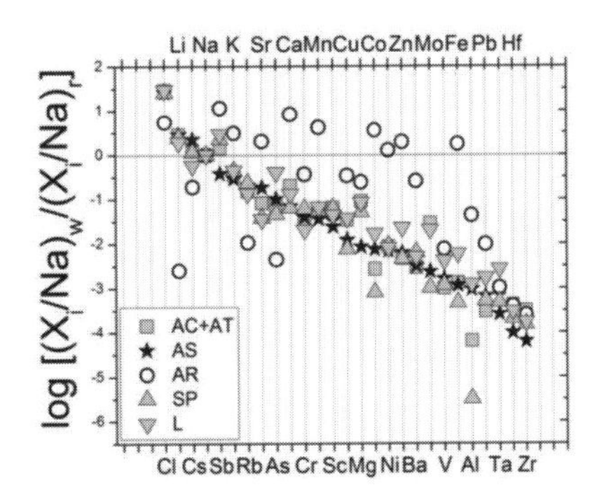

Figure 4. Na-normalized enrichment factors. X_i: concentration of I. I is an element present in both the water (w) and the rock (r). Na is used as reference for normalization due to its conservative behavior.

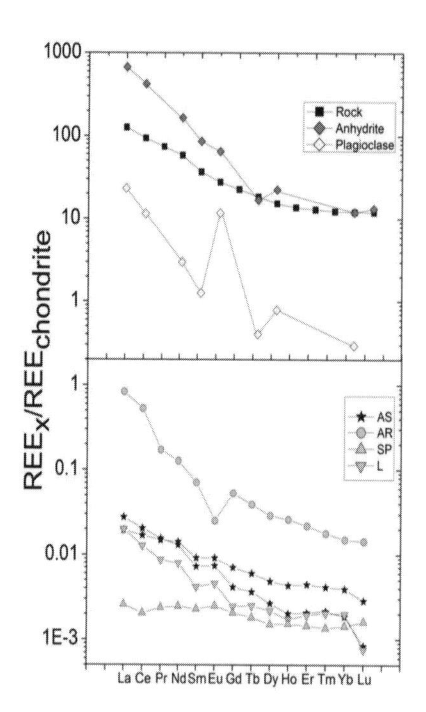

Figure 5. REE in water, El Chichon trachyandesite and phenocrysts of plagioclase and anhydrite normalized to chondrite.

Low pH observed in AS and lake water is attained as a result of H_2S oxidation close to the surface. As REE contents depend on pH, the REE profiles thus reflect water-rock interaction near the surface. No distinction can thus be made between Aq. 1 and Aq. 2 using the REE distribution.

Figure 5 shows our data on REE in El Chichón waters normalized to chondrite, as well as those of the bulk rock and phenocrysts of the 1982 eruption. Superficial and slightly acidic AR waters are more enriched, with a total REE concentration of 640 μg/l and LREE enrichment, while the AS springs and the lake water show lower REE concentrations that parallel concentrations in the El Chichón rock (Fig. 5). The AR profile is a result of the leaching of anhydrite from pyroclastic deposits as suggested by the similarity of the AR and anhydrite REE patterns.

AS and lake REE compositions seem to result from leaching of El Chichón volcanic rock. The lake profile shows a weak Eu enrichment that may be attributed to plagioclase dissolution. SP springs are characterized by a flat profile and some depletion in LREE comparing with the rock samples.

5 CONCLUSIONS

The existence of two distinct aquifers within the El Chichón volcano-hydrothermal system is clearly supported by two distinct trends on mixing plots for all thermal waters at El Chichón. All thermal springs located on SE to W volcano flanks are fed from a shallow aquifer Aq. 1. Springs located on the NW slopes belong to another, deep aquifer Aq. 2.

Ca/Sr ratios confirm the distinction between both aquifers but are not directly related to Ca/Sr ratios in the wall rocks. However, $^{87}Sr/^{86}Sr$ ratio data suggest that the shallower aquifer is composed of volcanic rocks while the deeper aquifer is composed of sedimentary rocks.

The REE distributions prevent confirmation of this evidence due to very low REE concentrations in neutral waters. Anyway, REE patterns of superficial AR waters clearly reflect the dissolution of anhydrite present in the last eruption pyroclastic deposits.

Better understanding of processes governing Ca/Sr ratios during water-rock interactions is also required. Dissolution of minerals such as celestite and strontianite in the sedimentary layers could control Ca/Sr ratios.

REFERENCES

Duffield, W.A., Tilling, R.I. & Cañul, R. 1984. Geology of El Chichón Volcano, Chiapas, Mexico. *J. Volcanol. Geotherm. Res*. 20, 117–132.

Giggenbach, W.F. 1988. Geothermal solute equilibria. Derivation of Na–K–Mg–Ca geoindicators. *Geochim. Cosmochim. Acta* 52, 2749–2765.

Ishikawa, H., Ohba, T. & Fujimaki, H. 2007. Sr isotope diversity of hot spring and volcanic lake waters from Zao volcano, Japan. *J. Volcanol. Geotherm. Res.* 166, 7–16.

Jones, D.A., Layer, P.W. & Newberry, R.J. 2008. A 3100-year history of argon isotopic and compositional variation at El Chichón volcano. *J. Volcanol. Geotherm. Res.* 175, 427–443.

Luhr, J.F., Carmichael, I.S.E. & Varekamp, J.C. 1984. The 1982 eruptions of El Chichón volcano, Chiapas, Mexico: mineralogy and petrology of the anhydrite-bearing pumices. *J. Volcanol. Geotherm. Res.* 23, 69–108.

Peterman, Z.E., Hedge C.E. & Tourtelot H.A. 1970. Isotopic composition of Sr in seawater throughout Phanerozoic time. *Geochim. Cosmochim. Acta* 34, 105–120.

Taran, Y., Fischer, T.P., Pokrovsky, B., Sano, Y., Armienta, M.A. & Macías, J.L. 1998. Geochemistry of the volcano–hydrothermal system of El Chichón Volcano, Chiapas, Mexico. *Bull. Volcanol.* 59, 436–449.

Taran, Y., Rouwet, D., Inguaggiato, S. & Aiuppa, A. 2008. Major and trace element geochemistry of neutral and acidic thermal springs at El Chichón volcano, Mexico. Implications for monitoring of the volcanic activity. *J. Volcanol. Geotherm. Res.* 178, 224–236.

Taran, Y.A. & Peiffer, L. 2009. Hydrology, hydrochemistry and geothermal potential of El Chichón volcano-hydrothermal system, Mexico. *Geothermics* 38, 370–378.

Tepley III, F.J., Davidson, J.P., Tilling, R.I. & Arth, J.G. 2000. Magma mixing, recharge and eruption histories recorded in plagioclase phenocrysts from El Chichon volcano, Mexico. *J. Petrol.* 41, 1397–1411.

Wood, S.A. 2006. Rare earth element systematics of acidic geothermal waters from the Taupo Volcanic Zone, New Zealand. *J. Geochem. Explor.* 89, 424–42.

Water-Rock Interaction – Birkle & Torres-Alvarado (eds)
© 2010 Taylor & Francis Group, London, ISBN 978-0-415-60426-0

Pliocene to present day evolution of the Larderello-Travale geothermal system, Italy

M. Puxeddu & A. Dini
CNR Istituto di Geoscienze e Georisorse, Pisa, Italy

ABSTRACT: The heat source of the Larderello-Travale geothermal system is a composite batholith undergoing a slow monotonic cooling since for ca. 4 Ma. Emplacement ages are 3.8–1.2 Ma. Three evolutionary stages were recognized: 1) early magmatic, 2) decompression, 3) late hydrothermal stage. Two geothermal systems intermittently coexisted for 3.8 Ma: a deeper one with T > 450°C, lithostatic pressures and high salinity brines, and a shallower one with T < 400°C, hydrostatic pressures and low salinity fluids. Cyclic rupture of the impermeable barrier between the two systems is documented. Stable isotopic data reveal prevalence of a magmatic component in the deeper system and of a meteoric one in the shallower system.

1 INTRODUCTION

A buried composite batholith emplaced between 3.8 and 1.2 Ma ago at shallow depth (3–6 km) is the heat source of the long lasting Larderello-Travale Geothermal System (LTGS). Teleseismic and tomographic studies (see Gianelli et al. 1997 and references therein) reveal the presence of a low velocity body extending down to at least 40 km depth (Moho depth: 20–25 km). Rough estimates of the volumes are 20,000 km^3 for the crustal part, overlying a soft, partially molten astenospheric mantle region (Cavarretta & Puxeddu 2001). The temperature anomaly is expected to last roughly another 30 million years on the basis of geological-geophysical modeling (Mongelli et al. 1998).

A first generation of granites named LAR1 (Dini et al. 2005), yielded Rb-Sr and ^{40}Ar-^{39}Ar ages between 3.8 ± 0.1 and 2.26 ± 0.06 Ma, a second gen-eration, LAR2, ages between 2.25 ± 0.04 and 1.23 ± 0.13 Ma (Dini et al. 2005, and related bibliography). All the granites are two-mica peraluminous (ASI in the range 1.12–1.26, normative corundum). LAR1 are monzogranites (SiO_2 mainly in the range 68–70 wt%) while LAR2 are tourmaline-rich syeno- to monzogranites (SiO_2: 74–79 wt%). Primary magmatic muscovite is stable at only 110–120 MPa owing to the dramatic lowering of the solidus temperatures for the Larderello granites due to their extremely high F, Li, and B contents. The occurrence of blue corundum and sanidine, with the assemblage tourmaline-biotite-calcic plagioclase, the stable isotope geothermometers (Qtz-Ms and Qtz-Bt) and the homogenization temperatures of fluid inclusions all indicate the attainment of temperatures higher than 560°–570°C and possibly up to 643°C in the innermost part of the contact aureole (Cavarretta & Puxeddu 1990, Cathelineau et al. 1994, Petrucci et al. 1994 and references therein). All radiometric, fluid inclusion, petrographic and geophysical data suggest a long life span of 4 Ma and a broadly slow monotonic cooling (Villa & Puxeddu 1994).

2 STAGES OF EVOLUTION

Fluid inclusion studies revealed a very complex history of the LTGS (Cathelineau et al. 1994). Three stages were identified:

1. Magmatic stage characterized by temperatures of 450°–600°C, lithostatic pressures, high salinity brines containing LiCl, NaCl, KCl, $CaCl_2$ (up to 50–60 wt% of NaCl equivalent and 30 wt% LiCl equivalent in Li-rich brines), and carbonic fluids produced by decarbonation reactions at temperatures of 550°C within carbonate-bearing layers of the metamorphic basement. The mineral assemblages consist mainly of biotite-tourmaline-calcic plagioclase.
2. Decompression stage, with temperatures of 400°–500°C, pressures intermediate between lithostatic and hydrostatic and the occurrence of Ca-Mg brines.
3. Late to present day stage, with temperatures up to 350°–400°C, hydrostatic to vaporstatic pressures, low salinity fluids and mixing between magmatic and meteoric waters.

The hydrothermal/metamorphic key minerals are K-feldspar (adularia), epidote, amphibole of the tremolite-actinolite solid solution series and albite.

3 COEXISTENCE OF LITHOSTATIC AND HYDROSTATIC SYSTEMS

Since initiation, LTGS has been characterized by the coexistence of two hydrothermal systems separated by an impermeable barrier:

1. a deeper hydrothermal-magmatic system with temperatures higher than 450°C, lithostatic pressures, F-, Li-, B-rich magmatic fluids, high salinity magmatic brines.
2. a shallower system with temperatures lower than 350°–400°C, hydrostatic pressures, low salinity waters with a dominant meteoric component.

Similar coexistence of an external meteoric hydrothermal system with hydrostatic pressure and a magmatic hydrothermal system with lithostatic pressures has been proposed for several types of magmatic-related ore deposits (e.g. Taylor 1974).

Cyclic rupture of the dividing impermeable barrier, mainly triggered by seismic-tectonic processes, periodically connected the two systems creating dramatic pressure drops from lithostatic to hydrostatic in the deeper system with a consequent hydrogen leakage and strong increase of oxygen fugacity. Then the uppermost part of the deeper system seals up again owing to the hydrothermal deposition in the newly created fractures (see Petrucci et al. 1994) while the fluid supply from the composite batholith recreates the original lithostatic pressure in the deeper system. This is similar to the cyclic opening and sealing of a fault-valve discussed by Sibson (1992).

The cyclic rupture of the impermeable barrier is supported in Larderello by two microprobe traverses from base to tip of acicular tourmaline crystals found in a drill core from the top of the barrier at 2389 m depth below ground level in the well San Pompeo 2 (Cavarretta & Puxeddu 1990, see Fig. 1 for the location of the wells quoted in this paper). These traverses show 3 to 4 peaks where Fe^{3+} replaces Al^{3+} (povondraite is replacing the schorldravite component of tourmaline) which indicate episodes of sudden hydrogen leakage and fO_2 increase associated with cyclic rupture of the barrier between the two systems. Similar cyclic hydrofracturing and tourmaline precipitation has been documented in a fossil magmatic-hydrothermal system at Elba Island (Dini et al. 2008).

The present day survival of the dichotomy—deep lithostatic-shallow hydrostatic systems—is demonstrated by the drilling history of the well San Pompeo 2. This well, during drilling, underwent

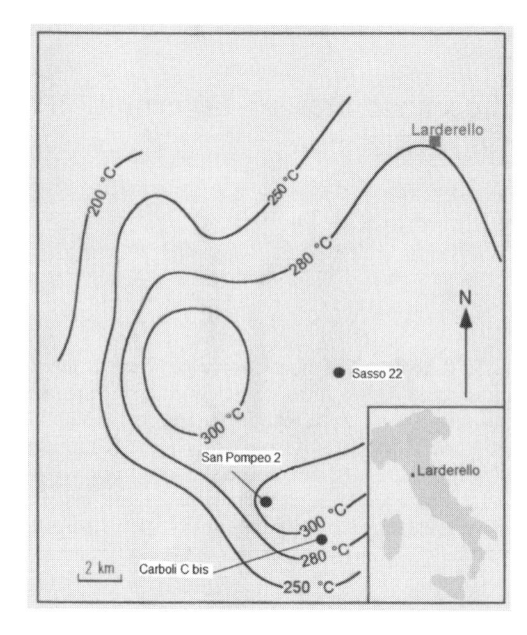

Figure 1. Location of wells discussed in the text. The isotherms at 2000 m depth below the ground level are also shown.

two dramatic blowouts episodes with outpouring of remarkable amounts of rock fragments from the well bottom. The estimated temperatures were at least 450°C and the pressures largely exceeded hydrostatic values.

Further indication of the relationship between the two systems is provided by oxygen isotopes (Petrucci et al. 1994). $\delta^{18}O$ values obtained on the most retentive minerals, i.e. secondary quartz (+12‰ to + 14.7‰ SMOW) and tourmaline (+ 9‰) indicates deposition from an ^{18}O-rich fluid. Muscovite yielded typical magmatic to metamorphic values of +8.2‰ and +10.3‰ while biotite shows two values, +1.5‰ and +1.9‰ consistent with an interaction with ^{18}O-depleted meteoric waters. In contrast, a value of +10.8‰ with a Qtz-Bt isotopic temperature of 643°C was obtained for biotite SA 22 4028 (bottom of Sasso 22, 4028 m b.g.l.) that still preserves the magmatic-metamorphic values of the contact aureole. As a reference, Panichi et al. (1974), for spring waters and well-head samples of fluids from the productive wells, found $\delta^{18}O$ values in the range +2 to −7‰ falling between magmatic values (>+5.5‰) and pure meteoric waters lower than −6‰ (see Petrucci et al. 1994).

This intermediate position more displaced towards the meteoric field demonstrates that, now and since the initiation of LTGS, the geothermal fluids derive from a mixing between magmatic fluids ascending from the batholith and the dominant meteoric waters percolating downward in the geothermal reservoirs.

Another important parameter for distinguishing between magmatic and meteoric fluids is the fluorine content in sheet silicates. During a late magmatic stage, the monzogranite underwent a strong albitisation process and a contemporaneous remarkable fluorine enrichment in the residual hydrothermal-magmatic fluids (see Cavarretta & Puxeddu 2001) testified to by extremely high F contents in primary biotite (2–3%) and muscovite (up to 2.93%) in a syeno-monzogranite crossed by the well Carboli C bis (4.2–4.3 km depth). 2.93% is the highest measured fluorine-concentration in muscovite published in the world from acidic igneous rocks. The following hydrothermal stage locally caused intense sericitisation of primary feldspars and strong chloritisation of primary biotite. All the microprobe analyses reveal the complete lack of fluorine in sericite and chlorite that are the most abundant hydrothermal alteration minerals in the depth range 0–3 km, mainly affected by circulation of meteoric fluids. This behaviour indicates that fluorine content of sheet silicates is an excellent indicator of the magmatic (high F contents) versus the meteoric nature (low to absent fluorine) of the very late fluids that affected the monzogranite and the surrounding regional to contact metamorphic rocks.

4 HYDROTHERMAL EVENTS REVEALED BY MUSCOVITES

The oldest evidence of hydrothermal activity is given by an inferred ^{40}Ar-^{39}Ar age of 1.87–2.02 Ma for a hydrothermal component of muscovites from

Figure 2. Photomicrograph of syeno-monzogranite from well Carboli C bis. Subhedral blades of primary muscovite (Ms) and biotite (Bt) on the left; late muscovite (Ms-H) and 5–20 μm-sized fluorite (Fl) crystals filling fractures, cleavages and cavities within plagioclase on the right.

well Radicondoli 26 at 4120 m depth (Villa et al. 1997). Two generations of hydrothermal muscovites were identified by Villa et al. (2001) on the 1.3 Ma old granite from well Carboli Cbis: one formed at 0.5 Ma, the second between 0.4 and 0.25 Ma. A gradual increase in Cl and Fe contents was observed from early to late hydrothermal muscovites. A hydrothermal event at 0.6–0.7 Ma was revealed by an ^{40}Ar-^{39}Ar study on feldspars from wells in the eastern part of the LTGS (Villa et al. 2006).

The distinction between the two muscovite generations is evident in petrographic thin sections: primary magmatic muscovite is subhedral, while the secondary hydrothermal muscovite is anhedral and forms an anastomosed network infiltrating along fractures, cleavages and grain boundaries (see Fig. 2).

5 CONCLUSIONS

The Larderello-Travale Geothermal System is characterized by:

1. The occurrence of a peraluminous, composite granitic batholith (estimated volume of 20,000 km^3) fed by multiple magma pulses during the last 4 Ma that sustained the long-lasting heat anomaly.
2. Cyclic rupture of the impermeable barrier between a deeper hydrothermal system with lithostatic pressures and a shallower one with hydrostatic pressures explains the hybrid nature of the present day and past geothermal fluids. The $\delta^{18}O$ values of high temperature hydrothermal minerals (Bt, Ms, Amp) vary from +1.5‰ to +10.8‰, those of the exploited gothermal fluids vary from +2 to −7‰, Both are intermediate between magmatic (>+5.5‰) and meteoric values (<−6‰) calculated considering mineral-water isotope fractionation.
3. The fluorine content of sheet silicates is a further parameter useful for distinguishing interactions with magmatic or meteoric fluids. Extremely high fluorine contents characterize magmatic muscovites and biotites. No fluorine was found in late sericite and chlorite produced by interaction with meteoric fluids.

REFERENCES

Cathelineau, M., Marignac, C., Boiron, M.C., Gianelli, G. & Puxeddu, M. 1994. Evidence for Li-rich brines and early magmatic fluid-rock interaction in the Larderello geothermal system. *Geochimica et Cosmochimica Acta* 58: 1083–1099.

Cavarretta, G. & Puxeddu, M. 1990. Schorl-dravite-ferridravite tourmalines deposited by hydrothermal magmatic fluids during early evolution of the Larderello geothermal field (Italy). *Economic Geology* 85: 1236–1251.

Cavarretta, G. & Puxeddu, M. 2001. Two-mica F-Li-B-rich monzogranite apophysis of the Larderello Batholith cored from 3.5 km depth. *Neues Jahrbuch für Mineralogie Abhandlungen* 177: 77–112.

Dini, A., Gianelli, G., Puxeddu, M. & Ruggieri G. 2005. Origin and evolution of Pliocene–Pleistocene granites from the Larderello geothermal field (Tuscan Magmatic Province, Italy). *Lithos* 81: 1–31.

Dini, A., Mazzarini, F., Musumeci, G. & Rocchi, S. 2008. Multiple hydro-fracturing by boron-rich fluids in the Late Miocene contact aureole of eastern Elba Island (Tuscany, Italy). *Terra Nova* 20: 318–326.

Gianelli, G., Manzella, A. & Puxeddu, M. 1997. Crustal models of the geothermal areas of southern Tuscany (Italy). *Tectonophysics* 281: 221–239.

Mongelli, F., Palumbo, F., Puxeddu, M., Villa, I.M. & Zito, G. 1998. Interpretation of the geothermal anomaly of Larderello, Italy. *Memorie della Società Geologica Italiana* 52: 305–318.

Panichi, C., Celati, R., Noto, P., Squarci, P., Taffi, L. & Tongiorgi, E. 1974. Oxygen and hydrogen isotope studies of the Larderello (Italy) geothermal system. In *Isotope Techniques in Groundwater Hydrology,* IAEA, Wien. vol. II: 4–28.

Petrucci, E., Gianelli, G., Puxeddu, M. & Iacumin, P. 1994. An oxygen isotope study of silicates in the Larderello geothermal field, Italy. *Geothermics* 23: 327–337.

Sibson, R.H. 1992. Implications of fault-valve behaviour for rupture nucleation and recurrence. *Tectonophysics* 211: 283–293.

Taylor, H.P., Jr. 1974. The application of oxygen and hydrogen isotope studies to problems of hydrothermal alteration and ore deposition. *Economic Geology* 69: 843–883.

Villa, I.M. & Puxeddu, M. 1994. Geochronology of the Larderello geothermal field: new data and the "closure temperature" issue. *Contributions to Mineralogy and Petrology* 115: 415–426.

Villa, I.M., Ruggieri, G. & Puxeddu, M. 1997. Petrological and geochronological discrimination of the two white-mica generations in a granite cored from the Larderello—Travale geothermal field (Italy). *European Journal of Mineralogy* 9: 563–568.

Villa, I.M., Ruggieri, G. & Puxeddu, M. 2001. Geochronology of magmatic and hydrothermal micas from the Larderello Geothermal Field. *Proc. 10th Intern. Symp. on Water-Rock Interaction, Villasimius, Cagliari 10–16 June 2001*. Rotterdam: Balkema, vol. 2: 1589–1592.

Villa, I.M., Ruggieri, G., Puxeddu, M. & Bertini, G. 2006. Geochronology and isotope transport systematics in a subsurface granite from the Larderello-Travale geothermal system (Italy). *Journal of Volcanology and Geothermal Research* 152: 20–50.

Water-Rock Interaction – Birkle & Torres-Alvarado (eds)
© 2010 Taylor & Francis Group, London, ISBN 978-0-415-60426-0

Evaluation of Tuwa geothermal system through water-rock interaction experiment

H.K. Singh & D. Chandrasekharam
Department of Earth Sciences, Indian Institute of Technology Bombay, Mumbai, India

ABSTRACT: The chemical evolution of the Tuwa thermal springs, together with the experimental results on the granite-water interaction at 100°C, indicate that Cenozoic sediments are the main contributors of chloride to the circulating thermal water. Chloride contribution by the granites seems to be slight. The Tuwa geothermal province is located within the triple junction which is the loci of Deccan volcanism. Granites and deep tectonic structures associated with the Deccan volcanism are the main contributors of high geothermal gradient and high heat flow in this province.

1 INTRODUCTION

A group of thermal springs, known as Tuwa thermal springs, located in the Panchmahal district of Gujarat state, was selected for this study. These springs are situated in the Cambay rift basin and forms part of Gujarat and Rajasthan geothermal provinces (Fig. 1). The Cambay basin is the foci of major alkaline magmatism (Sheth & Chandrasekharam 1997). Granite intrusives, such as Godhra granite, outcrop within the basin near Tuwa (Gopalan et al. 1979). The surface temperature of the thermal springs vary between 40 and 70°C, the thermal springs can be categorized into low enthalpy geothermal system. This study aims (i) to elucidate the geochemical evolution of the Tuwa thermal springs by using experimental leaching test of water rock interaction and geochemical signatures of the surface waters, groundwaters and springs and (ii) to estimate the reservoir temperatures by using geochemical thermometers.

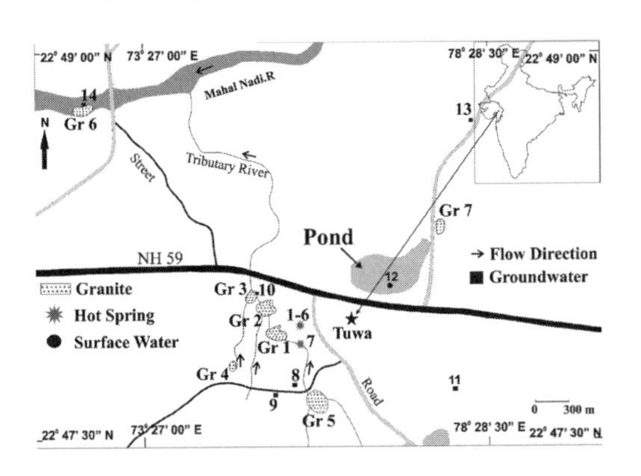

Figure 1. Location map of the Tuwa thermal spring.

2 METHOD OF STUDY

Water samples from seven thermal springs, three surface and four groundwater samples, were collected from the study area. Additionally seven granite samples from different locations were also collected. Water samples were analyzed for major cations, anions and total silica concentration. Cations and silica were analyzed by ICP-AES. Sulfate concentration was measured by spectrophotometer, alkalinity and chloride by titration (Table 1). The granite samples were crushed to about <1 mm. The experimental leaching tests of water-rock interaction were carried out in a glass chamber with fluid/solid ratio of 10:1 at 100°C. Rain water were utilized in the experiment as leaching fluids.

Table 1. Chemical analyses of water samples (in epm).

Sample	°C	pH	Na	K	Ca	Mg	Cl	HCO_3	SO_4
1	45	8	24.4	1.1	20.4	1.6	56.5	1.7	5.2
2	62	8	23.7	1.1	18.8	1.2	54.1	1.7	4.7
3	48	8	22.9	1.2	22.4	0.8	54.8	1.6	4.9
4	39	8	24.3	1.2	18.8	1.6	55.5	2.0	5.3
5	40	8	24.0	1.1	20.4	0.4	56.5	1.7	4.7
6	44	8	23.8	1.1	20.4	3.6	56.2	1.6	4.8
7	61	7	20.8	0.9	15.2	3.2	49.9	2.4	3.7
8	24	7	4.0	0.1	4.0	4.4	2.1	3.8	1.4
9	28	7	4.1	0.06	6.8	3.6	0.3	5.1	1.1
10	27	7	11.4	0.4	7.2	3.2	16.2	4.3	1.7
11	24	7	0.8	0.08	2.4	0.8	1.1	2.8	1.2
12	25	8	2.4	0.07	1.6	1.6	2.1	1.8	1.1
13	22	7	2.6	0.1	5.2	2.0	3.1	3.4	1.5
14	22	7	1.9	0.1	2.8	2.4	2.1	2.9	1.2

Sample no. 1–7: thermal waters; 8, 9, 11, and 13: Groundwaters; 10, 12, and 14: Surface waters.

Table 2. Geochemical data of experimental leaching test.

	A*	B**
pH	6.1	8.50
Na	0.03	1.37
K	0.005	0.52
Ca	0.15	0.68
Mg	0.078	0.00
Cl	0.09	1.30
SO_4	0.07	0.22
HCO_3	0.16	1.24

*A: meteoric water; **B: interacted water; Concentration in epm.

The experiments were conducted over a period of 30 days. Water sample were collected in an interval of every 10 days. Partial geochemical analysis of the reacted fluid is given in Table 2.

3 RESULT AND DISCUSSION

The major ion composition of the Tuwa thermal springs (Table 1), plotted in the Piper diagram (Fig. 2), indicate that the thermal springs are (Na-K/Ca)-Cl type dominated by Na and K over Ca and Cl > HCO_3 (Table 1). The geothermal system appear to be in chemical equilibrium with the rocks through which they circulate. This is apparent from the lack of temporal chemical variation in these springs when the data on these thermal springs are compared with that published earlier (Minissale et al. 2003).

The surface water samples also show similar chemical characteristics with exception of sample 10 whose Na, and Cl concentrations are higher than the other surface and groundwaters but less than the thermal springs. This sample was collected at a place where the thermal water springs discharge in to the river.

In order to compare the thermal springs issuing through similar geological setting, thermal water samples from Tattapani geothermal province (Chandrasekharam & Antu 1995) were plotted in Fig. 2. The Tattapani geothermal province lies along two faults parallel to the main E-W trending Narmada-Son lineament in the north and the Tapi lineament in the south. The thermal springs flow through granite gneiss of Precambrian age (Chandrasekharam & Antu 1995). Although the Tuwa thermal springs issue through granites, the chloride concentrations are much higher in comparison to the Tattapani thermal waters. The granites play an important role for the source of high chloride concentrations in aqueous system in Tattapani (Chandrasekharam & Antu 1995), it cannot

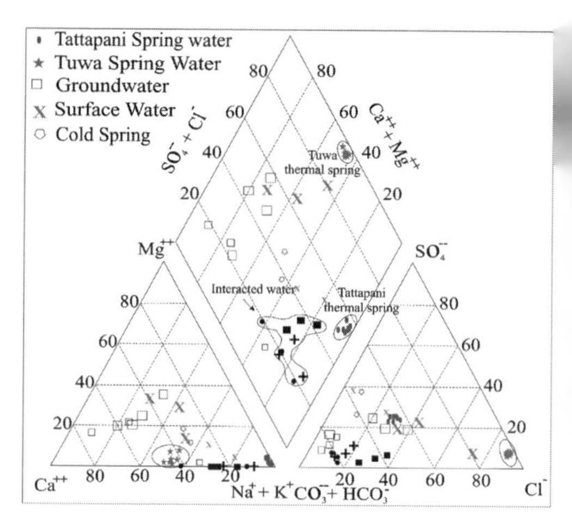

Figure 2. Piper diagram (Piper, 1944) showing the geochemical variation in different water types from Tuwa geothermal system. Tattapani thermal springs are from Chandrasekharam and Antu (1995) and Minissale et al. (2003).

be applicable for the Tuwa geothermal systems as evident from Fig. 2.

In experimental leaching tests, the water samples derived from the granites fall in the Na-HCO_3 field (close to Tattapani geothermal province) and can not support the view that granites are responsible for chloride concentrations in the Tuwa geothermal system (Table 2; Fig. 2).

Granite-water experimental test were conducted earlier to understand anomalous chloride concentration in hydrothermal solutions (Savage et al. 1985, 1987). In order to understand the source of high chloride concentration in these springs, subsurface lithological sequence of the Cambay basin, together with the leaching test results (Table 2), is considered which is shown in Fig. 3. The lithological sequence includes sandstone, limestone, shale and conglomerate representing sedimentation during Late Eocene to Paleocene.

The entire ~3.5 km thick sedimentary basin was deposited within continental margin basin environment (Biswas 1987). These sediments have locked old marine salts that are released to the circulating fluids. High salinity is common phenomena in all the shallow and deep aquifers located in the sedimentary rocks of the Cambay (Gupta & Deshpande 2003).

Thus, rain waters, as it circulated through the sedimentary sequence, becomes saline and enter the granites at depths >2 to 3 km before emerging at the surface as thermal waters. The high geothermal gradient (~70°C/km) and high heat flow (~93 mW/m^2) due to the Deccan volcanism as well as deep seated faults like the Narmada and

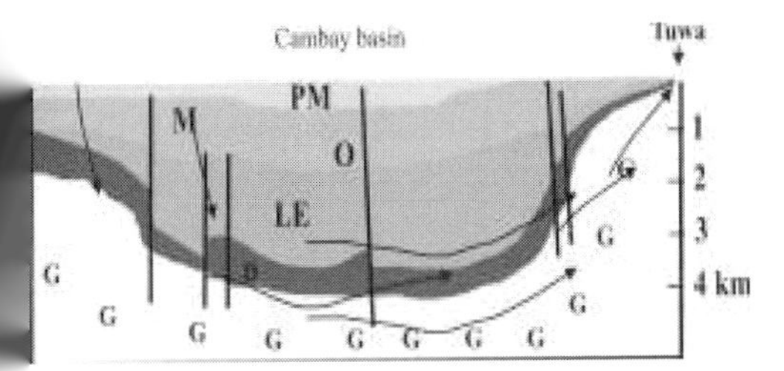

Figure 3. Subsurface structure of Cambay basin. G: Granite; D: Deccan basalt flows; LE: Late Eocene; O: Oligocene; M: Miocene; PM: Post Miocene-Early Pleistocene Dark thick lines: faults; Arrow indicate circulating water (modified from Chandrasekharam & Bundschuh 2008).

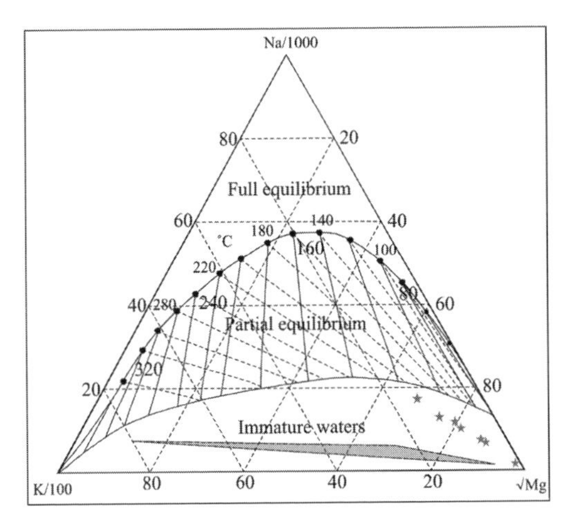

Figure 4. Na-K-√Mg diagram showing the position of Tuwa Thermal springs.

Cambay, and the radioactive decay heat from the Godhra granites is very typical for this region. The main Cambay rift and associated satellite faults are the main channel for the circulating thermal fluids.

The Tuwa thermal waters define an array towards the partial equilibrated field in the Giggenbach diagram (1988), indicating a steady state of flow of the circulating water, as indicated by the lack of temporal variation in their chemical signature, within the granite basement. The thermal waters define an array between the groundwater field and fully equilibrated field indicating mixing of near surface cold water before emerging to the surface. The samples fall between 140 (K-√Mg) and 200°C (Na-K) thermal tie lines indicating equilibrium temperature between this range (Fig. 4).

The Na-K-√Mg triangular diagram plot (Fig. 4) indicate that the thermal spring in the Tuwa region are fed by the geothermal aquifer whose temperature range from 150–160°C.

Thus the Tuwa geothermal waters is initially driven by the Cambay sedimentary sequence and heated by the Godhra granites.

REFERENCES

Biswas, S.K. 1987. Regional tectonic framework, structure and evolution of the western marginal basins of India. *Tectonophysics* 135: 307–327.

Chandrasekharam, D. & Antu, M.C. 1995. Geochemistry of tattapani thermal springs, Madhya Pradesh, India-field and experimental investigations. *Geothermics*. 24(4): 553–559.

Chandrasekharam, D. & Bundschuh, J. 2008. "Low Enthalpy Geothermal Resources for Power generation" *Taylor and Francis Pub., U.K.* 169.

Giggenbach, W.F. 1988. Geothermal solute equilibria. Derivation of Na-K-Mg-Ca geoindicators. *Geochemical et Cosmochimica Act.* 52: 2749–2765.

Gopalan, K. Trivedi, J.R. Merh, S.S. Patel, P.P. Patel, S.G. 1979. Rb-Sr age of Godhra and related granites, Gujarat, India. *Proceedings. Indian Academy of Sciences, 88 A, Part II, Number 1, March 1979.* 7–17.

Gupta, S.K. & Deshpande, R.D. 2003. Origin of groundwater helium and temperature anomalies in the Cambay region of Gujarat, India. *Chemical Geology*, 198(1–2): 33–46.

Minissale, A. Chandrasekharam, D. Vaselli, O. Magro, G. Tassi, F. Pansini, G.L. Bhramhabut, A. 2003. Geochemistry, geothermics and relationship to active tectonics of Gujarat and Rajasthan thermal discharges, India. *Journal of Volcanology and Geothermal Research.* 127: 19–37.

Minissale, A., Vaselli, O., Chandrasekharam, D., Magro, G., Tassi, F., Casiglia, A., 2000. Origin and evolution of 'intracratonic' thermal fluids from central-western peninsular India. *Earth Planet. Sci. Lett.* 181: 377–397.

Piper, A.M. 1944. A graphic procedure in the geochemical interpretation of water-analyses. *Am. Geophy. Union. Trans.* 25: 914–923.

Savage, D. & Mark, M. 1985. The origin of saline groundwater in granitic rocks: Evidence from hydrothermal experiments. *Materials Research Society Symposium Proceeding.* 50.

Savage, D. Mark, R.C. Antoni E. Milodowski & Ian George 1987. Hydrothermal alteration of granite by meteoric fluid: an example for the Carnmen ellies granite, United Kingdom. *Contrib. Mineral. Petrol.* 96: 391–405.

Sheth, H.C. & Chandrsekharam, D. 1997. Early alkaline magmatism in the Deccan traps: Implications for plume incubation and lithospheric rifting. *Physics of the Earth and Planetary Interiors.* 104(4): 371–376.

Water-Rock Interaction – Birkle & Torres-Alvarado (eds)
© 2010 Taylor & Francis Group, London, ISBN 978-0-415-60426-0

The origin of hot springs at Hengjing, south Jiangxi, China

W.-J. Sun & D.-L. Dong
College of Geoscience and Surveying Engineering, China University of Mining and Technology, Beijing, China

J. Jiao
College of Geological Science and Engineering, Shandong University of Science and Technology, Qingdao, China

ABSTRACT: Around the Hengjing area, south Jiangxi Province of China, 14 samples from hot springs and cold springs were collected and analyzed. The isotopic compositions of hydrogen and oxygen for hot springs are in accord with the local meteoric line, which implies that the geothermal waters in this area are of meteoric origin. It is also concluded that the hot springs came from the deep circulation of local precipitation. Meanwhile, the measurements include the composition of gas and helium isotope and carbon isotopic abundance of carbon dioxide of 4 gas samples from hot springs. The origin of gases from hot springs in this area seems to be the mantle CO_2. The $^3He/^4He$ ratios of gases show that the helium probably partly from the mantle source, and the percentage is at 16.8% to 26.2%.

1 INTRODUCTION

The de-gassing of the Earth makes natural gas to escape from different layers of both the crust and the mantle, reaching the earth's surface and subsequently the hydrosphere and atmosphere. The hot springs related to volcano-magma activity, that emerge to the surface through fracture systems as well as earthquakes are direct evidence of the de-gassing processes of the Earth (Liu et al. 2001; Xu et al. 1997; Giggenbach 1980; Giggenbach & Poreda 1993; Farley et al. 1997).

Hot water in the Hengjing area is of the typical middle-low temperature features in Jiangxi province of China. The authors study the isotopic makeup of carbon and helium in order to discuss the origin of the geothermal gas in the area.

2 GEOLOGICAL SETTING

Within the Southern China fold system, the Hengjing area (Figure 1) lies at the south of Wuyi mountain promontory and on the big Xunwu-Ruijin rupture. Rocks in the area consist of Sinian-Cambrian metamorphics, Upper Jurassic volcanic-sedimentary formations and Upper Cretaceous sedimentary formations. Granite of early Yanshan stage was well developed. Two sets of faults occur in the area, striking NE and EW respectively. There are also some associated faults. All the faults have the same characteristics in that they are large scale, steep dipping, and they are still active frequently, as evidenced by several small and medium earthquakes recorded during the last few decades. The heat flow for the southern Jiangxi province has been measured at 62.1 to 79.9 mW/m^2 with an average of 74.10 mW/m^2, which is higher than that (69.79 mW/m^2) for the whole province (Sun et al. 2004, Gao et al. 2006). From this we can see that the faults and the higher heat flow background favor the formation of the hot springs.

3 CHEMICAL AND ISOTOPIC ANALYSIS OF THE SAMPLES

The water and gas samples were collected from the main hot springs, and 8 water samples and 4 gas samples were collected. The results are given in Table 1 and Table 2.

Table 1 shows that the hydrochemical type of geothermal waters in the area is HCO_3-Na, and cold ground waters is HCO_3-Na or HCO_3-Ca.

4 RESULTS AND DISCUSSION

4.1 *Genetic type of hot springs*

According to the hydrogen and oxygen stable isotopes given in Table 1, the equation for the local meteoric water line was calculated approximatively (related coefficient was 0.97):

$$\delta D = 8.33\delta^{18}O + 8.52 (r = 0.97) \quad (1)$$

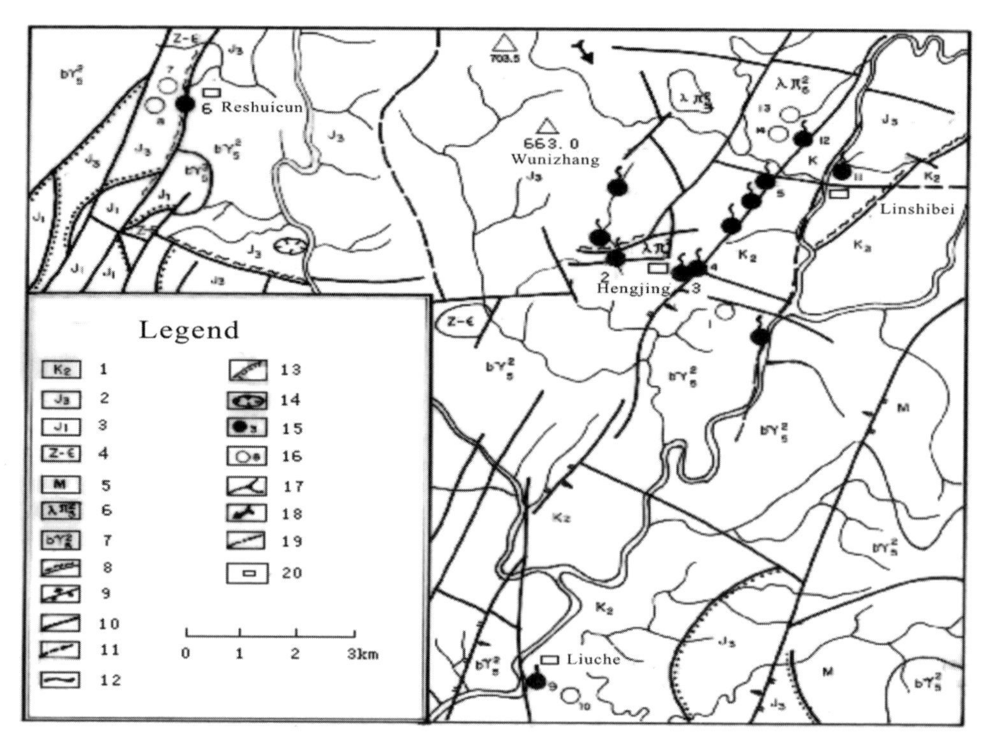

Figure 1. Geological map and sampling locations of hot springs in Hengjing area.
1-Upper Cretaceous sandstone and gravelstone, 2-Upper Jurassic sandstone and gravelstone, 3-lower Jurassic sandstone and gravelstone, 4-metamorphic sandstone Sinian-Cambrian, 5-migmatite, 6-Early Yanshan rhyolite porphyry, 7-Early Yanshan biotite-granite, 8-silicated fracture zone, 9-normal fault, 10-unknown faults, 11-inferred fault, 12-line of geological, 13-discordant contact line, 14-inferred volcanic crater, 15-thermal spring and number, 16-cold spring and number, 17-surface water system, 18-underground water flow direction, 19-surface watershed, 20-viliage.

Table 1. The isotopic compositions of hydrogen and oxygen and the chemical compositions of spring waters in Hengjing area.

Spring no.	Spring type	Water temperature (°C)	pH	δD (‰)	$\delta^{18}O$ (‰)	CO_2 (mg/L)	SiO_2 (mg/L)	Ca^{2+} (mg/L)	Mg^{2+} (mg/L)	K^+ (mg/L)	Na^+ (mg/L)	Cl^- (mg/L)	SO_4^{2-} (mg/L)	HCO_3^- (mg/L)
1	cold	20	–	−55	−7.4	–	–	–	–	–	–	–	–	–
2	hot	25	6.52	−57	−7.2	1263	99	138.5	15.73	84.60	698.6	81.18	325.0	1886.1
3	hot	48	6.67	−53	−6.9	725.3	94	117.1	10.95	71.80	679.1	70.19	764.9	1004.2
4	hot	40	–	−49	−6.7	–	–	–	–	–	–	–	–	–
5	hot	37	6.74	−53	−7.2	22.0	78	29.1	0.09	4.42	93.6	8.51	26.5	217.0
6	hot	73	7.30	−47	−6.1	8.8	81	9.9	0.09	3.09	71.5	8.51	17.0	106.1
7	cold	18	6.72	−46	−6.5	21.0	19	0.1	0.36	3.67	1.3	2.20	0.0	8.0
8	cold	21	6.92	−38	−5.6	16.0	17	0.0	0.01	1.49	2.0	2.20	0.0	25.2
9	hot	27	6.50	−48	−5.9	547.0	43	106.9	10.40	81.97	969.8	50.69	350.0	2253.1
10	cold	22	6.18	−49	−7.0	16.0	15	33.4	4.46	3.40	11.1	21.98	12.0	63.4
11	hot	48	6.77	−48	−5.9	722.0	135	107.8	6.74	43.17	711.2	35.10	300.0	1428.1
12	hot	44	6.63	−49	−7.0	250.0	82	52.3	0.33	6.85	154.4	28.71	115.0	276.8
13	cold	20	–	−48	−6.9	–	–	–	–	–	–	–	–	–
14	cold	19	–	−45	−6.6	–	–	–	–	–	–	–	–	–

Table 2. Chemical and isotopic compositions of geothermal gases in the Hengjing area.

Spring no.	Volume (%)					PDB (‰)		He	
	N_2	CO_2	CH_4	Ar	He	$\delta^{13}C_1$	$\delta^{13}C_{CO_2}$	$^3He/^4He$ (10^{-6})	R/R_a
2	–	99.84	0.04	0.115	0.0047	–27.69	–4.96	1.90 ± 0.06	1.36
3	–	97.96	1.86	0.117	0.0016	–34.86	–4.43	2.74 ± 0.12	1.96
9	2.01	96.47	1.28	0.211	0.0234	–59.31	–5.50	2.95 ± 0.10	2.11
11	1.91	97.92	0.06	0.095	0.0137	–45.30	–5.06	2.30 ± 0.08	1.64

Note: —not measured.

From equation (1) and the fitting line between δD and $\delta^{18}O$ in Hengjing area (Figure 2), the local meteoric water line basically agrees with the Craig (Craig 1961) meteoric water line ($\delta D = 8\delta^{18}O + 10$). The hydrogen and oxygen isotopic data in the area indicate that the data points of hot springs are close to the local meteoric water line. However, the thermal water came from the deep circulation of local precipitation.

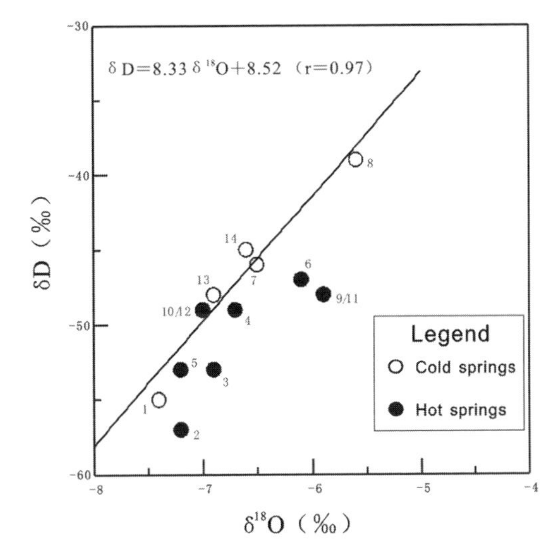

Figure 2. δD vs $\delta^{18}O$ relationship for natural waters in Hengjing area.

4.2 *Gas chemistry and helium and carbon isotopic compositions of hot springs*

4.2.1 *Gas chemical composition*

The gas analytic results indicate that the major gas in the hot springs in Hengjing area is CO_2, constituting 96.47 to 99.84% of the whole gas contents (Table 2). N_2 is very low, from 1.91 to 2.01%, and Ar is from 0.095 to 0.211%.

4.2.2 *Origin of the gases*

4.2.2.1 Origin of CO_2

The origin of CO_2 in different geothermal system can be estimated by using the carbon stable isotope method. In Table 2 it is seen that the $\delta^{13}C$ values for CO_2 in the Hengjing geothermal system are in the range of –5.50‰ to –4.43‰, with an average value of –4.98‰, which is consistent with a magma mantle-derived $\delta^{13}C$ value.

According to the standard, organic and inorganic CO_2 (Dai et al. 1997) can be estimated, the $\delta^{13}C$ values for CO_2 in this area are in the range of –5.50‰ to –4.43‰, so we can say that CO_2 in this study area is inorganic type. Just as we commented in the past, there are two ways to form inorganic CO_2 that comes from the deep Earth. One is metamorphic genesis, and the other is mantle genesis. The value of the carbon isotopes of CO_2 caused by carbonate metamorphic genesis is greater than the values caused by volcano-magma genesis, but usually the two values are in the same range (–1‰ to –8‰). Thus it is hard to distinguish the two kinds of CO_2 solely by the values of $\delta^{13}C$. There is no carbonate rock known in the Hengjing area and earthquakes occur frequently here. More than 30 earthquakes struck Hengjing from 1974 to 1992, reflecting that the geological structure is still active. This geological character shows that the inorganic CO_2 in Hengjing area is not derived from carbonate metamorphic genesis but rises directly from mantle.

4.2.2.2 Origin of helium

The earth's mantle is a region that is rich in ^3He. The ratio of $^3He/^4He$ is high (1.1×10^{-5} to 1.4×10^{-5}), while the crust is a region which is rich in radioactive ^4He. The ratio of $^3He/^4He$ in the crust is lower than the ratio of $^3He/^4He$ in air (1.4×10^{-6} Ra). If the ratio of $^3He/^4He$ is higher than that of air, helium can be regarded as coming from earth mantle (ShangGuan et al. 2000). For the gases of hot springs, the $^3He/^4He$ values are shown in Table 2. These are in the range of 1.90×10^{-6} to 2.95×10^{-6}, and the values of R/Ra are in the range of 1.36 to 2.11, which indicate that there are constituents coming from mantle.

If we take 0.02Ra and 8Ra as limits for $^3He/^4He$ of typical crust-derived and mantle-derived fluids respectively (Dai et al. 1995), R is 1.36Ra or 2.11Ra, according to the equation:

the proportion of mantle-derived helium

$$(\%) = \frac{R - R_1}{R_2 - R_1} \times 100 \qquad (2)$$

where R_1 is $^3He/^4He$ of crustal fluids; R_2 is $^3He/^4He$ of mantle fluids. According to this formula, we can calculate the percentage of mantle helium among total helium for hot springs gases that are in the range of 16.8% to 26.2%. Therefore the active faults in this area are good places for gases of mantle origin to reach the surface dissolved in hot spring waters.

5 CONCLUSIONS

From the discussion above, the hot springs in Hengjing area have the following characteristics:

1. The hot springs are of meteoric origin, and are formed by deep circulation of local precipitation along faults and fractures in Jiangxi Province.
2. The $\delta^{13}C$ values for CO_2 in Hengjing geothermal area are in the range of −5.50‰ to −4.43‰, and the average value is −4.98‰, which is consistent with magma-mantle inorganic derived $\delta^{13}C$ value. These indicate that CO_2 comes from inorganic mantle.
3. The characteristic values (R/Ra) of helium isotope of hot springs in the Hengjing area are all >1, and helium coming from mantle is in the range of 17% to 26% of the total helium.

ACKNOWLEDGEMENTS

This study was supported by the National Natural Science Foundation of China (40242018), which is gratefully acknowledged. As well, I would like to express my sincere gratitude to professor Z.-X. Sun and B. Gao for their instructions.

This study has also been supported by the National Basic Research Program of China, the research is on the basic theory about the mechanism of water inrush and its prevention in coalmines (No. 2007CB209401) and it has been supported by the Fundamental Research Funds for the Central Universities.

REFERENCES

Craig, H. 1961. Isotopic variation in meteoric waters. *Science*, 133: 436–468.

Dai, C.S., Dai, J.X., Song, Y., et al. 1995. Mantle Helium of natural gases from Huanghua depression in Bohai gulf basin. *Journal of Nanjing University (Natural Sciences Edition)* 31(2): 272–280. (In Chinese).

Dai, J.X., Song, Y., Dai, C.S., et al. 1997. *Abiogenic gas in east China and forming conditions of gas reservior*. Science Press, Beijing. (In Chinese).

Farley, K.A., Patterson, B.I. & McInnes, A. 1997. Helium isotopic composition of geothermal gases, phenocrysts and xenoliths from the Tabar-Lihir-Tanga-Feni Island arc, Papua New Guinea. *Geochimica et Cosmochimica Acta* 61: 2485–2496.

Gao, B., Sun, W.J., Zhang, W., et al. 2006. Identification of mantle origin of geothermal spring gases in Hengjing area in south Jiangxi province. *Journal of Guilin University of Technology* 26(1): 1–5. (In Chinese).

Giggenbach, W.F. & Poreda, R.J. 1993. Isotopic composition of helium, and CO_2 and CH_4 contents in gases produced along the New Zealand part of a convergent plate boundary. *Geochimica et Cosmochimica Acta* 57: 3427–3455.

Giggenbach, W.F. 1980. Geothermal gas equilibria. *Geochimica et Cosmochimica Acta* 44: 2021–2032.

Liu, D.L., Tao, S.Z. & Zhang, C.M. 2001. Gas geochemistry of warm springs in Tan-Lu fault belt. *Geological Journal of China Universities* 7(1): 81–86. (In Chinese).

ShangGuan, Z.G., Bai, C.H. & Sun, M.L. 2000. Release characteristics of modern mantle magmatic gas at Rehai area, Tengchong. *Science in China (Series D)* 30(4): 407–409. (In Chinese).

Sun, Z.X., Gao, B. & Liu, J.H. 2004. Geothermal gas geochemistry of the Hengjing hot springs area in Jiangxi province. *Geoscience* 18(1): 116–120. (In Chinese).

Xu, Y.C., Shen, P., Tao, M.X., et al. 1997. Geo-chemistry on mantle-derived volatiles in natural gases from eastern China oil/gas provinces (II). *Science in China (Series D)* 40(3): 315–321.

Water-Rock Interaction – Birkle & Torres-Alvarado (eds)

 ISBN 978-0-415-60426-0

Dissolved arsenic in shallow hydrothermal vents

R.E. Villanueva-Estrada, R.M. Prol-Ledesma & C. Canet

Departamento de Recursos Naturales, Instituto de Geofísica, Universidad Nacional Autónoma de México. Delegación Coyoacán, México D.F., Mexico

ABSTRACT: High arsenic concentrations are found in water discharged by hydrothermal vents in Bahía Concepción, Baja California Sur. The theoretical concentrations of the different As aqueous species, calculated with a speciation model, show that arsenic is present in reduced form. Mixing of a reduced fluid (hydrothermal fluid) with an oxidizing fluid (seawater) could form iron and manganese oxides that adsorb arsenic.

1 INTRODUCTION

Arsenic is a toxic element present in different natural environments. Arsenic chemical species occur in two oxidation states (III and V) and as oxyanions (arsenite and arsenate). In geothermal fluids, arsenic total concentrations range from less than 0.1 up to nearly 50 mg/L (Ballantyne & Moore 1988, Mc Carthy et al. 2005), although the majority falls in the range of 0.1 to 3 mg/L (Ballantyne & Moore 1988, Pichler & Veizer 1999, Prol et al. 2004). These concentrations are above the permissible limits for drinking water established by international organizations like USEPA (2002) and WHO (2001), which propose maximum values of about 0.01 mg/L.

However, the total concentration is not a good indicator for the toxicity of dissolved arsenic in a natural environment. On the other hand, the bioavailability is a function of the concentration and also of the chemical form of the element. The chemical speciation of arsenic is dependent on the pH and redox potential value of the fluid. As (III) is more toxic than As (V) (National Research Council 1977). At a pH range of 2 to 9, As (V) is found as $H_2AsO_4^-$ and $HAsO_4^{2-}$, meanwhile As (III) in form of H_3AsO_3. As (V) is the less mobile species. It can be incorporated into non-crystalline iron oxyhidroxides as $H_2AsO_4^-$ at pH less than 6 due to the positive charge surface, as the pHpzc is in the range from 6 to 10 (Cornell & Schwertmann 1996). On the other hand, the species $H_3AsO_3°$ is not attracted by the positive charge of $FeOH^+$ as is not charged (Maity et al. 2005).

2 ARSENIC IN A SHALLOW HYDROTHERMAL SYSTEM OF BAHÍA CONCEPCIÓN

Bahía Concepción is located in the northeastern part of the state of Baja California Sur. Hydrothermal activity occurs on the west side of Bahía Concepción. Prol-Ledesma et al. (2004) described two types of hydrothermal activity that occur in Bahía Concepción, intertidal springs and shallow submarine vents (Fig. 1).

Hydrothermal fluid is discharged by submarine vents at depths from 5 to 15 m and reaches a temperature of 87°C at 10 cm depth within the seafloor sediments. In the submarine vents, fluids have pH values between 5.95 and 6.2 and in the intertidal vents about 6.7 (Table 2). The water discharged in

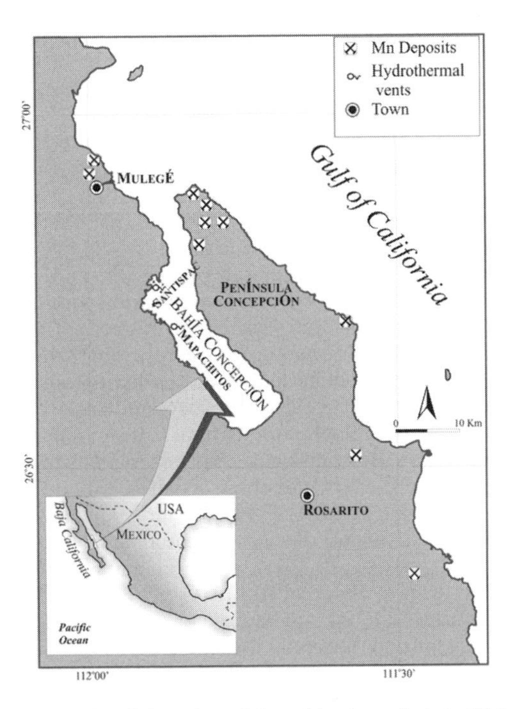

Figure 1. Submarine (Mapachitos) and intertidal (Santispac) hydrothermal vents in the Bahía Concepción.

Table 1. Manganese, iron and arsenic total concentrations (mg/L), and physical and chemical parameters measured in situ on submarine and intertidal hydrothermal vents. The data from samples BC1, BC4, and BC6 are from Prol-Ledesma et al. (2004).

	Temp. (°C)	pH	Mn	Fe	As
BC1	87	6.0	3.56	2.29	0.78
BC4	87	6.0	2.37	1.84	0.46
BC6	87	6.0	2.86	2.01	0.70
LP	61	6.5	1.58	0.27	0.36
SAN	66	6.2	2.8	1.11	0.27
MAN	40	7.0	0.11	0.05	0.15

Table 2. Abundances of chemical species of Mn, Fe, and As (in percentage), with respect to the total concentrations.

	Submarine discharges			Intertidal discharges		Mangrove
	BC1	BC4	BC6	LP	SAN	MAN
Mn^{2+}	83	81	81	94	94	96
$MnSO_4$	16	19	18	6	6	4
$FeCl^+$	69	69	69	44	51	29
$FeCl_2$	24	23	24	51	44	66
Fe^{2+}	6	6	6	3	4	3
H_3AsO_3	99.6	99.5	99.6	99.2	99.5	98.6

both type of manifestations is of Na-Cl type. The chemistry of the fluid shows enrichment in Ca, Mn, Si, Ba, B, As, Hg, I, Fe, Li, HCO_3 and Sr with respect to seawater.

The mineralogy in the submarine and intertidal manifestations was studied by Canet et al. (2005). Crusts of manganese oxides and detrital aggregates cemented by opal-A, barite and calcite occur around the intertidal discharge. In the submarine venting zone, the precipitates consist in millimeter-thick iron oxyhydroxide coatings on volcanic cobbles, boulders and seashells, with traces of cinnabar. X-Ray diffraction revealed the amorphous or poorly crystalline nature of the main components (Fe-oxyhydroxide, Mn-oxides and opal). The mineral assemblages and textural features suggest microbial activity influences the mineralizing processes. Vent precipitates show a significant enrichment in arsenic, with a concentration range from 962 to 5112 mg/kg. The absence of an arsenic mineral phase suggests that As is probably adsorbed onto the iron oxide substrates (Canet et al. 2005).

Forrest (2004) studied metal concentrations in sediments covered with flocculent material that could be incorporated in the tissue of the sea cucumber *Holothuria inhabilis* and its relation with the Bahía Concepción submarine vents. The author found elevated concentrations of Fe (4.92 mg/kg) and As (128.3 mg/kg) in vent sea cucumbers compared with the non-vent ones. The As concentration in the sediments was above 23.9 mg/kg (Valette-Silver et al. 1999). Also Fe and As are abundant in high concentration in the flocculent material (the mean values for Fe are 10,655 mg/kg and 546 mg/kg for As). This suggests that the enrichment of Fe and As in the tissue of sea cucumbers is due to feeding on the flocculent material around the vents. Arsenic concentrations in the tissue of the sea cucumbers in Bahía Concepción are low in comparison with bivalve tissues (about 14.5 mg/kg, Valette-Silver et al. 1999). According to Forrest (2004), the measured concentrations in organisms could be the result of arsenic bioaccumulation.

3 HYDROGEOCHEMISTRY

In this paper, we compare the geochemistry of submarine vents (BC1, BC4 and BC6), published by Prol-Ledesma et al. (2004) with the data obtained from the intertidal discharges (LP-La Posada and SAN-Santispac) and from a vent in a modern mangrove (MAN). The sample in the intertidal zone was collected using the air displacement method (Villanueva et al. 2006). For Mn and Fe analyses, the water sample was filtered using a filter paper of 0.45 micrometers pore size, and was acidified to pH = 2 using nitric acid (35% wt). The preservation of the samples for As analysis was achieved with acetic acid at pH conditions of 2. The samples were refrigerated prior to be analyzed in the laboratory. Trace metal concentrations were determined using inductively coupled plasma mass spectrometry (ICP-MS). The ICP analyses were carried out at the Activation Laboratories in Ancaster, Ontario, Canada. One blank sample was analyzed and it was below the detection limit for metal concentrations. Additionally, pH, temperature, conductivity and salinity were measured at the discharge zone using a multiparameter HACH SensION1.

4 CHEMICAL SPECIATION AND POURBAIX DIAGRAMS

Chemical speciation was calculated for Mn, Fe and As to estimate the distribution and the chemical form, in which they are being transported by the hydrothermal fluid in function of the pH and availability of ligands,

The React program of The Geochemist´s Workbench (GWB) software (release 6.0) was

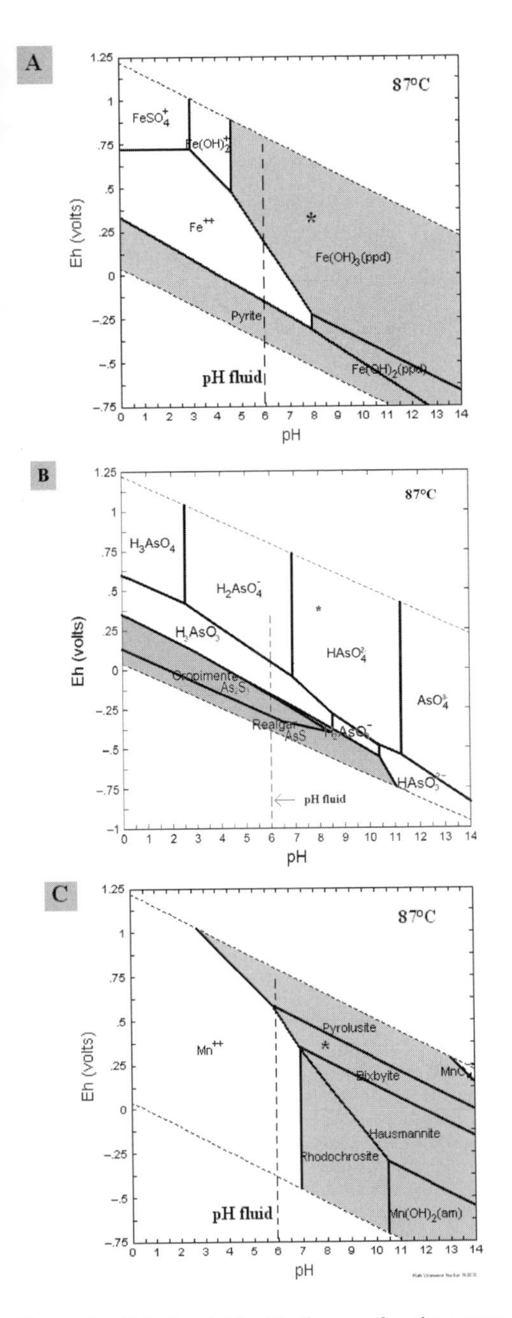

Figure 2. Calculated Eh-pH diagram for the system (A) As-H_2O-SO_4, (B) Fe-HCO_3-SO_4, and (C) Mn-CO_3 at the discharge temperature for the submarine vents. The Eh-pH diagram for intertidal and mangrove were not presented because they are similar to the submarine vent. The fields in white indicate the chemical species in solution and grey fields indicate the mineral phases. The dashed line shows the pH of the fluid measured at the discharge. The asterisk indicates pH (8.0) and redox potential (0.3 volts) values of the seawater.

used for modeling the equilibrium state of the aqueous solution. The assumptions made were that chemical equilibrium of the discharged thermal water exists, and the activities of the chemical species were calculated using the Debye-Hückel equation. The temperature and pH of discharge were taken into account to execute the equilibrium model.

Manganese, iron and arsenic are in reduced conditions in the hydrothermal fluid (Table 2). The complex formation with SO_4 and Cl with Mn and Fe respectively is due to the continuous interaction of the hydrothermal fluid with seawater. The chemical species of arsenic in the hydrothermal fluid is in form of H_3AsO_3.

In order to explain the chemical conditions (pH and Eh) for the deposition of manganese and iron oxides and the occurrence of arsenic, stability diagrams were constructed using the program Act2 from the GWB (Fig. 2).

The redox potential was not measured in field work. Therefore the estimation of the chemical species of arsenic was made assuming that the thermal fluid has not an oxidant character (Pichler 1999).

5 DISCUSSION

The arsenic concentration in the submarine and intertidal hydrothermal vents of Bahía Concepción is above the limits established by international organizations (Table 1). According to theoretical data from the speciation chemical model and from the stability diagram for As, up to the 98% of the As in the hydrothermal fluid is present in the chemical form of H_3AsO_3, (Fig. 2B). Moreover, when the hydrothermal fluid mixes with an oxidizing fluid as seawater (pH of about 8.3), As in H_3AsO_3 is oxide to form $H_2AsO_4^-$ or possibly to a deprotoned form such as $HAsO_4^{2-}$ (Fig. 2B). Alternatively, the Fe reduced from the hydrothermal process is oxide to ferric ion and form non-crystalline ferric hydroxide, Fe $(OH)_3$ (Fig. 2 A). During the mixing of hydrothermal fluid with seawater, an oxidation process was carried out and the adsorption of As onto the surface of non-crystalline Fe takes place.

The iron oxyhydroxides deposited in the submarine vents play an important role as arsenic scavengers (Hall 1998) because: (1) around the hydrothermal system, reduced chemical conditions locally prevail, (2) poorly crystalline minerals are precipitated, principally amorphous ferric hydroxide (3) As is present in high concentrations, and (4) the amount of As is too low in the organic matter to compete with the metal oxide (Rodríguez-Meza et al. 2009).

However, Mn is in reduced form Mn (II) at the pH of the fluid (Table 2 and Fig. 2C). Perhaps the biologic activity catalyzes the Mn (II) oxidation to form the non crystalline manganese oxide MnO_2 (Villalobos 2003). In addition, scanning electron microscope analysis showed us that arsenic is not adsorbed onto the Mn oxides (todorokite), deposited in the intertidal vents.

The adsorption of As onto non-crystalline Fe surface is the main mechanism to retain As in the most toxic form. The problem is that this iron oxyhydroxides material with adsorbed arsenic ($H_2AsO_4^-$ or $HAsO_4^{2-}$) could be present as flocculent material that can be incorporated to the tissue of some marine organism, such as *Holothuria inhabilis* in the Bahía Concepción.

6 CONCLUSIONS

According to the pH values, As in the vent fluid is mostly present as H_3AsO_3 (up to 98.6%) in the hydrothermal fluids. The mixing with seawater produces the oxidation of arsenic species ($H_2AsO_4^-$ or $HAsO_4^{2-}$). Mn and Fe are both in reduced form in solution. The mixing of hydrothermal fluid with seawater causes the oxidation of Fe to form the observed Fe-oxyhydroxides. The formation of Mn-oxides could occur by mediation of microorganisms.

Iron oxyhydroxides could adsorb As and thus prevent its release to seawater.

REFERENCES

Ballantyne, J.M. & Moore, J.N. 1988. Arsenic geochemistry in geothermal systems. *Geochimica et Cosmochimica Acta* 52: 475–483.

Canet, C., Prol-Ledesma, R.M., Proenza, J.A., Rubio-Ramos, M.A., Forrest, M.J., Torres-Vera, M.A. & Rodríguez-Díaz, A. 2005. Mn-Ba-Hg mineralization at shallow submarine hydrothermal vents in Bahía Concepción, Baja California Sur, México. *Chemical Geology* 224: 96–112.

Cornell, R.M. & Schwertmann, U. 1996. *The Iron Oxides*. Weinheim, Germany: VCH.

Forrest, M.J. 2004. *The geology, geochemistry and ecology of a shallow water submarine hydrothermal vent in Bahía Concepción, Baja California Sur, México.* California State University Monterey Bay. Thesis (in English), 112 pp.

Hall, G.E.M. 1998. Analytical perspective on trace element species of interest in exploration. *Journal of Geochemical Exploration* 61: 1–9.

Maity, S., Chakravarty, S., Bhattacharjee, S. & Roy, B.C 2005. A study on arsenic adsorption on polymetallic sea nodule in aqueous medium. *Water Research* 39 2579–2590.

McCarhty, K.T., Pichler, T. & Price, R.E. 2005. Geochemistry of Champagne Hot Springs shallow hydrothermal vent field and associated sediments, Dominica, Lesser Antilles. *Chemical Geology* 224: 55–68.

Pichler, T., Veizer, J. & Hall, G.E.M. 1999. Natural input of arsenic into a coral-reef ecosystem by hydrothermal fluids and its removal by Fe (III) oxyhydroxides. *Environmental Science & Technology* 33: 1373–1378.

Prol-Ledesma, R.M., Canet, C., Torres-Vera, M.A., Forrest, M.J. & Armienta, M.A. 2004. Vent fluid chemistry in Bahía Concepción coastal submarine hydrothermal system, Baja California Sur, México. *Journal of Volcanology and Geothermal Research* 137: 311–328.

Rodríguez-Meza, G.D., Shumilin, E., Saposhnikov, D., Méndez-Rodríguez, L. & Acosta-Vargas, B. 2009. Evaluación geoquímica de elementos mayoritarios y oligoelementos en los sedimentos de Bahía Concepción (B.C.S., México). *Boletín de la Sociedad Geológica Mexicana* 61: 57–72.

USEPA, 2002. *Implementation guidance for the arsenic rule- drinking water regulations arsenic and clarifications to compliance and new source contaminants monitoring* (EPA-816-K-02-018).

Valette-Silver, N.J., Riedel, G.F., Crecelius, E.A., Windom, H., Smith, R.G. & Dolvin, S.S. 1999. Elevated arsenic concentrations in bivalves from south east coasts of the USA. *Marine and Environmental Research* 48: 311–333.

Villalobos, M., Toner, B., Bargar, J. & Sposito G. 2003. Characterization of the manganese oxide produced by *Pseudomonas putida* strain MnB1. *Geochimica et Cosmochimica Acta* 67: 2649–2662.

Villanueva, R.E., Prol-Ledesma, R.M., Torres-Vera, M.A., Canet, C., Armienta, M.A. & de Ronde, C.E.J. 2006. Comparative study of sampling methods and in situ and laboratory analysis for shallow-water submarine hydrothermal systems. *Journal of Geochemical Exploration* 89: 414–419.

Villanueva-Estrada, R.E. 2008. *Procesos geoquímicos en las manifestaciones hidrotermales ubicadas en zonas intermareales y submarinas de las costas de Bahía Concepción (Baja California Sur) y Punta Mita (Nayarit).* Universidad Nacional Autónoma de México. Thesis (in Spanish). 123 pp.

WHO 2001. *Arsenic Compounds, Environmental Health Criteria,* 2a. ed. World Health Organization, Géneva.

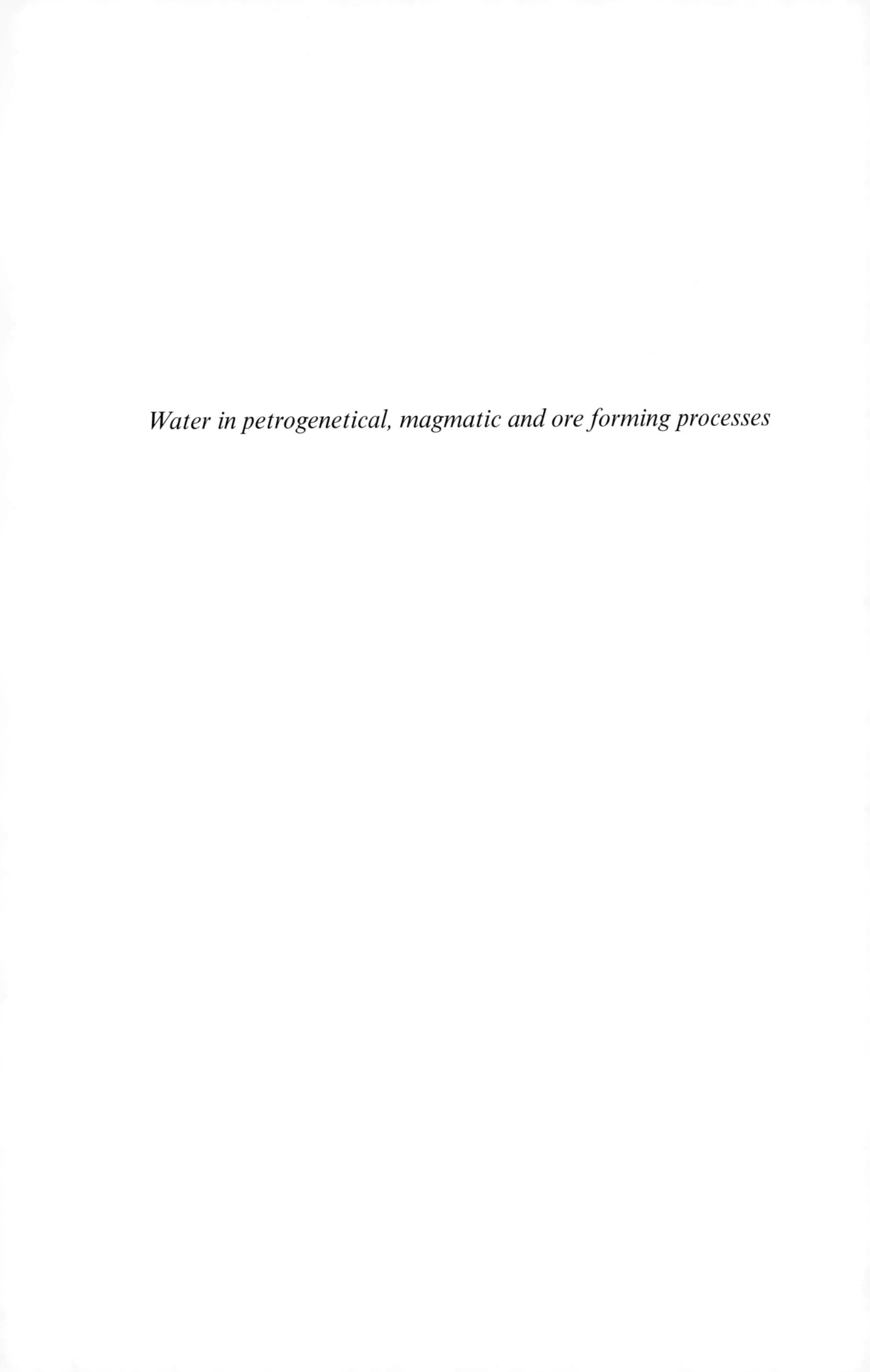

Water in petrogenetical, magmatic and ore forming processes

Water-Rock Interaction – Birkle & Torres-Alvarado (eds)

Carbon isotope composition of CO_2 at Cascade Arc volcanoes

W.C. Evans, R.H. Mariner & D. Bergfeld
U.S. Geological Survey, Menlo Park, California, USA

K.M. Revesz & J.P. McGeehin
U.S. Geological Survey, Reston, Virginia, USA

ABSTRACT: CO_2 from gas vents or dissolved inorganic carbon (DIC) in spring waters can constrain the isotopic composition of magmatic CO_2 at all but one (Medicine Lake) of the 12 major volcanoes of the Cascade Arc in western USA. A compilation of new and published carbon-isotope data obtained from these features shows that only Baker and Newberry have MORB-like $\delta^{13}C$ values near −7‰. Glacier Peak, Rainier, Adams, St. Helens, Hood, South Sister, Crater Lake, Shasta, and Lassen have $\delta^{13}C$ values in the −8.5 to −12‰ range. Inclusion of the DIC data fills in areal gaps and verifies the discharge of isotopically light CO_2, likely resulting from subduction of organic-rich sediments, along nearly the entire length of the arc.

1 INTRODUCTION

The Cascade Volcanic Arc south of the USA-Canada border hosts a dozen large volcanoes ranked in the USGS National Volcano Early Warning System as high or very high threat features (Ewert et al. 2005), where studies of magmatic degassing are needed. Most of the volcanoes have no summit fumaroles, and several lack gas vents of any kind, but dissolved magmatic He and C can be identified in the local spring waters based on isotopic analyses. Data from springs can be problematic in the study of carbon isotopes because groundwater can also acquire CO_2 from several non-magmatic sources. Subsequent fractionation of carbon isotopes between CO_2 and $HCO_3^- + CO_3^{2-}$, followed by either loss of CO_2 to the atmosphere or loss of CO_3^{2-} to calcite, can complicate efforts to constrain the $\delta^{13}C$ value of the magmatic component. With some consideration of spring characteristics, these problems are largely avoidable for the Cascade Arc volcanoes discussed here. We find good agreement between $\delta^{13}C$ values derived from CO_2 in gas vents and DIC ($CO_{2(aq)} + HCO_3^- + CO_3^{2-}$) in springs at those volcanoes with both types of features, implying that the $\delta^{13}C$ value of magmatic CO_2 can be constrained for the volcanoes that have only one type of feature. Our $\delta^{13}C$ compilation thus covers nearly the entire arc. This coverage allows us to better assess the provenance of the carbon dissolved in the arc magmas and will aid future efforts to quantify the magmatic CO_2 output of the arc volcanoes and track any changes in CO_2 output that might signal magmatic unrest.

2 SPRING TYPES

Three categories of C-enriched springs were considered in this study: hot springs, CO_2-rich soda springs, and dilute springs. Springs of all three types occur throughout the arc, but only springs discharging on or at the base of a volcanic edifice were used in this study. The spring waters should thus drain only volcanic rocks, ruling out dissolution of shallow marine carbonate rocks as a source of DIC. The $^3He/^4He$ ratios have been determined for many of these springs, and ratios of 2–8 R_A demonstrate a magmatic gas component.

Hot springs considered here have high DIC concentrations (2–25 mM) and pH values up to 7, reflecting partial conversion of CO_2 to $HCO_3^- + CO_3^{2-}$ through silicate hydrolysis. Although these springs are moderately mineralized (2.5–5 mS/cm) and typically saturated with calcite, they are immature waters that have not fully equilibrated with the host rocks. As long as little calcite precipitation has occurred, the $\delta^{13}C$ of DIC in these waters should be near the $\delta^{13}C$ of the magmatic CO_2 source.

Soda springs are cold, moderately mineralized (1.3 to 5 mS/cm), have very high DIC concentrations (50–150 mM), and nearly all produce bubbles of gas rich in CO_2. They are commonly saturated with calcite, and some deposit travertine. Carbon isotope fractionation is a serious concern for soda spring waters because of both CO_2 and CO_3^{2-} loss, but selection criteria were applied to minimize these problems. Soda springs were not used if they deposited travertine or issued at a pH >6.3, limiting

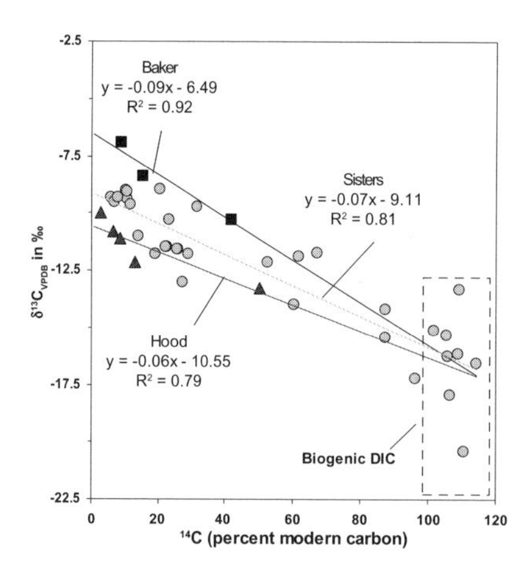

Figure 1. Carbon isotopes of DIC in dilute springs at three Cascade Arc volcanoes: Baker (squares); Three Sisters (circles); Hood (triangles). Biogenic DIC includes a small atmospheric component, both characterized by ^{14}C values ≥ 100% modern carbon. The $\delta^{13}C$ of the magmatic CO_2 component is given by the intercept of the regression line through each set of springs. Regression lines are similarly constrained at St. Helens, Shasta, and Lassen.

the equilibrium fractionation factor between CO_2 in the gas bubbles and DIC to 4‰.

Dilute springs (<1 mS/cm) are present on all of the volcanoes, but not all such springs contain magmatic carbon. Those that do are often slightly more mineralized and issue at slightly higher temperatures than typical cold springs at similar elevations (e.g., Nathenson et al. 2003). These springs range widely in pH, up to values of 8.5, where nearly all of the dissolved CO_2 has been converted to $HCO_3^- + CO_3^{2-}$, but they are generally unsaturated with calcite, eliminating the possibility of fractionation due to CO_3^{2-} loss. Most have low enough pCO_2 values that CO_2 loss is likely to be minor. However, the amount of magmatic C in this type of spring is usually small, and biogenic (including atmospheric) C can constitute a substantial fraction of the DIC. We use ^{14}C-DIC values to determine the biogenic fraction, following James et al. (1999), and calculate $\delta^{13}C$ of the magmatic end-member from regression analysis (e.g., Fig. 1).

3 DATA SOURCES

Our compilation incorporates published summaries of $\delta^{13}C$ values by Werner et al. (2009) for Baker, by Bergfeld et al. (2008) for St. Helens, by Janik & McLaren (2010) for Lassen, and by Symonds et al. (2003) for several volcanoes. Other $\delta^{13}C$ values are unpublished, or are from the on-line database (http://hotspringchem.wr.usgs.gov) of Mariner et al. (2006), which also gives methods information for unpublished data. Numerous other sources have reported some fumarolic $\delta^{13}C$-CO_2 values (e.g., Zimbelman et al. 2000), or carbon isotopic values in spring waters (e.g., James et al. 1999) at Cascade Arc volcanoes.

4 SITES

Baker is one of the few Cascade Arc volcanoes to host superheated summit fumaroles. Werner et al. (2009) point out that fumarolic $\delta^{13}C$-CO_2 values have decreased slightly (from −6‰ to −7.2‰) over the past 15 years and attribute this shift to Rayleigh degassing of a fresh pulse of magma in 1975. The high-pH, low-DIC hot spring (Baker) located on the edifice is not used here out of concern over excessive CO_3^{2-} loss, but $\delta^{13}C$-DIC data from three dilute cold springs on the southern flank of the edifice (Fig. 1) supplement the fumarolic CO_2 data.

Glacier Peak has no known gas vents. Its $\delta^{13}C$ value is derived from a DIC-rich, eastern flank hot spring (Gamma) and from bubbling gas and DIC in a warm spring (Kennedy) at its western base.

Rainier has very weak, air-contaminated vents (<5% CO_2) at its summit. Our $\delta^{13}C$-CO_2 values from two samples (−11.6 and −11.8‰) agree with the values (−11.8 ± 0.7‰) reported by Zimbelman et al. (2000). We also use the gas-bubble CO_2 and the DIC from a hot spring (Ohanapecosh) and a soda spring (Longmire) near the southern base.

Adams has a summit gas vent first sampled in 2005. Even though the discharge from this low-temperature feature is weak and slightly air contaminated, the gas is 70% CO_2 and provides a $\delta^{13}C$ value for this volcano.

St. Helens has erupted in two cycles during the past 30 years and is the only Cascade Arc volcano to emit true high-T magmatic gases such as SO_2 and HCl. Fumarolic CO_2 has been repeatedly analyzed over the years, as have bubbling hot springs in Loowit and Step Canyons, which drain the crater breach zone. As at Baker, Bergfeld et al. (2008) point out gradual shifts of ~1.5‰ in $\delta^{13}C$ over time that likely relate to Rayleigh degassing of fresh magma. A component of magmatic CO_2 has also been detected in several dilute cold springs around the flanks.

Hood has many fumaroles near its summit, some of which are vigorous but none superheated. Several dilute springs, one warm (Swim), on its southern flank contain DIC with a strong magmatic component (Fig. 1) and were used in the compilation.

Three Sisters are located close together and are considered here as a group. No gas vents are known on any of these volcanoes, but crustal uplift near South Sister beginning in 2001 fueled an extensive study of the DIC in dilute cold springs (Evans et al. 2004). A magmatic component was detected in numerous springs, giving exceptional definition to the mixing line between magmatic and biogenic carbon (Fig. 1). This line along with DIC and bubbling CO_2 in a soda spring near the SE base of South Sister were used to derive the magmatic $\delta^{13}C$ value.

Newberry has a weak summit gas vent. The CO_2 from this vent and gas bubbles from two areas of hot spring discharge were used to derive the magmatic $\delta^{13}C$ value.

Crater Lake has one soda spring near its western base. Both gas bubbles and DIC were collected from this feature.

Shasta has vigorously bubbling acid-sulfate pools (drowned fumaroles) at its summit, a soda spring (McCloud) at its southern base, and several dilute cold springs that contain a magmatic carbon component.

Lassen has weak gas vents near the summit and stronger vents high on the NW side, but all these features are heavily air-contaminated (10–20% CO_2). Several dilute cold springs on its NW flank are rich in magmatic DIC (Evans et al. 2002), and a huge hydrothermal system discharges on its southern flank (Janik & McLaren 2010). The exact connection between this system and the magma that feeds the Lassen edifice is not completely clear, but we group them together because of similarity in their carbon isotopes.

5 RESULTS

Four types of $\delta^{13}C$ data are distinguished in Figure 2: $\delta^{13}C$-CO_2 derived from fumaroles and gas vents (diamonds); $\delta^{13}C$-DIC for soda and hot spring waters (squares); $\delta^{13}C$-CO_2 for gas bubbles from soda and hot springs (circles); $\delta^{13}C$-DIC from regression-line intercepts for dilute springs as shown in Figure 1 (pluses). Some of the points for St. Helens and Lassen represent large groups of samples with similar values.

The spread in data points at most volcanoes is about 2–4‰, but despite this spread, some clear differences in $\delta^{13}C$ values can be seen amongst the volcanoes. For volcanoes with both gas-vent and spring DIC data, the values are generally in good agreement, whether the $\delta^{13}C$-DIC is from hot springs, soda springs, or from the dilute spring intercepts and whether the gas vents are vigorous fumaroles or weak, air-contaminated features. In the pH range of the soda springs considered, $\delta^{13}C$-CO_2 in gas bubbles should be between −4‰ and +1‰ different from the $\delta^{13}C$-DIC, assuming near-equilibrium fractionation, and this matches the observations (Fig. 2). Accounting for this shift and using qualitative field observations of gas/water discharge rates allows us to reduce the spread in the data and pick a best estimate for the characteristic $\delta^{13}C$ of each volcano, shown as filled triangles in Figure 2.

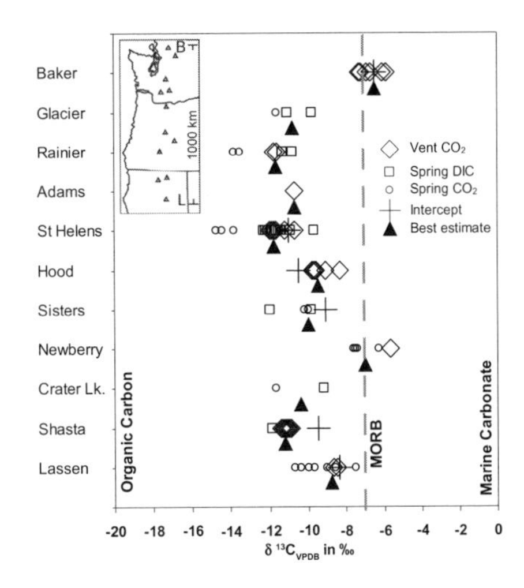

Figure 2. Compilation of $\delta^{13}C$ data from gas vents and springs at Cascade Arc volcanoes (see text for symbol explanation). Filled triangles are the best estimates for $\delta^{13}C$ of magmatic CO_2 at each volcano and are, from N to S: Baker (−6.5), Glacier Peak (−10.8), Rainier (−11.7), Adams (−10.7), St. Helens (−11.8), Hood (−9.5), South Sister (−10), Newberry (−7.0), Crater Lake (−10.4), Shasta (−11.2), and Lassen (−8.8).

Uncertainties in these best estimates are about ±1–1.5‰ and vary from volcano to volcano. The data compilation includes samples collected over a period of >30 years. Any long-term shifts in $\delta^{13}C$, such as at Baker and St. Helens (discussed above), fit within these uncertainties.

6 DISCUSSION

The $\delta^{13}C$ values at many volcanic arcs (Taran et al. 1992) differ from normal mantle values near −7‰ (MORB), but the predominance of light $\delta^{13}C$ values at Cascade Arc volcanoes is striking. Based on data from a subset of these volcanoes, Symonds et al. (2003) proposed that the carbon isotopes in large part reflect contamination from shallow crustal sources. The spring data reported here provide complete arc coverage, and based on that coverage, we argue that crustal controls on $\delta^{13}C$ are essentially limited to the small (~1‰) temporal changes linked to Rayleigh degassing of recently intruded magma. The $\delta^{13}C$ values show no obvious correlation to variations in basement-rock type, or to the predominant Holocene lava composition, which ranges from basalt/andesite to rhyolite (Fig. 3). These factors together imply that the Cascade Arc carbon isotopic compositions are mainly imparted from the deep basaltic parent magma, and the prevalence of light $\delta^{13}C$ values reflects the fact that organic-rich sediments contribute a large fraction of the carbon subducted at this arc (Hilton et al. 2002). Although

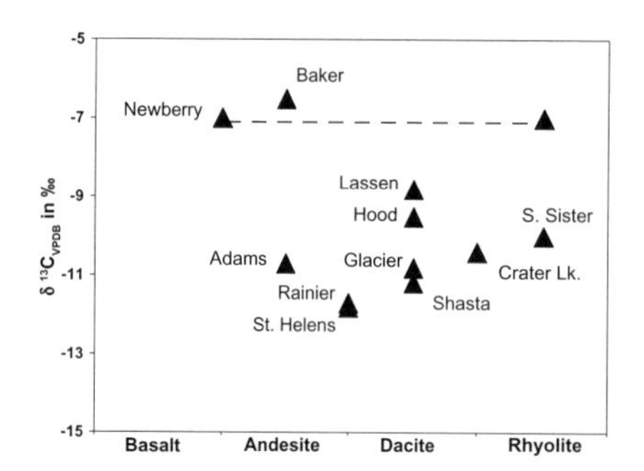

Figure 3. The $\delta^{13}C$ value for each Cascade Arc volcano vs. predominant postglacial lava composition (calculated from Hildreth 2007). Dashed line represents bimodal lava composition at Newberry.

the subducting slab is hot, and substantial dewatering is thought to occur beneath the forearc (e.g., Hurwitz et al. 2005), sediment-derived organic carbon has apparently persisted to the arc axis.

The 11 volcanoes are scattered over a 1000-km length of arc, characterized in the north by isolated stratovolcanoes, in the middle by rifting and closely-spaced vent lineaments, and in the south by increasing interaction with Basin and Range faulting. Their $\delta^{13}C$ values show no clear pattern linked to N-S position (Fig. 2). The MORB-like $\delta^{13}C$ value at Newberry might reflect its position substantially behind the arc axis, a hypothesis that could be tested by obtaining $\delta^{13}C$ for another volcano far behind the axis, like Medicine Lake. The MORB-like $\delta^{13}C$ at Baker seems enigmatic.

Conceivably the MORB-like $\delta^{13}C$ values at Newberry and Baker simply reflect parent magmas uncontaminated by sedimentary carbon. MORB-like lavas (olivine tholeiites) have erupted in the Cascade Arc and could be invoked as the gas source at Newberry, but they are uncommon at the arc's northern end (Hildreth 2007). Old MORB-like lava was recently described near Baker (Moore & DeBari 2009), raising the possibility that the 1975 unrest reflects intrusion of a magma type that hasn't erupted in this area during Holocene time.

ACKNOWLEDGEMENTS

The authors thank Yousif Kharaka, Manny Nathenson, Wes Hildreth, Dmitri Rouwet, and Yuri Taran for helpful comments on the manuscript.

REFERENCES

Bergfeld, D., Evans, W.C., McGee, K.A. & Spicer, K.R. 2008. Pre- and post-eruptive investigations of gas and water samples from Mount St. Helens, Washington, 2002 to 2005. In D.R. Sherrod, W.E. Scott, & P.H. Stauffer (eds.), *A volcano rekindled: the renewed eruption of Mount St. Helens, 2004–2006*. U.S. Geol. Surv. Prof. Pap. 1750: 523–542.

Evans, W.C., Sorey, M.L., Cook, A.C., Kennedy, B.M., Shuster, D.L., Colvard, E.L., White, L.D. & Huebner, M.A. 2002. Tracing and quantifying magmatic carbon discharge in cold groundwaters: lessons learned from Mammoth Mountain, USA. *J. Volcanol. Geotherm. Res.* 114: 291–312.

Evans, W.C., van Soest, M.C., Mariner, R.H., Hurwitz, S., Ingebritsen, S.E., Wicks, C.W. Jr. & Schmidt, M.E. 2004. Magmatic intrusion west of Three Sisters, central Oregon, USA: The perspective from spring geochemistry. *Geology* 32: 69–72.

Ewert, J.W., Guffanti, M. & Murray, T.L. 2005. An assessment of volcanic threat and monitoring capabilities in the United States: Framework for a National Volcano Early Warning System. *U.S. Geol. Surv. Open-File Rep.* 2005–1164: 62 p.

Hildreth, W. 2007. Quaternary magmatism in the Cascades – geologic perspectives. *U.S. Geol. Surv. Prof. Pap.* 1744: 125 p.

Hilton, D.R., Fischer, T.P. & Marty, B. 2002. Noble gases and volatile recycling at subduction zones. *Rev. Mineral. Geochem.* 47: 319–370.

Hurwitz, S., Mariner, R.H., Fehn, U. & Snyder, G.T. 2005. Systematics of halogen elements and their radioisotopes in thermal springs of the Cascade Range, Central Oregon, western USA. *Earth Planet. Sci. Lett.* 235: 700–714.

James, E.R., Manga, M. & Rose, T.P. 1999. CO_2 degassing in the Oregon Cascades. *Geology* 27: 823–826.

Janik, C.J. & McLaren, M.K. 2010. Seismicity and fluid geochemistry at Lassen Volcanic National Park, California: Evidence for two circulation cells in the hydrothermal system. *J. Volcanol. Geotherm. Res.* 189: 257–277.

Mariner, R.H., Venezky, D.Y. & Hurwitz, S. 2006. Chemical and isotopic database of water and gas from hydrothermal systems with an emphasis on the western United States. *U.S. Geol. Surv. Data Ser. Rep.* DS-169.

Moore, N.E. & DeBari, S.M. 2009. Origin and geochemical evolution of mafic magmas from Mount Baker in the northern Cascade Arc, Washington: Probes into Mantle and crustal processes. *GSA Abstr. Prog.* 41: 191.

Nathenson, M., Thompson, J.M. & White, L.D. 2003. Slightly thermal springs and non-thermal springs at Mount Shasta, California; Chemistry and recharge elevations. *J. Volcanol. Geotherm. Res.* 121: 137–153.

Symonds, R.B., Poreda, R.J., Evans, W.C., Janik, C.J. & Ritchie, B.E. 2003. Mantle and crustal sources of carbon, nitrogen, and noble gases in Cascade-Range and Aleutian-Arc volcanic gases. *U.S. Geol. Surv. Open-File Rep.* 03–436. 26p.

Taran, Y., Pilipenko, V.P., Rozhkov, A.M. & Vakin, E.A. 1992. A geochemical model for fumaroles of the Mutnovsky volcano, Kamchatka, USSR. *J. Volcanol. Geotherm. Res.* 49: 269–283.

Werner, C., Evans, W.C., Poland, M., Tucker, D.S. & Doukas, M. 2009. Long-term changes in quiescent degassing at Mount Baker volcano, Washington, USA; evidence for a stalled intrusion in 1975 and a connection to a deep magma source. *J. Volcanol. Geotherm. Res.* 186: 379–386.

Zimbelman, D.R., Rye, R.O. & Landis, G.P. 2000. Fumaroles in ice caves on the summit of Mount Rainier – preliminary stable isotope, gas, and geochemical studies. *J. Volcanol. Geotherm. Res.* 97: 457–473.

Water-Rock Interaction – Birkle & Torres-Alvarado (eds)
© 2010 Taylor & Francis Group, London, ISBN 978-0-415-60426-0

Porosity, density and chemical composition relationships in altered Icelandic hyaloclastites

H. Franzson, G.H. Guðfinnsson & H.M. Helgadóttir
Iceland GeoSurvey, Reykjavík, Iceland

J. Frolova
Geological Department, Moscow State University, Russia

ABSTRACT: Basaltic hyaloclastite tuffs play an important role in hosting groundwater and geothermal systems in Iceland. Their porosity is high and may exceed 60%. Alteration starts by palagonitization of the glass during eruption and is usually complete at relatively low temperatures. Petrophysical measurements show that grain density decreases with increasing alteration but increases again when alteration has reached the chlorite-epidote facies. Porosity changes during alteration, with macro-pores filling at the same time as micro-porosity increases. When glass alters, most chemical components are released from the rock. This study shows, however, that chemical mobility during this process is very limited, even at the highest alteration state. Only Na_2O shows high mobility and CaO, K_2O, P_2O_5, Rb and V to a lesser extent. Other elements show no apparent mobility.

1 INTRODUCTION

Icelandic volcanic rocks host vast water resources ranging from cold groundwater to high-temperature systems, and are utilized through wells drilled into these reservoirs. The type and longevity of these reservoirs are highly dependent on the petrophysical character of the rocks. Representative rock samples are needed for the study of these characteristics in the laboratory. However, because of the lack of cores taken during drilling, a research project, financed by National Energy Authority and Reykjavik Energy, was undertaken in 1993 to study the petrophysical character of Icelandic rocks of all types and degrees of alteration from surface samples to deeply eroded crustal sections. About 500 samples were collected and studied for these purposes (Stefansson et al. 1997, Sigurdsson et al. 2000). An additional 120 samples were taken in 1999 to study variations in petrophysical characteristics within a single fresh lava flow of olivine tholeiitic composition (Franzson 2001).

The present project, which is ongoing, focuses on basaltic hyaloclastite tuffs, ranging from relatively fresh to totally altered samples. An additional 100 samples of tuffs were collected in 2002, bringing the total up to some 140 samples. Duplicate samples were analyzed at Moscow State University for petrophysical properties (Frolova 2005). This paper summarizes the analytical methods and the relationship between permeability, porosity, density, alteration and chemistry of hyaloclastite tuffs.

2 SUMMARY OF DATA

Hyaloclastite tuffs are fragmental rocks formed during sub-glacial volcanic eruptions where magma contacts glacial meltwater.

Palagonitization is the first stage of glass alteration and is postulated to start during the eruption. Three time-dependent types of palagonite have been recognized as shown in Figures 2 and 3; rind palagonite, isolated spherical palagonite and layers of spherical palagonites, all of which we recognize as hydrated glass. When palagonite has altered into smectite, the glass becomes vulnerable to further alteration, and deposition of alteration minerals start to fill voids in the rock and enhance consolidation as the alteration proceeds (Helgadottir 2005, Thorseth et al. 1992, Stroncik & Schminke 2001, Schiffman et al. 2000).

Each thin section was point counted (2000 points) to establish quantitatively the changes occurring in the rock during the gradual alteration, and for comparison to geochemical and petrophysical data. This includes the proportion of glass, primary crystals, rock alteration, open voids and mineral deposition. SEM images were also collected to study the porosity structures of the tuff.

All samples were measured for total and effective porosity, gas and klinkenberg permeability, and grain density. Several other parameters were measured at Moscow State University, some of which have been published (Frolova 2005).

All samples were analyzed for major and basic trace elements, along with LOI, CO_2 and FeO analysis.

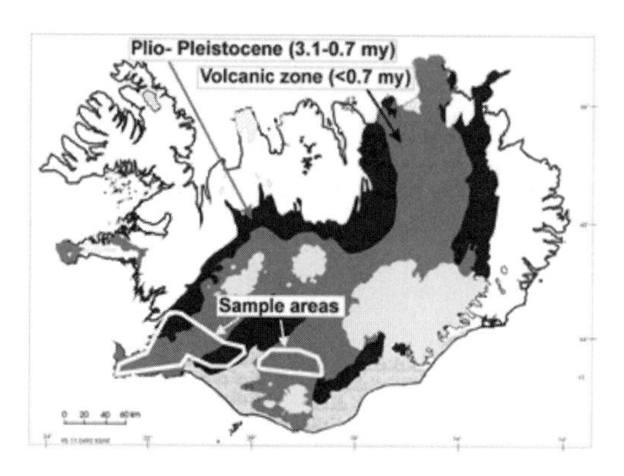

Figure 1. Geological map of Iceland with the sample areas outlined.

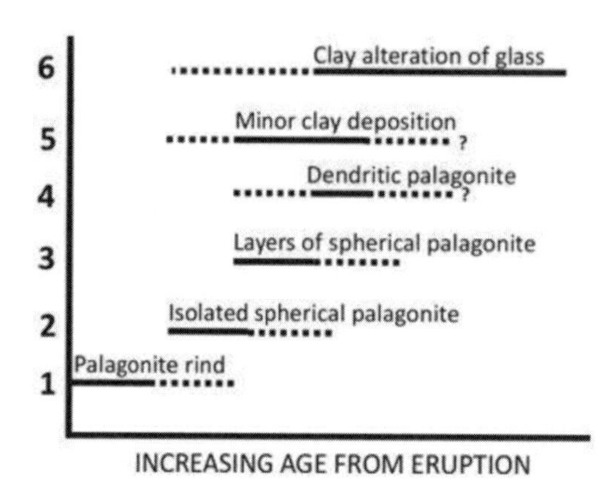

Figure 2. Time relation of the progressive alteration of basaltic glass.

Ten samples were analyzed for H_2O^+ to establish the relation between LOI and CO_2 analyses.

The porosity of tuff was determined by direct measurement in air and petrographically by point counting. A comparison of the results of the two methods is shown in Figure 4. It shows that the measured porosity is significantly higher than the thin- section porosity. Even in samples where thin-section inspection shows all pores to be filled with secondary minerals, measured porosity is up to a third of the bulk rock. The minimum size of pores observed in thin sections is related to the thickness of the section, indicating that the method only includes pores larger than 30 μm while the measured porosity includes all pores. This allows a distinction of pores into macro and micro-pores (<30 μm). Several lines of evidence suggest that, as the alteration of the glass proceeds and macro-pores fill with minerals, micro-pores are created within the alteration minerals, such as low-temperature clays. The creation of micro-pores is clearly seen at the intersection of glass and spherical palagonite in Figure 3.

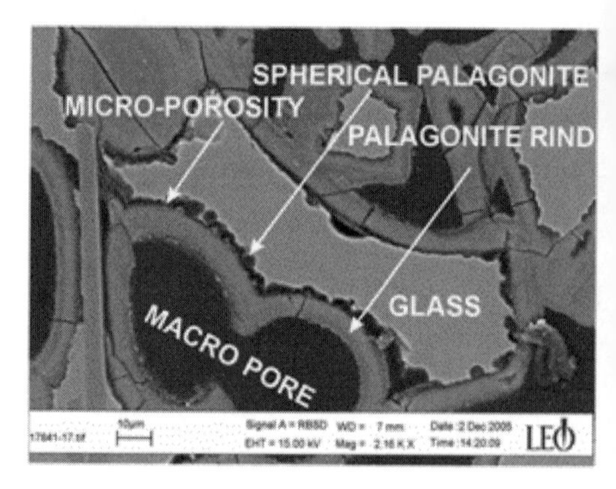

Figure 3. A backscatter-electron image of tuff showing rind palagonite, irregular spherical palagonite fresh glass, primary pores and secondary micro porosity at the intersection of the glass and palagonite.

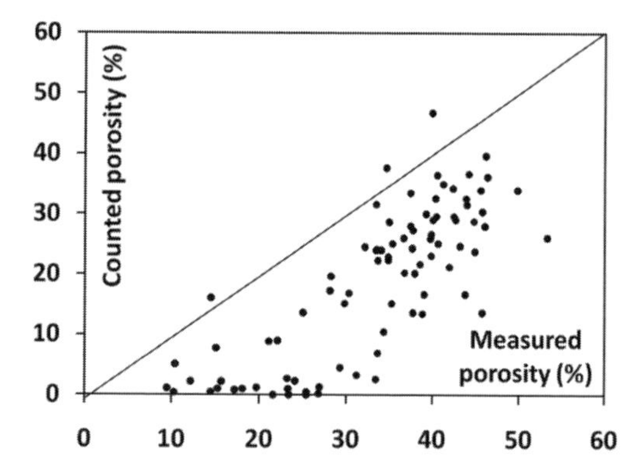

Figure 4. A comparison between measured porosity and porosity assessed from petrography (counted porosity).

Hydrous minerals form as glass is altered, thus water in the rock increases with increased alteration as shown in Figure 5, where water increases from less than 2% at 30% alteration to over 10% at intense alteration. An interesting feature is the marked loss of water in samples belonging to the chlorite-epidote alteration zone, which is explained by the transformation from smectite to less hydrous chlorite, and the disappearance of zeolites. It is interesting to note that the lower boundary of the alteration indicates the minimum rock consolidation needed to acquire core samples from the tuffaceous rocks (around 30% alteration as seen in Fig. 4).

Rock grain density (g/cm^3) is the average density of the minerals in the rock, excluding pores. Grain density of basaltic glass ranges from 2.7 to 2.8. When alteration starts, lower density palagonite and clays replace the glass along with zeolite precipitation in the voids, resulting in an overall decrease of density. Figure 6 shows this relation where the density gets progressively lower with increasing content of water. However, a distinct

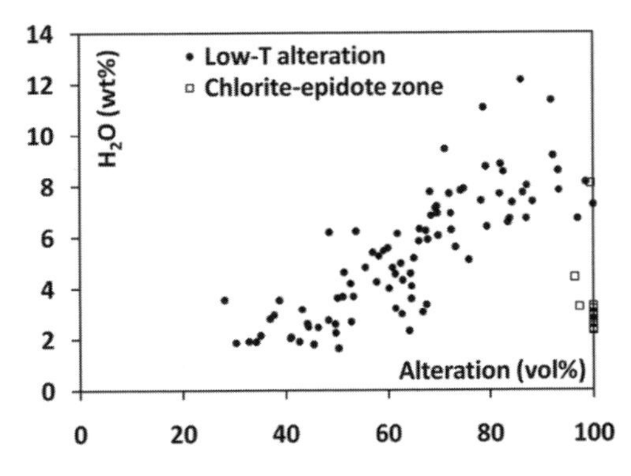

Figure 5. Graph of water content (LOI–CO_2) versus alteration volume% (determined form point counts).

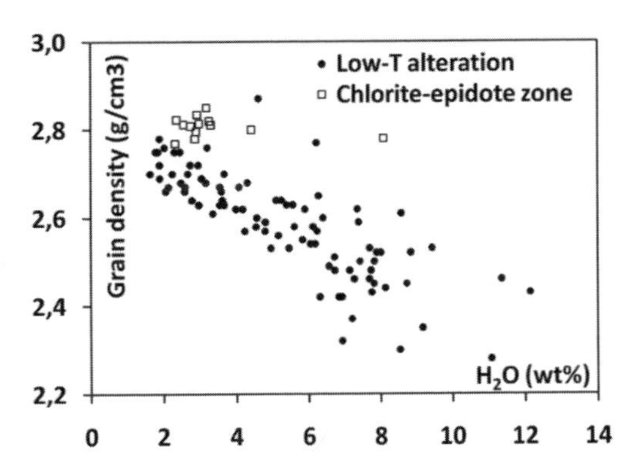

Figure 6. Graph of grain density versus water content.

density increase occurs in samples containing chlorite-epidote alteration, coinciding with a drastic decrease in water content. The decrease in the water content is caused by the disappearance of the more hydrous alteration minerals (clays, zeolites) along with the formation of higher density alteration minerals such as epidote.

Hydrothermal alteration is a change in the mineral content as a result of reaction of the rock with hydrothermal fluids. The rate of change depends to some extent on the stability of the minerals in the fresh rock. Basaltic volcanic glass is very sensitive to alteration, and made more vulnerable to alteration by the fur extreme porosity and permeability of the tuff it is contained in. Chemical transport during hydrothermal alteration in basaltic lavas and intrusions shows different mobility of the elements in the rocks partly related to the primary mineralogy of the rocks (Franzson et al. 2008). In basalt glass, the element mobility is not constrained by the different stabilities of crystal lattices, and the elements should all become mobile as the glass alters. The distance of element transport from glass thus depends more on the particular alteration minerals formed. A comparison of palagonite composition to fresh glass (Fig. 7) shows a very small difference between them with the exception of Na_2O, which is clearly highly mobile during palagonitization. A small loss of CaO also occurs. Other elements, such as FeO and TiO_2, are highly resistant to mobilization as shown in other studies of palagonitization (e.g. Thorseth et al. 1992, Crovisier et al. 1992), which support the finding of limited mobility of base cations on a microscale.

Another method of evaluating element mobility in the tuffs is to compare them with chemical trends produced by magmatic processes. If elements are mobile, alteration will lead to deterioration of such correlation, especially if the elements that produce the trends behave differentially during the hydrothermal alteration. Figures 8 and 9 show examples of such a comparison.

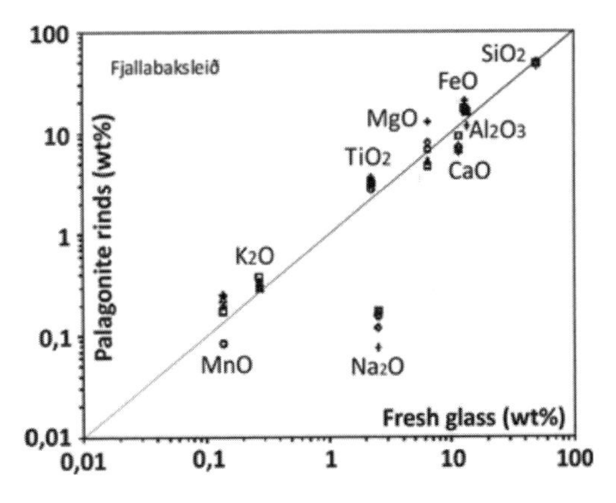

Figure 7. Palagonite composition (corrected for LOI) in comparison with fresh basaltic glass showing substantial leaching of Na from glass as palagonite forms.

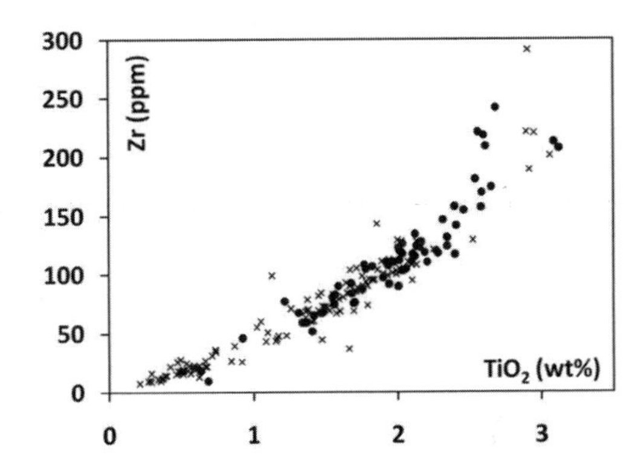

Figure 8. Variation diagram showing bulk concentration of Zr and TiO_2 in fresh rock samples from the Reykjanes-Langjökull volcanic zone (x) and in fresh to altered tuff from this study (filled circles). The compositional overlap shows that Zr and Ti are not significantly changed by alteration.

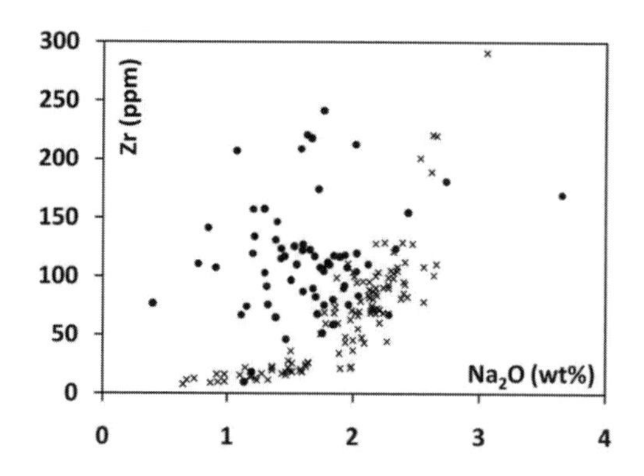

Figure 9. Variation diagram showing bulk rock concentrations of Zr and Na_2O in fresh rock samples from the Reykjanes-Langjökull volcanic zone (x) and fresh to altered tuff from this study (o).

The relation between TiO_2 and Zr shows that they conform closely with the magmatic evolutionary trend of the Reykjanes-Langjökull volcanic zone as seen in Figure 8, whereas Na_2O shows increasing loss with progressive alteration (Fig. 9).

Other elements that show signs of mobility are CaO, P_2O_5, K_2O and the trace elements of Rb and V.

3 CONCLUSIONS

Volcaniclastic tuffs are unusual in that the rocks have large porosity and permeability, but in nature they behave more like aquicludes where they form cap rocks for many high-temperature systems in Iceland. This study of about 140 fresh to totally altered tuff samples shows the following:

1. Porosity values are up to 60%.
2. Palagonitization starts during the eruption and progresses to total alteration at relatively low temperatures.
3. While macropores are filled during alteration, secondary microporosity is formed.
4. Grain density diminishes with increasing alteration, but increases again at the chlorite-epidote alteration state.
5. Only Na_2O shows strong mobility during palagonitization, while other elements remain relatively stable.
6. A comparison of fresh rock equivalent with tuffs shows that most elements are immobile within a core sample, with the exception of Na_2O which is highly mobile, and CaO, P_2O_5, K_2O, Rb and V to a lesser extent.

ACKNOWLEDGEMENT

This project has mainly been financed by the National Energy Authority and Reykjavik Energy. A NATO grant from the Icelandic Research Council and Russian Foundation for basic research (grant 05-03-64842) is also acknowledged. The manuscript benefitted greatly by the review of Halldór Armannsson and Mark Reed.

REFERENCES

Crovisier, J.-L., Honnorez, J., Fritz, B. & Petit, J.-C. 1992. Dissolution of Subglasial Volcanic Glasses from Iceland: Laboratory Study and Modelling, *Applied Geochemistry* 7(1): 55–81.

Franzson, H., Gudlaugsson, S.Th & Fridleifsson, G.O. 2001. Petrophysical properties of Icelandic rocks. *Proc. 6th Nordic Symp. on Petrophysics.* Throndheim, Norway: 1–14.

Franzson, H., Zierenberg, R. & Schiffman, P. 2008. Chemical transport in geothermal systems in Iceland. Evidence from hydrothermal alteration. *Journal of Volcanology and Geothermal Research* 173: 217–229.

Frolova, J., Ladygin, V., Franzson, H., Sigurdsson, O., Stefansson, V. & Sustrov, V. 2005. Petrophysical properties of fresh to mildly altered hyaloclastite tuffs. *Proc. World Geothermal Congress, Antalya, Turkey*: 15p.

Helgadottir, H.M. 2005. Formation of Palagonite. Petrographical analysis of hyaloclastite tuffs from the Western Volcanic Zone in Iceland. *BSc. Dissertation,* 40 p.

Schiffman, P., Spero, H.J., Southard, R.J. & Swanson, D.A. 2000. Control on palagonitization versus pedogenic weathering of basaltic tephra: Evidence from the consolidation and geochemistry of the Keanakako'i Ash Member, Kilauea Volcano. *Geochemistry Geophysics Geosystems*, vol. 1, paper no. 2000GC000068.

Sigurdsson, O., Gudmundsson, A., Fridleifsson, G.O., Franzson, H., Gudlaugsson, S.Th. & Stefansson, V. 2000. Database on igneous rock properties in Icelandic geothermal systems, status and unexpected results. *Proc. World Geothermal Congress, Kyushu – Tohuku, Japan*: 2881–2886.

Stefansson, V., Sigurdsson, O., Gudmundsson, A., Franzson, H., Fridleifsson, G.O. & Tulinius, H. 1997. Core measurements and geothermal Modelling. *Proc. Second Nordic Symp. on Petrophysics. Fractured reservoir. Nordic Petroleum Series: One*: 198–220.

Stroncik, N.A. & Schmincke, H.U. 2001. Evolution of Palagonite: Crystallization, chemical changes, and element budget. *Geochemistry, Geophysics, Geosystems* 2(7), 1017, doi: 10.1029/2000GC000102.

Thorseth, I.H., Furnes, H., & Tumyr, O. 1992. A textural and chemical study of Iceland palagonite of varied composition and its bearing on the mechanism of the glass-palagonite transformation. *Geochimica et Cosmochimica Acta* 56(2): 845–850.

Water-Rock Interaction – Birkle & Torres-Alvarado (eds)
 ISBN 978-0-415-60426-0

Aqueous phase composition of the early Earth

E.M. Galimov, B.N. Ryzhenko & E.V. Cherkasova
Vernadsky Institute of Geochemistry and Analytical Chemistry RAS, Russia

ABSTRACT: By computer thermodynamic simulation of carbonaceous CI chondrite-water system at 25°C, it is shown that the aqueous phase composition of the primary Earth at partial $P\ CO_2 = 10^{-7\pm1}$ and $P\ CH_4 = 10^{-6\pm1}$ bar is the following: aqueous K is 0.1–0.3 *m*, aqueous Na is 0.05–0.1 *m*, aqueous Mg ≤ 0.01 *m*, aqueous Ca 0.05–0.1 *m*, aqueous (NH_4^++NH_3) is about 2.10^{-3} *m*, aqueous P(V) ≤ 5×10^{-9} *m*, aqueous Cl is n × 10^{-2} *m*, pH = 8–9, Eh = –450 ± 50 mV. The K/Na ratio of the aqueous solution ranges from 4 to 14, which is similar to that in the cellular liquid of organisms.

1 INTRODUCTION

According to an hypothesis (Natochin 2005) the composition of the cellularliquid of organisms (K/Na > 1) should correspond to the primary Earth aqueous phase. The K/Na ratio of Baltic shield clay (>20–30) supports this idea (Natochin & Ahkmedov 2005). On the contrary for the contemporaneous crust aqueous solutions the K/Na ratio is less than 1 (Shvartsev 1998, Kraynov et al. 2004).

According to Vinogradov (1967), the primary ocean composition resulted from interaction of degassing mantle volatiles and the primary crust matter. The results of computer simulation of basic and ultra basic matter with H_2O show that aqueous solution with the K/Na ratio > 1 is formed at temperature higher than 700°C (Ryzhenko et al. 1997, 2000). Cooling of this aqueous fluid lowers the K/Na ratio to 0.n that is expected for aqueous solution at surface temperature and pressure. The formation of carbon dioxide–methane atmosphere in the above water-rock systems at high temperatures and pressures is determined by simulation also. It corresponds to the evidences about primary atmosphere composition of the Earth (Galimov 1967, 2000, 2005).

There is another hypothesis that the primary hydrosphere formed from comet or/and carbonaceous chondrite matter (Galimov 1967, Rauchfuss 2008). In any way the primary aqueous solution composition results from interaction of ultra basic matter and water at equilibrium state of liquid water. To investigate these hypotheses, the computer simulation technique is used.

2 METHODOLOGY

The simulated (H-O-K-Na-Ca-Mg-Fe-Al-Si-Ti-P-C-S-N) system includes 113 solids, 8 gases, and 227 aqueous species. The simulated system composition is the CI carbonaceous chondrite matter (Anders & Greevesse 1989). The simulation is performed by HCh code with stepwise flow technique (Shvarov 1999, Grichuk 2000). The Gibbs Free Energy values of minerals, aqueous species and gases are taken from (Borisov & Shvarov 1976). When equilibrium is reached the system is marked by the formation of aqueous solution and balanced mineral association of secondary (and primary) minerals to which the aqueous solution is saturated. All the calculated characteristics of the system (pH, Eh, concentrations of species) result from chemical interaction between water, gases and rock.

The following particular systems have been considered: (1) closed system, (2) the system opened to CO_2 only, (3) the system with various rock/water ratios, (4) the system opened to CO_2 and CH_4 with various partial pressure (CO_2/CH_4) ratios.

3 RESULTS

Results of simulations are shown as $P\ CH_4$– $P\ CO_2$ plain sections of three dimensional diagrams of K/Na (K, Na, Mg, Ca, N, P, pH, Eh) – $P\ CH_4$– $P\ CO_2$ (Figure 1). For calculation of the hydrogen pressure, the following relationships is used at 25°C: $1\,gP\ H_2 = 0.5(-83.1 - 1\,gP\ O_2)$; $1\,gP\ CH_4 = -143.3 + 1\,gP\ CO_2 - 21\,gP\ O_2$ (P in bars).

The aqueous principal component concentrations versus partial pressure of CO_2 and CH_4 are depicted in Figure 1 below. At $P\ CO_2 = 10^{-7\pm1}$ bar, $P\ CH_4 = 10^{-6\pm1}$ bar the K/Na ratio ranges 4–14. Aqueous K is 0.1–0.3 *m*, aqueous Na is 0.05–0.1 *m*, aqueous Mg ≤ 0.01 *m*, aqueous Ca 0.05–0.1 *m*, aqueous (NH_4^++NH_3) is about 2×10^{-3} *m*, aqueous P(V) ≤5×10^{-9} *m*, aqueous Cl is n × 10^{-2} *m*, pH = 8–9, Eh = -450 ± 50 mV at 25°C (Table 1).

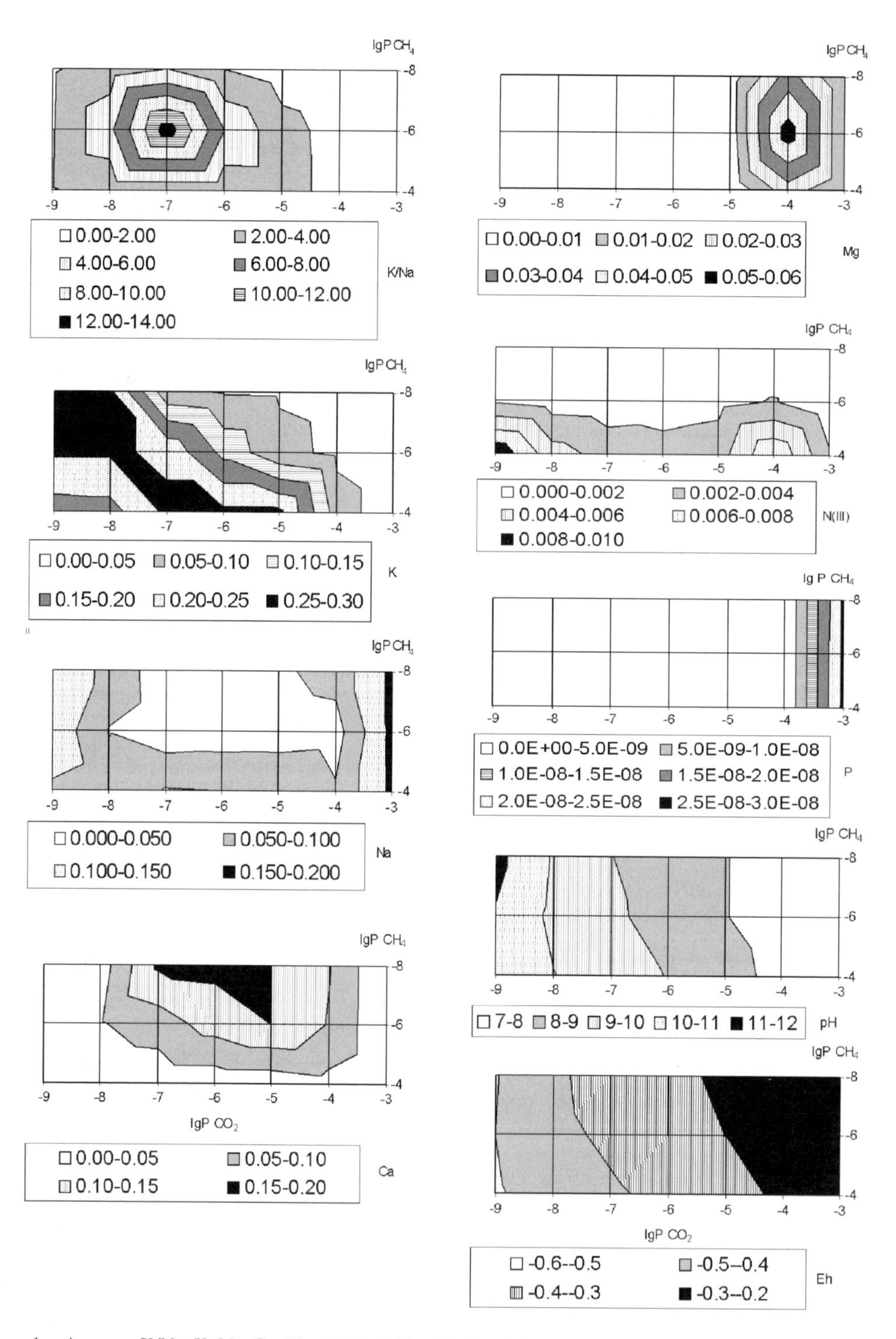

Figure 1. Aqueous K/Na, K, Na, Ca, Mg, N(III), P(V), pH, Eh of the water-CI chondrite system *versus* partial pressure of CO_2 and CH_4, 25°C.

Table 1. Aqueous phase composition and mineral assemblage water-CI chondrite system, 25°C (sections by plain l gP CO_2-l gP CH4, see Fig. 1).

Partial pressure of methane and carbon dioxide			
lg P CH_4	–4	–6	–8
lg P CO_2	–8	–7	–6
Aqueous solution (the element total concentration), m			
K/Na	2.54E+00	1.31E+01	2.02E+00
K	1.81E–01	2.39E–01	4.66E–02
Na	7.13E–02	1.83E–02	2.31E–02
Mg	1.85E–06	4.90E–05	3.17E+02
Ca	1.33E–02	7.30E–02	1.63E–01
N(III)	4.93E–03	8.34E–04	2.13E–04
P	2.33E–13	3.56E–13	1.08E–12
Cl	2.79E–01	4.04E–01	3.97E–01
Principal aqueous species, m			
pH	10.028	9.188	8.52
Eh	–0.454	–0.382	–0.32
H_2O (*)	9.90E–01	9.87E–01	9.89E–01
CO_3^{2-}	3.33E–06	8.28E–07	4.04E–07
HCO_3^-	2.35E–06	3.52E–06	7.64E–06
CO_2	3.43E–10	3.44E–09	3.43E–08
Ca^{2+}	1.30E–02	7.14E–02	1.60E–01
$CaOH^+$	8.20E–06	5.87E–06	2.75E–06
$CaCl^+$	2.02E–04	1.38E–03	2.89E–03
$CaCl_2$	3.12E–05	2.83E–04	5.70E–04
$CaCO_3$	6.76E–06	6.78E–06	6.77E–06
$CaHCO_3^+$	9.81E–08	7.03E–07	3.29E–06
Cl^-	2.77E–01	4.01E–01	3.93E–01
Fe^{2+}	1.61E–09	4.02E–08	3.79E–07
K^+	1.80E–01	2.39E–01	4.64E–02
KOH	4.55E–06	8.24E–07	3.40E–08
KCl	4.06E–04	7.01E–04	1.30E–04
KSO_4^-	2.96E–11	4.95E–10	2.89E–09
KCO_3^-	2.12E–06	5.72E–07	5.12E–08
$KHCO_3$	9.03E–08	1.64E–07	6.75E–08
Mg^{2+}	1.24E–06	2.97E–05	4.13E–04
$MgCl^+$	5.26E–07	1.64E–05	2.16E–04
$MgCl_2$	6.93E–08	2.87E–06	3.63E–05
$MgCO_3$	3.52E–10	1.61E–09	1.01E–08
$MgHCO_3^+$	7.56E–12	2.47E–10	7.27E–09
NH_4^+	9.74E–04	5.37E–04	1.91E–04
NH_3	3.95E–03	2.97E–04	2.22E–05
N_2	6.57E–04	6.60E–04	6.58E–04
Na^+	6.96E–02	1.77E–02	2.24E–02
NaOH	8.58E–07	3.03E–08	8.16E–09
NaCl	1.64E–03	5.51E–04	6.66E–04
$NaSO_4^-$	1.57E–11	5.14E–11	1.95E–09
$NaCO_3^-$	1.18E–06	6.21E–08	3.62E–08
$NaHCO_3$	4.38E–08	1.55E–08	4.17E–08
NaH_3SiO_4	4.12E–07	4.41E–08	2.28E–08
HS^-	2.28E–06	2.15E–07	3.04E–08

(Continued)

Table 1. (Continued).

SO_4^{2-}	1.20E–10	1.89E–09	6.07E–08
$H_2SiO_4^{2-}$	2.81E–08	2.12E–09	1.99E–10
$H_3SiO_4^-$	2.19E–05	9.90E–06	4.12E–06
SiO_2	5.76E–06	1.76E–05	3.36E–05
Equilibrium mineral assemblage, mole			
Antigorite	6.78E+00	6.70E+00	6.21E+00
Apatite-OH	1.03E+00	1.03E+00	1.03E+00
Calcite	1.81E+01	1.79E+01	1.70E+01
Clinochlore	1.36E+01	1.42E+01	1.57E+01
Daphnite	3.05E+00	2.43E+00	0
Goethite	0.00E+00	2.10E+02	2.22E+02
Magnetite	6.93E+01	0	0
Pyrite	5.16E+01	5.16E+01	5.16E+01
Riebeckite	1.20E+01	1.22E+01	1.22E+01
Saponite-K	0.00E+00	5.43E–02	5.22E–01
Saponite-Ca	0	0	4.22E–01
Sphene	8.76E–01	8.76E–01	8.76E–01

Table 2. Inner cell liquid composition (mm)

Na^+	K^+	Ca^{2+}	Mg^{2+}
10	140	0.0001	12
Cl^-	HCO_3^-	PO_4^{3-}	SO_4^{2-}
4	10	75	2

Simulation results correspond to aqueous K, Na, Mg concentrations of cellular liquid (Table 2, Trifonov 2009)). Discordance for aqueous Ca and aqueous PO_4 is the topic of our further research.

4 CONCLUSIONS

If the CI chondrite matter containing H_2O in amounts greater than 10% forms the crust of the early Earth and if the atmosphere contains nitrogen, CO_2 and CH_4, so the primary aqueous phase with the certain component concentrations is possible. The plain sections of three dimensional diagrams of K/Na (K, Na, Mg, Ca, N, P, pH, Eh) – P CH_4 – P CO_2 at P $CO_2 = 10^{-5} \rightarrow 10^{-8}$ bar, P $CH_4 = 10^{-8} \rightarrow 10^{-5}$ bar are as shown in Figure 1.

ACKNOWLEDGEMENTS

Authors thank Yu.V. Natochin, A.V. Ivanov, M.V. Mironenko and V.A. Dorofeeva for consultations, and Russian Academy of Sciences for the Project (N24) support.

REFERENCES

Anders, E. & Grevesse, N. 1989. Abundances of the elements: Meteoritic and solar. *Geochimica et Cosmochimica Acta* 53(2): 197–214.

Borisov, M.V. & Shvarov, Yu.V. 1976. *Termodynamics of geochemical processes*. Moskow: MSU Publ. House.

Galimov, E.M. 1967. The evolution of carbon of the Earth. *Geokhimiya* (5): 530–536.

Galimov, E.M. 2005. Prerequisites and conditions of the origin of life: research problems. *Geochemistry International* 43(5): 421–437.

Galimov, E.M. & Ryzhenko, B.N. 2008. Solution of the K/Na biogeochemical paradox. RAS Proceedings. *Doklady Earth Sciences* 421(3): 911–913.

Grichuk, D.V. 2000. Thermodynamic models of submarine systems. Nauchny Mir Publ. House (in Russian).

Kraynov, S.R., Ryzhenko, B.N. & Shvetz, V.M. 2004. *Geochemistry of groundwaters: Theoretical, applied, and Ecological Aspects.* Moskow: Nauka (in Russian).

Natochin, Yu.V. 2005. The role of sodium ions as a stimulus for the evolution of cells and multicellular animals *Paleontology Journal* 39: 345–357.

Natochin, Yu.V. & Akhmedov, S.M. 2005. Physiological and Paleogeochemical Arguments for a New Hypothesis of the Stimulus for Evolution of Eukaryotes and Multicellular Animals. *RAS Proceedings. Biological Sciences* 400: 31–34.

Rauchfuss H. 2008. *Chemical evolution and the origin of life*. Springer.

Ryzhenko, B.N., Barsukov, V.L. & Knazeva, S.N. 1997. Chemical characteristics (composition, pH, and Eh) of the water-rock systems: diorite (andesite)-water and gabbro (basalt)-water systems. *Geochemistry International* 35(12): 1089–1115.

Ryzhenko B.N., Barsukov V.L., Knazeva S.N. 2000. Chemical characteristics (composition, pH, and Eh) of the wayer-rock systems: pyroxinite-water and dunite-water systems. *Geochemistry International* 38(6): 560–583.

Shvarov, Y.V. 1999. Algorithmization of numeric equilibrium modeling of dynamic geochemical processes. *Geochemistry International* 37(6): 571–576.

Shvartsev, S.L. 1998. *Hydrogeochemistry of hypergenese zone.* Moskow: Nedra.(in Russian).

Tryfonov E.V. 2009. *Physiology of human being. Russian*-English Encyclopedy. 13th Edition.

Vinogradov, A.P. 1967. *Introduction to marine geochemistry.* M.: Nauka.

Water-Rock Interaction – Birkle & Torres-Alvarado (eds)
 ISBN 978-0-415-60426-0

Geochemistry and classification of cold groundwater in Iceland

H. Kristmannsdóttir
University of Akureyri, Faculty of Science and Business, Institute of Environmental Sciences, Akureyri, Iceland

S. Arnórsson & Á.E. Sveinbjörnsdóttir
University of Iceland, Division of Science and Engineering, Reykjavík, Iceland

H. Ármannsson
Ísor, Reykjavík, Iceland

ABSTRACT: In a comprehensive study of Icelandic groundwater special attention was paid to the chemical types of cold groundwaters and the correlation with the local geological and meteorological conditions. Before the chemistry of cold groundwater has been poorly investigated, even though over 90% of potable water is groundwater. The main factors affecting chemistry of the groundwater are age and type of the reservoir rocks, distance from sea and local meteorological conditions. Typical water from the Tertiary basaltic reservoir rocks has a pH of about 8.5–8.9, TDS of 60–100 mg/L and chloride concentration about or below 10 mg/L. Water from the recent basaltic bedrock of the active zones of rifting and volcanism is distinctly more alkaline with pH above 9. TDS is generally lower, 30–60 mg/L. Chloride concentration and TDS are higher nearer the coast. Water from the older Quaternary rock formations shows similarities to groundwater in the Tertiary rock formations.

1 INTRODUCTION

Water is one of the main natural resources in Iceland, both regarding the enormous quantities and its clean and unpolluted nature. The amount of precipitation is very high in Iceland, averaging 1800 mm/year (Jónsdóttir 2008). Total average river discharge in the country is about 5200 m^3/s, which is a European record. Since large areas of Iceland are covered with post-glacial lava, groundwater plays an important role in the hydrology of Iceland (Jónsdóttir 2008). Due to the young and highly permeable aquifers the quantity of groundwater is very high as compared to other European countries. The relation of groundwater to surface drainage varies considerably in different parts of the country as the bedrock permeability is highly dependent on the age and type of bedrocks. Due to geological and meteorological conditions in Iceland being considerable different from most other countries the properties of the groundwater are significantly different. There is however a great variation in the chemical composition of the groundwater depending mainly on the type of rocks, composition, structure and age, location in the country, elevation and distance from the sea. Potable water in Icelandic waterworks is mostly groundwater, but in a few places in Eastern and Western Iceland surface water is used. Due to the different nature of potable water in Iceland it is difficult to use information and experience from foreign countries for the choice of piping and construction of waterworks. Also the international references regarding permissible and admissible concentration of substances may be quite inappropriate. As an example the European regulations formerly required potable water to have pH below 8.5 as higher pH may indicate decay of organic substances. However the majority of potable water in Iceland greatly exceeds this limit due to the dominant basaltic bedrock and the standards had thus to be changed. Until now data for the Icelandic cold water, including groundwater have been rather poor. For the majority of Icelandic waterworks the chemical data have been insufficient, especially for the smaller ones. This study had as one aim to improve this situation.

2 GEOLOGY AND WATER RESOURCES

Iceland is an island located on the Mid-Atlantic Ridge which creates quite unique geological and hydrological conditions, greatly influencing the properties and quantity of the groundwater resources of the country.

2.1 *Geology of Iceland*

Iceland is formed on the Mid-Atlantic Ridge where a hot spot of volcanic activity is located below the divergent North American-Eurasian Plate boundaries and has created an anomalous elevated basalt plateau (Fig. 1). Crustal growth through volcanism and extensional faulting occur along the NE-SW active rift through the country.

In simple terms the geological formations of Iceland can be split into three main formations (Saemundsson 1979). The oldest rocks are the Tertiary basalts up to 16 Ma old (Fig. 2). Secondly there is a bedrock formation 0.8–3.3 Ma old composed of hyaloclastite basalt and interglacial lavas. The third formation is the volcano-tectonic zone younger than 0.8 Ma (Saemundsson 1979).

2.2 *Permeability of different rock formations*

The older Tertiary formation has a much lower permeability than the young recent lava formation. There the permeability is mainly fracture dependent as veins and vugs in the Teriary basalt lavas have to a great extent been filled by secondary minerals (Walker 1960). The hyaloclastites in the younger formations can act both as impermeable acquicludes and be quite permeable, depending on their structure and alteration state. The main permeable zones are also in those formations mainly confined to fault and fracture zones. In the zones covered

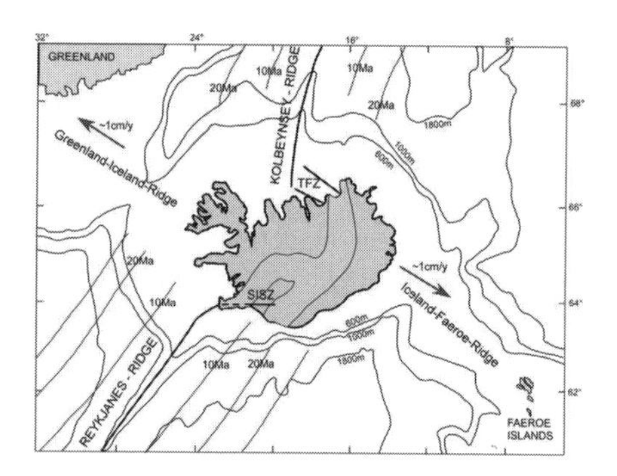

Figure 1. Location of Iceland on the Mid-Atlantic-Ridge at the plate boundaries of the American and the Eurasian plate (Björnsson et al. 2007).

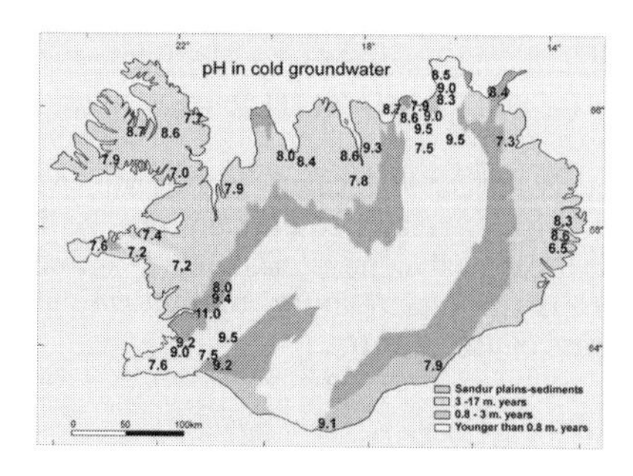

Figure 2. A simplified geological map of Iceland (Saemundsson 1979) with added values of typical pH of cold groundwater resources at the locations.

with recent lava fields younger than the last glaciation (<10.000 years) there are almost no creeks or surface waters as the rain seeps directly down to the groundwater table (Sigurdsson et al. 1988).

2.3 *Meteorological conditions*

The average precipitation in the whole country is about 1800 mm/year (Jónsdóttir 2008). But precipitation is very variable, with the south and southwest parts of Iceland getting the highest total precipitation, whereas the northern highlands being the driest parts. As the temperature in the Icelandic highlands remains below zero for some months during winter precipitation gets stored until the spring before draining of the surface and filtering down into the groundwater aquifers (Jónsdóttir 2008). Those facts influence also the groundwater composition as well as the water reservoirs. The salinity of the icelandic groundwaters depends directly on the distance from the sea (Sigurdsson et al. 1988).

3 SAMPLING AND ANALYSIS

For this project about 150 samples of cold water were collected all over the country, the majority being groundwater. Three labs located at the University of Iceland, University of Akureyri and the Iceland Geosurvey joined forces in the project. Sampling was conducted according to best geochemical practice (Arnórsson 2000, Kristmannsdóttir 2008). The samples were analyzed for the main dissolved solids (SiO_2, Na, K, Ca, Mg, CO_{2t}, SO_4, Cl, F), pH and 35 trace elements (Ag, Al, As, Au, B, Ba, Br, Cd, Co, Cr, Cs, Cu, Fe, Ga, Ge, Hg, I, Li, Mn, Mo, Ni, P, Pb, Rb, Sb, Se, Sn, Sr, Th, Ti, Tl, U, V, W, Zn). Nutrients were also analyzed in many of the samples, as well as Rn, electrical conductivity and dissolved oxygen. The stable isotopes δ^2H, $\delta^{18}O$ and $\delta^{13}C$ were also measured and ^{14}C (pMC) was measured in about 2/3 of the samples. The chemical analyses were performed according to standard methods for the examination of water and wastewaters (Franson 1998). The stable isotopes of oxygen, hydrogen and carbon were analyzed on a Finnegan MAT 251 mass-spectrometer at the Institute of Earth Sciences, University of Iceland. ^{14}C measurements were performed at the AMS Centre at Aarhus University, Denmark. (Sveinbjörnsdóttir et al. 1992).

4 RESULTS

The main chemical characteristics of the waters are high pH and alkalinity, low TDS and very little environmental contamination. This compares well with earlier studies of cold groundwater and potable water in Iceland (Arnórsson et al. 2002,

Kristmannsdóttir 2004). In Figure 3 pH is shown relative to temperature of the cold groundwater, clearly demonstrating the alkaline nature of the waters. In Figure 2 a simplified geological map of Iceland is shown along with the typical pH of cold groundwater in the formations. A significant difference in pH is demonstrated for waters originating in the Tertiary bedrock as compared to waters originating in the active zones of rifting and volcanism. The groundwater in the older Tertiary bedrock have typically a pH of 8.4–8.7, except when there is a substantial surface water derived component. Waters with very high pH are found at the borders of the volcanic zones, especially in waters originating in hyaloclastite formations. Some of the waters originating in the volcanic zone have lower pH due to influence of geothermal gases. Figure 4 shows a Schoeller classification diagram for the cold groundwater demonstrating that the waters are mostly classified as alkali-calcium-bicarbonate or chloride waters. The Mg concentration varies, mostly depending on salinity. A Piper classification diagram (Fig. 5) demonstrates a similar classification as well as the great variation in composition.

A graph showing the Cl plotted against total carbonate shows that only very few of the waters have chloride concentration exceeding 20 mg/L, whereas the total carbonate concentration varies much more widely. A small number of more saline and carbonated waters have been omitted from the graph, not to distort the scale.

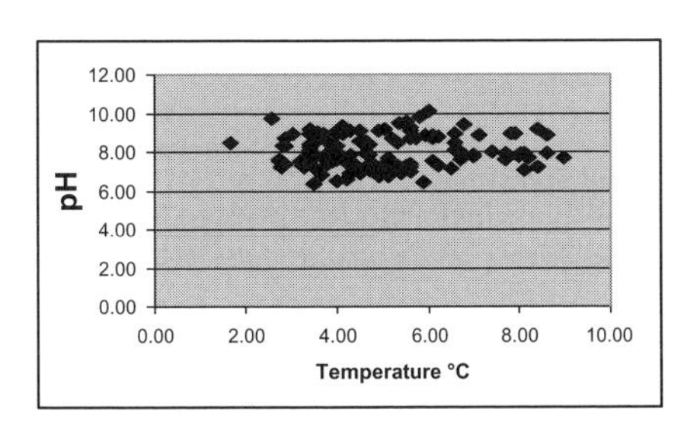

Figure 3. pH of the cold waters relative to temperature.

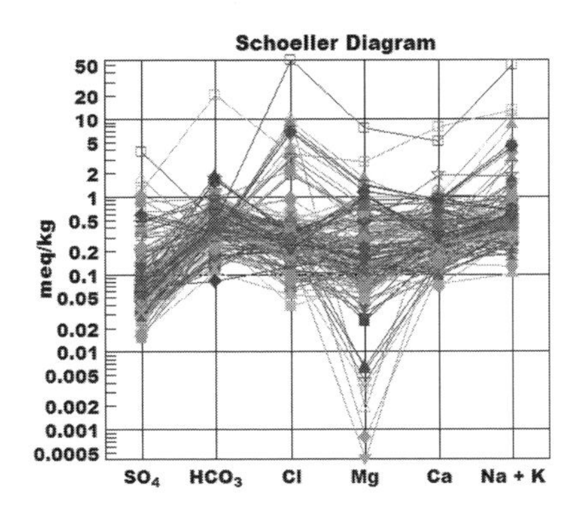

Figure 4. Schoeller diagram of the chemical properties of the cold groundwater samples studied in the project.

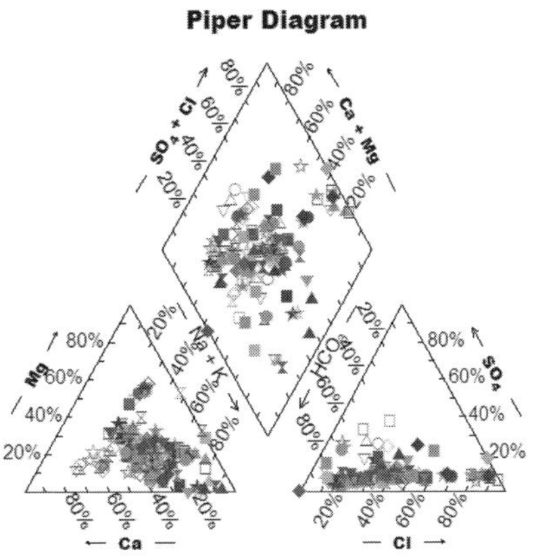

Figure 5. A Piper classification diagram of the waters.

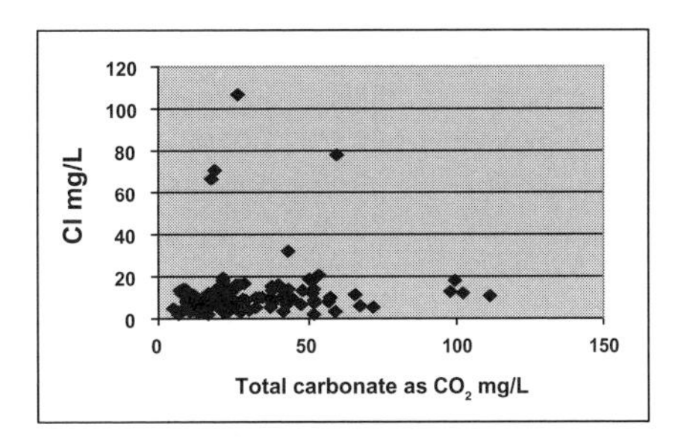

Figure 6. Chloride against total carbonate (as CO_2) for the cold groundwater.

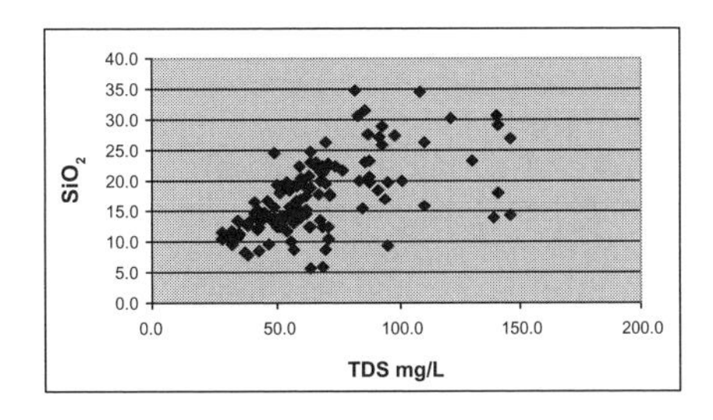

Figure 7. The concentration of silica against total dissolved solids (TDS) in the cold groundwater.

The relation of silica concentration to total dissolved solids (TDS) is shown in Figure 7. Most of the waters have TDS in the range 30–100 mg/L and silica concentrations vary from 5–35 mg/L. Table 1 shows analyses of typical groundwater from the various bedrock formations.

Table 1. Chemical analysis showing selected elements in typical cold groundwater samples.

Type	Teriary NE	Active Z SW	Flank Z SW	Recent Reykjanes
Temp.°C	2.9	4.5	5.2	5.0
pH/°C	8.6/21	9.1/24	10.9/20	7.6/24
SiO_2 mg/L	16.5	13.9	34.8	13.8
B mg/L	0.002	0.006	0.06	0.011
Na mg/L	4.5	11.1	29.8	33.7
K mg/L	2.9	0.4	0.28	1.35
Ca mg/L	4.8	4.4	7.6	7.6
Mg mg/L	0.728	0.814	0.005	7.0
CO_2 mg/L	16.5	17.0	8.6	17.6
SO_4 mg/L	0.7	2.3	2.8	9.2
Cl mg/L	2.3	10.1	5.7	67
F mg/L	0.04	0.05	0.22	0.07
P μg/L	44	16.4	2.2	20
TDS mg/L	41	51	86	149
Fe μg/L	1.0	2.7	0.4	2.8
Al μg/L	2.2	24	196	8
As μg/L	0.20	<0.05	0.10	0.35
Ba μg/L	0.03	0.03	0.01	0.46
Br μg/L	14.8	32.5	23.4	209.0
Cd μg/L	<0.002	<0.002	<0.002	0.005
Hg μg/L	<0.002	<0.002	<0.002	<0.002
Se μg/L	0.129	0.292	–	0.418
Zn μg/L	0.83	0.68	0.35	7
δ^2H ‰	–89.4	–58.6	–66.3	–50.7
δ^{18}O ‰	–12.53	–8.57	–9.28	–7.79
^{14}C age BP	681	470	2721	–

5 DISCUSSION AND CONCLUSION

The analysis of the groundwater have been classified after water types, correlated to geographical location, precipitation load, geological bedrock formation and chemical equilibrium with minerals was assessed. Groundwater in the active zones of volcanism and rifting has typically a high pH, about or exceeding 9, and very low mineralization as the basement rocks are highly permeable, unaltered and glassy. In the areas near to the sea in SW and NE Iceland, like in Reykjanes and Öxarfjördur the water has higher salinity (Table 1). In the Tertiary basement in the Western, Northern and Eastern part of the country the groundwater has somewhat lower pH, 8.3–8.7, and higher mineralization due to lower permeability and more altered reservoir rocks. The cold groundwaters are generally undersaturated with respect to calcium carbonate in contrast to geothermal waters which are generally in equilibrium with calcite (Kristmannsdóttir 2004). The groundwater from the Tertiary bedrock formation is often near to equilibrium with respect to chalcedony whereas the ones from the recent bedrock within the active zones are mostly undersaturated.

The groundwaters mostly originate in local mountain complexes, but some derive from further away in the highlands. Some places have waters with surface water influx or a very shallow and/or short influx time resulting in waters of lower pH, higher carbonate and stable isotopes as the local precipitation. The apparent ^{14}C age of the waters is mostly less than 1000 years, but there is a great variation. The concentration of heavy metals and trace elements is very low and often near the detection limits for conventional analytical methods. The waters are very different in composition and chemical properties from waters in most other countries and care has to be taken in the design of waterworks and selection of pipe and construction material.

REFERENCES

Arnórsson, S., 2000. Sampling of geothermal fluids: On-site measurements and sample trearment. In S. Arnórsson (ed), *Isotopic and Chemical techniques in Geothermal Exploration, Development and Use*: 84–142. Vienna: International Energy Agency.

Arnórsson, S., Gunnarsson, I., Stefánsson, A., Andrésdóttir, A. & Sveinbjörnsdóttir, Á.E. 2002. Major element chemistry of surface and ground waters in basaltic terrain, N-Iceland. *Geochim. Cosmochim. Acta* 66: 4015–4046.

Björnsson, A., Saemundsson, K., Sigmundsson, F., Halldórsson, P., Sigbjörnsson, R. & Snaebjörnsson, J.Th. 2007. *Geothermal Projects in NE Iceland at Krafla, Bjarnarflag, Gjástykki and Theistareykir. Assessment of geo-hazards affecting energy production and transmission systems emphasizing structural design criteria and mitigation of risk*. LV-2007/075. Reykjavík: Theistareykir Ltd, Landsnet, Landsvirkjun.

Franson, M.A.H. 1998. *Standard Methods for the examination of water and wastewater*, American Public Health Association, DC, USA: 2005–2605.

Jónsdóttir, J.F. 2008. A runoff map based on numerically simulated precipitation and a projection of future runoff in Iceland. *Hydrological Sciences Journal* 53: 100–111.

Kristmannsdóttir, H. 2008. *Sampling of hot and cold water for geochemical analysis*. Institute of Natural resources working paper: University of Akureyri.

Kristmannsdóttir, H. 2004. Chemical characteristics of potable water and water used in district heating systems in Iceland. *Proceedings 13th Scandinavian Corrosion Congress*: 1–6.

Saemundsson, K. 1979: Outline of the geology of Iceland. *Jökull* 29: 11–28.

Sigurðsson, F. & Einarsson, K. 1988. Groundwater resources of Iceland. Availability and demand. *Jökull* 38: 35–54.

Sveinbjörnsdóttir, Á.E., Heinemeier, J., Rud, N. & Johnsen, S.J. 1992. ^{14}C anomalies observed for plants growing in Icelandic geothermal waters. *Radiocarbon* 34: 696–703.

Walker, G.P.L. 1960. Zeolite zones and dike distribution in relation to the structure of the basalts of eastern Iceland, *Journal of Geology* 68: 515–528.

Water-Rock Interaction – Birkle & Torres-Alvarado (eds)
© 2010 Taylor & Francis Group, London, ISBN 978-0-415-60426-0

Fluid mixing and boiling during latest stage orogenic gold mineralization at Brusson, NW Italian Alps

G. Lambrecht & L.W. Diamond
Rock–Water Interaction Group, Institute of Geological Sciences, University of Bern, Switzerland

ABSTRACT: The orogenic gold deposit at Brusson, NW Italian Alps, consists of quartz-carbonate-sulphide veins, which are known to have formed from a CO_2–H_2O–NaCl hydrothermal fluid. Our recent discovery of an aqueous type of fluid inclusion, previously unknown at this location, raises the question of which processes triggered gold deposition. New petrographic evidence and fluid inclusion analyses show that the aqueous inclusions represent pore waters from the local wall rocks. These pore waters seeped into the vein structures and mixed with the main ore-bearing fluid, which at that time happened to be boiling. However, timing constraints suggest that this mixing event played no part in gold deposition. All the observations of gold and associated fluid inclusions suggest that cooling, wall-rock reactions and boiling of the main carbonic fluid were the principal causes of gold deposition. The aqueous inclusions simply record a lull in hydrothermal activity between gold-deposition events.

1 INTRODUCTION

A current question in the genesis of orogenic gold deposits is the influence on gold solubility of reactions between the ore-bearing fluid and the wall rocks of the deposits. In this context, we are investigating fluid inclusions in hydrothermal quartz-carbonate-sulphide veins from the abandoned Fenilia Mine at Brusson, northwestern Italian Alps.

A previous study (Diamond 1990) concluded that gold precipitated during main-stage vein growth from a homogeneous CO_2–H_2O–NaCl liquid, in response to wall-rock reactions, cooling and decompression of the liquid from ~300°C, 130 MPa to ~230°C, 60 MPa, driven by uplift and extension of the Western Alps. In a final stage of vein growth the continued cooling and decompression caused the hydrothermal fluid to boil (unmix/effervesce) and deposit free gold.

Using modern optical equipment we recently found a type of primary fluid inclusion that was previously unknown at this deposit: aqueous liquid + vapour ($L_{aq}V$) inclusions in which the low-density vapour contains variable amounts of N_2, CO_2 and CH_4. These inclusions were trapped within quartz near the end of its growth history, raising questions about (1) whether they record a fluid that was present during the deposition of free gold, and (2) whether this fluid played an active role in triggering gold deposition within the vein system. Both questions are addressed in this paper, based on petrographic relationships and on detailed fluid inclusion analyses.

2 SAMPLES AND METHODS

Quartz crystals were sampled from the main hydrothermal vein where it cuts through mafic amphibole-rich gneiss, albite-biotite paragneiss and calcite marble wall rocks. Fluid inclusion petrography was performed on polished sections using a 100X objective and a relatively new SEM technique known as charge-contrast imaging (Griffin 2000).

Phase transitions in the inclusions were measured by microthermometry to ± 0.2°C. Volume fractions of the carbonic phase [$\varphi_{(car)}$] were estimated using a spindle stage (Bakker & Diamond 2006) with a relative uncertainty of ±4%.

The compositions of the non-aqueous phases were determined by laser Raman spectroscopy. Peak integrals were inserted into the Placzek equation (1934; Dubessy et al. 1989) using scattering efficiencies (σ) of 2.5, 1 and 7.5 for CO_2, N_2 and CH_4 respectively. These values are valid for 1 bar pressure. In the absence of appropriate calibrations we assumed the constants are still valid for the high internal pressures of the fluid inclusions. This assumption, and the knowledge that the constants depend on the gas mixture (e.g. Seitz et al. 1996), gives rise to relative uncertainties of ~ ±15% in the molar ratios of the gases. Similarly, the dependence of spectral positions on gas composition precludes useful estimates of internal pressure. All these uncertainties propagate into the calculations of salinity of the $L_{aq}V$ inclusions based on clathrate modelling (Diamond 1994, Bakker 1997).

3 RESULTS

3.1 *Relative timing of fluid inclusions and gold*

The samples studied contain early primary and pseudosecondary inclusions that can be approximated by the CO_2–H_2O–NaCl system. Two types are observed: first, abundant low–XCO_2 $L_{aq}L_{car}V$ inclusions in which the homogenized carbonic phases contain 97.0 mol.% CO_2, 2.9 mol.% N_2 and 0.1 mol.% CH_4; second, high–XCO_2 $L_{aq}L_{car}$ inclusions which are present only in some late pseudosecondary assemblages. The two contrasting inclusion types were described by Diamond (1990) and interpreted to have formed during boiling (phase separation) of the low–XCO_2-type ore-bearing solution responsible for main-stage mineralisation.

New charge-contrast imaging has revealed fine overgrowths of quartz on the main-stage crystals. These host late-primary inclusion assemblages that consist of rare low–XCO_2 $L_{aq}L_{car}V$ inclusions (identical to those described above) coexisting with abundant $L_{aq}V$ inclusions. The latter have variable volume fractions of their carbonic phases [$\varphi_{(car)}$] ranging from 0.075 to 0.47 (±4%) (Fig. 1a). In quartz-vein samples hosted by albite-biotite paragneiss and amphibole-rich gneiss, solid inclusions of ankerite are associated with these late primary assemblages. In samples hosted by calcite marbles, solid inclusions of dolomite sometimes accompany some of the assemblages.

The quartz sample in Figure 2 exhibits two generations of gold: small inclusions within a late growth zone (Early gold in zone 2, Fig. 2c); and free gold mostly overlying the outer crystal surface (Late gold, Fig. 2c). Charge-contrast imaging reveals that this late gold is younger than the fine crystal growth zone 3 (Fig. 2c) that hosts the late primary $L_{aq}V$ inclusion assemblages. The late gold is also in direct contact, and therefore coeval with healed fractures (zone no. 4 in Fig. 2c) that host a secondary assemblage of variable–XCO_2 inclusions (including both low–XCO_2 $L_{aq}L_{car}V$ and high–XCO_2 $L_{aq}L_{car}$ types). No $L_{aq}V$ inclusions are present in zone 4.

3.2 *Gas compositions*

Raman analysis revealed CO_2, CH_4 and N_2 within the carbonic phases of all $L_{aq}V$ inclusions, but the molar ratios of the gases vary systematically. In a triangular diagram (Fig. 1b) these analyses reveal a linear array stretching from compositions close to those of low–XCO_2 $L_{aq}L_{car}V$ and high–XCO_2 $L_{aq}L_{car}$ inclusions (end-member no. 1 in Fig. 1) to compositions of about 4.4 mol.% CO_2, 91.4 mol.% N_2 and 4.2 mol.% CH_4 (end-member no. 2 in Fig. 1) for inclusions in vein-quartz hosted by mafic amphibole-rich gneiss and albite-biotite paragneiss. Quartz-vein samples hosted by calcite marbles contain $L_{aq}V$ inclusions with proportionally more methane in their carbonic phases (end-member no. 2b in Fig. 1b).

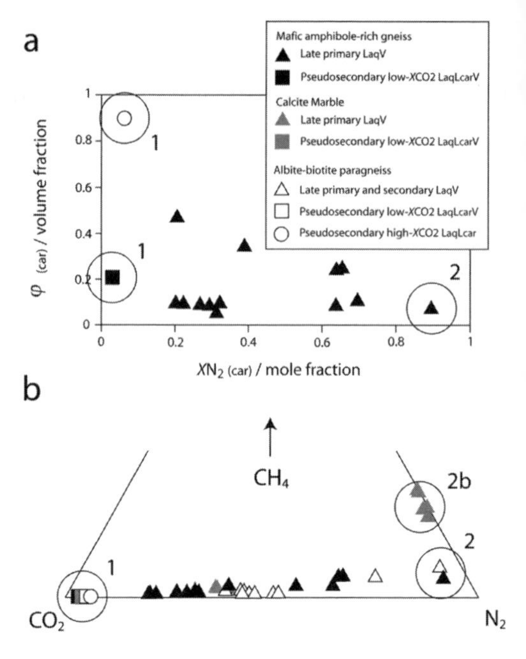

Figure 1. (a) Relationship between volume fraction of the carbonic phase [$\varphi_{(car)}$] and mole fraction of N_2 in the carbonic phase [$XN_{2(car)}$] and (b) composition of the carbonic phase in $L_{aq}V$, low–XCO_2 $L_{aq}L_{car}V$, and high–XCO_2 $L_{aq}L_{car}$ inclusions from quartz-vein samples hosted by different types of wall rock (types are listed in the text box). Compositions 1, 2 and 2b are interpreted to be end-members of a mixing line. See text for further details.

3.3 *Microthermometry*

Available microthermometric data correlate with the trends observed in Figure 1. In inclusions closer to end-member no. 1, clathrate is the last solid stable upon progressive heating whereas in inclusions closer to end-member no. 2 and 2b, ice is the last solid to melt. The low–XCO_2 $L_{aq}L_{car}V$ inclusions have an aqueous salinity of 3.7 mass% $NaCl_{equiv.}$ and they homogenize to liquid at ~230°C. The high–XCO_2 $L_{aq}L_{car}$ inclusions behave similarly upon heating and have an aqueous salinity of 2.2 mass% $NaCl_{equiv}$. For reasons given above, the salinity of the end-member $L_{aq}V$ inclusions (no. 2 in Fig. 1) could be constrained only roughly between 0.3 and 7 mass% $NaCl_{equiv}$. These inclusions homogenize at ~180°C via the transition $L_{aq}V \rightarrow L_{aq}$.

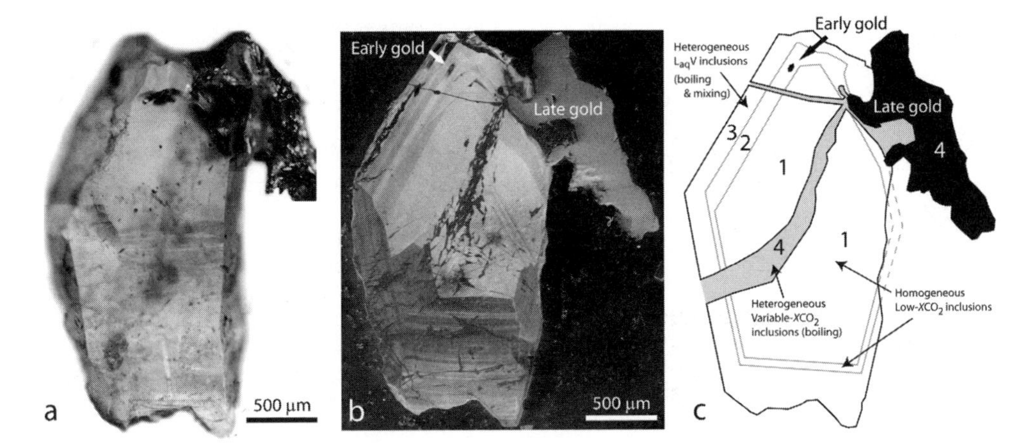

Figure 2. (a) Photograph of quartz crystal with free gold deposited on the surface. (b) SEM charge-contrast image of a polished section through this crystal showing quartz growth zones and solid gold inclusions (Early gold) as well as free gold (Late gold) (c) Schematic interpretation of the charge-contrast image in b: (1) main-stage quartz, (2) earliest growth zone with solid inclusions of gold (Early gold), (3) fine quartz overgrowths hosting late primary, $L_{aq}V$ inclusions, (4) healed fractures hosting low–XCO_2 $L_{aq}L_{car}V$ and high–XCO_2 $L_{aq}L_{car}$ inclusions (secondary assemblages), in direct contact with late gold.

3.4 *LA-ICP-MS*

Quantitative LA-ICP-MS analyses show that inclusions at and near the end-member no. 1 in Figure 1 have higher As and lower Sr contents than those nearer end-member 2. The Sr content in $L_{aq}V$ inclusions is higher where the vein is hosted by calcite marble rather than other wall rocks.

4 DISCUSSION

The variable volume fractions of the carbonic phases [$\varphi_{(car)}$] of the late primary $L_{aq}V$ inclusion assemblages (Fig. 1a) indicate heterogeneous entrapment. This suggests boiling (phase separation) of a single parent fluid. If boiling were the sole cause of this variability then the carbonic vapour phase in these inclusions would be enriched in the more volatile gas species. In our case, inclusions with larger $\varphi_{(car)}$ values should contain relatively more N_2. However, as visible in Figure 1a, this is not the case. Therefore, we conclude that the late primary $L_{aq}V$ assemblages are not a product of boiling alone.

The liquid end-members 1, 2 and 2b identified in Figure 1 are mutually miscible at the entrapment temperature of the low–XCO_2 liquid, 230°C. Therefore, the linear compositional trend of the late primary $L_{aq}V$ inclusions in Figure 1b can be interpreted as a mixing line. The mixture formed at 230°C when the low–XCO_2 liquid and the high–XCO_2 vapour end-members mixed with random amounts of liquid end-member 2 or 2b. Combining this interpretation with our conclusion above, it appears that boiling and mixing occurred simultaneously in the vein system during precipitation of quartz zone 3 (Fig. 2c).

Figure 1b shows that end-member 2b (hosted by calcite marble wall rocks) has relatively more methane than end-member 2 (hosted by gneisses). This suggests that the composition of $L_{aq}V$ inclusions is wall-rock dependent. Therefore, this fluid probably originated from within the wall rocks around the vein, rather than from an external source. This hypothesis is further supported by our LA-ICP-MS data that show higher Sr content in $L_{aq}V$ inclusions from vein-quartz hosted by calcite marbles.

The well-known geochemical coherency of Au and As in hydrothermal gold deposits, together with our observation that the low–XCO_2 inclusions are enriched in As, supports the earlier deduction (Diamond 1990) that this CO_2–H_2O–NaCl liquid (no. 1 in Fig. 1) was responsible for gold transport. A positive correlation between As and CO_2 was also observed by Yardley et al. (1993). Boiling of this liquid, as demonstrated by the fluid inclusions in zone 4 of Figure 2c, is clearly associated with deposition of free gold. Hence, its active role in inducing gold precipitation, as suggested previously (Diamond 1990), is confirmed by our new observations.

The question now is whether the admixture of pore water from the wall rocks also played a role in gold deposition. To shed light on this the new temporal and compositional deductions are used to reconstruct the history of quartz growth in the Fenilia vein (Fig. 3). First, a low–XCO_2-type

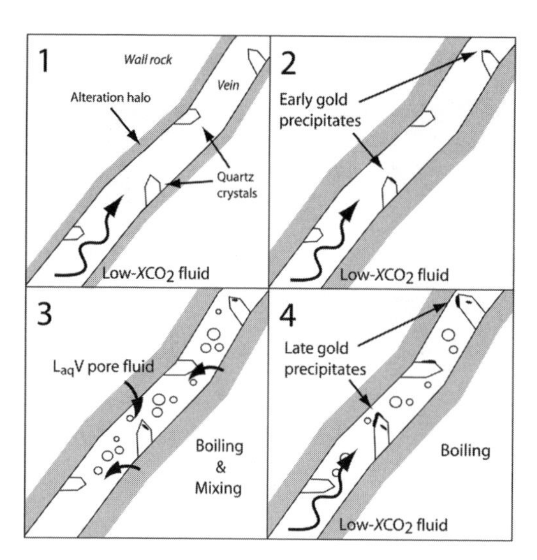

Figure 3. History of quartz veins at Fenilia (numbers coincide with numbers in Fig. 2c). (1) Quartz crystals form in a low–XCO_2 carbonic fluid. Wall rocks get altered (shaded area). (2) Early gold precipitates on quartz. Wall rocks are altered further. (3) The influx of vein fluid wanes, pressure drops and the fluid starts boiling. Pore fluid flows back into the vein and mixes with the boiling fluid. (4) A new influx of low-XCO_2 fluid triggers late gold precipitation.

carbonic liquid forms main-stage quartz (zone 1 in Fig. 2c) and alters the adjacent wall rock (Fig. 3.1). Subsequent simultaneous precipitation of gold and quartz from this homogeneous liquid (Fig. 3.2) is presumably triggered by cooling or by wall-rock reactions. Progressive cooling and reduction in fluid pressure owing to on-going uplift and extension of the Western Alps induces phase-separation (boiling) of the main-stage fluid (Diamond 1990, Fig. 3.3). The change in composition of the liquid end-member is minor, suggesting that boiling is gentle. Pore fluid from the wall rocks simultaneously seeps into the open vein, where it mixes with the small quantities of gently boiling carbonic fluid. At this moment the late primary $L_{aq}V$ inclusions are trapped (zone 3 in Fig. 2c). The drop in pressure of the main hydrothermal fluid may have driven this infiltration of local pore water. On the other hand, pore water may have been draining from the wall rock during the entire vein history, but it was swamped by the high flux of ore-bearing solution and therefore not recorded in fluid inclusions. Finally (Fig. 3.4), a new pulse of boiling, ore-bearing carbonic fluid enters the mine level, fracturing the existing quartz crystals and depositing the late gold on top of them (zone 4 in Fig. 2c). Without any evidence for wall-rock pore waters at this stage, it seems most likely that partitioning of gold-complexing ligands (aqueous HS^- and H_2S) from the liquid into the carbonic vapour triggered deposition of the late gold.

5 CONCLUSIONS

The newly discovered $L_{aq}V$ inclusions represent pore water from the wall rocks of the Fenilia vein. The pore water seeped into the vein and mixed with small amounts of boiling carbonic fluid during a lull in the main hydrothermal activity. Gold precipitated prior to this lull (probably triggered by cooling or by reactions with the wall rocks) and also after this lull (probably triggered by loss of gold-complexing ligands to the escaping vapour phase), but not during the lull. Thus, the influx of pore water from the wall rocks does not appear to have played a role in gold deposition. Evidence of other processes (mainly cooling and boiling) is sufficient to explain the observed gold.

REFERENCES

Bakker, R.J. 1997. Clathrates: Computer programs to calculate fluid inclusion V-X properties using clathrate melting temperatures. *Computers & Geosciences* 23(1): 1–18.

Bakker, R.J. & Diamond, L.W. 2006. Estimation of volume fractions of liquid and vapor phases in fluid inclusions, and definition of inclusion shapes. *American Mineralogist* 91(4): 635–657.

Diamond, L.W. 1990. Fluid inclusion evidence for P-V-T-X evolution of hydrothermal solutions in late-alpine gold-quartz veins at Brusson, Val-d'Ayas, Northwest Italian Alps. *American Journal of Science* 290(8): 912–958.

Diamond, L.W. 1994. Salinity of Multivolatile Fluid Inclusions Determined from Clathrate Hydrate Stability. *Geochimica et Cosmochimica Acta* 58: 19–41.

Dubessy, J., Poty, B. & Ramboz, C. 1989. Advances in C-O-H-N-S fluid geochemistry based on micro-Raman spectrometric analysis of fluid inclusions. *European Journal of Mineralogy* 1: 517–534.

Griffin, B.J. 2000. Charge contrast imaging of material growth and defects in environmental scanning electron microscopy – Linking electron emission and cathodoluminescence. *Scanning* 22(4): 234–242.

Placzek, G. 1934. *Handbuch der Radiologie*. Leipzig: Akademischer Verlag.

Seitz, J.C., Pasteris, J.D. & Chou, I.M. 1996. Raman spectroscopic characterization of gas mixtures 2. Quantitative composition and pressure determination of the CO_2-CH_4 system. *American Journal of Science* 296: 577–600.

Yardley, B.W.D., Banks, D.A., Bottrell, S.H. & Diamond, L.W. 1993. Post-metamorphic gold-quartz veins from NW Italy: the composition and origin of the ore fluid. *Mineralogical Magazine* 57(3): 407–422.

Water-Rock Interaction – Birkle & Torres-Alvarado (eds)
© 2010 Taylor & Francis Group, London, ISBN 978-0-415-60426-0

A new genetical copper deposit model from the East Pontic metallotect, NE Turkey: The Murgul type

N. Özgür
Groundwater and Mineral Resources, Research and Application Centre for Geothermal Energy, Süleyman Demirel Üniversitesi, Isparta, Turkey

ABSTRACT: The Murgul deposit in the East Pontic metallotect is assigned to a subvolcanic formation related to an Upper Cretaceous island arc temporally developed under subaerial conditions, whereas the deposits at Madenköy and Lahanos, in the western part, are genetically associated with a submarine hydrothermal activity. There is a Cu-Mo belt in the northern part of the Anatolian microplate. The Cu deposit of Murgul can be interpreted as a transition from Kuroko-type volcanogenic massive sulfides (VMS) to porphyry copper deposits and might be viewed as a new ore deposit model (Murgul type).

1 INTRODUCTION

Murgul is one of the principal copper deposits of Turkey, and is located in the eastern part of the East Pontic metallotect (Fig. 1), which hosts a great number of base metal deposits and represents an island arc system developed during Lias to Miocene time period (Özgür 1985; Dieterle 1986). The metallotect consists of a 2.000 to 3.000 m thick volcanic sequence with thin intercalations and lenses of marine sediments. The East Pontic volcanic sequence has been divided into three volcanic cycles (i.e. Akin 1979; Özgür 1985; Özgür & Schneider 1988; Schneider et al. 1988; Fig. 2):

i. The first cycle comprises a volcanic pile deposited between Jurassic and Upper Cretaceous and is characterized by a basal sequence of basaltic volcanic rocks that changes to felsic lava flows and thick pyroclastic rocks in the middle and top part of the cycle.
ii. the second cycle starts with volcanic breccias, tuffs and minor intercalations of marine sediments overlain by andesitic and rhyolitic flows. The volcanic sequence is in turn overlain by Maastrichtian limestones.
iii. The late cycle consists of a basal sequence of marine sediments of Paleocene age which are overlain by andesitic and basaltic lava flows representing Tertiary volcanic activity.

The base metal ratios of the most important deposits in the strongly altered pyroclastic rocks of Senonian age progressively change along the general strike of the metallotect, from E (Cu>>Pb+Zn) to W (Pb–Zn >>Cu). In the western part, the sulfide are deposits are predominantly stratiform, *e.g.* Madenköy (Cagatay & Boyle, 1980; Dieterle 1986) and Lahanos (Tugal 1969), whereas in the eastern part they are generally strata-bound (stockworks, veins, and disseminations, *e.g.* Murgul (Özgür 1985).

The aim of this paper is to present the copper deposit of Murgul, generated under subaerial conditions, as a new genetical ore deposit model which represents a transition from Kuroko-type VMS to porphyry copper deposits in the East Pontic metallotect.

2 GEOLOGIC SETTING

The copper deposit of Murgul occurs within the upper part of the first volcanic cycle and is associated with a 250 m thick group of felsic pyroclastic rocks whose upper contact is marked by a thin layer of marine sediments, and is characterized by intense erosion and weathering (Özgür 1985; Schneider et al. 1988). The pyroclastic host rocks can be classified into Senonian age according to paleontological observations (Buser & Cvetic 1973) and overlain by interbedded tuffs, sandstones and limestones, and by 200 to 500 m thick barren dacitic lava flows. The pyroclastic host rocks consist of altered breccias and tuffs. Their primary minerals can be observed only as relicts. In the less altered samples, the fluidal groundmass contains phenocrysts of plagioclase (An_{28-35}) and quartz and plagioclase microlites (An_{12-30}), relicts of hornblende and biotite, and minor apatite, sphene, and haematite (Özgür & Schneider 1988; Schneider et al. 1988).

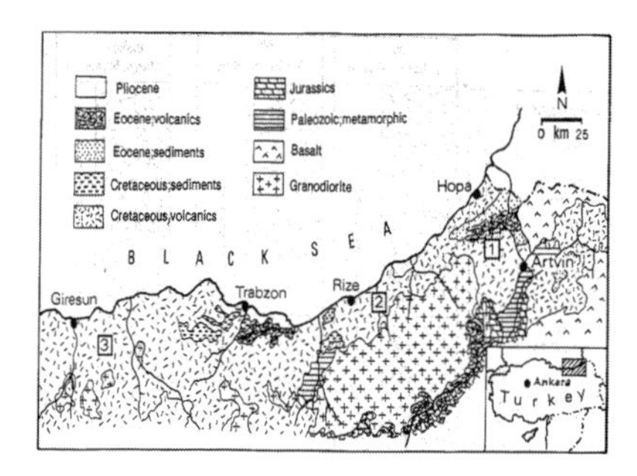

Figure 1. Geologic setting and location map of the East Pontic metallotect. Ore deposits: 1: Murgul, 2: Madenköy, 3: Lahanos (Özgür 1993a).

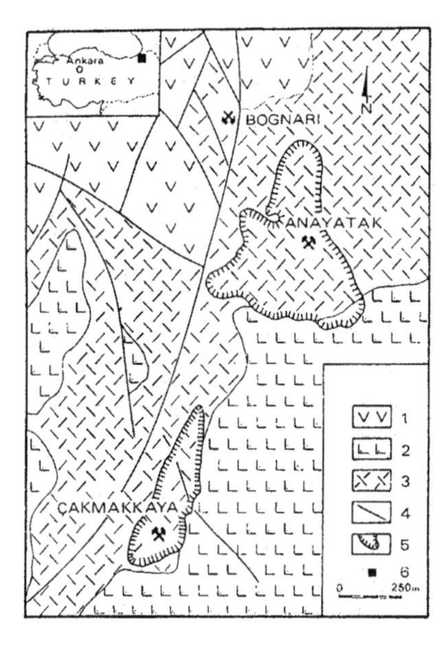

Figure 2. Geologic sketch map of the Murgul deposit. 1: Andesitic lava flows, 2: hanging dacitic lava flows, 3: dacitic pyroclastic rocks: host rock of the orebodies, 4: main faults, 5: boundary of the open pits, 6: study area (Özgür & Schneider 1988).

3 HYDOTHERMAL ALTERATION AND MINERALIZATION

The copper deposit of Murgul is mainly mined in the two open pits of Anayatak and Çakmakkaya. The mineralization is associated with a two-stage alteration of pyroclastic host rocks: (i) an initial stage of phyllic and argillic alteration, and a late stage characterized by silicification (Özgür 1985; Özgür & Schneider 1988; Schneider et al. 1988). The study of rare earth element (REE) distribution in the alteration zones supports a distinction between the both alteration stages and reveals a close correlation between increasing wall-rock alteration and depletion of total REE contents (Schneider et al. 1988).

The first stage of hydrothermal alteration led to the destruction of the primary paragenesis of the pyroclastic rocks and the replacement by quartz and pale green to greasy sericite. An extensive but poor mineralization of disseminated pyrite and chalcopyrite of type 1 (Schneider et al. 1988) took place during the first stage. There is a rather sharp contact between phyllic and argillic zones with 1 to 3 m wide transition spheres. The phyllic zone is surrounded by a pervasive argillic alteration zone in which the alteration assemblage consists of quartz, montmorillonite, dickite, illite, pyrite and chalcopyrite.

The late stage of hydrothermal alteration is characterized by silification which consists of quartz replacement of the volcanic host rocks producing cryptocrystalline jasper. The mineralized veins and veinlets (types 2 and 3; Schneider et al. 1988) crosscut the altered pyroclastic host rocks. Both ore occurrence types can be interpreted to be a late phase of ore remobilization (Özgür & Schneider 1988), which was generated during the last episode of volcanic activity. During this time period mechanical disintegration took place, opening faults and fissures for ascending ore-bearing solutions resulted in the formation of the stockworklike mineralization (type 2). Finally, the latest open-space fillings (type 3) suggest ore mineralization closely below the terrestrial surface.

The intense hydrothermal alteration shows a sharp contact with the upper part of the pyroclastic host rocks. Thereon, the sulfide ore mineralization discontinues in vertical profile as well. The hydrothermal alteration seems to be distributed in the upper part of the first volcanic cycle and can be considered, therefore, as standard stratum field geological.

The Murgul deposit consists of (i) a widespread disseminated ore with Cu contents ranging from 0,2 to 0,7 percent (type 1), (ii) a stockworklike ore with average Cu contents between 1,0 and 2,5 percent (type 2), and (iii) small ore lodes with Cu contents from 5,0 to 10,0 percent (type 3).

The ore mineral assemblage of the Murgul deposit consists of pyrite, chalcopyrite and minor contents of sphalerite, galena, fahlore, aikinite, hessite, and tetradymite (Özgür 1985; Özgür & Schneider 1988; Willgallis et al. 1990). For the fist time, native gold was detected (Özgür 1985; Özgür & Schneider 1988).

4 A NEW ORE DEPOSIT MODEL

The base metal deposits of the western part of the East Pontic metallotect, *e.g.* Madenköy and Lahanos, are formed by submarine-hydrothermal activity in a volcano-sedimentary sequence under temporarily subaquatic conditions (Özgür 1993a), and represent Kuroko-type VMS deposits. In comparison, the copper deposit of Murgul can be assigned to a subvolcanic-hydrothermal deposit in an island arc under subaerial conditions (Fig. 3). Moreover, the presence of a Cu-Mo belt in the northern part of the Anatolian microplate (Taylor & Fryer 1980; Taylor 1981) can be considered as a genetical and paragenetical indications of plate tectonic position of the East Pontic island arc development accompanied by base metal deposits. Therefore, we interpret the copper deposit of Murgul as a transition between Kuroko-type VMS deposits and porphyry copper deposits (Murgul type).

The following observations allow us to establish the subvolcanic character of the deposit:

i. The scarce, locally intercalated marine sediments in the thick volcanic sequence indicate a shallow-water depositional environment, at least for the upper part of the first volcanic cycle. In Murgul, there are no stratiform orebodies that are bedded within the pyroclastic sequence. Therefore, the mineralization is endogenetic with respect to the volcanic pile (Schneider et al. 1988). In contrast, there are synsedimentary ores in the submarine tuffs of the East Pontic metallotect.
ii. The pyroclastic sequence in the upper part of the first volcanic cycle was altered and mineralized during a late stage of the volcanic activity by ascending hydrothermal fluids.
iii. The formation of the orebodies of Anayatak and Cakmakkaya must have took place before the short period of subaerial erosion and weathering (Özgür, 1985), which is represented by a thin strongly kaolinized layer of reworked pyroclastics and sediments (Özgür 1985; Dieterle 1986). This layer is considered a regional marker and is evidence for the existence of a temporal subareal formation conditions.
iv. The mineralized and strongly altered host rocks are overlain by a series of relatively weakly altered and barren volcanic rocks. The mineralization does not traverse the marker bed. This emphasizes that the deposits were formed prior the erosional interval and the later eruption of the hanging wall volcanic rocks.
v. The altered and mineralized host rocks show structures similar to the "ore-related breccias" described by Sillitoe (1985). They suggest a local subsurface brecciation which might have been generated by repeated volcanic activity heating the system, because there is evidence for several nearly contemporaneous eruption centers at a distance of a few hundred meters. The Murgul deposit shows some similarities to porphyry ore deposits described by Lowell & Guilbert (1970), but there are some remarkable differences (Fig. 3; Schneider et al. 1988): (1) the high-grade ore is mainly concentrated in the center of the orebodies, (2) there is no potassic alteration and (3) the mineralization must have taken place relatively close to the surface. The $\delta^{34}S$ values of sulfides in Murgul and Lahanos range from 2,33 to 4,83 per mil (Özgür et al. 2008) and are comparable with the values of the Kuroko-type VMS deposits (Ohmoto et al. 1983).

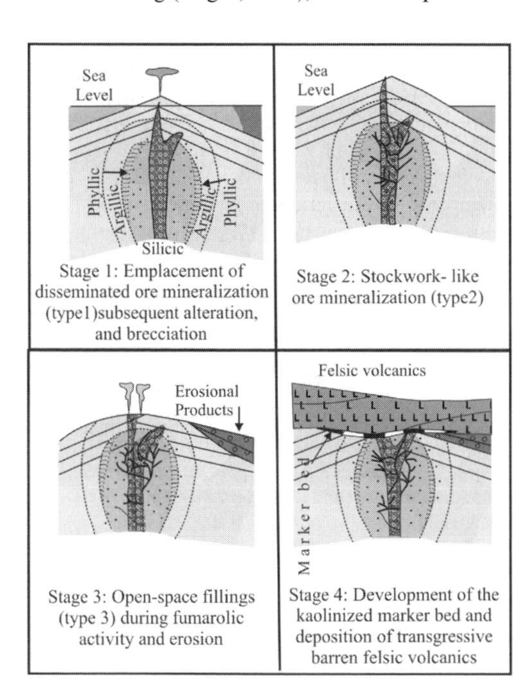

Figure 3. Formation of the copper deposit of Murgul in a schematic representation (Schneider et al. 1988).

From Jurassic to Lower Cretaceous, an island arc was generated and produced the spilites from Murgul and its environs (Özgür, 1985). In Senonian, the output and deposition of volcanic materials increased, the sea water was displaced from Murgul and its environs temporarily and deposited shallow water sediments at the border of island arc locally. During this time period, in which the first volcanic cycle was developed, the formation of the copper deposit of Murgul took place. Within a radius of approximately one to two km, the deposited pyroclastic rocks and small thick sediments were altered by hydrothermal fluids. The base metal deposits are generated in connection with two-stage alteration processes. Thereby, repeated tectonic events led increasely to the breakage of silicified host rocks increasingly and opened further fissures for an increasingly mineralization thereby.

Finally, the copper deposit of Murgul resulted from at least three stages of mineralization (Schneider et al. 1988). After the deposition of the thick dacitic to rhyolitic pyroclastic sequence, subsequent volcanic activity caused the disintegration of the pile at isolated small centers, thus producing optimal permeability. During this stage hydrothermal fluids spread upward forming the first stage

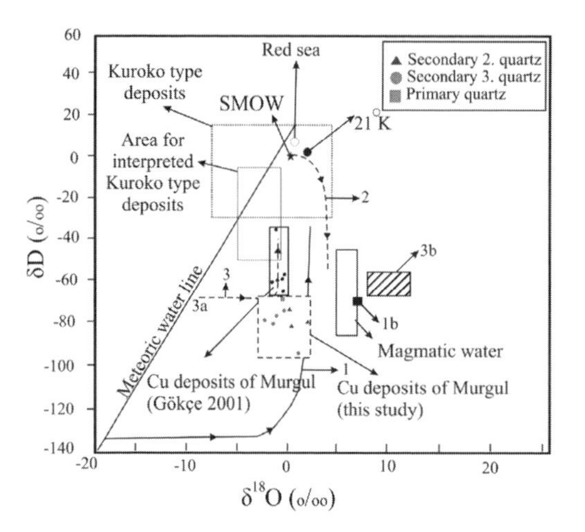

Figure 4. δD versus $\delta^{18}O$ in fluid inclusions of secondary quartz crystals of Anayatak and Çakmakkaya open pits of Cu deposit of Murgul.

of disseminated pyrite and chalcopyrite mineralization (type 1). The continuation of hydrothermal activity led to the formation vein mineralization (stockwork-like ore: type 2; open-space fillings: type 3).

Fluid inclusion measurements in quartz crystals (type ii and iii) associated with ore types 1, 2 and 3 show formation temperatures from 150 to 350°C (mean value: 225°C) and salinity ranging from 1,0 to 12,0 (mean value: 5,0–7,0) NaCl% equivalent (Özgür et al. 2008) which can be considered as an epithermal to mesothermal character (Özgür et al. 2008). Murgul differs from the Kuroko-type VMS deposits in the $\delta^{18}O$ and δD values, contents of anions and cations in fluid inclusions of secondary quartz crystals and due to $\delta^{32}S$ values in sulfide ore minerals obviously (Fig. 4; Özgür et al. 2008) and can be interpreted as a transition between Kuroko-type VMS deposits and porphyry copper deposits. Finally, the Cu deposit of Murgul can be considered as a hot spring-type ore mineralization due to the results of the $\delta^{18}O$ and δD, contents of anions and cations in secondary quartz crystals revealing meteoric origin of the hydrothermal fluids and $\delta^{32}S$ values in ore samples.

In Murgul, the enrichment of fluorine and the depletion of manganese and titanium in the pyroclastic host rocks are suitable for the exploration of blind ore deposits in the East Pontic metallotect, as pathfinder elements in connection with hydrothermal alteration (Özgür 1993b).

REFERENCES

Akin, H. 1979. Geologie, Magmatismus und Lagerstättenbildung im Ostpontischen Gebirge/Türkei aus der Sicht der Plattentektonik. *Geol. Rundschau* 68: 253–283.

Buser, S. & Cvetic, S. 1973. Geology of the environs from the Murgul copper deposit, Turkey. *Maden Tetkik ve Arama Enstitüsü Bull.* 81: 22–45 (in Turkish).

Cagatay, M.N. & Boyle, D.R. 1980. Geology, geochemistry and hydrothermal alteration of the Madenköy massive-sulphide deposit, eastern Black Sea region, Turkey. *Proc. 5th Quadrennial IAGOD Symp*: 653–677.

Dieterle, M. 1986. *Zur Geochemie und Genese der schichtgebundenen Buntmetall-Vorkommen in der Ostpontischen Metallprovinz/NE-Türkei.* Ph.D. thesis, Freie Universität Berlin.

Gökçe, A. 2001, Çakmakkaya ve Damarköy (Murgul-Artvin), bakır yataklarında sıvı kapanımı, oksijen ve hidrojen izotopları jeokimyası incelemeleri ve yatakların oluşumu açısından düşündürdükleri. *Türkiye Jeoloji Bülteni* 44: 1–37.

Lowell, J.D. & Guilbert, J.M. 1970. Lateral and vertical alteration-mineralization zoning in porphyry ore deposits. *Econ. Geol.* 65: 373–408.

Ohmoto, H., Mizukami, M., Drummond, S.E., Eldridge, C.S., Pisutha-Arnond, V. & Lenagh, T.C. 1983, Chemical process of Kuroko formation. *Econ. Geol. Mon.* 5: 570–604.

Özgür, N. 1985. *Zur Geochemie und Genese der Kupferlagerstätte Murgul, E-Pontiden/Türkei.* Ph.D. thesis, Freie Universität Berlin.

Özgür, N. 1993a. Volcanogenic massive sulfide deposits in the East Pontic metallotect, NE Turkey. *Resources Geology Special Issue* 16: 181–185.

Özgür, N. 1993b. Geochemical pathfinder elements of the Murgul copper deposit, NE Turkey. *Resources Geology Special Issue* 16: 164–169.

Özgür, N., Elitok, Ö. & Zerener, M. 2008, Doğu Karadeniz Bölgesi metalojenik kuşağında bulunan Murgul masif sülfit yatağının hidrotermal çözeltileri ve gelişimi. Süleyman Demirel Üniversitesi, Internal Final Report.

Özgür, N. & Schneider, H.-J. 1988. New metallogenetic aspects concerning the copper deposit of Murgul, NE Turkey. *Soc. Geology Applied to Mineral Deposits, Spec. Pub.* 6: 229–239.

Schneider, H.-J., Özgür, N. & Palacios 1988. Relationship between alteration, rare earth element distribution, and mineralization of the Murgul copper deposit, northeastern Turkey. *Econ. Geol.* 83: 1238–1246.

Sillitoe, R.H. 1985. Ore-related breccias in volcanoplutonic arcs. *Econ. Geol.* 80: 1467–1514.

Taylor, R.P. 1981. Isotope geology of the Bakircay porphyry copper prospect, northern Turkey. *Mineral. Deposita* 16: 375–390.

Taylor, R.P. & Fryer, B.J. 1980. Multiple-stage hydrothermal alteration in porphyry copper systems in northern Turkey: the temporal interplay of potassic, prophylitic, and phyllic fluids. *Can. J. Earth Sci.* 17: 901–927.

Tugal, H.T. 1969. *Pyrite sulphide deposits of the Lahanos mine area, eastern Black Sea region, Turkey*. Ph.D. thesis, Dunham University.

Willgallis, A., Özgür, N. & Siegmann, E. 1990. Se- and Te-bearing sulphides in copper ore deposits of Murgul, NE Turkey. *Eur. J. Mineral.* 2: 145–148.

Water-Rock Interaction – Birkle & Torres-Alvarado (eds)
© 2010 Taylor & Francis Group, London, ISBN 978-0-415-60426-0

Mineralogy and genesis of the Rayas ore shoot, and its geological relation with other ore bodies, Veta Madre, Guanajuato, Mexico

F.J. Orozco-Villaseñor
Centro de Geociencias, Universidad Nacional Autónoma de México, Juriquilla, Mexico

ABSTRACT: The Guanajuato mining district, located at the center of Mexico was discovered in 1548. The Veta Madre was found two years later. Its silver production (~36,000 t) has only been surpassed by Pachuca-Real del Monte. The Veta Madre strikes NW-SE for about 26 km dipping 37–55° to the SW. At the end of the 1960's and the beginnings of the 1970's the district seemed to be exhausted but a series of geological and mining discoveries revived it. The Las Torres and Cebada mines proved new reserves; the Rayas showed the first base metals ore body and at the El Cubo mine, several rich gold veins were found. Several ideas have been developed to explain the mineralization, based on diverse analysis techniques, specifically fluid inclusions and stable isotopes. Hypothesis on the origin of the ores have varied from magmatic (related to an intrusion or to volcanism), to lateral secretion, to mixing of fluids, etc., but the debate continues. Here, we will comment about new proposals.

1 INTRODUCTION

Guanajuato is located northwest of Mexico City, nearly at the center of Mexico. Its settlement was prompted by the discovery of the ore deposits of its surroundings, an event that occurred on July 11, 1548. The newly discovered vein was named "San Bernabé" which is part of the vein complex today known as "La Luz". La Luz is a part of the district which additionally encompasses the Veta Madre -discovered by Juan de Rayas, two years after the initial discovery in the district-, and the "vetas de La Sierra" group.

2 ECONOMIC GEOLOGY

The silver production of the Guanajuato district has only been surpassed by the vein system of Pachuca-Real del Monte, Hidalgo, being both Mexican districts above Cerro Rico del Potosí in Bolivia, which was once thought to be "the biggest silver ore deposit of the world". The Guanajuato precious metal production was estimated in 2002, to be around 36,000 t of silver and some 134 t of gold. The mineralization occurs mainly as filling veins although there are also stockworks, breccia chimneys, replacement zones and disseminated ores. The most important structure in the district is the Veta Madre vein, which strikes NW-SE for about 26 km and varies in width from a few cm to nearly 40 m. It dips between 37 to 55° to the SW.

Towards the end of the 1960's and the beginning of the 1970's, when the mining production seemed to be exhausted, a series of mining discoveries revitalized it: The "Las Torres" mine started operations after the discovery of 2,375,000 t of ore, with 353 g/Ag and 2.2 g/t Au. The Cebada mine reported reserves of 1,277,216 t of ore with 372 g/t Ag and 4.04 g/t Au, and the "clavo de Rayas" had 1,200,000 tons of ore with 350 g/t Ag, 4 to 5 g/t Au, 3% Cu and 1.8% Pb-Zn (combined). In 1981, the Cubo mine added to the list of discoveries a "bonanza" type ore shoot containing of 600 g/t Ag and 15 g/t Au.

These discoveries revolutionized concepts about the mining and geology of the district, which until then was considered as "only argentiferous". This revolution opened new perspectives and challenges for the exploration geologist. The existence of base metals at the Rayas mine caused Gross (1975) to revive and to modify the local concept about metallic zonation: at the shallowest, from the surface (~2,350 masl) to the 2,200 m elevation, there were the occurrence of native precious metals and sulfosalts; from there to the 1,700 elevation there was the silver-rich horizon and below this level a base metal sulfide zone.

Along the vein or at its width the ores are irregularly distributed; they are concentrated in ore shoots (*clavos*) that are discontinuous. These ore shoots show dip slip displacements plunging to the SW and their dimensions vary both vertically and horizontally. The clavo de Rayas is one of longest, being a little more than 100 m in length. This ore shoot is a breccia pipe in which, at least, two periods of economic mineralization have been identified.

3 STRUCTURE AND LOCAL STRATIGRAPHY

The only control to the mineralization is structural: the main structure that the ore fluids followed is a normal fault which has at its center (near Valenciana mine) its maximum slip dip of about 1,700 m and at the Las Torres of nearly 1,220 m (Aranda-Gómez & Vasallo 2007). Along its strike Veta Madre cuts nearly all the rock units in the district: the Esperanza Formation, the Conglomerado Rojo Formation, an unnamed andesitic dike, and the La Bufa rhyolite. The complete sequence in the district includes:

a. the Mesozoic basement with: the Cerro Pelón tonalite which according to Monod (1990) has a radiometric K-Ar age of 157 ± 8.8 Ma; the Tuna Mansa diorite dated by Ortiz-Hernández (1992) at 122.5 ± 5.6 Ma (K-Ar in amphibole) and at 112.5 ± 5.6 Ma (K-Ar, on a gabbroic phase), and the La Palma diorite -or Santa Ana Filonian Complex- dated by Ortiz-Hernández (1992) at 157.1 ± 8.8 (K-Ar, on a dioritic phase) and 143 ± 9.6 Ma (K-Ar, on a tonalitic phase)
b. a volcanic-sedimentary complex, composed of the La Luz-El Cubilete volcanic sequence (~108.4 Ma, according to Martínez-Hernánez 1992, who use the K-Ar method), the Esperanza Formation, consisting calcareous, argillaceous and siliciclastic sediments alternating with volcanic rocks (Valanginian-Turonian, according to Dávila-Alcocer & Martínez Reyes 1987, based on microfossils) and the Arperos Formation, composed of volcanic and volcaniclastic rocks (Dávila Alcocer & Martínez-Reyes 1987) assigned the age for the overlying strata of this unit as Valanginian-Turonian, based on several radiolaria species), and
c. a sedimentary and volcanic cover of the Conglomerado Rojo (late Eocene- Oligocene, according to Ferrusquía-Villafranca 1987, based on the study of two endemic rodent fossils) and the Losero (medium to late Eocene, according to its stratigraphic position), La Bufa, dated by Gross (1975) at 37.0 ± 3.0 Ma (K-Ar), Calderones (~35 Ma considering its stratigraphic position), the Cedro andesite referred by Labarthe-Hernández (1995) with age of 32.9 ± 1.6 Ma (K-Ar) and the Chichindaro rhyolites dated by Gross (1975) at 32 ± 1.0 Ma K-Ar), by Nieto-Samaniego (1995) at 30.8 ± 0.8 and 30.1 ± 0.8 Ma, and by Labarthe-Hernández (1982) at 30.0 ± 1.5 Ma.

The intrusive rocks in the area are: the Comanja granite which was dated by Mujica-Mondragón & Albarrán-Jacobo at 55.0 ± 4.0 and 58.0 ± 5.0 Ma (K-Ar in biotite), the intrusive Peregrina rhyolitic dome (Oligocene, considering its stratigraphic position) and several dikes of andesitic affiliation.

4 ORE AND ALTERATION MINERALOGY

The ore mineralogy is relatively simple: in the upper part, native gold and silver, electrum, cinnabar, embolite, cerargyrite and some other silver halides, selenides and sulfosalts. At the intermediate zone the selenide content diminishes, as sulfosalts and sulfides increase, specifically: acanthite (or argentite), the argentite-canfieldite, aguilarite-naumannite, polybasite-pearceite and pyrargyrite-proustite series, in addition to stephanite, tetrahedrite, guanajuatite, and paraguanajuatite. At the deepest zone there are acanthite (or argentite), chalcopyrite, sphalerite, and galena all of them together with diminishing quantities of silver sulfosalts.

Gangue minerals include: nontronite, halloysite and metahalloysite, chlorite, calcite, dolomite, siderite, rhodochrosite, quartz, adularia, sericite, fluorite, zeolites and alunite. The abundances of these minerals are variable along the different mines at the district, as well as in the different levels in each mine, showing a very marked vertical zoning. As it occurs in many ore bodies in vein complexes, those of the Guanajuato district show complicated associations of mineralization and alteration assemblages; the most prominent feature is the presence of an argillic alteration zone at the surface, associated with the existence of ore shoots at depth; apparently the intensity of this alteration is directly related to the magnitude of the associated ore shoots.

5 DISCUSSION

In general, the ore deposits of the district have been considered as epithermal and more specifically they are of the adularia-sericite type (or low sulfidation). Nearly all authors agree on the sequence of geological events that predated the mineralization: tensional stresses during the Laramide orogeny, followed by the rejuvenation of the region. As a consequence, erosion intensified and resulted in the Red Conglomerate; a relaxing period that caused the development or intense fracturing and normal faulting followed. After that, the emplacement of the Comanja granite contributed in preparing the country rocks that will contain the district's riches.

The origin of metals is still a subject of debate. Wandke & Martínez (1928) attributed them to the fluids coming from the Comanja granite. Taylor (1971) proposed that metals came from the volcanic sequence that, according to him, had concentrations sufficient enough to mineralize the fractures by means of lateral secretion; both, magma and metals coming from the differentiated mantle. Gross (1975) agreed with Taylor about the origin of the magma but was reluctant to accept his idea of the source of metals. He presented a series of reasons debating Taylor's ideas and based on isotopic studies concluded that the source of metals was the crust, more precisely the Esperanza Formation. Randall (1979) stated that the mineralizing fluids ascended along two breccia chimneys very similar in mineral content and alteration. Those zones are known as the Rayas and Sirena ore shoots. Buchanan (1979, 1980), by means of fluid inclusion studies, without addressing the origin of metals, established the mechanism of ore deposition: boiling, which -he suggested- occurred about 650 m below the present surface, at a mean temperature of 230 °C, from salinity fluids (<6 wt. % eq. NaCl). Mango (1988) refuted Buchanan's data. She suggested that deposition did not happen because of boiling but by means of mixing of two different fluids: one, the dominant, of meteoric origins and, the other one of magmatic origins. The result was a mixture of very low salinity (<1%, wt. % eq. NaCl). The metals came from the same cortical source. Randall (1989, 1990a, 1990b) insisted -citing Gross- that the gold mineralization is related to the Esperanza Formation and that the dikes associated with the Peregrina rhyolitic dome are spatially related to the gold mineralization found at the vicinity of El Cubo mine. Mango (1992) based on isotopic and fluid inclusion studies, established that ore deposition occurred at 200 to 300°C, from fluids with salinity varying only from 0 to 3 wt. % eq. NaCl, in the absence of boiling, although she admitted that this mechanism probably occurred in the El Cubo area. Her oxygen isotopic analysis suggested that meteoric water was not the only fluid involved in the mineralization; the carbon isotopic analysis indicated a reduced component for the isotopic signature, which was interpreted as originated from the Esperanza Formation. Lead, according to her interpretations was derived, by leaching, from the Tertiary volcanic sequence. She proposed a model for the mineralization in the central part of the Veta Madre: meteoric waters percolated through the country rocks to depth, becoming hotter due to an intrusive body. The meteoric waters leached Ag, Cu, and Zn from the country rocks meanwhile the magmatic body contributed Pb and other metals. Randall (1994), aided by the studies of Mango (1988), suggested that the boiling level was above the present-day topography and that this was the mechanism for the deposition of metals for the El Cubo mine and Vetas de la Sierra, but not for the other ore bodies in the Veta Madre. Additionally, he suggested that for all the district, the mineralizing fluids were a mixture of both, meteoric water and magmatic fluids, the latter coming from a plutonic-volcanic center that served as the source of the lavas (and/or tuffs) of the Oligocene sequence. He also proposed the development of a caldera that erosive processes destroyed. Labarthe et al. (1995, 1996) concluded (in a cartographic contribution submitted to the Las Torres mine), that they did not find evidence of any caldera. Abeyta (2003) mentioned, based on fluid inclusion studies from samples from the San Nicolás vein (El Cubo) that the depositional mechanism was boiling and that it "was absent or not preserved during the development of other vein systems in this mining district such as the Veta Madre and La Luz system". Orozco-Villaseñor (2003) mentioned that he observed indications of boiling in his Veta Madre samples and particularly in fluid inclusions in samples from deep levels of the Rayas mine (390 and 405 levels). In addition, he reported about the presence of bladed calcite (which according to several authors is evidence of boiling, e.g. Camprubí & Albinson 2006) not only at the surface, but also in the zone corresponding to the Clavo de Rayas at depth, and even in the deep mine workings. Finally, Orozco-Villaseñor (2010, in prep.) mentioned that according to preliminary fluid inclusion studies made on "original" samples from the Clavo de Rayas, his salinity data do not coincide with those presented by Mango. He found salinity values varying from 5 to 14 wt. % eq. NaCl. He also found temperature variation between 288 to 370°C, with a mean of 347°C, in primary inclusions in samples from the 435 level. Fluid inclusions studied in valencianite (a variety of adularia very common at the central part of Veta Madre) by the same author, showed a melting temperature mean of −7.93°C (which corresponds to 11.53 wt. % eq. NaCl), and a mean homogenization temperature of 277.9°C.

REFERENCES

Abeyta, R.L. 2003. Epithermal Gold Mineralization of the San Nicolas Vein, El Cubo Mine, Guanajuato, Mexico: Trace Element Distribution Fluid Inclusion Microthermometry and gas Chemistry. Unpublished Thesis. Socorro, New Mexico Institute of Mining and Technology.

Aranda-Gómez, J. & Vasallo, L.F. 2007. Geology of the Guanajuato mining district and sightseeing in the old town of Guanajuato. Field trip guidebook. *4th International Conference "GIS in Geology & Earth Sciences"*, Querétaro, México, October 22–26.

Buchanan, J.L. 1979. The Las Torres Mine, Guanajuato, Mexico. Ore controls of a fossil geothermal system. Unpublished Ph.D. dissertation. Golden, Colorado School of Mines.
Buchanan, J.L. 1980. Ore controls of vertically stacked deposits, Guanajuato, Mexico. Society of Mining Engineers, American Institute of Mining, Metallurgical and Petroleum Engineers, Preprint 80–82.
Camprubí, A. & Albinson, T. 2006. Depósitos epitermales en México: actualización de su conocimiento y reclasificación empírica. *Boletín de la Sociedad Geológica Mexicana, Vol. Conmemorativo del Centenario* LVIII(4): 27–81.
Gross, W.H. 1975. New Ore Discovery and Source of Silver-Gold Veins, Guanajuato, Mexico. *Economic Geology* 70: 1175–1189.
Labarthe-H., G. et al. 1995. Cartografía 1:25,000 de la Sierra de Guanajuato. Universidad Autónoma de San Luis Potosí, Instituto de Geología, Investigación para la Compañía Minera Las torres, Guanajuato, Gto.
Labarthe-H., G. et al. 1996. Cartografía 1:10,000 de la Sierra de Guanajuato, Sierra del Chorro y lote Villalpando. Universidad Autónoma de San Luis Potosí, Instituto de Geología, Investigación para la Cía Minera Las torres, Guanajuato, Gto.
Mango, H.N. 1988. A Fluid Inclusion and Isotope Study of the Las Rayas Ag-Au-Pb-Cu Mine, Guanajuato, Mexico. Unpublished Thesis. Hanover, New Hampshire, Darmouth College.
Mango, H.N. 1992, Origin of Epithermal Ag-Au-Cu-Pb-Zn Mineralization on the Veta Madre, Guanajuato, Mexico. Unpublished Ph.D. dissertation. Hanover, New Hampshire, Darmouth College.
Mango, H., Zantop, H. & Oreskes, N. 1991. A Fluid Inclusion and Isotope Study of the Rayas Ag-Au-Cu-Pb-Zn Mine, Guanajuato, Mexico. *Economic Geology* 86: 1554–1559.
Orozco-Villaseñor, F.J. 2003. Estudio metalogenético del clavo de Rayas en la veta Madre de Guanajuato, México. Unpublished Thesis. UNISON, Sonora.
Orozco-Villaseñor, F.J. 2010. Mineralogía y génesis del "Clavo de Rayas" y su relación geológica con otros cuerpos de mineral de la parte central de la Veta Madre de Guanajuato. Ph. D. Thesis, in prep. Centro de Geociencias, UNAM, Mexico.
Randall-R., J.A. 1979. Structural setting, emplacement of Veta Madre orebodies using the Sirena and Rayas mines as examples, Guanajuato, Mexico. Nevada Buereau of Mines and Geology, Report 33: 203–212.
Randall-R., J.A. 1990a. Gold overprint in silver veins, Guanajuato, Mexico. Proceedings of the International Association on the Genesis of Ore Deposits, Ottawa, Canada, A109.
Randall-R., J.A. 1990b. Geology of El Cubo mine area, Guanajuato, Mexico. *Society of Economic Geologists Guidebook* 6: 218–227.
Randall-R., J.A., Saldaña-A., E. & Clark, K.F. 1994. Exploration in a Volcano-Plutonic Center at Guanajuato, Mexico. *Economic Geology* 89: 1722–1751.
Taylor, P.S. 1971. Mineral Variation in the Silver Veins of Guanajuato, Mexico. Unpublished Thesis. Hanover, New Hampshire, Darmouth College.
Wandke, A. & Martínez 1928. The Guanajuato Mining District, Guanajuato, Mexico. *Economic Geology* 23: 1–44.

Water-Rock Interaction – Birkle & Torres-Alvarado (eds)
© 2010 Taylor & Francis Group, London, ISBN 978-0-415-60426-0

Alkali elements as geothermometers for ridge flanks and subduction zones

W. Wei & M. Kastner
Scripps Institution of Oceanography, La Jolla, CA, USA

R. Rosenbauer
U.S. Geological Survey, Menlo Park, CA, USA

L.H. Chan
Department of Geology and Geophysics, Louisiana State University, Baton Rouge, LA, USA

Y. Weinstein
Department of Geography and Environment, Bar-Ilan University, Ramat-Gan, Israel

ABSTRACT: Hydrothermal experiments were conducted on basaltic ash and smectite, with a water/rock mass ratio of ~5–8, at 600 bar pressure. The reaction temperature was gradually changed from 35 to 350°C at 20–50°C increments, and at each temperature steady-state was reached. The results indicate that each alkali metal has a distinct temperature-dependant partition profile between fluid and the solid phase. The data indicate significant fractionations among alkali metals from 35–350°C and a strong influence by the starting solids on the mobility of the alkali metals. The reactivity of the solid for Li isotopes increases dramatically with temperature. Cesium exhibits significant affinity for the fluid at temperatures as low as 150°C, suggesting that ridge flanks may be an important source of Cs to the ocean. The results also suggest that alkali element ratios together with Li isotope ratios may be used as approximate geothermometers for deep-sourced fluid at subduction zones.

1 INTRODUCTION

Laboratory experiments can provide important constraints for the understanding of hydrothermal reactions in natural systems. Early hydrothermal experiments (Bischoff & Dickson 1975; Seyfried & Bischoff 1977, 1979; Seyfried et al. 1998) focused on reproducing the observations at Mid Ocean Ridges (MOR) hydrothermal systems and were conducted at fixed P-T conditions. Nevertheless, they showed that seawater was significantly modified traveling through these systems. The mobility of elements and the temperature of deep sourced hydrothermal fluids are of major concern in subduction zone research. A few experiments were conducted with gradually changing temperature (e.g. You et al. 1996, 2001 on smectites; James et al. 2003 on both smectites and basalts), which better mimic processes at subduction zones.

The alkali metals (Li, K, Rb, Cs) are very good tracers of fluid-rock interaction because of their strong partitioning into the fluid phase, particularly at moderate to high temperatures. As such, they can provide information on the process of fluid recycling at the principal plate boundaries, the reaction site temperature and the nature of the reacting solid phases (i.e. the involvement of sediments in arc volcanoes). James et al. (2003) studied the mobility of alkali and other elements with temperature, and concluded that geochemical fluxes at ridge-flanks are important for the global budgets of some elements. In order to better understand the hydrothermal processes at subduction zones and their dependence on temperature and rock type, we conducted detailed experiments on the mobility of alkali elements under changing temperature and with two different solids, typical of subduction zones. Our experiments indicate that the reactivity of each alkali metal is distinct; each has a characteristic behavior with respect to partitioning into the fluid phase with temperature. The data thus suggest that alkali metal ratios may be used for geothermometry.

2 EXPERIMENT DESIGN

Basaltic ash from Lau Basin and smectite from offshore Barbados were reacted with artificial seawater under varying temperature using Dickson-type rocking autoclaves. A total of seven experiments were conducted with a water/rock mass ratio of ~5–8, all at 600 bars. The reaction temperature was gradually changed from 35 to 350°C at 20–50°C increments. Steady-state was reached at each step. Run-times continued until steady-state partitioning was observed. All results were normalized to chloride concentration.

3 RESULTS & IMPLICATIONS

In the basaltic ash-SW experiments, fluid K/Cl molar ratios increased slightly from 35–65°C, then decreased to a minimum of 1.10×10^{-2} at 275°C. At >275°C the ratio sharply increased to 6.53×10^{-2} at 350°C. In contrast, Li/Cl, Rb/Cl, and Cs/Cl ratios all increased from 35–350°C, but each at a distinct slope, indicating significant fractionation between the alkali metals. The Li/Cl ratio gently increased between 35–250°C, and sharply increased to a ratio of $\sim 1.20 \times 10^{-3}$ at 350°C (Fig. 1). Rb/Cl behaved similarly, except for a higher inflection temperature of ~300°C, and a steeper slope between 300–350°C. Cs/Cl behaved distinctly, with no change until 100°C and then a steady and sharp (10 fold) increase of ratios with temperature (i.e. no inflection point) to 1.8×10^{-8} at 350°C (Fig. 2). As a result of the different behavior of K and the other alkali elements, Li/K and Cs/K ratios exhibit

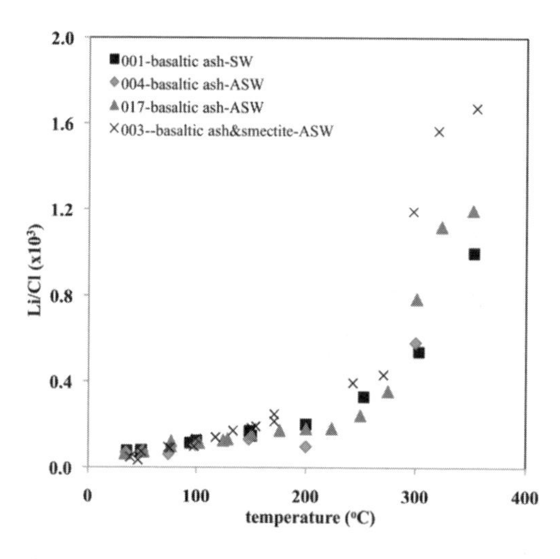

Figure 1. Cl-normalized Li concentrations in the basaltic ash-seawater experiments. SW stands for filtered seawater, while ASW is for artificial seawater experiments. In Experiment 003, the starting material was a mixture of basaltic ash and smectite (proportions of 2:1, respectively).

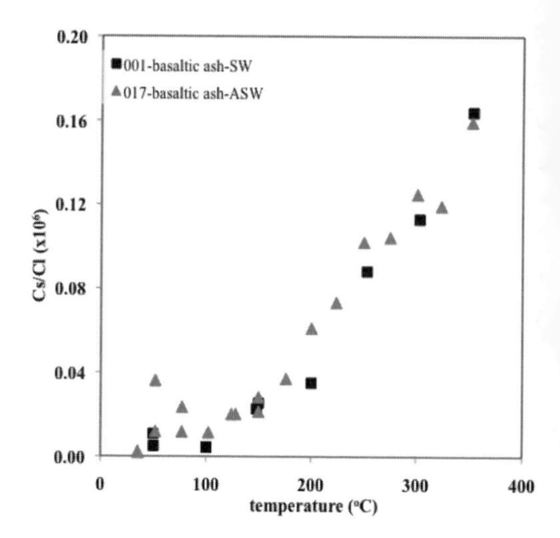

Figure 2. Cl-normalized Cs concentrations in the basaltic ash-seawater experiments.

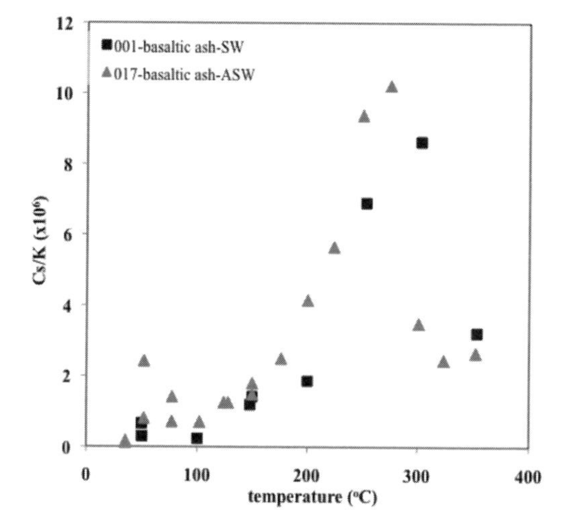

Figure 3. Cs/K ratios in the basaltic ash-seawater experiments.

sharp maxima at ~275°C (Fig. 3). In the smectite experiments, K/Cl ratio pattern is very different from the basalt experiment, with ratios steadily increase from 50–300°C. Rb/Cl inflection point is at 100°C rather than at 300°C as in the basaltic ash experiment, and on the other hand Cs/Cl ratios do not start incr easing until >100°C. Li/Cl is the only case, where patterns in both experiments are similar.

At 350°C the enrichment factors (ER) of Li/Cl and Cs/Cl are higher in the smectite than in the basaltic ash experiments (~70 & 55 vs. ~25 & 30, respectively). The ER of Rb/Cl is ~10 in both experiments, but that of K/Cl is higher in the basaltic ash than in the smectite experiments (3.5 & 1.8, respectively).

The relatively low inflection point of the Cs/Cl ratios in the basaltic ash experiment (<150°C) and the significant release of Li to the fluid already at 100°C (in both experiments) suggest that the exchange of Cs and Li between the solids and fluids is not confined to the high-temperature hydrothermal ridge crest setting, but also occurs at ridge flanks, which may be an important source of these elements to the ocean.

The fluid Li totally exchanges its isotopes with the ash or with the smectite. In the basaltic ash experiments, the fluid δ^7 Li fall on a SW-MORB mixing line, shifting from 30.95‰, the SW value, to 17.4‰ at 150°C, and 7.33‰ at 350°C. Hence, δ^7 Li values together with Li/K, Li/Rb and Li/Cs ratios constrain the reaction temperature (± 20–30°C) and the nature of the rock involved. Thus, these ratios can be used as approximate geothermometers for assessment of the temperature of deep-sourced fluids at subduction zones.

REFERENCES

Bischoff, J.L. & Dickson, F.W. 1975. Seawater-basalt interaction at 200 degrees C and 500 bars; implications for origin of sea-floor heavy-metal deposits and regulation of seawater chemistry. *Earth and Planetary Science Letters* 25: 385–397.

James, R.H., Allen, D.E. & Seyfried, Jr., W.E. 2003. An experimental study of alteration of oceanic crust and terrigenous sediments at moderate temperatures (51 to 350 degrees C); insights as to chemical processes in near-shore ridge-flank hydrothermal systems. *Geochimica et Cosmochimica. Acta* 67: 681–691.

Seyfried, Jr., W.E. & Bischoff, J.L. 1977. Hydrothermal transport of heavy metals by seawater; the role of seawater/basalt ratio. *Earth and Planetary Science Letters* 34: 71–77.

Seyfried, Jr., W.E. & Bischoff, J.L. 1979. Low temperature basalt alteration by seawater; an experimental study at 70 degrees C and 150 degrees C. *Geochimica et Cosmochimica. Acta* 43: 1937–1948.

Seyfried, Jr., W.E., Chen, X. & Chan, L.-H. 1998. Trace element mobility and lithium exchange during hydrothermal alteration of seafloor weathered basalt; an experimental study at 350 degrees C, 500 bars. *Geochimica et Cosmochimica. Acta* 62: 949–960.

You, C.F., Castillo, P.R., Gieskes, J.M., Chan, L.H. & Spivack, A.J. 1996. Trace element behavior in hydrothermal experiments; implications for fluid processes at shallow depths in subduction zones. *Earth and Planetary Science Letters* 140: 41–52.

You, C.F. & Gieskes, J.M. 2001. Hydrothermal alteration of hemi-pelagic sediments; experimental evaluation of geochemical processes in shallow subduction zones. *Applied Geochemistry* 16: 1055–1066.

Water-Rock Interaction – Birkle & Torres-Alvarado (eds)
© 2010 Taylor & Francis Group, London, ISBN 978-0-415-60426-0

Water in the Earth's interior: A planetary water-rock problem

A.E. Woermann
RWTH, Aachen, Germany(deceased)

B.G. Eriksson
GTT-Technologies, Herzogenrath, Germany

C.S.K. Saxena
Center for Study of Matter at Extreme Conditions, International University of Florida, FL, USA

D.G.C. Ulmer
Earth & Environmental Sciences, Temple University, Philadelphia, PA, USA

ABSTRACT: XRD and Ramaan data report hydrogen-contents in mantle wadsleyite, Mg-rich olivines. This hydrogen has been 'assigned' as 2 wt% H_2O (Kleppe 2006). The fluid-rock interaction aspects of this hydrogen need attention since statements are appearing: "the Earth's mantle with this water content has more overall water than the oceans of the world" (Smyth 1994). By mass balance, this is correct but disregards the thermodynamic "price-tags" that would come with a mantle with 2 wt% water (Woermann et al. 2008, Ulmer & Woermann 2008). Two major equations of state (MRK & Chemsage GTT-Aachen web site 2009) show that the oxidation state of the mantle with 2 wt% water would require the mantle to be at the redox value ~ (QFM). Much data show that this is not true (Ulmer et al. 1987, Frost & McCammon 2008). Thus, the problem of hydrogen in the mantle is not as simple as calling it "water".

1 BACKGROUND

Decades of reliable research from more than a dozen labs in many countries has been aimed at understanding the oxidation-reduction equilibria of the Earth's mantle. A review of these redox studies was made in Ulmer et al. (1987). Techniques and environments in these redox researches have included mantle studies of xenoliths from as deep as at least 400 km (Saufer et al. 1991, Virgo et al. 1988), deep seated igneous plutons like the Stillwater and Bushveld Complexes (emplacement depths up to 15 km; Flynn et al. 1978, Elliott et al. 1982), laboratory PT studies of various redox buffers to temperature and pressures of the geothermal gradient to at least depths of 300 km (reviewed in Ulmer & Barnes 1987), fluid inclusion analyses of diamonds and lab redox stability studies of both diamonds and Moissanite (SiC) (Ulmer et al. 1998).

2 A MINI-REVIEW OF REDOX BUFFERS

Since 1958, several oxidation-reduction buffers have been calibrated to measure critical points in fO_2-P-T space of geological environments. The work of Eugster and students at Johns Hopkins and of Muan, Osborn and students at Penn State are reviewed for example in Ulmer (1971), Barnes & Ulmer (1978) and Deines et al. (1974), wherein equations are given for the buffers to calculate log fO_2 at P and T. The mini-review below will refresh the reader's ability to follow acronyms used throughout the rest of this paper for a few of these important buffers:

(QFM) Quartz + Magnetite = Fayalite + oxygen
(WM) Magnetite = Wüstite + oxygen
(WI) Wüstite = Iron + oxygen
(CHO) $CO_2 + H_2 = CO + H_2O$
(CCO) $CO_2 = CO$ + oxygen
(H_2O) $H_2O = H_2$ + oxygen

3 DISCUSSION

A general trend of these redox studies has established that there is some redox gradient from the Earth's surface downward. The surface basaltic redox conditions both on the Earth and the Moon's interior are reported as being generally equivalent to the (FMQ) buffer (Basalt Volcanism Study Project 1981). The general data for terrestrial igneous plutons is more equivalent to the (WM) buffer (Ulmer et al. 1987). By contrast, the mantle-core boundary is at (WI). Even with the known iron oxidation state disproportionation ($4Fe^2 = Fe_3O_4 + Fe^{o}_{metallic}$) (e.g. Frost et al. 2004), the core is certainly more reduced than the planet's surface. Another example of lower redox state for the mantle is provided by Haggerty & Tomkins (1983). Stagno & Frost (2008) have reviewed carbon and carbonate equilibria and have demonstrated that log fO_2 values can be depressed at depth as much as $\Delta FMQ = -1.75$. Frost & Mc Cammon (2008) make the statement that "modeling of experimental data indicates that at approximately 8 GPa, [80 Kb], mantle fo_2 will be 5 log units below FMQ and at a level where Ni-Fe metal becomes stable." Moissanite (SiC) has been established as a mantle phase and, as shown by Ulmer et al. (1998), to be also as much as five orders of magnitude more reduced than (FMQ), both at 1 bar and at 90 Kbars at temperatures of 1525 K and 1775 K. All these observations of mantle mineralogy and equilibria are indications that redox conditions vary from the surface to depth from more oxidizing (~FMQ) to less oxidizing (ΔFMQ~ −5), and are accordingly certainly not in equilibrium with abundant water fluid at depth.

The buffering power of water according to many existing equations of state (EOS) varies, in its fine details, according to which P, T and EOS are employed. The overall total consensus is, that water at the current modeled weight percent (~2 wt%) would, particularly given over geologic time, have become the ubiquitous mantle redox buffer. It would have simply 'annihilated' by oxidation, any and all species more reduced than those in equilibrium with water or the (QFM buffer). As shown in Figure 2, an internally consistent set of calculations is presented from Factsage (2009). High molar volume of water is shown to equate with (QFM). Any lower oxidation state would have been overpowered by the claimed preponderance of water, if the mantle were to have had water, or still have water as a phase at the 2 wt% level. Note the inset box that shows that the oxygen isobars converge at high mole fraction of water with the (QFM) buffer. Thus, if there is still an existing 2 wt% of water as a phase in the mantle, it could only be because water is the dominate redox buffer. And there is too much evidence that high water content per se is not the redox situation throughout the mantle.

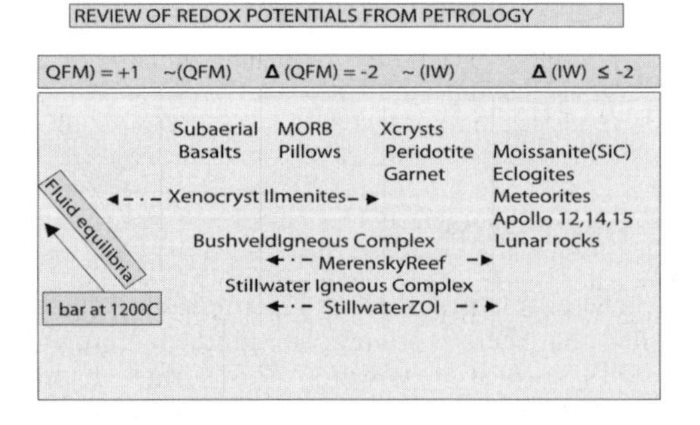

Figure 1. Review of the redox data *vs* published techniques. It is shown that there is an indication of more reducing conditions existing at depth than at the surface.

4 CONCLUSIONS

For many years, careful experimental studies of dense hydrous silicates in the magnesium olivine-water system (DHMS) have been performed (Ganskow et al. 2008). Many studies of potential sites for hydrogen or water in the mantle silicate mineralogies have been examined. A summary diagram by Saxena (2004) is shown in Figure 3.

However, the problem is to find out what speciation is correct for the observed hydrogen in the

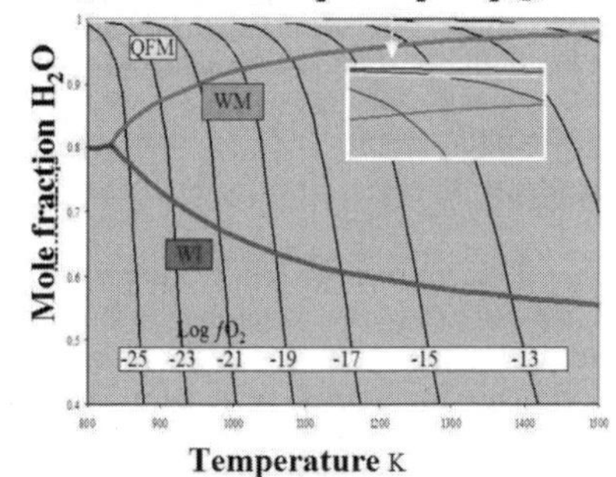

Figure 2. This computer compilation shows the log fO_2 versus P and T isopleths for the common redox buffers including fluids with large mole% of water for the dissociation of water at 3000 bars.

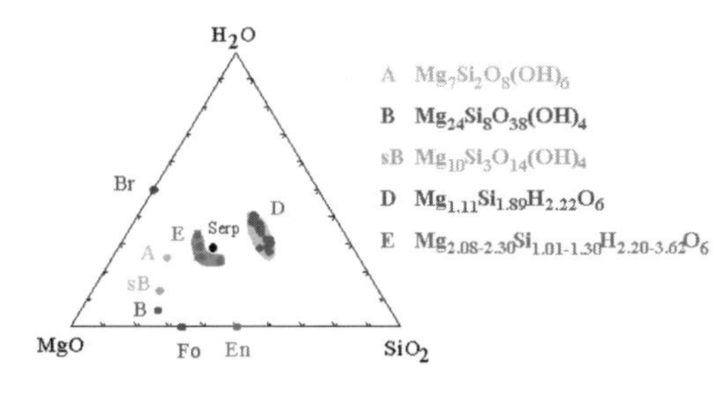

Figure 3. Potential sites for water in the mantle minerals such as (DHMS) (from Saxena 2004).

mantle. At least six additional suggestions/questions need to be examined in light of all the reliable XRD data that do suggest hydrogen in the wadsleyite olivines.

1. Is the hydrogen really there as water or is it interstitial hydrogen at the atomic level?
2. Is the water present as real hydoxyl species or as interstitial hydrogen at the atomic level?
3. Is the hydrogen present as hydrate water in the crystal structure, or again, can it be better described as interstitial hydrogen?
4. In their phase equilibria studies of the Mg-Fe-Si-O system, Nakamura & Schmalzreid (1983) and Muan (1958) have shown that the fO_2 isobars 'piled up' in an unusual manner on any given single bulk compositions of Fe-Mg olivine. All these authors suggested that this may be because the bulk concentration, with the defects present displayed a more reduced nature in these equilibria studies. When the defects were not present (by annealing?) in the same bulk composition, the olivine appeared orders of magnitude more oxidized in these lab studies. This shift in apparent oxidation potentials for a given Fe/Mg bulk composition of olivine was as much as four orders of magnitude in terms of the range of measured log fO_2 values. These older data now may suggest that the hydrogen observed by XRD and IR spectra in Wadsleyite could be present in olivine structure as defect-fillings or as interstitials, and not as water per se.
5. The hydrogen studies in more iron-rich olivine have, in fact, been reporting that experiments to equilibrate iron-containing olivine with water are more complicated, "because there is also present some ferric iron in the resulting material" (Frost et al. 2004). This last observation seems itself to be a tacit and quiet proof that copious water and ferrous iron-rich olivine will not exist in the mantle?
6. Another suggestion has been made by more than a dozen studies (as reviewed by Saxena et al. 2004) that there is the possibility of hydrogen being 'stored' in/as hydrides of iron. Their results indicate that at pressures corresponding to the earth's lower mantle, the hydride phase is stable, and that orthorhombic high-magnetite (h-Fe_3O_4) may also be stabilized in lieu of, or in addition to magnesio-wüstite. The stability of these phases opens up the possibility that water (as a component of a fluid phase or hydrous solids) may be present not only in the mantle but also in the core (as dissolved hydride and oxide). Note they do not claim water per se.

Thus the problem is that hydrogen in the mantle can not be simply ascribed to water as a phase. There are pervasive mantle equilibria buffered by the mineralogy of the rocks themselves that have been shown to be at log fO_2 values as much as five orders of magnitude more reduced than (FMQ). Water per se, particularly if still present at wt% amounts, is predicted by EOS to buffer the mantle at or very near to (FMQ).

ACKNOWLEDGEMENT

The coauthors of this manuscript wish to dedicate this paper to the memory and career of Dr. Eduard Woermann, formerly Professor Emeritus, RWTH Aachen, Germany, whose interest in applied mineralogy ranged for decades from the steel industry to ceramics to the Earth's Interior. His untimely death has been felt in many fields of materials science.

REFERENCES

Bale, C.W., Bélisle, E., Chartrand, P., Decterov, S.A., Eriksson, G., Hack, K., Jung, I.-H., Kang, Y.-B., Melançon, J., Pelton, A.D., Robelin, C. & Petersen, S. 2009. CALPHAD: FactSage thermochemical software and databases—recent developments: comput. coupling phase diagrams *Thermochem.* 33, 295–311. *www.gtt-technologies.de or at www.factsage.com.*

Basalt Volcanism Study Project 1981. *Basaltic Volcanism on the Terrestrial Planets.* New York: Pergamon Press, 1286 pp.

Deines, P., Nafziger, R.H., Ulmer, G.C. & Woermann, E. 1974. Temperature—Oxygen fugacity tables for selected gas mixtures in the system C-H-O at one atmosphere. *Bulletin 88 College of Mineral Sciences, The Pennsylvania State University.*

Elliott, W.C., Grandstaff, D.E., Ulmer, G.C., Buntin, T.J. & Gold, D.P. 1982. An intrinsic oxygen fugacity study of the platinum-carbon associations in layered intrusions. *Economic Geology* 77(7): 209–226.

Flynn, R., Ulmer, G.C. & Sutphen, C. 1978. Petrogenesis of the Bushveld Complex: Crystallization of the Middle Critical Zone. *Journal of Petrology* 19(1): 136–152.

Frost, D.J. &. McCammon, C.A. 2008. The redox state of Earth's mantle. *Annual Review of Earth and Planetary Sciences* 36: 389–420.

Frost, D.J., Lieske, C., Langenhorst, F.S., McCammon, C., Tronnes, R.G. & Rubie, D.C. 2004. Experimental evidence for the existence of iron-rich metal in the Earth's lower mantle. *Nature* 428: 409–41.

Ganskow, G., Langenhorst, F. & Frost, D.J. 2008. Bayerisches Forschungsinstitut (BGI) 95–96.

Haggerty, S.E. & Tompkins, L.A. 1983. Redox state of the Earth's upper mantle from kimberlitic ilmenites. *Nature* 303: 295–300.

Kleppe, A. 2006. High pressure Raman spectroscopic studies of hydrous wadsleyite II. *American Mineralogist* 91(7): 1102–1109.

Moats, M.A. & Ulmer, G.C. 1989. (CCO) and (FMQ) oxygen buffer values for upper mantle conditions: Implications for kimberlite, carbonatite and diamond genesis. *Workshop on Diamonds, 28th Intern. Geolog. Congress, Washington D.C.*

Muan, A. 1958. *The system FeO-MgO-SiO$_2$ in phase diagrams for ceramists*. vol. 1, American Ceramic Society Publications.

Nakamura, A. & Schmalzreid, H. 1983. On the nonstoichoiometry and point defects of olivine. *Physics and Chemistry of Minerals* 10(1): 27–37.

Saufer, V., Haggerty, S.E. & Field, S. 1991. Ultra deep (>300 km) ultramafic xenoliths from the transition zone. *Science* 262: 827–830.

Saxena, S.K. 2004. High pressure fluids in the Earth's Core. *Conference on Water: Lecture-presentation at Sendai University, Japan, November 15, 2004.*

Saxena, S.K., Liermann, H.P. & Shen, G.Y. 2004. Formation of iron hydride and high magnetite at high pressure and temperature. *Physics of the Earth and Planetary Interiors* 146: 313–317.

Smyth, J.R. 1994. A crystallographic model for hydrous Wadlsleyite: An ocean in the Earth's interior. *American Mineralogist* 79: 1021–1024.

Stagno, V. & Frost D.J. 2008. *Carbonatites, kimberlites and diamonds in the Earth's Mantle*. Annual Report of Bayerisches Forschungsinstitut (BGI) 51–53.

Ulmer, G.C. 1971. *Research techniques for high pressure and high temperature*. New York: Springer-Verlag, 367 pp.

Ulmer, G.C. & Barnes, H.L. 1978. *Hydrothermal experimental techniques*. New York: Wiley and Sons, 525 pp.

Ulmer, G.C. & Woermann, E. 2008. Thermodynamic price tags for a wet mantle. European Union of Geosciences, *Vienna, Abstract EUG 2008-A-12052.*

Ulmer, G.C., Grandstaff, D.E., Moats, M.A., Weiss, D., Buntin T.J., Gold, D.P., Hatten, C., Kadik, A., Koseluk, R.A. & Rosenhauer, M. 1987. Mantle redox: an unfinished story. *Geological Society of America Special Paper* 215: 383–392.

Ulmer, G.C., Grandstaff, D.E., Woermann, E., Schönitz, M., Goebbels, M. & Woodland, A. 1998. The redox stability of moisssonite compared to metal/metal oxide buffers. *Neues Jahrbuch für Mineralogie Abhandlungen* 172(2/3): 279–307.

Virgo, D.A., Luth, R.W., Moats, M.A. & Ulmer, G.C. 1988. Constraints on the oxidation state of the mantle: An electrochemical and ^{57}Fe Mossbauer study of mantle-derived ilmenite. *Geochemica et Cosmochemica Acta* 53: 1781–1794.

Woermann, E., Ulmer, G.C., Eriksson, G. & Saxena, S.K. 2008. Oxygen fugacity in the laboratory and in terrestrial systems. European Union of Geosciences, *Vienna. Abstract EUG 2008-A-12054.*

Water-Rock Interaction – Birkle & Torres-Alvarado (eds)

Characteristics of chloritization in granite-type uranium deposits in southern China

Z.S. Zhang, Z.P. Jiang, G.L. Guo & Y. Zhao
Key Laboratory of Nuclear Resources and Environment (East China Institute of Technology), Ministry of Education, Nanchang

R.M. Hua
State Key Laboratory for Mineral Deposit Research, Department of Earth Science, Nanjing University, Nanjing

ABSTRACT: Granite-type uranium mineralization is the most important uranium-ore-type in China. Chloritization was one of the most important alteration types associated with this type of mineralization. For better understand the genesis of the uranium mineralization, two typical granite-type uranium deposits had been selected to study the characteristics of chloritic alteration. Chloritized granitic rocks and chloritized ores have been systematically sampled and examined by scanning electrom microscopy and EPMA. These studies revealed that chlorite distributed in the rocks as either pseudomorphs of biotite or vermiform and flaky conglomerations in veins. Two main kinds of chlorite, brunsvigite and ripidolite had been classified on the basis of their chemistry. Formation temperatures of chlorite varied between 163 and 276 °C and the chlorite formed under reducting conditions. Dissolution-precipitation and hydrothermal precipitation were the formation mechanisms. Chloritization not only reflects the uranium distribution in granite, but it also contributed to a favorable precipitation environment.

1 INTRODUCTION

Chlorite is a common mineral associated with medium to low grade metamorphism, hydrothermal alteration, and diagenesis, and can be found in silicic to ultramafic rocks. Chloritization was one of the most important alteration types before and during granite-type uranium mineralization in southern China (Cheng 2000, Zhang et al. 2008). Granite-type uranium mineralization is still the most important uranium-ore-type in China. Thus, the study of its characteristics including the paragenetic association and the formation conditions would be helpfully to understanding uranium mineralization in this region. Two typical granite-type uranium deposits, No. 201 and No. 333 deposit were selected in this study. The former study revealed that potassic, sodic, sericitic and chloritic alterations were widespread in the granite in those two mining areas. The chloritic alteration was distributed either with potassic, sodic and sericitic alteration or just formed a wide chloritic only zone independently which reached hundred meters in width. The uranium ore-body formed at the transition zone of these two kinds of chloritic alteration zones and near the independent chloritic only zone (Meng & Xie 1990). The former studies were focused on the alteration types and their zoning. In this study, we focused on the chemical composition of the chlorite and its formation condition. Different kinds of chloritized granites and ores have been systematically sampled and examined by optical and scanning electron microscopy; their chemical compositions have been studied by EPMA. The chemical classification and the formation conditions including calculated temperature and paragenesis of the chloritization with the uranium mineralization are discussed.

2 METHODS

Chemical compositions of the chlorite were analyzed by JXA-8100 EPMA in the Key Laboratory of Nuclear Resources and Environment (East China Institute of Technology), Ministry of Education, under the conditions: accelerating voltage of 15 kV and current of 1×10^{-8} A, using the GB/T 15617–95 mineral standard. All the calculations of chlorite are based on 28 oxygen structural formula, with all the iron being considered as Fe^{2+}. Because the chemical composition of chlorite may have been affected by mineral inclusions, and fine-grained admixtures of associated phases, we judged the quality of the analyses as being acceptable with ($Na_2O + K_2O + CaO$) < 0.5%, and assumed that ($Na_2O + K_2O + CaO$) > 0.5% reflected chemical contamination of chlorite (Zang & Fyfe 1995).

3 RESULTS AND DISCUSSION

3.1 *Distribution and paragenetic association*

Four types of chlorite have been distinguished in the granite-type uranium ore field. The first type of chlorite, most of it shaped irregularly, has a paragenetic association with pitchblende or rutile which may indicate a close relationship with uranium mineralization (Fig. 1-A). These chlorites occur in quartz-pitchblende veins or in altered granite. The second type of chlorite is the secondary product of biotite, which usually displayed in pseudomorphing biotite (Fig. 1-B). The third type of the chlorite is the secondary product of feldspar alteration, which is usually distributed as very fine flaky grains (Fig.1-C). The fourth type of the chlorite occurs in chlorite veins and has worm-like or flaky forms (Fig.1-D).

3.2 *Composition and the classification of chlorite*

Chemical composition of the chlorite varied; the content of SiO_2 varied between 22.2 and 28.81 wt.%, the average was 25.27 wt.%; the content of Al_2O_3 varied between 15.79 and 22.38 wt.%, the average was 19.79 wt.%; the content of FeO varied between 26.18 and 39.74 wt.%, the average was 33.71 wt.%; the content of MgO varied between 3.47 and 15.45 wt.%, the average was 7.58 wt.%; Mg and Fe varied much more than Si and Al. The Fe-Si diagram of Foster (1962) was used for the classification of chlorite (Fig. 2). Chlorite types 1, 2 and 3 are located in the ripidolite and brunsvigite fields; there were belong to the ferroamesite and aphrosiderite region both in No. 201 and No. 333 deposits; type 4 chlorite in No. 201 deposit was also in the same areas, whereas this type of the chlorite in No. 333 deposit was spread over the ripidolite, brunsvigite and pycnochlorite fields, still some of them were located in sheridanite and disbantite fields. This revealed that this type of the chlorite is more enriched in silicon but deficient in iron. Hua et al. (2003) pointed out that the assemblage of brunsvgite and pycnochlorite in one sample may mean that cation partitioning in the crystal

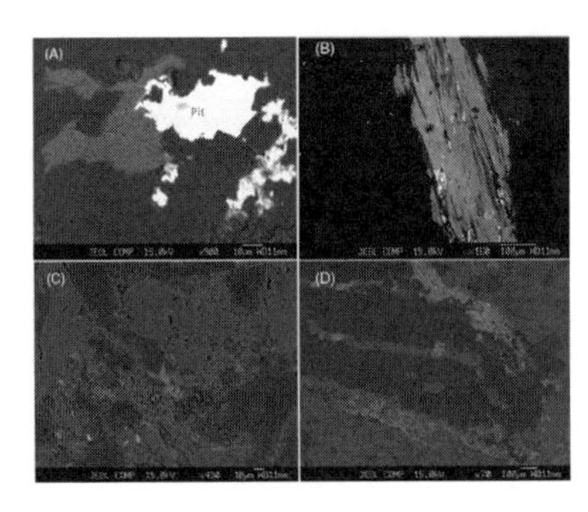

Figure 1. Morphology and paragenetic association of chlorite.

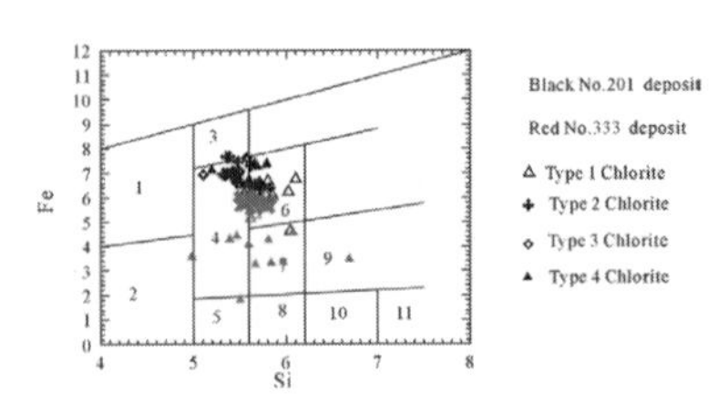

Figure 2. Classification diagram of chlorite in selected uranium deposits. Note: 1. Pseudothuringite; 2. Corundophilite; 3. Daphnite; 4. Ripidolite; 5. Sheridanite; 6. Brunsvigite; 7. Pycnochlorite; 8. Clinochlore; 9. Disbantite; 10. Penninite; 11. Talc-chlorite.

structure of chlorite did not reached equilibrium and may been formed in different physical-chemical conditions (Hua et al. 2003). So the chlorite assembles in the studied deposits might means there were formed in different physical-chemical conditions.

3.3 *The formation condition of chlorite*

3.3.1 *The protolith type of the chlorite*

The molar (Mg)/(Fe + Mg)–Al/(Al + Mg + Fe) diagram, which first used by Laird (1988), has been widely used to evaluate the protolith type of the chlorite. The chlorite derived from argillaceous rock had higher Al/(Al + Mg + Fe) values (>0.35) than those derived from femic rock. The Al/(Al + Mg + Fe) value of the chlorite in No. 201 deposit and No. 333 deposit varied between 0.31 and 0.43, most of the values (>93%) were greater than 0.35, which may infer that they were derived from argillaceous rock, which consistent with the genesis of the granite of those area. However, those chlorites that have Al/(Al + Mg + Fe) values less than 0.35 also revealed that they may been affected by the femic fluids during the formation process. The chlorite compositions are scattered in the (Mg)/(Fe + Mg)–Al/(Al + Mg + Fe) diagram (Fig. 3) and show an inconspicuous negative correlation.

3.3.2 *(Al^{IV}) and Al^{VI} value and (Al^{VI} + Fe) vs Mg diagram*

The molar (Al^{IV}) and Al^{VI} values can be used to illuminate the substitution schemes in chlorite. Xie et al (1997) studied the Barberton greenstone belt (BGB) chlorite and pointed out that there were two major cation substitution schemes in the BGB chlorite, the tschermakite substitution and the Fe-Mg substitution. The tschermakite substitution illustrates the 1:1 dependence of Al^{IV} and Al^{VI} (Xie et al. 1997). In our case the (Al^{IV}) value varied between 1.38 and 2.9, the (Al^{VI}) value varied between 2.19 to 3.46. Most of the samples had (Al^{IV}) < (Al^{VI}), and the (Al^{IV})/(Al^{VI}) value varied between 0.70 to 1.04. No. 201 deposit has $Al^{VI} = 0.5583\ Al^{IV} + 1.4783$, and No. 333 deposit has $Al^{VI} = 0.09\ Al^{IV} + 2.69$, which indicates in the studied deposit that there was not a

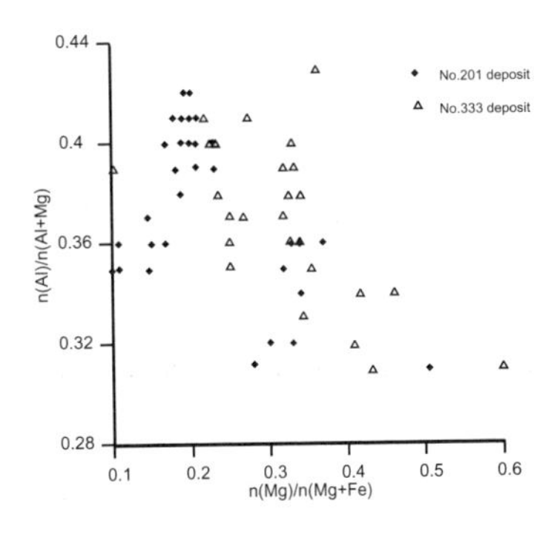

Figure 3. (Mg)/n(Fe + Mg)-nAl/n(Al + Mg + Fe) diagram of chlorite in selected deposit (after Laird 1988).

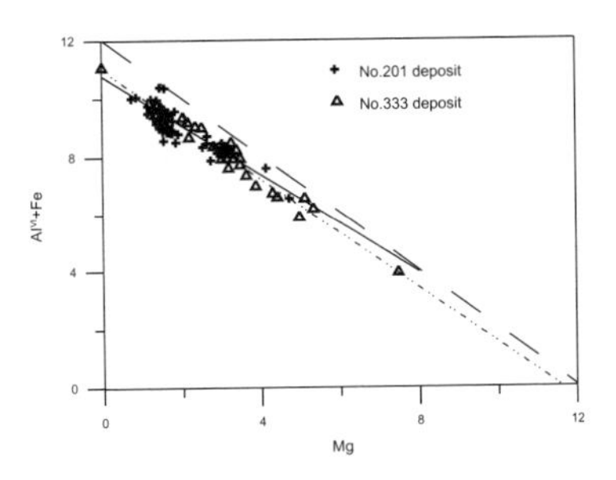

Figure 4. n(Al^{VI} + Fe) vs. Mg diagram of chlorite (after Xie 1997).

pure aluminum tschermakite substitution in tetrahedral site. It is also inferred that more Al^{IV} substituted for Si^4 in tetrahedral site, while Al^{VI} substituted for Fe or Mg in octahedral site to reach the charge balance. The range of the (Al^{IV})/(Al^{VI}) value also inferred that the sample had lower Fe^{3+}contents.

The (Al^{VI} + Fe) vs. Mg diagram has been used to show the substitution relation in octahedral site in chlorite (Xie et al. 1997). Both deposits exhibit a good negative correaltion in the (Al^{VI} + Fe) vs. Mg diagram (Fig. 3), which indicates that Al^{VI}, Fe, and Mg were the three main elements taken up the octahedral site, and also shows that both Al^{VI} and Fe could substitute for Mg in this site.

3.3.3 *Formation temperature and environment of chlorite*

Foster (1962) found that Al^{IV} is higher than Al^{VI} in metamorphic chlorite. Hillier and Velde (1991) reported that Al^{VI} is usually higher in diagenetic chlorite. McDowell and Elders (1980) and Jahren and Aagaard (1989) also reported that Al^{IV} increases with chlorite formation temperature. These observations are consistent with the data of Cathelineau and Nieva (1985), in which Al^{IV} increases and Al^{VI} decreases with chlorite formation temperature. Cathelineau and Nieva (1985) proposed an empirical chlorite geothermometer calibrated using the data from the geothermal system of Los Azufres, Mexico, and later modified by Cathelineau (1988) with new data from Salton Sea geothermal system. Klein calculated the formation temperature of chloritee in Cipoeiro gold deposit using the methods forward by Cathelineau (1988), Kranidiotis & MacLean (1987), Zang & Fyfe (1995) respectively but get very similarly results (Klein et al. 2007). Since then the relationship between the formation temperature and its structure, chemical composition, and polytype of chlorite has been widely studied. In this paper we calculated the interplanar spacing d_{001} value by the formulae d_{001}(0.1 nm) = 14.339–0.115n(Al^{IV}) – 0.0201 n(Fe^2), which was propsed by Raused-Colom then revised by Nieto. Using the formula of t°C = (14.379 – d_{001}(0.1 nm))/0.001, which was presented by Battaglia (1999), to calculate the formation temperature of chlorite. The formation temperature of chlorite in No.201 deposit varied between 220 and 276°C, the average value is 259°C; the formation temperature of chlorite in No. 333 deposit varied between 163 and 260°C, the average value is 214°C, slightly lower than No. 201 deposit.

The formation of the chlorite was a water-rock interaction process controlled by reaction kinetics; therefore, it could have been affected by factors such as temperature, pressure, water/rock ratio, and fluid and rock chemical composition. Inoue (1995) pointed out that in vein hydrothermal deposits, Mg-enriched chlorite favor forming under slight oxidizing and low pH conditions, whereas Fe-enriched chlorite form under reducing conditions. Most of the chlorites in the studied deposits are Fe-rich brunsvigites or ripidolites, expect for a few pycnochlorites.

3.4 *Formation mechanism of chlorite and the relationship with uranium mineralization*

As mentioned before, chloritization was one of the most important alteration types in the granite-type uranium deposits being investigated. They were brunsvigite and ripidolite, Fe-enriched chlorite, which imply (Fe^{2+}) > (Mg^{2+}) in the fluid. Chlorite either scattered in the rocks or occurring in veins of different scales from several centimeters in width to extra fine vein could be observed under the microscope or EMPA. Combining with the formation temperature information of this study, that the chlorite in the studied deposits varied in 163 and 276°C (average value 259°C for No. 201 deposit and 214°C for No. 333 deposit), we could deduced that there were two kinds of formation mechanism for chlorite in these deposits. Some chlorites formed as the alteration product of biotite and usually pseudomorphed biotite through dissolution-precipitation mechanism; the formation temperature varied between 220 and 276°C. The other chlorites occurred in veins of different scales or had the characteristics of fluid precipitation suggesting that they were formed directly from a hydrothermal precipitation mechanism; the formation temperature varied between 163 and 232°C. There were two kinds of observations on the relationship between chlorite and uranium mineralization. One was the chlorite formed before uranium mineralization; it was the chloritization of biotite reactive the uranium exist in biotite at isomorphism, so the chloritization contributes to the active uranium for uranium mineralization. That was confirmed by fission track studies (Cheng et al. 2000). The other one was chloritization in veins of different scale, which were formed during the uranium mineralization stage, and were reconfirmed by field and fission track studies. Therefore chlorite not only inherits the characteristics of fission tracks of uranium-bearing accessory mineral with major inclusions in biotite, but also overprints some reactive uranium during the chloritization through adsorption. This study revealed that the chlorite in studied deposits formed at low to moderate temperatures and under reducing condition that favored by uranium mineralization. We propose that the chloritization not only reflects the uranium distribution in granite but also contributed to a favorable depositional environment.

4 CONCLUSIONS

Through this study, we can conclude the following:

1. Four types of chlorite have been distinguished in the granite-type uranium ore field. On the basis of chemical classification, ripidolite and brunsvigite were the two main kinds of chlorite, with minor amounts of pycnochlorite, sheridanite, and disbantite.
2. The molar Al/(Al + Mg + Fe) values revealed that the chlorite may have been derived from argillaceous rock; a negative linear relationship in (Al^{VI} + Fe) vs. Mg indicated Al^{VI}, Fe, and Mg were the three main elements taken up the octahedral site; it also shows that both $Al^{VI}$and Fe could substitute for Mg in this site.
3. The formation temperature of chlorite varied between 163 and 276°C; the chlorite formed under reducing condition. Dissolve-precipitation and dissolve-transfer-precipitation were the two main formation mechanisms. Chloritization not only reflects the uranium distribution in the granite but also contributed to a favorable depositional environment.

ACKNOWLEDGEMENT

This study was financial supported by the National Natural Science Foundation of China (Grant No: 40972068 and No: 40772068). The authors wish to express their gratitude to two reviewers for their critical and suggestion comments and especially to Dr. Seal for help improving the English expressions.

REFERENCES

Battaglia, S. 1999. Applying X-ray geothermometer diffraction to a chlorite. *Clay and Clay Minerals* 47(1): 54–63.

Cathelineau, M. & Nieva, D. 1995. A chlorite solid solution geothermometer: the Los Azufres (Mexico) geothermal system. *Contribution to Mineralogy and Petrology* 91: 235–244.

Cathelineau, M. 1988. Cation site occupancy in chlorites and illite as a function of temperature. *Clay minerals* 23: 471–485.

Cheng, H.H., Ma, H.F. & Xiang, W.D. 2000. Study on changes of existing state of uranium during alkalic metasomatism using fission-track method. *Uranium Geology* 16(5): 291–296 (in Chinese with English abstract).

Foster, M.D. 1962. Interpretation of the composition and classfication for the chlorite. *US Geology Survey Prof. Paper. 414 A*, 33 p.

Hillier, S. & Velde, B. 1991. Octahedral occupancy and the chemical composition of diagenetic (low-temperature) chlorites. *Clay Minerals* 26:149–168.

Hua, R.M., Li, X.F., Zhang, K.P., Ji, J.F. & Zhang, W.L. 2003. Characteristics of clay minerals derived from hydrothermal alteration in Jinshan gold deposit: implication for the environment of water-rock interaction. *Acta Mineralogical Sinica* 23(1): 23–30 (in Chinese with English abstract).

Inoue, A. 1995. Formation of clay minerals in hydrothermal environments. In Viede B. (ed.), *Origin and Mineralogy of clays*. Berlin: Springer.

Jahren, J.S. & Aagaard, P. 1989. Compositional variations in diagenetic chlorites and illites, and relationships with formation-water chemistry. *Clay Minerals* 24:157–170.

Klein, E.V., Harris, C., Giret, A. & Moura, C.A.V. 2007. The Cipoeiro gold deposit, Gurupi belt, Brazil: Geology, chlorite geochemical, and stable isotope study. *Journal of South American Earth Sciences* 23: 242–255.

Laird, J. 1988. Chlorites: metamorphic petrology. In S.W. Bailey (ed.), *Hydrous Phyllosilicates. Reviews in Mineralogy* 19: 405–453.

MacDowell, S.D. & Elders, W.A. 1980. Authigenic layer silicate minerals in borehole Elmore Salton Sea geothermal field. California. USA. *Contribution to Mineralogy and Petrology* 74: 293–310.

Men, X.L. & Xie, Q.P. 1990. The physhical chemistry conditions in which No. 201 ore deposit is formed and its relation to the concentrate uranium ore deposit. *Journal of ECGI* 13(2): 31–41(in Chinese with English abstract).

Nieto, F. 1997. Chemical composition of metapelitic chlorites: X-ray diffraction and optical property approach. *European Journal of Mineralogy* 9: 829–841.

Rausell-Colom, J.A., Wiewiora, A. & Matesanz, E. 1991. Relation between composition and d_{001} for chlorite. *American Mineralogist* 76: 1373–1379.

Xie, X.G. 1997. IIb trioctahedral chlorite from the Barberton greenstone belt: crystal structure and rock composition constraints with implications to geothermometry. *Contribution to Mineralogy and Petrology* 126: 275–291.

Zang, W. & Fyfe, W.S. 1995. Chloritization of the hydrothermally altered bedrock at the Igarape Bahia gold deposit, Carajas, Brazil. *Mineral deposita* 30: 30–38.

Zhang, Z.S., Liu, S. & Wu, J.H. 2008. Characteristic and the formation conditions of chlorite in Xiazhuang uranium ore-field, South China. *Geochemica et Cosmochmica Acta* 72 (12): 1092–1092.

Water-Rock Interaction – Birkle & Torres-Alvarado (eds)
 ISBN 978-0-415-60426-0

Global geological water exchange and evolution of the Earth

V.P. Zverev
Institute of Environmental Geosciences RAS, Moscow, Russia

ABSTRACT: From data on the masses of subsurface waters in the Earth's crust and their global mass fluxes, an estimated is made of the geological mass fluxes of free and connected subsurface water in the process of drifting lithospheres plates and subduction under the continental crust. These waters are involved in volcanic eruptions and create a top-down flux reaching the Earth's lower crust and upper mantle. The mass of this flux is ~0.95×10^{24} g, which is commensurate with the mass of water in the mantle (1×10^{24} g). It is suggested that the geological cycle of subsurface water is the mechanism for compensating mantle dehydration and supports the content of the water at a level sufficient for convection to take place.

1 INTRODUCTION

Water on Earth as well as on planets of the Solar system was formed as a result of accretion and accumulation of proto-planetary material. Heating up the interiors of planets and their internal differentiation was accompanied by degassing of volatile substances, the rate and extent of which was largely dependent on the mass of the planet and its distance to the Sun, which in turn determined the particular distribution and evolution of water on each of the terrestrial planets.

2 WATER IN THE EARTH

Water in a hypothetical primitive mantle equals ~0.1% of its mass, or approximately 4×10^{24} g. The water content in MORB glasses (190 g/t) equals the concentration of water in a depleted and degassed upper mantle (Ryabchikov 1999). On this basis, there should be about 1×10^{24} g of water in the mantle, of which about 3×10^{24} g should be degassed. The most part of that mass went into the formation of the surface and subsurface hydrosphere, which began, according to isotopic data, at the beginning of the Archean, about 4 billion years ago. The average rate of water provision during that time was about 0.75×10^{15} g/a. In the view of the majority of researchers, these rates were higher at the early stages of the evolution of the Earth, decreasing to a current rate of 0.25×10^{15} g/a (Sorochtin 2007).

Currently, the surface hydrosphere totals 1.37×10^{24} g. The amount of water in the Earth's crust is more problematic. Recently, (Zverev 2004, 2007a, b) has estimated this amount using data from Ronov et al. (1990). Currently, the sedimentary cover of the Earth's crust contains 0.285×10^{24} g of water, i.e., approximately 4.8 times less than in the recent ocean. Much more water (0.504×10^{24} g), is concentrated in the so-called 'granitic' and 'basaltic' layers of the crust. The total of all types of subsurface water of the crust is 0.79×10^{24} g, i.e., more than half the mass of surface hydrosphere (Zverev 2009a).

We can conclude that at this stage of the evolution of the Earth the masses of surface and subsurface hydrosphere are slightly greater than 2×10^{24} g, and there is ~1×10^{24} g of water in the mantle.

The most important feature of subsurface waters of the Earth's crust is that they are mobile and constantly transfer material within the crust. This transfer is controlled jointly by solar radiation, and the geothermal and gravitational fields of the Earth. Processes in which this transfer is involved include active water exchange within the continental crust; lithogenic-metamorphic reactions, lithosphere-mantle processes, the discharged in zones within island arcs and active continental-margin systems during subduction.

The geological role of free, physically and chemically connected waters during plate tectonic processes is very important, with a total mass flux of $1{,}294 \times 10^{15}$ g/a. They are part of hydrothermal systems, they participate in volcanic eruptions, and form descending water flow reaching the lowest layers of the crust and the uppermost mantle in amounts of ~0.38×10^{15} g/a.

Ryabchikov (1999) emphasizes the importance of these deep flows (up to 400 km) which can reach the transition zone and to control the start of melting mantle rocks.

3 WATER IN THE TERRESTRIAL PLANETS

On the basis of the common origin of the terrestrial planets (Mercury, Venus, Earth and Mars), it is assumed that the initial water was proportional to their masses (Table 1).

Interior heating was accompanied by release of mobile components, the rates, and scales of the degassing being considerably different for each planet. In turn, this defined the peculiarity of water distribution and evolution on these planets. Excluding the Earth, the absence or extremely low water content in their atmospheres and on their surfaces is common for these planets.

There is no atmosphere or hydrosphere on Mercury or on Earth's Moon. The atmosphere on Venus consists of carbon dioxide, helium, and water vapour. The latter dissociates into OH and H under the influence of sunlight in the upper atmosphere. These dissociation products leave the planet (along with helium), resulting in the formation of the carbon dioxide atmosphere.

There is water in the atmosphere and at the surface in parts of Mars, but comprises an insignificant part of the planet's initial mass, which included the presence of a surface hydrosphere and ocean. It is supposed that the Martian atmosphere, which controlled the water cycle under warm humid conditions, was destroyed 4.1 to 3.8 billion years ago and the planet has now only an insignificant part of the original hydrosphere. Currently, water on Mars appears to exist only in polar ice caps. The actual existence of water on Mars was confirmed recently by measurements by *Phoenix*.

Table 1. The Terrestrial planets

Parameters & processes	Mercury	Venus	Earth	Mars
Mass (10^{27} g)	0.3302	4.8685	5.974	0.6418
Volcanic activity (10^9 years ago)	until ~3.5	up to 1.0–0.5	until the present	up to 1.0–2.0
Continental drift (10^9 years ago)	did not exist	did not exist	Since 2.5	did not exist
Presence of surface hydrosphere (10^9 years ago)	did not exist	did not exist	Since 3.0	4.1–3.5
Initial water mass (10^{24} g)	~0.22	~3.2	~4.0	~0.42
Recent water mass (10^{24} g)	$<10^{-5}$	>0.8	~3.2	~0.12
Water mass in planets thickness (10^{24} g)	$<10^{-5}$	~0.8	~0.8	~0.1

It is estimated that the water mass concentrated in ice on the surface and near-surface of Mars reaches about 1.2×10^{21} g, which is three orders of magnitude less than on Earth.

It is supposed that the water content in their mantles is in accordance with the values typical for the depleted mantle of Earth and is in proportion to the respective mass of the planets. The tentative water masses of the planets calculated on this basis are given in Table 1.

4 CONSEQUENCE OF THE GLOBAL GEOLOGICAL WATER EXCHANGE

Due to the Earth's temperature regime the surface hydrosphere and water cycle in the crust and uppermost mantle have existed for more than 3 billion years. Volcanic activity has taking place during its whole history and, in contrast to other planets, there has also been the processes of plate tectonics, rifting, continental drift, and subduction.

The development of hydrothermal activity accompanying volcanism, rifting, and subduction is usually regarded as a secondary product of these processes. However, the following considerations allow one to come to a slightly different, and maybe unexpected and paradoxical conclusion.

The time of existence and the development peculiarities of volcanism, along with the absence of continental drift on the terrestrial planets, allows us to conclude that the amount of water and its distribution could be a reason limiting these processes. Indeed, volcanism on Mercury, Venus, Mars, and the Moon developed only in the first billion years after their formation, when the depths of planets contained water masses sufficient for conditions of magma formation to be maintained. As degassing progressed, the volcanism on Mercury, Venus, and Mars decreased. Moreover, the more water there was, the longer the process took. There is no rifting or subduction on these planets now because these processes are controlled by plate tectonics. In order for plate tectonics to take place there must be a permanent presence of sufficient water masses coming from the surface hydrosphere and water-containing rocks of the oceanic crust and moving into subduction zones in order to sustain the physical parameters of rocks for convection in the uppermost mantle.

On Earth, in the Mid-ocean Ridges, water transfer into the zone where magma is formed results from direct hydrothermal convection from oceanic waters. Lithogenic and geological cycles are responsible for water transfer in subduction zones (Table 2). The former results in physically connected waters of the oceanic crust's sedimentary and volcanogenic rocks sinking beneath the continental crust and changing into the free state hydrothermal

Table 2. Mass fluxes of subsurface waters of the geological cycle for the past 2.5 billion years

Mass fluxes components	For time realization, (10^{24} g)
Hydrothermal systems	2.000
Volcanic eruptions	0.29
Descending fluxes	0.947
Total mass flux	3.237

waters, and in volcanism of island arcs and active continental borders. The influx of chemically connected water contained in the basaltic oceanic crust and uppermost mantle (Fyfe et al. 1978), where it turns into the free state during continental drift, is the geological cycle which continually sustains the appropriate physical parameters of rocks and magmatic melts that are needed for convection in the uppermost mantle. Trubitsyn (2004) notes that the influence of water on the viscosity of materials is comparable to that of temperature. It should be noted that on Earth the processes of rifting and subduction, and consequently plate tectonics, began approximately 2.5 billion years ago, after the formation of the oceans and simultaneously with the formation of hydrated volcanogenic and sedimentary rocks of the oceanic crust.

The water mass transferred into the deep Earth's crust and the uppermost mantle during the time that the geological cycle has existed, and the progress of continental drift, amount to 0.947×10^{24} g (Table 3), which is comparable to the water mass contained in the mantle (1×10^{24} g).

Thus one can assume that the geological cycle of subsurface water, implemented during the drift of lithospheric plates, may serve as mechanism compensating degassing of the upper mantle, and supports the content of the water level necessary for the process of convection (Zverev 2009b).

It is likely that on Mars, when surface hydrosphere and oceans were present that the processes of rifting could have begin (similar structures on Mars have been discovered recently), but because of the short-term existence of ocean water plate tectonics did not occur. It should be emphasized that most of Earth's continental rifts, where there are no required amount of water to create powerful hydrothermal circulation systems, eventually stop their development.

Thus it follows from all the above that the geological cycle of subsurface water on Earth to a certain extent may compensate for degassing and dehydration of the Earth, and in contrast to other terrestrial planets, sustain the water content at the level which is enough for mantle convection and tectonic activity for a longtime and up to the present.

Table 3. Masses of water compensation degassing the upper mantle

Parameters of hydrosphere	Mass of water, 10^{24} g
Mass of water in depletion and degassing the upper mantle	~1
Mass of dissipation waters	~1
Descending mass fluxes of the geological cycle of subsurface water for 2.5 billion years	0.95

Finally, because biological activity and life can't arise without water, life can only develop on planets with tectonic activity.

5 CONCLUSION

In its communication, the author has not claimed the final decision of the problem. The main aim was precisely the statement of the problem of the need to learn more about the role of water in the evolution of the Earth and its covers.

REFERENCES

Fyfe, W.S., Price, N.J. & Thompson, A.B. 1978. *Fluids in the Earth's Crust.* Elsevier.

Ronov, A.B., Jaroschevsky, A.A. & Migdisov, A.A. 1990. *A Chemical Structure of the Earth's Crust and Geochemical Balance of Major Elements.* Moscow: Nauka (in Russian).

Ryabchikov, I.D. 1999. Fluid regime of Earth's mantle. *Vestnic BGGGMS RAS* (in Russian).

Sorochtin, O.G. 2007. *Life of the Earth.* Moscow-Izhevsk (in Russian).

Trubitsyn, V.P. 2004. Single global tectonic evolution of Mars, Earth and Venus and tectonic stage continents evolution on the Earth. Evolution tectonic processes at the Earth's history. *Proceeding XXXYII Tectonic conference*, Novosibirsk, 220–223 (in Russian).

Zverev, V.P. 2004. New data on masses and mass flows of subsurface water in the Earth's crust. *Doklady Earth Sciences,* Vol. 397 A (6), 836–839.

Zverev, V.P. 2007a. *Subsurface waters of the Earth's crust and geological processes.* Second Edition. Scientific World (in Russian).

Zverev, V.P. 2007b. New Data on Masses and Mass Fluxes of Subsurface water in the Earth's Crust and Rock-Water Ratio in Geological Processes. In T.D. Bullen & Y.X. Wang (eds.), *Water Rock Interaction-Proceedings of WRI-12:* 1521–1524.

Zverev, V.P. 2009a. *The Water in the Earth. The Introduction in Doctrine of Subsurface Waters.* Scientific World (in Russian).

Zverev, V.P. 2009b. Peculiarity and Consequence of Geological Subsurface Water Circulation. *Doklady Earth Sciences* 425 A(3): 367–370.

Water-rock interactions in watersheds

Water-Rock Interaction – Birkle & Torres-Alvarado (eds)
© 2010 Taylor & Francis Group, London, ISBN 978-0-415-60426-0

Assessing surface water—groundwater connectivity using hydraulic and hydrochemical approaches in fractured rock catchments, South Australia

E.W. Banks & A.J. Love
School of the Environment, Flinders University, Adelaide, Australia

C.T. Simmons
School of the Environment & National Centre for Groundwater Research and Training, Flinders University, Adelaide, Australia

P. Shand
CSIRO Land and Water, Glen Osmond, Australia

ABSTRACT: In Australia, native vegetation clearance has had considerable impacts on surface water and groundwater salinities. The impact on surface water-groundwater connectivity is less understood. A hydraulic, hydrochemical, and tracer-based study was conducted at two contrasting fractured rock catchments in South Australia. Results indicate that connectivity was variable across each of the catchments. The influence of the fractured rock aquifer was minimal in the pristine, uncleared Rocky River catchment, whereas in the cleared, mixed land-use Cox Creek catchment, the fractured rock aquifer played a more significant role. The results emphasise the need to understand the importance that the impacts of land-use change (particularly vegetation clearance) can have on surface water-groundwater connectivity.

1 INTRODUCTION

The importance of surface water-groundwater interactions has received greater attention in the last decade, in response to water resource allocation needs and the impacts on groundwater dependent ecosystems. It is becoming increasingly evident that there are often strong hydraulic connections between these two resources (Winter et al. 1998; Kalbus et al. 2006; Banks et al. 2009). Considerable research has been undertaken in sedimentary aquifer systems but very few studies are reported for fractured bedrock systems (Haria and Shand 2006; Kahn et al. 2008). The latter are substantially more complex owing to the geological heterogeneity of the fractured rock aquifer (FRA).

Rates of groundwater flow and connectivity in FRA are difficult to determine, and methods commonly used for porous media are often not applicable (Cook et al. 1996). Multi-tracer approaches have been used in fractured rock systems (e.g. Genereux et al. 1993; Shand et al. 2007) and have proven invaluable in constraining contributing sources of solutes, preferential flow pathways and residence times in these types of systems.

This study examines the results of two contrasting catchments in the greater Mount Lofty Ranges (MLR) of South Australia using a multi-tracer approach. The aim was to improve understanding of surface water-groundwater connectivity in fractured rock catchments and evaluate potential differences between cleared and uncleared catchments.

1.1 *Study sites*

1.1.1 *Cox Creek catchment*

The Cox Creek catchment (CCC) is situated approximately 20 km east of Adelaide. The catchment covers an area of 28.8 km^2 and land use is a mix of agriculture, small farms and forest. Average annual rainfall is 1189 mm/y, the majority of which falls between May and October.

The geology of CCC includes several stratigraphic sequence associated with the Adelaide Geosyncline. The Neoproterozoic Emeroo Subgroup (quartzite, sandstone and dolomite) dominates the north and south of the catchment and is separated by the Archean Barossa Complex (metamorphic rocks with retrograde metamorphism), which lies across the middle of the catchment (Fig. 1).

1.1.2 *Rocky River*

The Rocky River Catchment (RRC) is located within Flinders Chase National Park on Kangaroo

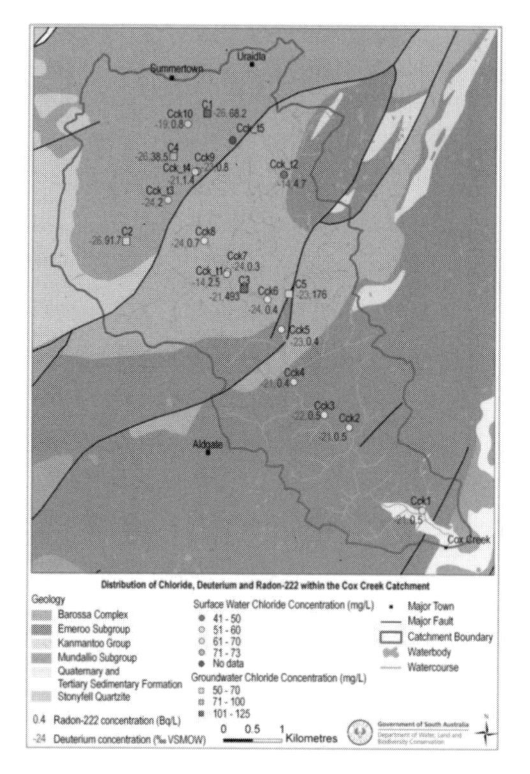

Figure 1. Spatial distribution of ^{222}Rn (Bq/L), deuterium (‰ relative to Vienna Standard Mean Ocean Water-VSMOW) and chloride (mg/L) concentrations of surface water and groundwater in the Cox Creek Catchment, December 2005.

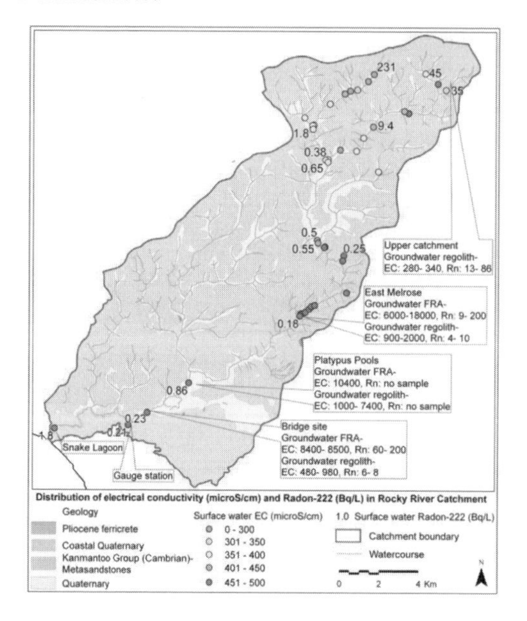

Figure 2. Spatial distribution of electrical conductivity (EC μS/cm) and ^{222}Rn (Bq/L) of surface water and groundwater in the Rocky River Catchment.

Island, South Australia. The catchment covers an area of about 216 km^2 and is one of very few remaining catchments in South Australia still covered by native vegetation (Fig. 2). The Rocky River is semi-perennial, flowing all year round in the mid to upstream part of the catchment whilst the lower part usually ceases to flow during the summer months. The average annual rainfall is 780 mm/y.

The geology is dominated by a laterite plateau underlain by Cambrian and Neoproterozoic metasediments. Towards the bottom of the catchment, the river has incised through Quaternary calc-arenite and limestone deposits.

2 METHODS

2.1 *Sampling and analytical techniques*

Surface water and groundwater sampling was conducted along the length of the creeks during a 12 month period. A YSI® multi-parameter meter was used to measure the pH, specific electrical conductance (SEC), dissolved oxygen (DO), redox potential (Eh) and temperature in the creek and in groundwater bores. The alkalinity (as HCO_3^-) was measured in the field using a HACH titration kit.

Surface water and groundwater samples were analysed for major elements, stable isotopes, ^{222}Rn, and $^{87}Sr/^{86}Sr$. All analyses were done using standard analytical techniques.

Manual flow gauging was conducted using a pigmy flow meter (OTT) and YSI® flow tracker at sampling locations in both catchments.

3 RESULTS AND DISCUSSION

3.1 *Cox Creek*

Significant changes in solute concentrations and environmental tracers along the length of Cox creek may indicate locations of groundwater discharge to the creek and the influence of the underlying geology and/or fault zones that traverse the creek (Fig. 1). The trends in the ^{222}Rn activity during the sampling period were similar, and suggest that there is connectivity between the groundwater system and Cox Creek (Figure 3). The locations of high ^{222}Rn activity indicate localised groundwater influx to the stream (in this investigation we assume that the groundwaters have a constant high ^{222}Rn activity and reflect the lithology of the major aquifers sampled). The ^{222}Rn activity downstream of influx declines rapidly due to its short half-life (3.82 days) and loss to the atmosphere by gas exchange. The constant ^{222}Rn activities along some creek reaches suggests a balance between a consistent groundwater inflow and the de-gassing and radioactive decay of ^{222}Rn in creek water. During baseflow conditions (Dec-Mar), ^{222}Rn activities in the stream varied from 0.6 Bq/L at Cck7 to 3.4 Bq/L at Cck2, compared to winter (Jul) where the activity only varied from 0.39 Bq/L at Cck7 and 0.43 Bq/L at Cck2 (Fig. 3). The lower values in winter are likely to be a result of dilution by rainfall and surface runoff.

The ^{222}Rn activities in the groundwater were an order of magnitude higher than in the surface water samples and reflect the mineralogy of the aquifer (Love et al. 2002). Groundwater from the FRA in the sandstone, quartzite and dolomite units had lower ^{222}Rn activities (37.9–87.1 Bq/L) compared to groundwater from the FRA in the metamorphosed gneisses and schists (220–489 Bq/L).

The chloride concentrations (Cl) in Cox Creek showed a decreasing trend in the downstream direction from sample site Cck9 to Cck8/Cck7, and then gradually increased to Cck1. This suggests discharge of a low salinity groundwater end member in the top part of the catchment. This was supported by the stable isotopes of the water molecule which showed the groundwater had a more depleted isotopic composition than Cox Creek (Fig. 1). Downstream of sample site Cck8 the δ^2H became progressively more positive indicating surface evaporation. The higher Cl of groundwater sample C1 compared to the other samples in the top part of the catchment indicates that the contribution of groundwater from this area of the top part of the catchment would have to be small given that the surface water samples have a much lower Cl.

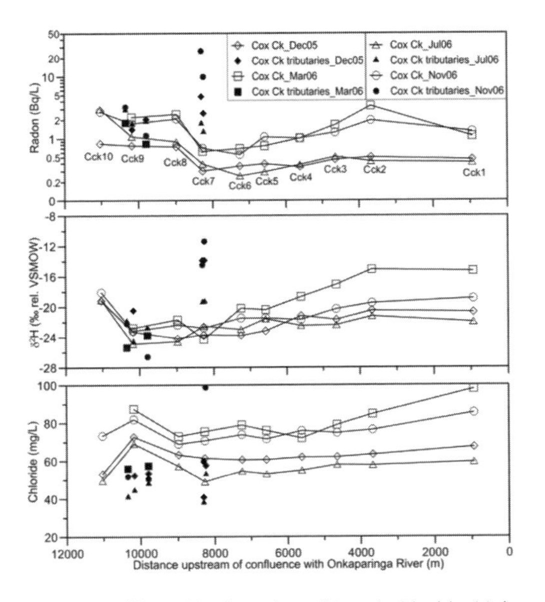

Figure 3. ^{222}Rn (a), deuterium (b) and chloride (c) in Cox Creek in December 2005, March 2006, July 2006 and November 2006. Distances are measured upstream of confluence with the Onkaparinga River (m).

The plot of the ^{87}Sr/^{86}Sr ratio versus the reciprocal of the strontium concentration (1/Sr) shows that the surface water samples trend between local rainwater and groundwater (Fig. 4). A more detailed analysis suggests that groundwater from aquifers located in the top part of the catchment dominate the hydrochemical signature in Cox Creek and are likely to be the main source of groundwater to the creek. Evaluation of the ^{87}Sr/^{86}Sr ratio versus δ^{18}O provided further evidence, which indicated that the majority of the samples from Cox Creek had an evaporated groundwater signature.

3.2 *Rocky River*

Hydraulic and hydrochemical results showed both gaining and losing sections of the Rocky River and in some locations the river is disconnected from the aquifers beneath. The increase in EC from the top of the catchment down to Snake Lagoon is a result of evapotranspiration and/or groundwater discharge (Figure 2). Continuous water level monitoring showed that at the study sites (East Melrose, Platypus Pools and the Bridge) the river was losing at these locations and therefore, the increase in EC can only be a result of evapotranspiration. In the upper region of the catchment, the river must be gaining to sustain perennial flow at and above East Melrose. Below East Melrose during summer periods flow has sometimes ceased (Fig. 5).

Gaining conditions in the upper region of the catchment indicates a subsurface fresh water source to the river that is either from shallow sedimentary and/or fractured rock aquifer systems. There is no substantial winter surface runoff to sustain perennial river flow. Groundwater samples from the fractured rock aquifer had much higher salinities (>6000 μS/cm) than the river (<500 μS/cm). Soil chloride profiles taken across the catchment also indicated that there are large stores of salt present at shallow depth. However, groundwater sampling in shallow perched sand aquifers in the upper reaches of the catchment had similar salinities and ^{222}Rn activities to the river, suggesting that these shallow aquifers are the subsurface source to Rocky River (Figure 2). The ^{222}Rn activity in the river reduced to less than 0.2 Bq/L by the time it

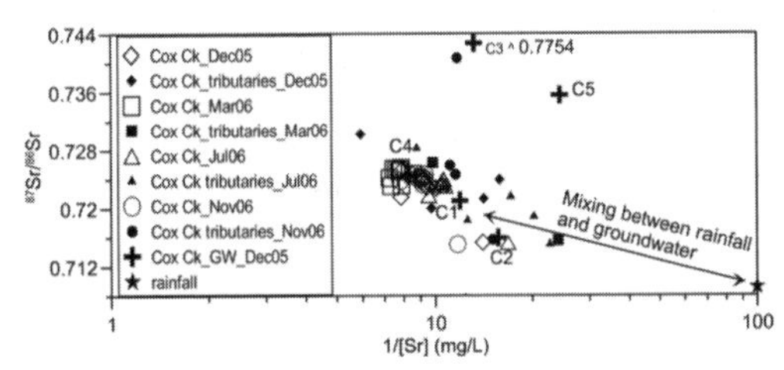

Figure 4. ^{87}Sr/^{86}Sr ratio versus 1/Sr in Cox Creek and the sampled groundwater bores.

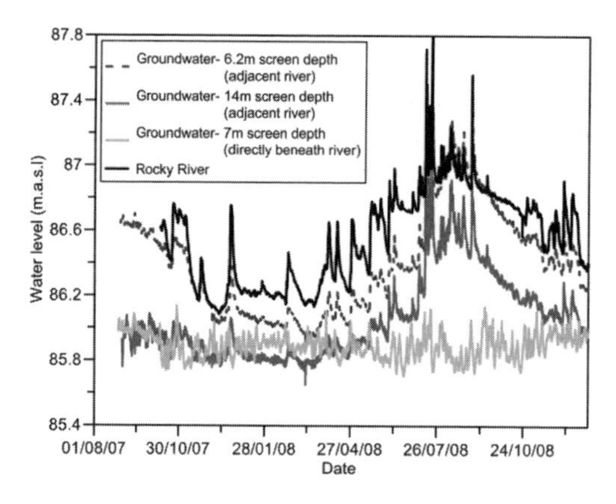

Figure 5. Continuous water level data (m above sea level) from monitored piezometers and Rocky River at East Melrose from August 2007 until December 2008.

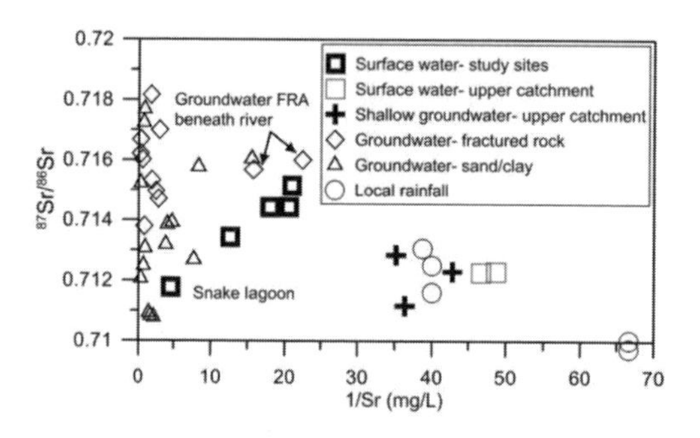

Figure 6. $^{87}Sr/^{86}Sr$ ratio versus 1/Sr for surface water, groundwater and local rainfall in Rocky River Catchment.

reached East Melrose suggesting minimal to no groundwater contribution below this point.

The plot of $^{87}Sr/^{86}Sr$ ratio versus 1/Sr shows the shallow groundwater and surface water from the upper region of the catchment plot close to the rainfall samples, whilst the surface water samples at and below East Melrose have a higher ratio and Sr concentration. The groundwater samples from the fractured rock have a large variation in the $^{87}Sr/^{86}Sr$ ratio however, the groundwater samples from the fractured rock sampled beneath the river plot closely to the surface water samples supporting the losing stream conceptual model (Fig. 6).

The emerging conceptual model of this system is that winter rainfall replenishes the shallow perched sand aquifers located in the upper reaches of the catchment which gradually drain and discharge into the tributaries of Rocky River. The contribution from the FRA is minimal. A high evapotranspiration rate from the native vegetation in a dominantly losing system is likely to maintain a fresh river system because in this state, deeper regional saline groundwater from the FRA cannot discharge into it as baseflow.

4 CONCLUSION

This study investigated surface water-groundwater connectivity in fractured rock catchments at two sites in South Australia. Results indicate that connectivity was variable across each of the catchments. The influence of the fractured rock aquifer was minimal in the pristine, uncleared RRC whereas in the cleared, mixed land-use CCC, the fractured rock aquifer played a more significant role. This result calls for further research investigating the impacts of land-use change (particularly vegetation clearance) on surface-water groundwater connectivity.

REFERENCES

Banks, E., Simmons, C., Love, A., Cranswick, R., Werner, A., Bestland, E., Wood, M. & Wilson, T. 2009. Fractured bedrock and saprolite hydrogeologic controls on groundwater/surface-water interaction: a conceptual model (Australia). *Hydrogeology Journal* 17(8): 1969–1989.

Cook, P.G., Solomon, D.K., Sandford, W.E., Busenberg, E., Plummer, L.N. & Poreda, R.J. 1996. Inferring shallow groundwater flow in saprolite and fractured rock using environmental tracers. *Wat Resorces Res* 32(6): 1501–1509.

Genereux, D.P., Hemond, H.F. & Mulholland, P.J. 1993. Spatial and temporal variability in streamflow generation on the West Fork of Walker Branch Watershed. *Journal of Hydrology* 142(1–4): 137–166.

Haria, A.H. & Shand, P. 2006. Near-stream soil water-groundwater coupling in the headwaters of the Afon Hafren, Wales: Implications for surface water quality. *Journal of Hydrology* 331: 567–579.

Kahn, K., Ge, S., Caine, J. & Manning, A. 2008. Characterization of the shallow groundwater system in an alpine watershed: Handcart Gulch, Colorado, USA. *Hydrogeology Journal* 16(1): 103–121.

Kalbus, E., Reinstorf, F. & Schirmer, M. 2006. Measuring methods for groundwater—surface water interactions: a review. *Hydrology and Earth System Sciences* 10(6): 873–887.

Love, A.J., Cook, P.G., Harrington, G.A. & Simmons, C.T. 2002. *Groundwater Flow in the Clare Valley*, Department for Water Resources.

Shand, P., Darbyshire, D.P.F., Gooddy, D. & H. Haria, A. 2007. $^{87}Sr/^{86}Sr$ as an indicator of flowpaths and weathering rates in the Plynlimon experimental catchments, Wales, U.K. *Chemical Geology* 236(3–4): 247–265.

Winter, T.C., Harvey, J.W., Franke, O.L. & Alley, W.M. 1998. *Grounwater and Surfacewater: A Single Resource*, U.S. Geological Survey Circular 1139.

Water-Rock Interaction – Birkle & Torres-Alvarado (eds)

Depth profiles in a tropical, volcanic critical zone observatory: Basse-Terre, Guadeloupe

H.L. Buss & A.F. White
U.S. Geological Survey, Water Resources Discipline, Menlo Park, CA, USA

C. Dessert & J. Gaillardet
Institut de Physique du Globe de Paris, Paris, France

A.E. Blum
U.S. Geological Survey, Water Resources Discipline, Boulder, CO, USA

P.B. Sak
Department of Geology, Dickinson College, Carlisle, PA, USA

ABSTRACT: The Bras David watershed on the French island of Basse-Terre, Guadeloupe in the Lesser Antilles is located on a late Quaternary volcaniclastic debris flow of dominantly andesitic composition. The bedrock is mantled by more than 12 m of highly leached regolith. The regolith is depleted with respect to most primary minerals and weathering is dominated by the dissolution and precipitation of clays. Mineral nutrient cations such as Mg, K, and P are largely present adsorbed to, or co-precipitated with, clays and iron oxides. Surface soils (< 0.3 m depth) are enriched in feldspar, quartz, cristobalite, and Fe(II), Ca, K, and Mg relative to the underlying regolith, likely reflecting atmospheric deposition, possibly related to volcanic activity.

1 INTRODUCTION

The Bras David watershed is located in a rugged, humid, tropical environment with a mean annual temperature of 25°C and a mean annual precipitation of 4500 mm yr^{-1} (Météo-France 2008). Thin soils top very thick (>12 m) regolith, which is exposed at roadcuts and excavations. The regolith is interpreted as highly weathered volcanic debris flows containing rocky clasts at various stages of weathering. Volcanic flows in the immediate vicinity were dated by Ar/Ar as having been emplaced at 900 ka (Samper et al. 2007). Here we document depth profiles (density, elemental concentrations, mineralogy) in Bras David to investigate weathering and mineral nutrient processes in a deep, tropical, volcanic regolith.

2 METHODS

Vadose zone pore waters were collected approximately every month for 2 years from 5-cm diameter nested porous-cup suction water samplers (Soil Moisture Inc., Santa Barbara, CA) that were installed in hand-augered holes at depths from 0.15 to 12.5 m. Pore waters were filtered to 0.45 μm and analyzed by ICP-MS and ion-chromatography. A 12.5 m solid core was collected and used for bulk density measurements, quantitative mineralogy by XRD using RockJock (Eberl 2003), and bulk chemical analysis by ICP-OES (SGS, Canada).

3 RESULTS

Augered core samples contained a number of weathered rocky clasts of varying hardness and color. Similarly, the regolith matrix material exhibited a variety of colors and textures that were visible during sample collection. Bulk densities of augered samples are extremely low, on average 0.96 g cm^{-3} (Fig. 1). A large clast, with a relatively unweathered core, collected from a nearby roadcut has a bulk density of 2.6 g cm^{-3} in the core and 1.4 g cm^{-3} in the rind (Sak et al. 2010).

Clays, dominantly halloysite, comprise about 75 wt% of the mineralogy at all depths (Table 1). Non-clays are almost entirely Fe(III)-hydroxides and quartz/cristobalite. The only distinct depth trends in mineralogy are an inverse relationship between halloysite and gibbsite. Feldspars are nearly absent except at the top and bottom of the augered core, consistent with the observation of

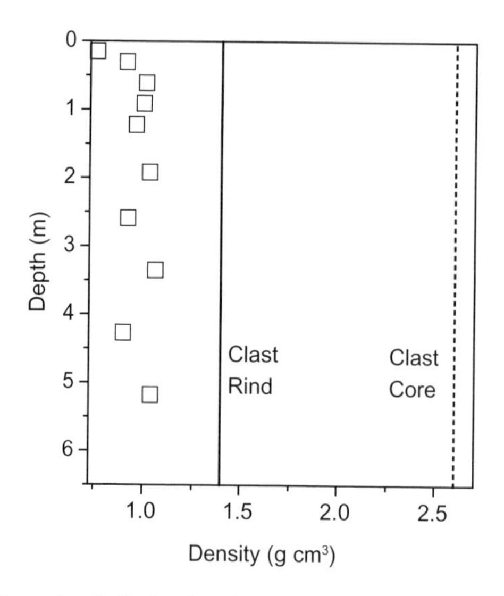

Figure 1. Bulk density of the augered profile (□). The densities of a clast and its weathering rind from a nearby roadcut (Sak et al. 2010) are shown as vertical lines.

Sak et al. (2010) of feldspar dissolving completely at a clast-rind interface within the same formation nearby. Clasts recovered during augering are weathered and differ from the matrix mineralogy only in the proportion of specific clays (augered clast data not shown). More feldspar (mostly microcline) is present in the upper 0.3 m than anywhere else in the profile (Table 1); quartz and cristobalite also increase at the surface.

Solid state chemical composition of the bulk regolith is dominated by Si and Al with Na and Ca near or below detection limits at most depths (Table 2). However, Fe(II), Ca, K, and Mg are enriched at the surface (< 0.3 m depth) relative to the underlying regolith. Roadcut clast rind compositions are comparable to regolith compositions with the exception that the rinds are slightly enriched in P. (Sak et al. 2010; Table 2). Augered clasts (data not shown) are similar in composition to the surrounding regolith.

Pore waters are dominated by sea salts, as is typical in tropical island watersheds (e.g., White et al. 1998). With the exception of Si, cations show little

Table 1. Mineralogy by Quantitative XRD, in weight percent.

Depth (m)	Quartz	K-spar[1]	Albite[2]	Magnetite	Goethite	Maghemite	Cristo-balite	Kaolinite[3]	Gibbsite	Halloysite
0.15	5.2	4.3	0.1	0.7	4.5	9.8	6.0	12.0	11.4	45.9
0.30	3.7	2.1	0.0	0.7	5.6	9.3	5.6	13.7	10.3	49.1
0.61	0.4	0.1	0.0	2.0	4.3	14.4	4.0	12.0	8.3	54.4
0.91	0.6	0.0	0.0	0.7	5.5	8.5	2.7	24.2	4.7	53.1
1.22	0.7	0.0	0.0	3.1	3.9	14.3	2.8	13.6	6.8	54.8
1.52	0.5	0.0	0.0	2.7	2.9	16.0	3.2	11.5	4.8	58.4
1.83	0.5	0.0	0.0	3.3	3.2	13.5	5.5	9.9	6.8	57.2
2.13	0.5	0.0	0.0	3.8	3.2	13.6	3.8	12.4	6.4	56.4
2.44	1.1	0.0	0.1	2.6	3.2	15.1	3.4	11.6	4.1	58.8
2.74	1.4	0.0	0.0	2.1	3.5	10.9	2.4	26.3	2.5	50.8
3.05	6.1	0.7	0.0	1.1	6.6	8.4	3.4	16.7	8.1	48.9
3.66	1.4	0.0	0.0	3.8	4.1	13.6	5.4	11.1	8.7	51.8
4.27	8.3	1.4	0.0	1.3	5.1	9.3	2.0	7.0	17.8	47.7
4.88	2.8	0.0	0.0	1.8	4.8	17.5	1.4	3.2	41.3	27.1
5.49	1.0	0.0	0.0	2.8	4.1	17.5	2.8	10.7	10.0	51.2
6.10	1.8	0.0	0.0	2.3	5.1	11.6	2.4	15.4	4.4	57.1
6.71	3.3	0.0	0.0	1.8	2.8	13.4	2.1	5.9	5.1	65.6
7.32	4.6	0.0	0.0	1.9	2.8	9.2	2.2	5.3	64.3	9.6
7.92	5.5	0.5	0.0	0.9	3.1	8.6	0.6	1.7	0.4	78.7
8.53	1.9	0.0	0.0	0.9	6.2	10.4	2.5	15.3	7.1	55.5
9.14	0.7	0.0	0.0	2.0	4.0	12.9	3.1	13.9	0.9	62.4
9.75	0.8	0.0	0.0	2.8	4.4	12.3	6.7	13.5	4.2	55.3
10.36	0.8	0.0	0.0	2.9	3.6	11.1	4.1	17.9	1.3	58.4
10.97	1.0	0.0	0.0	1.2	5.7	12.8	2.6	14.7	0.1	61.9
11.58	0.3	0.0	0.0	3.4	4.3	17.7	2.2	14.2	2.0	55.9
12.19	5.1	1.7	0.0	1.4	7.0	12.5	2.0	9.0	0.7	60.5
12.50	5.8	1.8	0.0	2.1	7.6	8.4	3.3	13.8	0.7	56.5

[1]Intermediate microcline
[2]Albite var. cleavelandite
[3]Disordered kaolinite.